Advances in High-Order Predictive Modeling

Continuing the author's previous work on modeling, this book presents the most recent advances in high-order predictive modeling. The author begins with the mathematical framework of the 2nd-BERRU-PM methodology, an acronym that designates the "second-order best-estimate with reduced uncertainties (2nd-BERRU) predictive modeling (PM)." The 2nd-BERRU-PM methodology is fundamentally anchored in physics-based principles stemming from thermodynamics (maximum entropy principle) and information theory, being formulated in the most inclusive possible phase-space, namely the combined phase-space of computed and measured parameters and responses.

The 2nd-BERRU-PM methodology provides second-order output (means and variances) but can incorporate, as input, arbitrarily high-order sensitivities of responses with respect to model parameters, as well as arbitrarily high-order moments of the initial distribution of uncertain model parameters, in order to predict best-estimate mean values for the model responses (i.e., results of interest) and calibrated model parameters, along with reduced predicted variances and covariances for these predicted responses and parameters.

Advances in Applied Mathematics

Series Editor: Daniel Zwillinger

Delay Ordinary and Partial Differential Equations
Andrei D. Polyanin
Vsevolod G. Sorkin
Alexi I. Zhurov

Clean Numerical Simulation
Shijun Liao

Multiplicative Partial Differential Equations
Svetlin Georgiev and Khaled Zennir

Engineering Statistics
A Matrix-Vector Approach with MATLAB®
Lester W. Schmerr Jr.

General Quantum Numerical Analysis
Svetlin Georgiev and Khaled Zennir

An Introduction to Partial Differential Equations with MATLAB®
Matthew P. Coleman and Vladislav Bukshtynov

Handbook of Exact Solutions to Mathematical Equations
Andrei D. Polyanin

Introducing Game Theory and its Applications, Second Edition
Elliott Mendelson and Dan Zwillinger

Modeling Operations Research and Business Analytics
William P. Fox and Robert E. Burks

Decision Analysis through Modeling and Game Theory
William P. Fox

Advances in High-Order Predictive Modeling
Methodologies and Illustrative Problems
Dan Gabriel Cacuci

https://www.routledge.com/Advances-in-Applied-Mathematics/book-series/CRCADVAPPMTH?pd=published,forthcoming&pg=1&pp=12&so=pub&view=list

Advances in High-Order Predictive Modeling

Methodologies and Illustrative Problems

Dan Gabriel Cacuci

CRC Press
Taylor & Francis Group
Boca Raton London New York

CRC Press is an imprint of the
Taylor & Francis Group, an **informa** business
A CHAPMAN & HALL BOOK

Cover credit: TPCX/shutterstock

First edition published 2025
by CRC Press
2385 Executive Center Drive, Suite 320, Boca Raton, FL 33431

and by CRC Press
4 Park Square, Milton Park, Abingdon, Oxon, OX14 4RN

CRC Press is an imprint of Taylor & Francis Group, LLC

Library of Congress Cataloging-in-Publication Data
Names: Cacuci, Dan Gabriel, author.
Title: Advances in high-order predictive modeling / authored by Dan Gabriel Cacuci.
Other titles: Advances in high-order predictive modeling
Description: First edition. | Boca Raton, FL : CPC Press, 2025. |
Includes bibliographical references.
Identifiers: LCCN 2024024626 | ISBN 9781032740560 (hardback) |
ISBN 9781032763538 (paperback) | ISBN 9781003478119 (ebook)
Subjects: LCSH: Prediction theory. | Sensitivity theory (Mathematics) |
Maximum entropy method. | Newton-Raphson method. |
Uncertainty (Information theory)
Classification: LCC QA279.2 .C33 2025 | DDC 519.2/87—dc23/eng20240823
LC record available at https://lccn.loc.gov/2024024626

ISBN: 9781032740560 (hbk)
ISBN: 9781032763538 (pbk)
ISBN: 9781003478119 (ebk)

DOI: 10.1201/9781003478119

Typeset in Times
by Apex CoVantage, LLC

Contents

Preface

Continuing the author's previous work on predictive modeling, this book presents the most recent advances in high-order predictive modeling, beginning with the mathematical framework of the 2nd-BERRU-PM methodology, an acronym that designates the "2nd-order best-estimate with reduced uncertainties (2nd-BERRU) predictive modeling (PM)." The 2nd-BERRU-PM methodology is fundamentally anchored in physics-based principles stemming from thermodynamics (maximum entropy principle) and information theory, being formulated in the most inclusive possible phase-space, namely the combined phase-space of computed and measured parameters and responses. The 2nd-BERRU-PM methodology provides second-order output (means and variances) but can incorporate, as input, arbitrarily high-order sensitivities of responses with respect to model parameters, as well as arbitrarily high-order moments of the initial distribution of uncertain model parameters, in order to predict best-estimate mean values for the model responses (i.e., results of interest) and calibrated model parameters, along with reduced predicted variances and covariances for these predicted responses and parameters.

As discussed in Chapter 2, the 2nd-BERRU-PM methodology comprises two variants, depending on whether the model under consideration is incorporated deterministically or probabilistically. It is shown that the results produced by the 2nd-BERRU-PM methodology include and extend the results produced by the extant data assimilation and/or data adjustment procedures while overcoming the fundamental limitations of these procedures. The 2nd-BERRU-PM methodology eliminates the need for "least-square minimizing a user-defined functional" meant to represent the discrepancies between computations and experiments and is significantly more efficient computationally than the extant data assimilation and/or data adjustment procedures since it requires the inversion of a single matrix of the size given by the total number of simultaneously considered responses, which is considerably smaller than the matrices needed to be inverted in data assimilation procedures, which have dimensions equal to the number of model parameters (much larger in practice than the number of model responses).

The capabilities of the 2nd-BERRU-PM methodology are illustrated in Chapter 3 by considering the polyethylene-reflected plutonium (acronym: PERP) OECD/NEA reactor physics benchmark (Valentine, 2006). This benchmark is modeled using the neutron transport Boltzmann equation comprising 21976 uncertain parameters, the solution of which is representative of "large-scale computations." The response of interest for this benchmark is the leakage of neutrons through the outer sphere of the plutonium benchmark. The application of the 2nd-BERRU-PM methodology is illustrated by considering apparently consistent and inconsistent measurements and computations: "consistent" means that the standard deviations of the measured responses overlap with the standard deviations of the computed responses, while "inconsistent" means that the standard deviation of the measured response does not initially overlap with the standard deviation of the computed response. It is shown that the predicted inconsistency diminishes when applying the 2nd-BERRU-PM methodology as higher-order sensitivities are progressively included, thus highlighting the significant impact of sensitivities of orders two through four. The impact of the precision ("high," "typical," "low") of the measured responses and of the model parameters is also underscored, considering various combinations thereof.

Chapter 3 presents a novel generic 4th-order moment-constrained maximum entropy distribution, which is used in Chapter 4 to construct the 4th-order 4th-BERRU-PM methodology, which generalizes to 4th-order the 2nd-BERRU-PM methodology, yielding 4th-order expressions for the following best-estimate posterior quantities: (i) mean values and covariances for the predicted responses; (ii) mean values and covariances for the calibrated model parameters; (iii) correlations between the predicted responses and calibrated model parameters; (iv) triple correlations among predicted responses and calibrated model parameters; (v) quadruple correlations among predicted responses and calibrated model parameters; (vi) a quantitative indicator of consistency among computed and measured responses and parameters.

In Chapter 5, the 4th-BERRU-PM methodology is applied to the PERP reactor physics benchmark, underscoring the methodology's computational efficiency for obtaining hitherto unavailable high-order expressions for the previously mentioned predicted quantities while highlighting the impact of high-order sensitivities in conjunction with the impact of the precision of the measured responses and model parameters.

For completeness and convenient referencing, the mathematical framework of the 4th-order comprehensive adjoint sensitivity analysis methodology is presented in Chapter 6, and the particular equations and expressions produced by applying this methodology to the neutron transport equation that models the PERP benchmark are presented in Chapter 7.

The author gratefully acknowledges the constructive reviewing and recommendations provided by Dr. Mihaela Ionescu-Bujor, a scientist in the Fusion Programme at the Karlsruhe Institute of Technology, and the essential computational contributions by Dr. Ruixian Fang, Research Associate Professor at the University of South Carolina. Last, but certainly not least, the publication of this book would not have been possible without the patience and unwavering support of Mr. Robert Ross and the assistance of Ms. Elizabeth Hawkins of Taylor and Francis.

About the Author

Dan Gabriel Cacuci is Distinguished Professor Emeritus in the Department of Mechanical Engineering at the University of South Carolina and the Karlsruhe Institute of Technology, Germany. He received his PhD in applied physics, mechanical, and nuclear engineering from Columbia University. He is also the recipient of many awards including four honorary doctorates, the Ernest Orlando Lawrence Memorial award from the US Department of Energy, and the Arthur Holly Compton, Eugene P. Wigner, and the Glenn Seaborg Awards from the American Nuclear Society. He was named an "Inaugural Highly Ranked Scholar" by Scholar GPS, being ranked #2 in the world in the field of *Uncertainty Analysis*, #5 in the world in the field of *Sensitivity Analysis*, and ranked in the top 0.05% of all scholars worldwide.

This is Dr. Cacuci's fifth book for CRC Press. The others include *The Second-Order Adjoint Sensitivity Analysis Methodology* (2018); *Computational Methods for Data Evaluation and Assimilation* with Ionel Michael Navon and Mihaela Ionescu-Bujor (2013), *Sensitivity and Uncertainty Analysis, Volume I Applications to Large-Scale Systems* (2003) and *Volume II* (2005) also with Mihaela Ionescu-Bujor and Ionel Michael Navon.

1 2nd-BERRU-PM *2nd-Order Maximum Entropy Predictive Modeling Methodology for Reducing Uncertainties in Predicted Model Responses and Parameters*

1.1 INTRODUCTION

Scientific progress stems from the judicious combination of experimental information with results produced by computational models, neither of which are perfectly accurate. On the one hand, computations are afflicted by errors stemming from numerical procedures, uncertain model parameters, boundary/initial conditions and/or imperfectly known physical processes. On the other hand, results of measurements are afflicted by experimental errors which means that around any reported experimental value there always exists a range of values that may also be plausibly representative of the true but unknown value of the measured quantity. Extracting "best-estimate" values for model parameters and predicted results, together with "best-estimate" uncertainties for these parameters and results, requires reasoning from incomplete, error-afflicted, and occasionally discrepant information while combining experimental and computational data and their uncertainties. Under ideal circumstances, the result of predictive modeling is a probabilistic description of possible future outcomes based on all recognized errors and uncertainties.

The earliest systematic activities aimed at extracting best-estimate values for "model parameters" and thereby also improving the "model responses" (i.e., results computed using the respective models) were initiated in the 1960s in the field of nuclear reactor physics (see, e.g., Humi et al., 1964; Cecchini et al., 1964; Usachhev, 1964). The computational models used in these activities were time-independent solvers of the neutron transport and/or diffusion equation, comprising various imprecisely known neutron cross-sections as model parameters; the model responses were "integral quantities" such as reaction rates and multiplication factors. The improvements, called "adjustments," in the neutron cross-sections under consideration arose from the combination of *measured* and *computed* values of the "integral quantities."

DOI: 10.1201/9781003478119-1

These activities reached conceptual maturity under the name of "cross-section adjustment" in the 1970s (see, e.g., Rowlands et al., 1973; Gandini and Petilli, 1973). The cross-section adjustment procedure uses a weighted least-square functional that combines uncertainties in the computed model response with uncertainties in the experimentally measured response, subject to the constraint imposed by the linearized neutron transport and/or diffusion model. The weights in this least-square functional are the 1st-order functional derivatives (called "response sensitivities") of the model's response with respect to the model parameters. The resulting "adjusted" neutron cross-sections (model parameters) and their "adjusted" uncertainties were subsequently employed in the respective computational model to predict improved model responses (e.g., improved reaction rates and reaction rate ratios, reactor multiplication factors, Doppler coefficients) in an extended application domain (e.g., a new or improved reactor core design). By the late 1970s, adjoint neutron fluxes were introduced and used (see, e.g., Kuroi and Mitani, 1975; Dragt et al., 1977; Weisbin et al., 1978) to compute efficiently the 1st-order response sensitivities which served as weighting functions in the least squares adjustment procedure. It is important to note that all of these works dealt with the time-independent linear neutron transport or diffusion equation, as encountered in reactor physics and reactor shielding, for which the corresponding adjoint equations were already known and readily available. For *nonlinear* systems, the "adjoint method" for computing efficiently response sensitivities with respect to model parameters was generally formulated by Cacuci (1981a, 1981b), who also pioneered its introduction into the earth sciences (see, e.g., Prăger and Kelemen, 2013; Luo et al., 2020). Since then, Cacuci (2022, 2023a) has generalized his adjoint method of sensitivity analysis to enable the computation of sensitivities of arbitrarily-high order while applying this method to many problems, in various fields.

A pioneering formalism for performing data adjustment in the context of time-dependent nonlinear system was presented by Cacuci's group at Oak Ridge National Laboratory (Barhen et al., 1982). However, this advanced (for its time) methodology stagnated in the field of nuclear engineering although it may have inspired later developments in the geophysical sciences. In the late 1980s and during the 1990s, the fundamental concepts underlying "data adjustment" seem to have been rediscovered in the geophysical sciences while developing the so-called "data assimilation" procedure, in that the concepts underlying data assimilation are the same as those underlying the previously developed "data adjustment" procedure. Since then, numerous works have been published on data assimilation; the most representative can be found cited in the books by Lewis et al. (2006), Lahoz et al. (2010), and Cacuci et al. (2014). Fundamentally, both the data adjustment and the data assimilation procedures rely on the "least-squares" minimization of a user-defined quadratic functional that represents the weighted errors between measured and computed responses.

The predictive modeling methodology developed by Cacuci and Ionescu-Bujor (2010) uses the maximum entropy (MaxEnt) principle originally formulated by Jaynes (1957) to dispense with the need to minimize an a priori chosen "cost functional" thus generalizing and significantly extending the data adjustment and data assimilation procedures. This methodology was called "Best-Estimate Results with Reduced

Uncertainties Predictive Modeling (BERRU-PM)" and can be used for both "forward predictions" and "inverse predictions" within an a priori selected convergence criterion, without user intervention and/or introduction of arbitrary "regularization parameters," as the conventional procedures for analyzing inverse problems need to do. The BERRU-PM methodology was extended by Cacuci (2014) from a single multi-physics system to coupled multi-physics systems, such that the systems can be considered sequentially rather than simultaneously while preserving exactly the same results as if the systems were treated simultaneously. Consequently, very large coupled systems, which could perhaps exceed available computational resources if treated simultaneously, can be treated sequentially using the BERRU-PM methodology, without any loss of generality or information, while requiring only the computational resources that would be needed for treating each system separately. The mathematical framework of the BERRU-PM methodology and applications to both direct and inverse models of large-scale installations (e.g., OECD/NEA ICSBEP reactor physics benchmarks, Savannah River National Laboratory cooling towers, etc.) are presented in the book by Cacuci (2019).

The BERRU-PM methodology (Cacuci, 2019) has recently been extended by Cacuci (2023b, 2023c) to include the full second-order representation of the computed responses, thereby creating the "*2nd-BERRU-PM*" methodology, where the attribute "2nd" indicates that this methodology incorporates second-order uncertainties (means and covariances) and 2nd-order sensitivities of computed model responses to model parameters, and where the acronym BERRU stands, as before, for "Best-Estimate Results with Reduced Uncertainties." The 2nd-BERRU-PM methodology is fundamentally based on the MaxEnt principle. There are two variants of the 2nd-BERRU-PM methodology, depending on whether the computational model is being combined deterministically or probabilistically with the experimental information. The deterministic incorporation of the computational model gives rise to the "2nd-BERRU-PMD" methodology (where the letter "*D*" in the acronym indicates "deterministic" inclusion of the computational model responses), while the probabilistic incorporation of the computational model gives rise to the "2nd-BERRU-PMP" methodology (where the letter "*P*" in the acronym indicates "probabilistic" inclusion of the computational model responses). The material presented in this chapter follows closely this original work (Cacuci, 2023b, 2023c).

This chapter is structured as follows: Section 1.2 presents the mathematical modeling of the physical system under consideration. Section 1.3 presents the generic construction of the "minimally discrepant maximum entropy distribution," which is subsequently used in Section 1.4 to construct the "2nd-order minimally discrepant maximum entropy distribution" for experimentally measured responses and parameters." The generic 2nd-order methodology presented in Section 1.4 is then used in Sections 1.5 and 1.6 to construct the 2nd-BERRU-PMD and, respectively, the 2nd-BERRU-PMP methodologies. The 2nd-BERRU-PMP and 2nd-BERRU-PMD are compared to each other in Section 1.7 and are compared to the current versions of the data adjustment and data assimilation procedures in Section 1.8. This chapter is concluded by a summarizing discussion presented in Section 1.9.

1.2 GENERIC MATHEMATICAL MODELING OF A PHYSICAL SYSTEM

The generic modeling of a physical system and/or of a result of an indirect experimental measurement requires consideration of the following modeling components:

1. A mathematical model comprising linear and/or nonlinear equations that relate the system's independent variables and parameters to the system's state (i.e., dependent) variables;
2. Statistical information (e.g., means, covariances, triple correlations, etc.), inequality and/or equality constraints that characterize the ranges of the system's parameters;
3. One or several computational results, customarily referred to as system or model *responses* (or "objective functions", or "indices of performance"), which are computed using the mathematical model; and
4. Experimentally measured responses, characterized by corresponding nominal (mean) values and uncertainties (variances, covariances, skewness, kurtosis, etc.).

A nonlinear physical system can be generally modeled by means of coupled equations which can be represented in operator form as follows:

$$\mathbf{N}\left[\mathbf{u}(\mathbf{x}),\boldsymbol{\alpha}\right]=\mathbf{Q}(\mathbf{x},\boldsymbol{\alpha}),\ \mathbf{x}\in\Omega_x(\boldsymbol{\alpha}), \tag{1.1}$$

Boundary and/or initial conditions must also be provided if differential operators appear in Eq. (1.1). In operator form, these boundaries and/or initial conditions are represented as follows:

$$\mathbf{B}\left[\mathbf{u}(\mathbf{x});\boldsymbol{\alpha};\mathbf{x}\right]-\mathbf{B}^s(\mathbf{x},\boldsymbol{\alpha})=\mathbf{0},\ \mathbf{x}\in\partial\Omega_x(\boldsymbol{\alpha}), \tag{1.2}$$

where the vector $\mathbf{B}^s(\mathbf{x},\boldsymbol{\alpha})$ represents "source terms on the model's boundary." In this chapter and throughout the book, matrices will be denoted using capital bold letters while vectors will be denoted using either capital or lower-case bold letters. The symbol "$\triangleq$" will be used to denote "is defined as" or "is by definition equal to." Transposition will be indicated by a dagger $(\dagger)$ superscript. The equalities in this work are considered to hold in the weak ("distributional") sense. The right sides of Eqs. (1.1) and (1.2), and various equations to be derived in this work, may contain "generalized functions/functionals," particularly Dirac distributions and derivatives thereof. The quantities which appear in Eqs. (1.1) and (1.2) are defined in what follows.

1. The quantity $\boldsymbol{\alpha}\triangleq(\alpha_1,\ldots,\alpha_{TP})^\dagger$ is defined to be a *TP*-dimensional column vector, with components $\alpha_1,\ldots,\alpha_{TP}$, which denote the model's imprecisely known parameters. The quantity "*TP*" denotes the "total number of model parameters." Without loss of generality, the model parameters $\boldsymbol{\alpha}\triangleq(\alpha_1,\ldots,\alpha_{TP})^\dagger\in D_\alpha\in\mathbb{R}^{TP}$ can be considered to be real scalars defined

over a domain D_α, which is included in a *TP*-dimensional subset of the $\mathbb{R}^{TP}$. These model parameters usually stem from processes that are external to the physical system under consideration and their precise value is seldom, if ever, known. The components of the *TP*-dimensional column vector $\boldsymbol{\alpha} \in \mathbb{R}^{TP}$ are considered to also include imprecisely known geometrical parameters that characterize the physical system's boundaries in the phase space of the model's independent variables. In practice, the model parameters can be considered to be quasi-random scalar-valued quantities which follow an unknown multivariate distribution denoted as $p_\alpha(\boldsymbol{\alpha})$. The first four moments of $p_\alpha(\boldsymbol{\alpha})$ are defined as follows:

(i) The nominal (or expected, or mean) values of the model parameters are denoted as α_i^0 and are formally defined as follows:

$$\alpha_i^0 \triangleq \int_{D_\alpha} \alpha_i p_\alpha(\boldsymbol{\alpha}) d\boldsymbol{\alpha}, \; i = 1, ..., TP. \tag{1.3}$$

The nominal values α_i^0 are considered to be the components of the *TP*-component column "vector of nominal parameter values" defined as follows: $\boldsymbol{\alpha}^0 \triangleq \left(\alpha_1^0, ..., \alpha_{TP}^0\right)^\dagger$.

(ii) The covariance, $\text{cov}(\alpha_i, \alpha_j) \triangleq c_{ij}^\alpha$, of two model parameters, α_i and α_j, is defined as follows:

$$\begin{aligned} &\text{cov}(\alpha_i, \alpha_j) \triangleq c_{ij}^\alpha \triangleq \int_{D_\alpha} (\delta\alpha_i)(\delta\alpha_j) p_\alpha(\boldsymbol{\alpha}) d\boldsymbol{\alpha}; \\ &\delta\alpha_i \triangleq \alpha_i - \alpha_i^0; \; i, j = 1, ..., TP. \end{aligned} \tag{1.4}$$

The covariances $\text{cov}(\alpha_i, \alpha_j) \triangleq c_{ij}^\alpha$ are considered to be the components of the parameter covariance matrix, denoted as $\mathbf{C}_{\alpha\alpha} \triangleq \left[\text{cov}(\alpha_i, \alpha_j)\right]_{TP \times TP}$ $\triangleq \left[c_{ij}^\alpha\right]_{TP \times TP}$. The standard deviation of a parameter α_i is defined as follows: $\sigma_i^\alpha \triangleq \sqrt{c_{ii}^\alpha}$, $i = 1, \dots, TP$.

(iii) The triple correlations of three model parameters α_i, α_j, and α_k, are denoted as t_{ijk}^α, $i, j, k = 1, \dots, TP$, and are defined as follows:

$$t_{ijk}^\alpha \triangleq \int_{D_\alpha} (\delta\alpha_i)(\delta\alpha_j)(\delta\alpha_k) p_\alpha(\boldsymbol{\alpha}) d\boldsymbol{\alpha}; \quad i, j, k = 1, \dots, TP; \tag{1.5}$$

In particular, the triple self-correlation for a parameter, α_i, is defined as follows:

$$t_i^\alpha \triangleq \int_{D_\alpha} (\delta\alpha_i)^3 p_\alpha(\boldsymbol{\alpha}) d\boldsymbol{\alpha}; \quad i = 1, \dots, TP; \tag{1.6}$$

(iv) The quadruple correlations of four model parameters α_i, α_j, α_k, and α_ℓ, are denoted as $q_{ijk\ell}^\alpha$ and are defined as follows for $i, j, k, \ell = 1, ..., TR$:

$$q_{ijk\ell}^{\alpha} \triangleq \int_{D_\alpha} (\delta\alpha_i)(\delta\alpha_j)(\delta\alpha_k)(\delta\alpha_\ell)\, p_\alpha(\boldsymbol{\alpha})\, d\boldsymbol{\alpha};\; i,j,k,\ell = 1,\ldots,TP. \quad (1.7)$$

In particular, the quadruple self-correlation for a parameter, α_i, is defined as follows:

$$q_i^{\alpha} \triangleq \int_{D_\alpha} (\delta\alpha_i)^4\, p_\alpha(\boldsymbol{\alpha})\, d\boldsymbol{\alpha};\quad i = 1,\ldots,TP; \quad (1.8)$$

2. The generic nonlinear model represented by Eqs. (1.1) and (1.2) are considered to comprise TI independent variables which are denoted as x_i, $i = 1,...,TI$ and considered to be real-valued quantities. These independent variables are considered to be the components of a TI-dimensional column vector denoted as $\mathbf{x} \triangleq (x_1,\ldots,x_{TI})^\dagger \in \mathbb{R}^{TI}$, where the sub/superscript "TI" denotes the "total number of independent variables." The vector $\mathbf{x} \in \mathbb{R}^{TI}$ of independent variables is considered to be defined on a phase-space domain which will be denoted as $\Omega(\boldsymbol{\alpha})$ and which is defined as follows: $\Omega(\boldsymbol{\alpha}) \triangleq \{-\infty \le \lambda_i(\boldsymbol{\alpha}) \le x_i \le \omega_i(\boldsymbol{\alpha}) \le \infty;\, i = 1,...,TI\}$. The lower boundary-point of an independent variable is denoted as $\lambda_i(\boldsymbol{\alpha})$ and the corresponding upper boundary-point is denoted as $\omega_i(\boldsymbol{\alpha})$. The boundary $\Omega(\boldsymbol{\alpha})$ is also considered to be imprecisely known since it may depend on both geometrical parameters and material properties. A typical example of boundaries that depend on both geometrical parameters and material properties, are the "boundaries facing vacuum" in models based on diffusion theory, where conditions are imposed on the "extrapolated boundary" of the respective spatial domain. The "extrapolated boundary" depends both on the imprecisely known physical dimensions of the problem's domain and also on the medium's imprecisely known material properties, such as atomic number densities and microscopic transport cross-sections. The boundary of $\Omega(\boldsymbol{\alpha})$, denoted as $\partial\Omega(\boldsymbol{\alpha}) \triangleq \{\lambda_i(\boldsymbol{\alpha}) \cup \omega_i(\boldsymbol{\alpha}),\, i = 1,...,TI\}$, includes the set of the endpoints $\lambda_i(\boldsymbol{\alpha})$, $\omega_i(\boldsymbol{\alpha})$, $i = 1,...,TI$,of the respective intervals on which the components of $\mathbf{x}$ are defined.
3. The vector $\mathbf{u}(\mathbf{x}) \triangleq [u_1(\mathbf{x}),\ldots,u_{TD}(\mathbf{x})]^\dagger$ is a TD-dimensional column vector with components $u_i(\mathbf{x})$, $i = 1,...,TD$, which represent the model's dependent variables (also called the model's "state functions"). The abbreviation "TD" denotes "total number of dependent variables." Without loss of generality, it can be considered that $\mathbf{u}(\mathbf{x}) \in \mathscr{E}_u$, where $\mathscr{E}_u$ is a normed linear space over the scalar field $\mathscr{F}$ of real numbers.
4. The quantity $\mathbf{N}[\mathbf{u}(\mathbf{x});\boldsymbol{\alpha}] \triangleq [N_1(\mathbf{u};\boldsymbol{\alpha}),\ldots,N_{TD}(\mathbf{u};\boldsymbol{\alpha})]^\dagger$ is a TD-dimensional column vector comprising the left sides of the equations underlying the mathematical/computational model. The components $N_i(\mathbf{u};\boldsymbol{\alpha})$, $i = 1,...,TD$, are operators (including differential, difference, integral, distributions, and/or finite or infinite matrices) acting nonlinearly on the dependent variables $\mathbf{u}(\mathbf{x})$, the independent variables $\mathbf{x}$ and the model parameters $\boldsymbol{\alpha}$.

The mapping $\mathbf{N}(\mathbf{u};\boldsymbol{\alpha})$ is defined on the combined domains of the model's parameters and state functions.

5. The vector $\mathbf{Q}(\mathbf{x},\boldsymbol{\alpha}) \triangleq [q_1(\mathbf{x};\boldsymbol{\alpha}),....,q_{TD}(\mathbf{x};\boldsymbol{\alpha})]^\dagger \in \mathscr{E}_Q$ is a TD-dimensional column vector which represents inhomogeneous source terms (i.e., the right sides of the equations underlying the model), which usually depend nonlinearly on the uncertain parameters $\boldsymbol{\alpha}$. The vector $\mathbf{Q}(\mathbf{x},\boldsymbol{\alpha})$ is defined on a normed linear space denoted as $\mathscr{E}_Q$.
6. In Eq. (1.2), the vector $\mathbf{B}(\mathbf{u};\boldsymbol{\alpha}) \triangleq [B_1(\mathbf{u};\boldsymbol{\alpha}),...,B_{TB}(\mathbf{u};\boldsymbol{\alpha})]^\dagger$, where the subscript "$TB$" denotes "total number of boundary conditions," comprises components $B_i(\mathbf{u};\boldsymbol{\alpha})$, $i = 1,...,TB$, which operate nonlinearly on $\mathbf{u}(\mathbf{x})$ and $\boldsymbol{\alpha}$, and are defined on the boundary $\partial\Omega_x(\boldsymbol{\alpha})$ of the model's domain $\Omega_x(\boldsymbol{\alpha})$. The vector $\mathbf{B}^s(\mathbf{x},\boldsymbol{\alpha}) \triangleq [B_1^s(\mathbf{x};\boldsymbol{\alpha}),...,B_{TB}^s(\mathbf{x};\boldsymbol{\alpha})]^\dagger$ comprises inhomogeneous boundary sources (indicated by the superscript "s") which are nonlinear functions of $\boldsymbol{\alpha}$. The column vector $\mathbf{0}$ has TB components, all of which are identically zero.

Solving Eqs. (1.1) and (1.2) at the nominal parameter values, $\boldsymbol{\alpha}^0$, provides the "nominal solution" $\mathbf{v}^0(\mathbf{x})$, in other words, the vectors $\mathbf{u}^0(\mathbf{x})$ and $\boldsymbol{\alpha}^0$ satisfy the following equations:

$$\mathbf{N}[\mathbf{u}^0(\mathbf{x});\boldsymbol{\alpha}^0] = \mathbf{Q}(\mathbf{x},\boldsymbol{\alpha}^0),\ \mathbf{x} \in \Omega_x, \tag{1.9}$$

$$\mathbf{B}[\mathbf{u}^0(\mathbf{x});\boldsymbol{\alpha}^0;\ \mathbf{x}] - \mathbf{B}^s(\mathbf{x},\boldsymbol{\alpha}^0) = \mathbf{0},\ \mathbf{x} \in \partial\Omega_x(\boldsymbol{\alpha}^0). \tag{1.10}$$

The results computed using a mathematical model are customarily called "model responses" (or "system responses" or "objective functions" or "indices of performance"). It is considered that there are a total number of TR such model responses, each of which can be considered to be a component of the "vector of model responses" $\mathbf{r} \triangleq (r_1,...r_k,...,r_{TR})^\dagger$. Each of these model responses is formally a function (implicit and/or explicit) of the dependent variables and model parameters $\boldsymbol{\alpha}$, which can be represented formally as follows:

$$r_k = r_k[\mathbf{u}(\mathbf{x});\boldsymbol{\alpha}];\ \ k = 1,...,TR. \tag{1.11}$$

In particular, a scalar-valued response, such as a measurement of a physical quantity that depends on the model's state functions and parameters, can be represented generically in the following integral form:

$$R[\mathbf{u}(\mathbf{x});\boldsymbol{\alpha}] \triangleq \int_{\lambda_1(\alpha)}^{\omega_1(\alpha)} ... \int_{\lambda_{TI}(\alpha)}^{\omega_{TI}(\alpha)} S(\mathbf{u},\boldsymbol{\alpha};\mathbf{x})\,dx_1...dx_{TI}, \tag{1.12}$$

where $S(\mathbf{u},\boldsymbol{\alpha};\mathbf{x})$ is a suitably defined differentiable function of $\mathbf{u}(\mathbf{x})$ and $\boldsymbol{\alpha}$. The model response $R[\mathbf{u}(\mathbf{x});\boldsymbol{\alpha}]$ will also be affected by the uncertainties that affect the endpoints $\lambda_i(\boldsymbol{\alpha})$, $\omega_i(\boldsymbol{\alpha})$, $i = 1,...,TI$, of $\partial\Omega(\boldsymbol{\alpha})$.

For an actual measurement of a quantity that involves the model's dependent variables, the function in the integrand of the response would take on the following particular form: $S(\mathbf{u},\boldsymbol{\alpha};\mathbf{x}) \triangleq F[\mathbf{u}(\mathbf{x});\boldsymbol{\alpha};\mathbf{x}]\delta[x_1 - x_1^p(\boldsymbol{\alpha})]\times\ldots\times\delta[x_{TI} - x_{TI}^p(\boldsymbol{\alpha})]$ where the function $F[\mathbf{u}(\mathbf{x});\boldsymbol{\alpha};\mathbf{x}]$ comprises the mathematical dependence of the measurement device on the model's dependent variable(s), where $\mathbf{x}^p(\boldsymbol{\alpha}) \triangleq [x_1^p(\boldsymbol{\alpha}),\ldots x_k^p(\boldsymbol{\alpha}),\ldots,x_{TI}^p(\boldsymbol{\alpha})]^\dagger$ denotes the location in phase-space of the specific "measurement point" and where the quantity $\delta[x_i - x_i^p(\boldsymbol{\alpha})]$ denotes the Dirac delta function. The measurement's location in phase space, $\mathbf{x}^p(\boldsymbol{\alpha})$, may itself be afflicted by measurement (experimental) uncertainties. Hence, the components of $\mathbf{x}^p(\boldsymbol{\alpha})$ must be included among the components of the vector $\boldsymbol{\alpha}$ of model parameters, even though the quantity $\mathbf{x}^p(\boldsymbol{\alpha})$ appears only in the definition of the response but does not appear in Eqs. (1.1) and (1.2) which mathematically define the physical model. Thus, the physical "system" is to be understood as being defined to comprise both the system's computational model and the system's responses. In most cases, the coordinates $x_k^p(\boldsymbol{\alpha})$, $k = 1,\ldots,TI$, will simply be independent (albeit uncertain) model parameters included among the components of the vector $\boldsymbol{\alpha}$, in which case $\partial x_k^p(\boldsymbol{\alpha})/\partial\alpha_n = 1,\ if\ \alpha_n \equiv x_k^p$ and $\partial x_k^p(\boldsymbol{\alpha})/\partial\alpha_n = 0,\ if\ \alpha_n \neq x_k^p$.

Each computed response, $r_k(\boldsymbol{\alpha})$, can be formally expanded in a multivariate Taylor series around the parameters' mean values. In particular, the 4th-order Taylor series of a model response around the expected (or nominal) parameter values $\boldsymbol{\alpha}^0$ has the following formal expression:

$$
\begin{aligned}
r_k(\boldsymbol{\alpha}) &= r_k(\boldsymbol{\alpha}^0) + \sum_{j_1=1}^{TP}\left\{\frac{\partial r_k(\boldsymbol{\alpha})}{\partial\alpha_{j_1}}\right\}_{\boldsymbol{\alpha}^0}\delta\alpha_{j_1} + \frac{1}{2}\sum_{j_1=1}^{TP}\sum_{j_2=1}^{TP}\left\{\frac{\partial^2 r_k(\boldsymbol{\alpha})}{\partial\alpha_{j_1}\partial\alpha_{j_2}}\right\}_{\boldsymbol{\alpha}^0}\delta\alpha_{j_1}\delta\alpha_{j_2} \\
&+ \frac{1}{3!}\sum_{j_1=1}^{TP}\sum_{j_2=1}^{TP}\sum_{j_3=1}^{TP}\left\{\frac{\partial^3 r_k(\boldsymbol{\alpha})}{\partial\alpha_{j_1}\partial\alpha_{j_2}\partial\alpha_{j_3}}\right\}_{\boldsymbol{\alpha}^0}\delta\alpha_{j_1}\delta\alpha_{j_2}\delta\alpha_{j_3} \qquad (1.13) \\
&+ \frac{1}{4!}\sum_{j_1=1}^{TP}\sum_{j_2=1}^{TP}\sum_{j_3=1}^{TP}\sum_{j_4=1}^{TP}\left\{\frac{\partial^4 r_k(\boldsymbol{\alpha})}{\partial\alpha_{j_1}\partial\alpha_{j_2}\partial\alpha_{j_3}\partial\alpha_{j_4}}\right\}_{\boldsymbol{\alpha}^0}\delta\alpha_{j_1}\delta\alpha_{j_2}\delta\alpha_{j_3}\delta\alpha_{j_4} + \varepsilon_k.
\end{aligned}
$$

In Eq. (1.13), the quantity $r_k(\boldsymbol{\alpha}^0)$ indicates the computed value of the response using the expected/nominal parameter values $\boldsymbol{\alpha}^0 \triangleq (\alpha_1^0,\ldots,\alpha_{TP}^0)^\dagger$. The notation $\{\ \}_{\boldsymbol{\alpha}^0}$ indicates that the quantities within the braces are also computed using the expected/nominal parameter values. The quantity ε_k in Eq. (1.13) comprises all quantifiable errors in the representation of the computed response as a function of the model parameters, including the truncation errors $O(\delta\alpha_j)^5$ of the Taylor series expansion, possible bias errors due to incompletely modeled physical phenomena, and possible random errors due to numerical approximations. The radius/domain of convergence of the series in Eq. (1.13) determines the largest values of the parameter variations $\delta\alpha_j$ which are admissible before the respective series becomes divergent. In turn, these maximum admissible parameter variations limit, through Eq. (1.13), the largest parameter covariances/standard deviations which can be considered for using the

Taylor expansion for the subsequent purposes of computing moments of the distribution of computed responses.

As is well-known, and as indicated by Eq. (1.13), the Taylor series of a function of TP-variables (e.g., $r_k(\boldsymbol{\alpha})$) comprises TP 1st-order derivatives, $TP(TP+1)/2$ distinct 2nd-order derivatives, and so on. The computation by conventional methods of the nth-order functional derivatives (called "sensitivities" in the field of sensitivity analysis) of a response with respect to the TP-parameters (on which it depends) would require at least $O(TP^n)$ large-scale computations. The exponential increase –with the order of response sensitivities– of the number of large-scale computations needed to determine higher-order sensitivities is the manifestation of the "curse of dimensionality in sensitivity analysis," by analogy to the expression coined by Bellman (1957) to express the difficulty of using "brute-force" grid search when optimizing a function with many input variables. The effects of the "curse of dimensionality" are already apparent when computing second-order sensitivities (Cacuci, 2018). The "n^{th}-order Comprehensive Adjoint Sensitivity Analysis Methodology for Response-Coupled Forward/Adjoint Linear Systems" (n^{th}-CASAM-L) conceived by Cacuci (2022) and the "n^{th}-order Comprehensive Adjoint Sensitivity Analysis Methodology for Nonlinear Systems" (n^{th}-CASAM-N) conceived by Cacuci (2023a) are currently the only methodologies that enable the exact and efficient computation of arbitrarily high-order sensitivities of model responses directly with respect to the model parameters while overcoming the curse of dimensionality. Both the nth-CASAM-L and the nth-CASAM-N are constructed on the same fundamental principles. The nth-CASAM-L aims specifically at sensitivity analysis of linear model involving responses that depend on (e.g., are linear or bilinear functionals of) both the forward and the adjoint variables (see, e.g., Lewins, 1965; Stacey, 1974), as is the case, for example, for eigenvalues in linear systems. Such situations cannot happen for nonlinear systems since only linear operators admit bona fide adjoint operators. If the response of a linear model depends only on the model's forward state functions (i.e., forward dependent variables) then the nth-CASAM-L can be treated as a particular (linear) case within the nth-CASAM-N. This particular situation applies to the OECD/NEA Polyethylene Plutonium (acronym: PERP) reactor physics benchmark (Valentine, 2006), which will be used as a paradigm model for illustrating numerically the methodologies presented in this work (book); the "neutron leakage response" is a function of just the benchmark's forward neutron flux (forward dependent variables), so its sensitivity analysis can be performed by using either the nth-CASAM-L or the nth-CASAM-N. As will be shown in the remainder of this chapter (and book), the 4th-order Taylor series expansion presented in Eq. (1.13) provides the starting point for constructing the expressions underlying the computations of the response expected values, variances/covariances, skewness, and kurtosis, which will subsequently be used for performing uncertainty quantification (UQ) of the effects of uncertainties in the model parameters on the uncertainty in the computed responses, and predictive modeling (including data assimilation and model calibration) to obtain best-estimate values for the predicted model responses and calibrated model parameters while reducing their accompanying predicted uncertainties. For convenient referencing, therefore, the mathematical underpinnings of the 4th-CASAM-N, which underlies the computation of the 1st-, 2nd-, 3rd-, and 4th-order sensitivities of model responses that involve just forward dependent variables, will be summarized in Chapter 6 of this book, while the specific equations that arise by applying the 4th-CASAM-N to the PERP benchmark will be presented in Chapter 7.

Uncertainties in the model's parameters will give rise to uncertainties in the computed model responses $r_k(\boldsymbol{\alpha})$. The computed model responses are considered to be distributed according to an unknown distribution denoted as $p_c(\mathbf{r})$. The unknown joint probability distribution of model parameters and responses will be denoted as $p_c(\boldsymbol{\alpha},\mathbf{r}) \triangleq p_\alpha(\boldsymbol{\alpha})\, p_c(\mathbf{r})$; this joint probability distribution is formally defined on a domain $D \triangleq D_\alpha \cup D_r$, where D_α denotes the domain of definition of the parameters and D_r denotes the domain of definition of the model responses. The approximate moments of the unknown distribution of $r_k(\boldsymbol{\alpha})$ are obtained by using the so-called "propagation of errors" methodology originally proposed by Tukey (1957), which entails the formal integration over $p_c(\boldsymbol{\alpha},\mathbf{r}) \triangleq p_\alpha(\boldsymbol{\alpha})\, p_c(\mathbf{r})$ of various expressions involving the truncated Taylor series expansion of the response provided in Eq. (1.13). Tukey's results were explicitly generalized to the 6th-order by Cacuci (2022).

The expectation value, $E_c(r_k)$, of a computed response $r_k(\boldsymbol{\alpha})$ is obtained by integrating formally Eq. (1.13) over the joint distribution $p_c(\boldsymbol{\alpha},\mathbf{r})$, which yields the following expression:

$$
\begin{aligned}
E_c(r_k) &\triangleq \int_D r_k(\boldsymbol{\alpha})\, p_c(\boldsymbol{\alpha},\mathbf{r})\, d\boldsymbol{\alpha}\, d\mathbf{r} \\
&= r_k(\boldsymbol{\alpha}^0) + \frac{1}{2}\sum_{j_1=1}^{TP}\sum_{j_2=1}^{TP}\left\{\frac{\partial^2 r_k(\boldsymbol{\alpha})}{\partial \alpha_{j_1}\partial \alpha_{j_2}}\right\}_{\boldsymbol{\alpha}^0} c^{\alpha}_{j_1 j_2} \\
&+ \frac{1}{6}\sum_{j_1=1}^{TP}\sum_{j_2=1}^{TP}\sum_{j_3=1}^{TP}\left\{\frac{\partial^3 r_k(\boldsymbol{\alpha})}{\partial \alpha_{j_1}\partial \alpha_{j_2}\partial \alpha_{j_3}}\right\}_{\boldsymbol{\alpha}^0} t^{\alpha}_{j_1 j_2 j_3} \\
&+ \frac{1}{24}\sum_{j_1=1}^{TP}\sum_{j_2=1}^{TP}\sum_{j_3=1}^{TP}\sum_{j_4=1}^{TP}\left\{\frac{\partial^4 r_k(\boldsymbol{\alpha})}{\partial \alpha_{j_1}\partial \alpha_{j_2}\partial \alpha_{j_3}\partial \alpha_{j_4}}\right\}_{\boldsymbol{\alpha}^0} q^{\alpha}_{j_1 j_2 j_3 j_4} + \ldots
\end{aligned}
\tag{1.14}
$$

The expectation values $E_c(r_k)$, $k = 1,\ldots,TR$, are considered to be the components of a vector defined as follows: $\mathbf{E}_c(\mathbf{r}) \triangleq \left[E_c(r_1),\ldots,E_c(r_k),\ldots,E_c(r_{TR})\right]^\dagger$. Notably, the expected value, $E_c(r_k)$, of the expected computed response, differs from the value, $r_k(\boldsymbol{\alpha}^0)$, of the model response computed at the nominal/mean value of the model parameters.

The expression of the correlation between a computed response and a parameter variance, which will be denoted as $\operatorname{cor}(\alpha_i, r_k)$, is presented in what follows:

$$
\begin{aligned}
\operatorname{cor}(\alpha_i, r_k) &\triangleq \int_D (\delta\alpha_i)(\delta r_k)\, p_c(\boldsymbol{\alpha},\mathbf{r})\, d\boldsymbol{\alpha}\, d\mathbf{r} = \sum_{j=1}^{TP}\left\{\frac{\partial r_k(\boldsymbol{\alpha})}{\partial \alpha_j}\right\}_{\boldsymbol{\alpha}^0} c^{\alpha}_{ij} \\
&+ \frac{1}{2}\sum_{j_1=1}^{TP}\sum_{j_2=1}^{TP}\left\{\frac{\partial^2 r_k(\boldsymbol{\alpha})}{\partial \alpha_{j_1}\partial \alpha_{j_2}}\right\}_{\boldsymbol{\alpha}^0} t^{\alpha}_{i,j_1 j_2} + \frac{1}{6}\sum_{j_1=1}^{TP}\sum_{j_2=1}^{TP}\sum_{j_3=1}^{TP}\left\{\frac{\partial^3 r_k(\boldsymbol{\alpha})}{\partial \alpha_{j_1}\partial \alpha_{j_2}\partial \alpha_{j_3}}\right\}_{\boldsymbol{\alpha}^0} q^{\alpha}_{i,j_1 j_2 j_3} + \ldots
\end{aligned}
\tag{1.15}
$$

where

$$\delta r_k \triangleq \left[r_k(\boldsymbol{\alpha}) - E_c(r_k) \right], \; k = 1,...,TR. \tag{1.16}$$

The parameter-response correlations, $\text{cor}(\alpha_i, r_k) = \text{cor}(r_k, \alpha_i)$, $i = 1,...,TP$, $k = 1,...,TR$, are considered to be the components of a "parameter-response computed correlation matrix" denoted as $\mathbf{C}^c_{\alpha r}$ and defined as follows:

$$\mathbf{C}^c_{\alpha r} \triangleq \begin{pmatrix} \text{cor}(\alpha_1, r_1) & \cdots & \text{cor}(\alpha_1, r_{TR}) \\ \vdots & \ddots & \vdots \\ \text{cor}(\alpha_{TP}, r_1) & \cdots & \text{cor}(\alpha_{TP}, r_{TR}) \end{pmatrix} = \left(\mathbf{C}^c_{r\alpha} \right)^{\dagger}. \tag{1.17}$$

The covariance between two responses r_k and r_ℓ, denoted as $\text{cov}(r_k, r_\ell)$, has the following expression:

$$\begin{aligned} \text{cov}(r_k, r_\ell) &\triangleq \int_D (\delta r_k)(\delta r_\ell)\, p_c(\boldsymbol{\alpha}, \mathbf{r})\, d\boldsymbol{\alpha} d\mathbf{r} = \sum_{i=1}^{TP}\sum_{j=1}^{TP} \left\{ \frac{\partial r_k(\boldsymbol{\alpha})}{\partial \alpha_i} \frac{\partial r_\ell(\boldsymbol{\alpha})}{\partial \alpha_j} \right\}_{\boldsymbol{\alpha}^0} c^{\alpha}_{ij} \\ &+ \frac{1}{2}\sum_{j_1=1}^{TP}\sum_{j_2=1}^{TP}\sum_{j_3=1}^{TP} \left\{ \frac{\partial^2 r_k(\boldsymbol{\alpha})}{\partial \alpha_{j_1} \partial \alpha_{j_2}} \frac{\partial r_\ell(\boldsymbol{\alpha})}{\partial \alpha_{j_3}} + \frac{\partial r_k(\boldsymbol{\alpha})}{\partial \alpha_{j_1}} \frac{\partial^2 r_\ell(\boldsymbol{\alpha})}{\partial \alpha_{j_2} \partial \alpha_{j_3}} \right\}_{\boldsymbol{\alpha}^0} t^{\alpha}_{j_1 j_2 j_3} \\ &+ \frac{1}{4}\sum_{j_1=1}^{TP}\sum_{j_2=1}^{TP}\sum_{j_3=1}^{TP}\sum_{j_4=1}^{TP} \left\{ \frac{\partial^2 r_k(\boldsymbol{\alpha})}{\partial \alpha_{j_1} \partial \alpha_{j_2}} \frac{\partial^2 r_\ell(\boldsymbol{\alpha})}{\partial \alpha_{j_2} \partial \alpha_{j_3}} \right\}_{\boldsymbol{\alpha}^0} \left(q^{\alpha}_{j_1 j_2 j_3 j_4} - c^{\alpha}_{j_1 j_2} c^{\alpha}_{j_3 j_4} \right) \\ &+ \frac{1}{6}\sum_{j_1=1}^{TP}\sum_{j_2=1}^{TP}\sum_{j_3=1}^{TP}\sum_{j_4=1}^{TP} \left\{ \frac{\partial^3 r_k(\boldsymbol{\alpha})}{\partial \alpha_{j_1} \partial \alpha_{j_2} \partial \alpha_{j_3}} \frac{\partial r_\ell(\boldsymbol{\alpha})}{\partial \alpha_{j_4}} + \frac{\partial r_k(\boldsymbol{\alpha})}{\partial \alpha_{j_1}} \frac{\partial^3 r_\ell(\boldsymbol{\alpha})}{\partial \alpha_{j_2} \partial \alpha_{j_3} \partial \alpha_{j_4}} \right\}_{\boldsymbol{\alpha}^0} q^{\alpha}_{j_1 j_2 j_3 j_4} + \ldots \end{aligned} \tag{1.18}$$

The covariances $\text{cov}(r_k, r_\ell)$, $k, \ell = 1,...,TR$, are considered to be the components of a "computed-responses covariance matrix" denoted as $\mathbf{C}^c_{rr}$ and defined as follows:

$$\mathbf{C}^c_{rr} \triangleq \begin{pmatrix} \text{cov}(r_1, r_1) & \cdots & \text{cov}(r_1, r_{TR}) \\ \vdots & \ddots & \vdots \\ \text{cov}(r_{TP}, r_1) & \cdots & \text{cov}(r_{TP}, r_{TR}) \end{pmatrix}. \tag{1.19}$$

The triple correlations, denoted as $\mu_3(r_k, r_\ell, r_m)$, among three responses r_k, r_ℓ and r_m, are defined as follows:

$$\begin{aligned} \mu_3(r_k, r_\ell, r_m) &\triangleq \int_D (\delta r_k)(\delta r_\ell)(\delta r_m)\, p_c(\boldsymbol{\alpha}, \mathbf{r})\, d\boldsymbol{\alpha} d\mathbf{r} \\ &= \sum_{j_1=1}^{TP}\sum_{j_2=1}^{TP}\sum_{j_3=1}^{TP} \left\{ \frac{\partial r_k(\boldsymbol{\alpha})}{\partial \alpha_{j_1}} \frac{\partial r_\ell(\boldsymbol{\alpha})}{\partial \alpha_{j_2}} \frac{\partial r_m(\boldsymbol{\alpha})}{\partial \alpha_{j_3}} \right\}_{\boldsymbol{\alpha}^0} t^{\alpha}_{j_1 j_2 j_3} \\ &+ \frac{1}{2}\sum_{j_1=1}^{TP}\sum_{j_2=1}^{TP}\sum_{j_3=1}^{TP}\sum_{j_4=1}^{TP} \left\{ \frac{\partial r_k(\boldsymbol{\alpha})}{\partial \alpha_{j_1}} \frac{\partial r_\ell(\boldsymbol{\alpha})}{\partial \alpha_{j_2}} \frac{\partial^2 r_m(\boldsymbol{\alpha})}{\partial \alpha_{j_3} \partial \alpha_{j_4}} \right. \left(q^{\alpha}_{j_1 j_2 j_3 j_4} - c^{\alpha}_{j_1 j_2} c^{\alpha}_{j_3 j_4} \right) \end{aligned} \tag{1.20}$$

$$+\frac{\partial r_k(\boldsymbol{\alpha})}{\partial \alpha_{j_1}}\frac{\partial^2 r_\ell(\boldsymbol{\alpha})}{\partial \alpha_{j_2}\partial \alpha_{j_3}}\frac{\partial r_m(\boldsymbol{\alpha})}{\partial \alpha_{j_4}}\left(q^{\alpha}_{j_1 j_2 j_3 j_4}-c^{\alpha}_{j_1 j_4}c^{\alpha}_{j_2 j_3}\right)\Bigg\}_{\boldsymbol{\alpha}^0}+\ldots$$

In particular, setting $k=\ell=m$ in Equation (1.20) yields the expression of the (3rd-order) triple self-correlation, $\mu_3(r_k)$, for a response r_k, which is defined as follows:

$$\mu_3(r_k)\triangleq\int_D(\delta r_k)^3\,p_c(\boldsymbol{\alpha},\mathbf{r})\,d\boldsymbol{\alpha}\,d\mathbf{r},\quad k=1,\ldots,TR. \tag{1.21}$$

The expressions of the triple correlations $\mu_3(\alpha_i,r_k,r_\ell)$ and $\mu_3(\alpha_i,\alpha_j,r_k)$ among parameters and responses will not be considered in this work but are provided by Cacuci (2022).

The expression of the quadruple correlations of the distribution of responses denoted as $\mu_4(r_k,r_\ell,r_m,r_n)$ among four responses, r_k, r_ℓ, r_m, and r_n, is defined as follows:

$$\mu_4(r_k,r_\ell,r_m,r_n)\triangleq\int_D(\delta r_k)(\delta r_\ell)(\delta r_m)(\delta r_n)\,p_c(\boldsymbol{\alpha},\mathbf{r})\,d\boldsymbol{\alpha}\,d\mathbf{r}$$
$$=\sum_{j_1=1}^{TP}\sum_{j_2=1}^{TP}\sum_{j_3=1}^{TP}\sum_{j_4=1}^{TP}\left\{\frac{\partial r_k(\boldsymbol{\alpha})}{\partial \alpha_{j_1}}\frac{\partial r_\ell(\boldsymbol{\alpha})}{\partial \alpha_{j_2}}\frac{\partial r_m(\boldsymbol{\alpha})}{\partial \alpha_{j_3}}\frac{\partial r_n(\boldsymbol{\alpha})}{\partial \alpha_{j_4}}\right\}_{\boldsymbol{\alpha}^0}q^{\alpha}_{j_1 j_2 j_3 j_4}+\ldots \tag{1.22}$$

In particular, setting $k=\ell=m=n$ in Equation (1.22) yields the 4th-order self-correlation, $\mu_4(r_k)$, for a response r_k, which is defined as follows:

$$\mu_4(r_k)\triangleq\int_D(\delta r_k)^4\,p_c(\boldsymbol{\alpha},\mathbf{r})\,d\boldsymbol{\alpha}\,d\mathbf{r},\quad k=1,\ldots,TR. \tag{1.23}$$

The expressions of the quadruple correlations among parameters and responses will not be considered in this work but are provided by Cacuci (2022).

1.3 CONSTRUCTION OF THE MINIMALLY DISCREPANT MAXIMUM ENTROPY DISTRIBUTION

When an unknown distribution, denoted as $p(\mathbf{z})$, needs to be reconstructed from a finite number of its known moments, the principle of maximum entropy (MaxEnt) originally formulated by Jaynes (1957) provides the optimal compatibility with the available information, while simultaneously ensuring minimal spurious information content. The generic construction of the constrained "minimum discrepancy distribution with a given function" aims at obtaining a mathematical expression for a distribution $p(\mathbf{z})$, defined over a domain $\mathbf{z}\in D$, which is normalized to unity and satisfies K independent constraints of the following form:

$$\int_D g_k(\mathbf{z})\,p(\mathbf{z})\,d\mathbf{z}=c_k<\infty,\quad k=1,\ldots,K; \tag{1.24}$$

with

$$\int_D p(\mathbf{z})\,d\mathbf{z} = 1. \tag{1.25}$$

In addition, it is required that the normalized distribution $p(\mathbf{z})$ be "minimally discrepant from" (i.e., "fit as well as possible") a known function $f(\mathbf{z})$, which is also defined over the same domain $\mathbf{z} \in D$.

The *discrepancy* (i.e., "lack of fit") between an *approximate distribution* $f(\mathbf{z})$ and a *true distribution* $p(\mathbf{z})$, both defined over the same domain $\mathbf{z} \in D$, is denoted as $\delta(f|p)$, and is defined as follows:

$$\delta(f|p) \triangleq \int_D p(\mathbf{z}) \log \frac{p(\mathbf{z})}{f(\mathbf{z})} d\mathbf{z}. \tag{1.26}$$

The discrepancy $\delta(f|p)$ is also called (see e.g., Bernardo and Smith, 2006) the "Kulback-Leibler divergence" or "relative entropy." The normalized distribution $p(\mathbf{z})$ which minimizes the discrepancy defined in Eq. (1.26) while satisfying the constraints expressed by Eqs. (1.24) and (1.25) are obtained by minimizing the Lagrangian functional $F(p)$ defined in (1.27):

$$\begin{aligned} F(p) &\triangleq \int_D p(\mathbf{z}) \log \frac{p(\mathbf{z})}{f(\mathbf{z})} d\mathbf{z} + \sum_{k=1}^{TR} \theta_k \left[\int_D g_k(\mathbf{z}) p(\mathbf{z}) d\mathbf{z} - c_k \right] \\ &+ c \left[\int_D p(\mathbf{z}) d\mathbf{z} - 1 \right]. \end{aligned} \tag{1.27}$$

In Eq. (1.27), the quantities c and $\theta_1, \ldots, \theta_k, \ldots, \theta_{TR}$ are Lagrange multipliers which remain to be determined. The stationary value of $F(p)$ is provided by the probability $p(\mathbf{z})$ where the first variation of $F(p)$ vanishes, in other words, $p(\mathbf{z})$ is the solution of the following equation

$$\begin{aligned} \delta F(p) &\triangleq \frac{d}{d\varepsilon} \left\{ F\left[p(\mathbf{z}) + \varepsilon \tau(\mathbf{z}) \right] \right\}_{\varepsilon=0} \\ &= \int_D \left[\log \frac{p(\mathbf{z})}{f(\mathbf{z})} + \sum_{k=1}^{TR} \theta_k g_k(\mathbf{z}) + (c+1) \right] \tau(\mathbf{z}) d\mathbf{z} = 0. \end{aligned} \tag{1.28}$$

The desired posterior distribution is the normalized solution of Eq. (1.28), which has the following expression:

$$p(\mathbf{z}) = \frac{1}{Z} f(\mathbf{z}) \exp\left[-\sum_{k=1}^{TR} \theta_k g_k(\mathbf{z}) \right]; \; Z \triangleq \int_D f(\mathbf{z}) \exp\left[-\sum_{k=1}^{TR} \theta_k g_k(\mathbf{z}) \right] d\mathbf{z}. \tag{1.29}$$

It is important to note that in the limiting case when the function $f(\mathbf{z})$ is the uniform distribution $f(\mathbf{z}) = \text{constant}$, the distribution $p(\mathbf{z})$ becomes a MaxEnt distribution. In other words, the distribution which is least discrepant from (i.e., "best-fits") the uniform distribution is the MaxEnt distribution.

1.4 CONSTRUCTION OF THE 2ND-ORDER MINIMALLY DISCREPANT MAXIMUM ENTROPY DISTRIBUTION OF EXPERIMENTALLY MEASURED RESPONSES AND PARAMETERS

The imprecisely known model parameters $\alpha_1,\ldots,\alpha_{TP}$, where "*TP*" denotes the "total number of model parameters," can be considered to be quasi-random scalar-valued quantities that follow an unknown multivariate distribution denoted as $p_\alpha(\boldsymbol{\alpha})$. The mean values (which in the context of computational modeling are called "nominal" values) of the model parameters will be denoted as α_i^0; the superscript "0" will be used throughout this monograph to denote "nominal values." These nominal values are formally defined as follows:

$$\alpha_i^0 \triangleq \int_{D_\alpha} \alpha_i p_\alpha(\boldsymbol{\alpha})\, d\boldsymbol{\alpha}, \quad i = 1,\ldots,TP. \tag{1.30}$$

The expected values of the measured parameters will be considered to constitute the components of a vector denoted as $\boldsymbol{\alpha}^0 \triangleq \left[\alpha_1^0,\ldots,\alpha_i^0,\ldots,\alpha_{TP}^0\right]^\dagger$. The covariance, $\text{cov}(\alpha_i,\alpha_j)$, of two model parameters, α_i and α_j, is defined as follows:

$$\text{cov}(\alpha_i,\alpha_j) \triangleq \int_{D_\alpha} \left(\alpha_i - \alpha_i^0\right)\left(\alpha_j - \alpha_j^0\right) p_\alpha(\boldsymbol{\alpha})\, d\boldsymbol{\alpha} \triangleq \rho_{ij}\sigma_i\sigma_j;\ i,j = 1,\ldots,TP. \tag{1.31}$$

The quantity σ_i denotes the standard deviation of parameter α_i; the quantity σ_j denotes the standard deviation of parameter α_j; the quantity ρ_{ij} denotes the *correlation* between the parameters α_i and α_j. In particular, the variance, $\text{var}(\alpha_i)$, of a parameter α_i, is formally defined as follows:

$$\text{var}(\alpha_i) \triangleq \int_{D_\alpha} (\delta\alpha_i)^2\, p_\alpha(\boldsymbol{\alpha})\, d\boldsymbol{\alpha} = \sigma_i^2; \quad i = 1,\ldots,TP, \tag{1.32}$$

where $\delta\alpha_i \triangleq \alpha_i - \alpha_i^0$ The covariances $\text{cov}(\alpha_i,\alpha_j)$ are considered to constitute the components of the covariance matrix for the model parameters. which will be denoted as $\mathbf{C}_{\alpha\alpha} \triangleq \left[\text{cov}(\alpha_i,\alpha_j)\right]_{TP\times TP}$.

Consider that the total number of *experimentally measured* responses are *TR*. These responses are subject to uncertainties and can be considered to be quasi-random scalar-valued quantities that follow an unknown multivariate normalized distribution denoted as $p_r(\mathbf{r})$, defined on a domain denoted as D_r. The 1st-order moments (mean values) of these experimentally measured responses will be denoted as r_i^e, $i = 1,\ldots,TR$, and are formally defined as follows:

$$r_i^e \triangleq \int_{D_r} r_i p_r(\mathbf{r}) d\mathbf{r}, \quad i = 1,\ldots,TR. \tag{1.33}$$

The letter "e" will be used either as a superscript or a superscript to indicate "experimentally measured" quantities. The expected values of the experimentally measured responses are considered to constitute the components of a vector denoted as $\mathbf{r}^e \triangleq \left(r_1^e,\ldots,r_{TR}^e\right)^\dagger$.

The 2nd-order moments (variances/covariances) of the experimentally measured responses will be denoted as $\text{cov}\left(r_i, r_j\right)_e$, $i, j = 1,\ldots,TR$, and are formally defined as follows:

$$\text{cov}\left(r_i, r_j\right)_e \triangleq \int_{D_r} \left(r_i - r_i^e\right)\left(r_j - r_j^e\right) p_r(\mathbf{r}) d\mathbf{r}; \quad i, j = 1,\ldots,TR. \tag{1.34}$$

The covariances of the measured responses are considered to be components of the $TR \times TR$-dimensional covariance matrix of measured responses, which will be denoted as $\mathbf{C}_{rr}^e \triangleq \left[\text{cov}\left(r_i, r_j\right)_e\right]_{TR\times TR}$.

In principle, it is also possible to obtain correlations between some measured responses and some model parameters. When such correlations between measured responses and measured model parameters are available, they will be denoted as $\text{cor}\left(\alpha_i, r_j\right)_e$, $i = 1,\ldots,TP; j = 1,\ldots,TR$, and are formally defined as follows:

$$\text{cor}\left(\alpha_i, r_j\right)_e \triangleq \int_{D_r} \left(\alpha_i - \alpha_i^0\right)\left(r_j - r_j^e\right) p_\alpha(\boldsymbol{\alpha}) p_r(\mathbf{r}) d\boldsymbol{\alpha}\, d\mathbf{r};$$
$$i = 1,\ldots,TP; j = 1,\ldots,TR. \tag{1.35}$$

The correlations $\text{cor}\left(\alpha_i, r_j\right)_e$, $i = 1,\ldots,TP; j = 1,\ldots,TR$, can formally be considered to be elements of a rectangular correlation matrix which will be denoted as $\mathbf{C}_{\alpha r}^e \triangleq \left[\text{cor}\left(\alpha_i, r_j\right)_e\right]_{TP\times TR}$.

When only the 1st-order and 2nd-order moments of the joint distribution of model parameters and responses are known, the expression of the corresponding MaxEnt distribution $p(\boldsymbol{\alpha}, \mathbf{r}) \triangleq p_\alpha(\boldsymbol{\alpha}) p_r(\mathbf{r})$, defined on a domain $D \triangleq D_\alpha \cup D_r$, is constructed by applying the principles illustrated in Section 1.3, which comprise the following requirements:

1. Require that $p(\boldsymbol{\alpha}, \mathbf{r})$ maximizes the Shannon information entropy (Shannon, 1948), denoted as$S$ and defined here:

$$S = -\int_D p(\boldsymbol{\alpha}, \mathbf{r}) \ln\left[p(\boldsymbol{\alpha}, \mathbf{r})\right] d\boldsymbol{\alpha} d\mathbf{r}, \tag{1.36}$$

2. Require that $p(\boldsymbol{\alpha},\mathbf{r})$ satisfies the following known 1st- and 2nd-order "moments constraints" defined in Eqs. (1.30)–(1.35).
3. Require that $p(\boldsymbol{\alpha},\mathbf{r})$ satisfies the normalization condition:

$$\int_D p(\boldsymbol{\alpha},\mathbf{r})\,d\boldsymbol{\alpha}\,d\mathbf{r}=1. \tag{1.37}$$

It follows from the generic considerations that led to the result shown in Eq. (1.29) that the expression of the MaxEnt distribution $p(\boldsymbol{\alpha},\mathbf{r})$ will be obtained as the solution of the constrained variational problem $\partial H(p)/\partial p=0$, where the entropy (Lagrangian functional) $H(p)$ is defined as follows:

$$\begin{aligned}
H(p)&=-\int_D p(\boldsymbol{\alpha},\mathbf{r})\ln\left[p(\boldsymbol{\alpha},\mathbf{r})\right]d\boldsymbol{\alpha}\,d\mathbf{r}-\lambda_0\left[\int_D p(\boldsymbol{\alpha},\mathbf{r})\,d\boldsymbol{\alpha}\,d\mathbf{r}-1\right]\\
&-\sum_{k=1}^{TR}\lambda_k^{(1)}\left[\int_D r_k\,p(\boldsymbol{\alpha},\mathbf{r})\,d\boldsymbol{\alpha}\,d\mathbf{r}-r_k^0\right]-\sum_{i=1}^{TP}\lambda_i^{(2)}\left[\int_D \alpha_i\,p(\boldsymbol{\alpha},\mathbf{r})\,d\boldsymbol{\alpha}\,d\mathbf{r}-\alpha_i^0\right]\\
&-\frac{1}{2}\sum_{k=1}^{TR}\sum_{\ell=1}^{TR}\lambda_{k\ell}^{(11)}\left[\int_D r_k r_\ell\,p(\boldsymbol{\alpha},\mathbf{r})\,d\boldsymbol{\alpha}\,d\mathbf{r}-cov(r_k,r_\ell)-r_k^0 r_\ell^0\right]\\
&-\sum_{k=1}^{TR}\sum_{i=1}^{TP}\lambda_{ki}^{(12)}\left[\int_D r_k\alpha_i\,p(\boldsymbol{\alpha},\mathbf{r})\,d\boldsymbol{\alpha}\,d\mathbf{r}-cov(r_k,\alpha_i)-r_k^0\alpha_i^0\right]\\
&-\frac{1}{2}\sum_{i=1}^{TP}\sum_{j=1}^{TP}\lambda_{ij}^{(22)}\left[\int_D \alpha_i\alpha_j\,p(\boldsymbol{\alpha},\mathbf{r})\,d\boldsymbol{\alpha}\,d\mathbf{r}-\mathrm{cov}(\alpha_i,\alpha_j)-\alpha_i^0\alpha_j^0\right].
\end{aligned} \tag{1.38}$$

In Eq. (1.38), the quantities $\lambda_k^{(1)}$, $\lambda_k^{(2)}$, $\lambda_{k\ell}^{(11)}$,$\lambda_{k\ell}^{(12)}$, and $\lambda_{k\ell}^{(22)}$ denote the respective Lagrange multipliers, and the factors 1/2 have been introduced for subsequent computational convenience.

Solving the equation $\partial H(p)/\partial p=0$ yields the following expression for the resulting MaxEnt distribution $p(\mathbf{z})$:

$$p_c(\mathbf{z})=\frac{1}{Z(\mathbf{b},\boldsymbol{\Lambda})}\exp\left(-\mathbf{b}^\dagger\mathbf{z}-\frac{1}{2}\mathbf{z}^\dagger\boldsymbol{\Lambda}\mathbf{z}\right), \tag{1.39}$$

where the various matrices and vectors are defined as follows:

$$\boldsymbol{\Lambda}\triangleq\begin{pmatrix}\boldsymbol{\Lambda}^{(11)} & \boldsymbol{\Lambda}^{(12)}\\ \left[\boldsymbol{\Lambda}^{(12)}\right]^\dagger & \boldsymbol{\Lambda}^{(22)}\end{pmatrix}; \tag{1.40}$$

$$\mathbf{z}\triangleq[\mathbf{r},\boldsymbol{\alpha}]^\dagger;\ \mathbf{z}\triangleq\left[\boldsymbol{\lambda}^{(1)},\boldsymbol{\lambda}^{(2)}\right]^\dagger;\boldsymbol{\lambda}^{(i)}\triangleq\left[\lambda_1^{(i)},\ldots,\lambda_{TR}^{(i)}\right]^\dagger;\ i=1,2. \tag{1.41}$$

with

$$\mathbf{\Lambda}^{(11)} \triangleq \begin{pmatrix} \lambda_{11}^{(11)} & \bullet & \lambda_{1,TR}^{(11)} \\ \bullet & \bullet & \bullet \\ \lambda_{TR,1}^{(11)} & \bullet & \lambda_{TR,TR}^{(11)} \end{pmatrix}; \mathbf{\Lambda}^{(12)} \triangleq \begin{pmatrix} \lambda_{11}^{(12)} & \bullet & \lambda_{1,TP}^{(12)} \\ \bullet & \bullet & \bullet \\ \lambda_{TR,1}^{(12)} & \bullet & \lambda_{TR,TP}^{(12)} \end{pmatrix}; \mathbf{\Lambda}^{(22)} \triangleq \begin{pmatrix} \lambda_{11}^{(22)} & \bullet & \lambda_{1,TP}^{(22)} \\ \bullet & \bullet & \bullet \\ \lambda_{TP,1}^{(22)} & \bullet & \lambda_{TP,TP}^{(22)} \end{pmatrix}.$$

The normalization constant $Z(\mathbf{b},\mathbf{\Lambda})$ which appears in Eq. (1.39) is defined as follows:

$$Z(\mathbf{b},\mathbf{\Lambda}) = \int_D \exp\left(-\mathbf{b}^\dagger \mathbf{z} - \frac{1}{2}\mathbf{z}^\dagger \mathbf{\Lambda} \mathbf{z}\right) d\mathbf{z}; \quad d\mathbf{z} \equiv d\boldsymbol{\alpha} d\mathbf{r}. \tag{1.42}$$

In statistical mechanics, the normalization constant Z is called the partition function (or sum over states) and carries all of the information available about the possible states of the system, while the MaxEnt distribution $p(\mathbf{z})$ is called the canonical Boltzmann-Gibbs distribution. The integral in Eq. (1.42) can be evaluated explicitly by conservatively extending the computational domain D to the entire multidimensional real vector space $\mathbb{R}^N$, where $N \triangleq TR + TP$, to obtain the following expression:

$$Z_c(\mathbf{b},\mathbf{\Lambda}_c) = \int_{\mathbb{R}^N} \exp\left(-\mathbf{b}^\dagger \mathbf{z} - \frac{1}{2}\mathbf{z}^\dagger \mathbf{\Lambda} \mathbf{z}\right) d\mathbf{z} = \frac{(2\pi)^{N/2}}{\sqrt{Det(\mathbf{\Lambda})}} e^{\left(\frac{1}{2}\mathbf{b}^\dagger \mathbf{\Lambda}^{-1} \mathbf{b}\right)}. \tag{1.43}$$

The Lagrange multipliers are determined in terms of the known information (means and covariances of parameters and responses) by differentiating the "free energy" $F(\mathbf{b},\mathbf{C}) \triangleq -\ln Z(\mathbf{b},\mathbf{C})$ with respect to the components of the vector $\mathbf{b} \triangleq \left(\boldsymbol{\lambda}^{(1)},\boldsymbol{\lambda}^{(2)}\right)^\dagger$ to obtain the following expressions:

$$\frac{\partial F(\mathbf{b},\mathbf{C})}{\partial \lambda_k^{(1)}} = \frac{1}{Z}\int_D r_k e^{\left(-\mathbf{b}^\dagger \mathbf{z} - \frac{1}{2}\mathbf{z}^\dagger \mathbf{\Lambda} \mathbf{z}\right)} d\mathbf{z} = \int_{D_c} r_k p(\boldsymbol{\alpha},\mathbf{r}) d\mathbf{z} = r_k^e; \; k = 1,\ldots,TR; \tag{1.44}$$

$$\begin{aligned} \frac{\partial F(\mathbf{b},\mathbf{C})}{\partial \lambda_i^{(2)}} &= \frac{1}{Z}\int_D \alpha_i e^{\left(-\mathbf{b}^\dagger \mathbf{z} - \frac{1}{2}\mathbf{z}^\dagger \mathbf{\Lambda} \mathbf{z}\right)} d\mathbf{z} \\ &= \int_D \alpha_i p(\boldsymbol{\alpha},\mathbf{r}) d\boldsymbol{\alpha} d\mathbf{r} = \alpha_i^0; \; i = 1,\ldots,TP. \end{aligned} \tag{1.45}$$

The results obtained in Eqs. (1.44) and (1.45) can be collectively written in vector-matrix form as follows:

$$\frac{\partial F(\mathbf{b},\mathbf{C})}{\partial \mathbf{b}} = \mathbf{z}_e; \quad \mathbf{z}_e \triangleq \left(\mathbf{r}^e,\boldsymbol{\alpha}^0\right)^\dagger. \tag{1.46}$$

On the other hand, it follows from Eq. (1.43) that:

$$\frac{\partial F(\mathbf{b},\mathbf{C})}{\partial \mathbf{b}} = -\frac{\partial \left[\ln Z(\mathbf{b},\mathbf{C}) \right]}{\partial \mathbf{b}} = -\boldsymbol{\Lambda}^{-1}\mathbf{b}. \tag{1.47}$$

The relations obtained in Eqs. (1.46) and (1.47) imply the following relation:

$$\mathbf{b} = -\boldsymbol{\Lambda}\mathbf{z}_e. \tag{1.48}$$

Differentiating a second time the relation provided in Eq. (1.44) or (1.45) yields the following relations:

$$\frac{\partial^2 F(\mathbf{b},\mathbf{C})}{\partial \lambda_j^{(1)} \partial \lambda_k^{(1)}} = -\frac{1}{Z^2}\frac{\partial Z}{\partial \lambda_j^{(1)}} r_k^e + \frac{1}{Z}\int_D (-r_j) r_k e^{\left(-\mathbf{b}^\dagger \mathbf{z} - \frac{1}{2}\mathbf{z}^\dagger \boldsymbol{\Lambda}\mathbf{z}\right)} d\mathbf{z}$$
$$= -\operatorname{cov}(r_j, r_k);\, j,k = 1,\ldots,TR. \tag{1.49}$$

$$\frac{\partial^2 F(\mathbf{b},\mathbf{C})}{\partial \lambda_i^{(2)} \partial \lambda_k^{(1)}} = -\operatorname{cov}(\alpha_i, r_k);\quad i = 1,\ldots,TP;\quad k = 1,\ldots,TR; \tag{1.50}$$

$$\frac{\partial^2 F(\mathbf{b},\mathbf{C})}{\partial \lambda_i^{(2)} \partial \lambda_j^{(2)}} = -\operatorname{cov}(\alpha_i, \alpha_j);\quad i,j = 1,\ldots,TP. \tag{1.51}$$

The results obtained in Eqs. (1.49)–(1.51) can be collectively written in vector-matrix form as follows:

$$\frac{\partial^2 F(\mathbf{b},\mathbf{C}_c)}{\partial \mathbf{b}\partial \mathbf{b}} = -\mathbf{C}_e;\ \mathbf{C}_e \triangleq \begin{pmatrix} \mathbf{C}_{rr}^e & \mathbf{C}_{r\alpha}^e \\ \mathbf{C}_{\alpha r}^e & \mathbf{C}_{\alpha\alpha} \end{pmatrix}. \tag{1.52}$$

On the other hand, it follows from Eq. (1.47) that:

$$\frac{\partial^2 F(\mathbf{b},\mathbf{C})}{\partial \mathbf{b}\partial \mathbf{b}} = -\boldsymbol{\Lambda}^{-1}. \tag{1.53}$$

The relations obtained in Eqs. (1.53) and (1.52) imply the following relation:

$$\boldsymbol{\Lambda}^{-1} = \mathbf{C}_e. \tag{1.54}$$

Introducing the results obtained in Eqs. (1.54) and (1.48) into Eq. (1.39) and (1.43) yields the following expression for the "2nd-order" MaxEnt joint distribution $p_e(\mathbf{z}|\mathbf{z}_e,\mathbf{C}_e)$ of the experimentally measured responses and parameters:

$$p_e(\mathbf{z}|\mathbf{z}_e,\mathbf{C}_e) = \frac{(2\pi)^{-(TR+TP)/2}}{\sqrt{Det(\mathbf{C}_e)}} \exp\left[-\frac{1}{2}(\mathbf{z}-\mathbf{z}_e)^\dagger \mathbf{C}_e^{-1} (\mathbf{z}-\mathbf{z}_e) \right], \tag{1.55}$$

where $Det(\mathbf{C}_e)$ denotes the determinant of the matrix $\mathbf{C}_e$.

1.5 2ND-BERRU-PMD: 2ND-ORDER MAXENT PREDICTIVE MODELING METHODOLOGY WITH DETERMINISTICALLY INCLUDED COMPUTED RESPONSES

This section presents the mathematical and physical considerations leading to the development of the "2nd-order MaxEnt predictive modeling methodology for obtaining best-estimate results with reduced uncertainties" by deterministically incorporating the computational model. This methodology is designated using the acronym 2nd-BERRU-PMD, where the attribute "2nd" indicates that this methodology incorporates second-order uncertainties (means and covariances) and 2nd-order sensitivities of computed model responses to model parameters, and where the acronym BERRU stands, as before, for "Best-Estimate Results with Reduced Uncertainties;" the last letter ("*D*") in the acronym indicates "deterministic," referring to the deterministic inclusion of the computational model responses. The 2nd-BERRU-PMD methodology is fundamentally based on the MaxEnt principle. The posterior 1st-order moments (best-estimate predicted mean values) and posterior 2nd-order moments (best-estimate predicted correlations) produced by the 2nd-BERRU-PMD methodology will be compared with the results produced by the 1st-order BERRU-PM methodology developed by Cacuci (2019). The 2nd-BERRU-PMD methodology will also be compared to the results produced by data assimilation (DA) procedures, which are formulated either just in the phase-space of measured responses ("observation-space formulation") or just in the phase-space of the model's dependent variables ("state-space formulation"). Consequently, even the so-called "2nd-order DA" can only calibrate initial conditions as "direct results" but cannot directly calibrate any other model parameters. Furthermore, 2nd-order DA methods fail fundamentally if the experiments are perfectly well-known and/or if the response measurements happen to coincide with the computed value of the response. Although such situations are not expected to occur frequently in practice, there are no negative consequences (should such a situation occur) if the 2nd-BERRU-PM methodology is used, in contradistinction to using the DA methodology. Furthermore, the 2nd-BERRU-PM methodology is significantly more efficient computationally than DA since it requires the inversion of a single matrix of the size given by the total number of simultaneously considered responses, in contradistinction with DA methods, where the smallest possible matrix to be inverted has the size of the number of discretization nodes/points of the grid designed for computing numerically the model's dependent variables. These advantages of the 2nd-BERRU-PM methodology over the DA methodology stem from the fact that the 2nd-BERRU-PM methodology is fundamentally anchored in physics-based principles (thermodynamics and information theory) formulated in the most inclusive possible phase-space (namely the combined phase-space of computed and measured parameters and responses), whereas the DA methodology is fundamentally based on the minimization of a subjective user-chosen functional.

Consistent with the consideration that only mean values and covariances (i.e., 1st- and 2nd-order distributional moments) are available, the 2nd-BERRU-PMD methodology considers that only 1st- and 2nd-order sensitivities of the *computed* responses with respect to the model parameters are available. In this case, the vector of model responses can be represented by the following 2nd-order Taylor series written in "vector/matrix" form:

$$\begin{gathered}\mathbf{r}=\mathbf{r}^{c}\left(\boldsymbol{\alpha}^{0}\right)+\left[\mathbf{S}\left(\boldsymbol{\alpha}^{0}\right)\right]\left(\boldsymbol{\alpha}-\boldsymbol{\alpha}^{0}\right)+\mathbf{p}\left(\boldsymbol{\alpha};\boldsymbol{\alpha}^{0}\right); \\ \mathbf{p}\left(\boldsymbol{\alpha};\boldsymbol{\alpha}^{0}\right)\triangleq\left[p_{1}\left(\boldsymbol{\alpha};\boldsymbol{\alpha}^{0}\right),\ldots,p_{TR}\left(\boldsymbol{\alpha};\boldsymbol{\alpha}^{0}\right)\right]^{\dagger}, \\ p_{k}\left(\boldsymbol{\alpha};\boldsymbol{\alpha}^{0}\right)\triangleq\frac{1}{2}\left(\boldsymbol{\alpha}-\boldsymbol{\alpha}^{0}\right)^{\dagger}\left[\mathbf{Q}_{k}\left(\boldsymbol{\alpha}^{0}\right)\right]\left(\boldsymbol{\alpha}-\boldsymbol{\alpha}^{0}\right),\end{gathered} \tag{1.56}$$

where the components of the vector $\mathbf{r}^{c}\left(\boldsymbol{\alpha}^{0}\right)\triangleq\left[r_{1}^{c}\left(\boldsymbol{\alpha}^{0}\right),\ldots,r_{TR}^{c}\left(\boldsymbol{\alpha}^{0}\right)\right]^{\dagger}$ represent the values of the model responses computed at the nominal parameter values $\boldsymbol{\alpha}^{0}$, while the other vectors and matrices contain the 1st-order and 2nd-order sensitivities of responses with respect to the model parameters, evaluated at the nominal parameter values $\boldsymbol{\alpha}^{0}$, and are defined here:

$$\begin{gathered}\mathbf{S}(\boldsymbol{\alpha})\triangleq\begin{pmatrix} s_{11} & \cdots & s_{1,TR} \\ \vdots & \ddots & \vdots \\ s_{TR,1} & \cdots & s_{TR,TR}\end{pmatrix}\triangleq\begin{pmatrix}\frac{\partial r_{1}(\boldsymbol{\alpha})}{\partial\alpha_{1}} & \cdots & \frac{\partial r_{1}(\boldsymbol{\alpha})}{\partial\alpha_{TP}} \\ \vdots & \ddots & \vdots \\ \frac{\partial r_{TR}(\boldsymbol{\alpha})}{\partial\alpha_{1}} & \cdots & \frac{\partial r_{TR}(\boldsymbol{\alpha})}{\partial\alpha_{TP}}\end{pmatrix}; \\ \mathbf{Q}_{k}(\boldsymbol{\alpha})\triangleq\begin{pmatrix} q_{11}^{(k)} & \cdots & q_{1,TP}^{(k)} \\ \vdots & \ddots & \vdots \\ q_{TP,1}^{(k)} & \cdots & q_{TP,TP}^{(k)}\end{pmatrix}\triangleq\begin{pmatrix}\frac{\partial^{2} r_{k}^{c}(\boldsymbol{\alpha})}{\partial\alpha_{1}\partial\alpha_{1}} & \cdots & \frac{\partial^{2} r_{k}^{c}(\boldsymbol{\alpha})}{\partial\alpha_{1}\partial\alpha_{TP}} \\ \vdots & \ddots & \vdots \\ \frac{\partial^{2} r_{k}^{c}(\boldsymbol{\alpha})}{\partial\alpha_{TP}\partial\alpha_{1}} & \cdots & \frac{\partial^{2} r_{k}^{c}(\boldsymbol{\alpha})}{\partial\alpha_{TP}\partial\alpha_{TP}}\end{pmatrix}.\end{gathered} \tag{1.57}$$

The merging of the experimental information with the computational information using the mathematical model is accomplished by seeking a mathematical representation of a "best fit" probability distribution $p(\mathbf{z})$, $\mathbf{z}\triangleq(\mathbf{r},\boldsymbol{\alpha})^{\dagger}$, which incorporates all of the 2nd-order information provided in the foregoing about the model parameters, measured, and computed responses. This "best fit" probability distribution must satisfy the following characteristics:

1. $p(\mathbf{z})$ is normalized to unity;
2. $p(\mathbf{z})$ approximates as closely as possible the experimental distribution $p_{e}\left(\mathbf{z}\middle|\mathbf{z}^{e},\mathbf{C}_{e}\right)$;
3. $p(\mathbf{z})$ satisfies the following integral forms of the *deterministic relationships* among responses and parameters represented mathematically by Eq. (1.56):

$$\int_{Dz} g_k(\mathbf{z})\, p(\mathbf{z})\, d\mathbf{z} = 0; \quad k = 1,...,TR, \ \textit{where}:$$
$$g_k(\mathbf{z}) \triangleq r_k - r_k^c\left(\boldsymbol{\alpha}^0\right) - \sum_{i=1}^{TP} s_{ki}\delta\alpha_i - \frac{1}{2}\sum_{i=1}^{TP}\sum_{j=1}^{TP} q_{ij}^{(k)}\delta\alpha_i\delta\alpha_j = 0. \tag{1.58}$$

The "best fit" normalized distribution $p(\mathbf{z})$ which minimizes the discrepancy between it and the distribution $p_e\left(\mathbf{z}\middle|\mathbf{z}^e,\mathbf{C}_e\right)$ while satisfying the constraints expressed by Eq. (1.58) is obtained by applying the steps generally outlined in Section 1.3, which yields the following expressions:

$$p(\mathbf{z}) = \frac{p_e\left(\mathbf{z}\middle|\mathbf{z}^e,\mathbf{C}_e\right)}{Z}\exp\left\{-\sum_{k=1}^{TR}\theta_k g_k(\mathbf{z})\right\}; \tag{1.59}$$

$$Z \triangleq \int_{Dz} p_e\left(\mathbf{z}\middle|\mathbf{z}^e,\mathbf{C}_e\right)\exp\left\{-\sum_{k=1}^{TR}\theta_k g_k(\mathbf{z})\right\} d\mathbf{z}. \tag{1.60}$$

In Eqs. (1.59) and (1.60), the quantities $\theta_1,...,\theta_k,...,\theta_{TR}$ are Lagrange multipliers which remain to be determined. In statistical mechanics, the normalization integral Z defined in Eq. (1.60) is called the partition function (or sum-over-states) and carries all of the information available about the possible states of the system. The quantity $(-\log Z)$ is called the "free energy" and plays an important role in determining the expressions of the Lagrange multipliers and hence of the probability distribution $p(\mathbf{z})$. In the present case, it follows from Eqs. (1.60) and (1.58) that:

$$\frac{\partial[-\ln Z]}{\partial\theta_k} = \frac{1}{Z}\int_{Dz} g_k(\mathbf{z})\, p_e\left(\mathbf{z}\middle|\mathbf{z}^e,\mathbf{C}_e\right)\exp\left\{-\sum_{k=1}^{TR}\theta_k g_k(\mathbf{z})\right\} d\mathbf{z} = 0, \quad k = 1,...,TR. \tag{1.61}$$

On the other hand, alternative expressions for the derivatives $\partial[-\ln Z]/\partial\theta_k$ of the free energy with respect to the Lagrange multipliers can be obtained by evaluating the integral which defines the partition function Z in Eq. (1.60). Equating these alternative expressions to zero, in view of Eq. (1.61), will provide equations which are to be solved to obtain the explicit expressions of the Lagrange multipliers.

Integrals such as those represented by Eq. (1.61) can be quantified to a high degree of accuracy (even exactly) using the well-known saddle-point (Laplace expansion) method (see, e.g., Shun and McCullagh, 1995; Evangelou et al., 2011; Tang and Reid, 2023). For a ratio of integrals of the form

$$I = \left[\int \exp[-g(\mathbf{z})]\, d\mathbf{z}\right]^{-1}\int f(\mathbf{z})\exp[-g(\mathbf{z})]\, d\mathbf{z}, \tag{1.62}$$

the saddle-point (Laplace) method yields the following result:

$$I = \hat{f} - \frac{1}{2}\hat{f}_{i_1}\hat{g}_{i_2i_3i_4}\hat{g}^{i_1i_2}\hat{g}^{i_3i_4} + \frac{1}{2}\hat{f}_{j_1j_2}\hat{g}^{j_1j_2} + O\left(\hat{f}_{j_1j_2j_3}\right), \tag{1.63}$$

where:

1. The derivative of a function with respect to a component of $\mathbf{z}$ is denoted using a subscript, for example, $f_i \triangleq \partial f / \partial z_i, f_{ij} \triangleq \partial^2 f / \partial z_i \partial z_j$, $i, j = 1,...,TI$, where TI denotes the total number of independent variables;
2. The superscripts denote the respective component of the inverse Hessian of the respective function, for example, f^{ij} denotes the (i, j)-element of the inverse Hessian matrix $\left[f_{ij}^{-1}\right]$;
3. An index that appears as a subscript and a superscript implies a summation over all possible values of that index;
4. The "hat" denotes that the respective quantity is evaluated at the saddle point of $\exp\left[-g(\mathbf{z})\right]$, which is defined as the point at which the gradient $g_{\mathbf{z}}(\mathbf{z}) \triangleq (\partial g / \partial z_1,...,\partial g / \partial z_{TI})$ of $g(\mathbf{z})$ vanishes, that is, $g_{\mathbf{z}}(\mathbf{z}) = \mathbf{0}$.

Examining the exponent of the function entering the normalization integral Z defined in Eq. (1.60) readily indicates that if the term involving the 2nd-order sensitivities of the response with respect to the parameters is retained in the exponent, the expression of the saddle-point will involve the vector of Lagrange multipliers, $\boldsymbol{\theta} \triangleq (\theta_1,...,\theta_k,...,\theta_{TR})^\dagger$, in a nonlinear equation that cannot be solved analytically to obtain a closed-form expression for these Lagrange multipliers. Furthermore, solving numerically this nonlinear equation in $\boldsymbol{\theta}$ would require the inversion of the (very large) parameter-covariance matrix $\mathbf{C}_{\alpha\alpha}$, which is impractical. On the other hand, as will be shown in what follows, a closed-form expression for the saddle-point and, consequently, for the Lagrange multipliers can be obtained if the terms containing the 2nd-order response sensitivities are considered by expanding the respective exponential term in a Taylor series in powers of $\boldsymbol{\theta} \triangleq (\theta_1,...,\theta_k,...,\theta_{TR})^\dagger$ or, equivalently, in a Taylor series around the nominal parameter values. Performing this Taylor expansion and retaining the 1st- and 2nd-order terms make it possible to recast the normalization integral Z defined in Eq. (1.60) into the following form:

$$Z \triangleq \int_{Dz} f(\mathbf{r},\boldsymbol{\alpha}) e^{-h(\mathbf{r},\boldsymbol{\alpha};\boldsymbol{\theta})} d\mathbf{r} d\boldsymbol{\alpha}, \tag{1.64}$$

where $\boldsymbol{\theta} \triangleq (\theta_1,...,\theta_k,...,\theta_{TR})^\dagger$ and where:

$$\begin{aligned} h(\mathbf{r},\boldsymbol{\alpha};\boldsymbol{\theta}) &\triangleq \frac{1}{2}\begin{pmatrix} \mathbf{r}-\mathbf{r}^e \\ \boldsymbol{\alpha}-\boldsymbol{\alpha}^0 \end{pmatrix}^\dagger \begin{pmatrix} \mathbf{C}_{rr}^e & \mathbf{C}_{r\alpha}^e \\ \mathbf{C}_{\alpha r}^e & \mathbf{C}_{\alpha\alpha} \end{pmatrix}^{-1} \begin{pmatrix} \mathbf{r}-\mathbf{r}^e \\ \boldsymbol{\alpha}-\boldsymbol{\alpha}^0 \end{pmatrix} \\ &+\boldsymbol{\theta}^\dagger \left[-\mathbf{r}+\mathbf{r}^c(\boldsymbol{\alpha}^0)+\mathbf{S}(\boldsymbol{\alpha}-\boldsymbol{\alpha}^0)\right], \end{aligned} \tag{1.65}$$

$$f(\mathbf{r},\boldsymbol{\alpha}) \triangleq 1 - \frac{1}{2}\sum_{k=1}^{TR}\theta_k (\boldsymbol{\alpha}-\boldsymbol{\alpha}^0)^\dagger \left[\mathbf{Q}_k(\boldsymbol{\alpha}^0)\right](\boldsymbol{\alpha}-\boldsymbol{\alpha}^0) + hots, \tag{1.66}$$

where:

$$\mathbf{Q}_k(\boldsymbol{\alpha}) \triangleq \begin{pmatrix} q_{11}^{(k)} & \cdots & q_{1,TP}^{(k)} \\ \vdots & \ddots & \vdots \\ q_{TP,1}^{(k)} & \cdots & q_{TP,TP}^{(k)} \end{pmatrix} \triangleq \begin{pmatrix} \dfrac{\partial^2 r_k^c(\boldsymbol{\alpha})}{\partial\alpha_1\partial\alpha_1} & \cdots & \dfrac{\partial^2 r_k^c(\boldsymbol{\alpha})}{\partial\alpha_1\partial\alpha_{TP}} \\ \vdots & \ddots & \vdots \\ \dfrac{\partial^2 r_k^c(\boldsymbol{\alpha})}{\partial\alpha_{TP}\partial\alpha_1} & \cdots & \dfrac{\partial^2 r_k^c(\boldsymbol{\alpha})}{\partial\alpha_{TP}\partial\alpha_{TP}} \end{pmatrix}, \tag{1.67}$$

and where the acronym "hots" denotes "higher-order terms" in the expansion of the exponential $\exp\left\{-\sum_{k=1}^{TR}\theta_k g_k(\mathbf{z})\right\}$." Neglecting these higher-order terms, it follows from Eqs. (1.64)–(1.66) that

$$Z = I_1 - \frac{1}{2}\sum_{k=1}^{TR}\theta_k J_k, \tag{1.68}$$

where

$$I_1 \triangleq \int_{Dz} e^{-h(\mathbf{r},\boldsymbol{\alpha};\boldsymbol{\theta})} d\mathbf{r}d\boldsymbol{\alpha}, \tag{1.69}$$

$$J_k \triangleq \int_{Dz} (\boldsymbol{\alpha}-\boldsymbol{\alpha}^0)^\dagger \left[\mathbf{Q}_k(\boldsymbol{\alpha}^0)\right](\boldsymbol{\alpha}-\boldsymbol{\alpha}^0) e^{-h(\mathbf{r},\boldsymbol{\alpha};\boldsymbol{\theta})} d\mathbf{r}d\boldsymbol{\alpha}. \tag{1.70}$$

The saddle-point, denoted as $\mathbf{z}^{sp} \triangleq (\mathbf{r}^{sp},\boldsymbol{\alpha}^{sp})^\dagger$, of the function $h(\mathbf{r},\boldsymbol{\alpha};\boldsymbol{\theta})$ is defined as the point where the following partial gradients vanish:

$$\nabla_{\mathbf{r}} h(\mathbf{r},\boldsymbol{\alpha};\boldsymbol{\theta}) = \mathbf{0}, \quad \nabla_{\boldsymbol{\alpha}} h(\mathbf{r},\boldsymbol{\alpha};\boldsymbol{\theta}) = \mathbf{0}, \; at\; \mathbf{z} = \mathbf{z}^{sp}. \tag{1.71}$$

The conditions represented by Eq. (1.71) yield the following equation:

$$\begin{pmatrix} \mathbf{C}_{rr}^e & \mathbf{C}_{r\alpha}^e \\ \mathbf{C}_{\alpha r}^e & \mathbf{C}_{\alpha\alpha} \end{pmatrix}^{-1} \begin{pmatrix} \mathbf{r}^{sp} - \mathbf{r}^e \\ \boldsymbol{\alpha}^{sp} - \boldsymbol{\alpha}^0 \end{pmatrix} = \begin{pmatrix} \boldsymbol{\theta} \\ -\mathbf{S}^\dagger\boldsymbol{\theta} \end{pmatrix}. \tag{1.72}$$

Solving Eq. (1.72) yields the following expressions for the components of the saddle-point $\mathbf{z}^{sp} \triangleq (\mathbf{r}^{sp},\boldsymbol{\alpha}^{sp})^\dagger$ of the function $h(\mathbf{r},\boldsymbol{\alpha};\boldsymbol{\theta})$:

$$\mathbf{r}^{sp} = \mathbf{r}^e + \left(\mathbf{C}_{rr}^e - \mathbf{C}_{r\alpha}^e\mathbf{S}^\dagger\right)\boldsymbol{\theta}, \tag{1.73}$$

$$\boldsymbol{\alpha}^{sp} = \boldsymbol{\alpha}^0 + \left(\mathbf{C}_{\alpha r}^e - \mathbf{C}_{\alpha\alpha}\mathbf{S}^\dagger\right)\boldsymbol{\theta}. \tag{1.74}$$

Using the saddle-point method (which is exact for the integral I_1) or "completing the square in the exponent of I_1" yields the following exact expression for the integral I_1:

$$I_1 = (2\pi)^{(TR+TP)/2} \left|C_e\right|^{1/2} \exp\left[-\boldsymbol{\theta}^\dagger\mathbf{d} + \frac{1}{2}\begin{pmatrix} \boldsymbol{\theta} \\ -\mathbf{S}^\dagger\boldsymbol{\theta} \end{pmatrix}^\dagger C_e \begin{pmatrix} \boldsymbol{\theta} \\ -\mathbf{S}^\dagger\boldsymbol{\theta} \end{pmatrix}\right], \tag{1.75}$$

where $|C_e|$ denotes the determinant of the matrix C_e and where $\mathbf{d}$ denotes the "vector of deviations between the nominal values of computed and measured responses" and is defined here:

$$\mathbf{d} \triangleq \mathbf{r}^c\left(\boldsymbol{\alpha}^0\right)-\mathbf{r}^e; \quad \mathbf{d} \triangleq \left(d_1,...,d_{TR}\right)^\dagger; \quad d_1 \triangleq r_i^c\left(\boldsymbol{\alpha}^0\right)-r_i^e; \quad i=1,...,TR. \tag{1.76}$$

The expression of the response in terms of the computational model, in other words, Eq. (1.76), can be used to obtain the following representation for $\mathbf{d}$ in terms of the computed responses:

$$\mathbf{d} \triangleq \mathbf{r}-\mathbf{r}^e-\mathbf{S}\left(\boldsymbol{\alpha}-\boldsymbol{\alpha}^0\right)-\mathbf{p}\left(\boldsymbol{\alpha};\boldsymbol{\alpha}^0\right). \tag{1.77}$$

Thus, the "vector of deviations" $\mathbf{d}$ can be considered to be a (quasi-) random variable. The expressions of the "vector of mean values of deviations" $\overline{\mathbf{d}} \triangleq \left[\overline{d}_1,...,\overline{d}_{TR}\right]^\dagger$ and the covariance matrix $\mathbf{D} \triangleq \left[D_{ij}\right]_{TR\times TR}$ of the vector of deviations $\mathbf{d}$ are derived:

$$\overline{\mathbf{d}} \triangleq \int_{Dz} \left[\mathbf{r}-\mathbf{r}^e-\mathbf{S}\left(\boldsymbol{\alpha}-\boldsymbol{\alpha}^0\right)-\mathbf{p}\left(\boldsymbol{\alpha};\boldsymbol{\alpha}^0\right)\right] p_e\left(\mathbf{z}\middle|\mathbf{z}^e,\mathbf{C}_e\right)d\mathbf{z} = \left[\overline{d}_1,...,\overline{d}_{TR}\right]^\dagger, \tag{1.78}$$

$$\overline{d}_k \triangleq \sum_{i=1}^{TP}\sum_{j=1}^{TP} q_{ij}^{(k)} \operatorname{cov}\left(\alpha_i,\alpha_j\right); \quad k=1,...,TR; \tag{1.79}$$

$$\begin{aligned}\mathbf{D} &\triangleq \int_{Dz} \left(\mathbf{d}\mathbf{d}^\dagger\right) p_e\left(\mathbf{z}\middle|\mathbf{z}^e,\mathbf{C}_e\right)d\mathbf{z} \\ &= \int_{Dz} p_e\left(\mathbf{z}\middle|\mathbf{z}^e,\mathbf{C}_e\right)\left[\mathbf{r}-\mathbf{r}^e-\mathbf{S}\left(\boldsymbol{\alpha}-\boldsymbol{\alpha}^0\right)-\mathbf{p}\left(\boldsymbol{\alpha};\boldsymbol{\alpha}^0\right)\right] \\ &\times\left[\mathbf{r}-\mathbf{r}^e-\mathbf{S}\left(\boldsymbol{\alpha}-\boldsymbol{\alpha}^0\right)-\mathbf{p}\left(\boldsymbol{\alpha};\boldsymbol{\alpha}^0\right)\right]^\dagger d\mathbf{z} \cong \mathbf{C}_{rr}^e+\mathbf{S}\mathbf{C}_{\alpha\alpha}\mathbf{S}^\dagger-\mathbf{C}_{r\alpha}^e\mathbf{S}^\dagger-\mathbf{S}\mathbf{C}_{\alpha r}^e.\end{aligned} \tag{1.80}$$

The 3rd- and 4th-order correlations were neglected in the final expression obtained for the covariance matrix $\mathbf{D} \triangleq \left[D_{ij}\right]_{TR\times TR}$ in Eq. (1.80). Notably, the result obtained in Eq. (1.78) indicates that *if the 2nd-order sensitivities of the response with respect to the model parameters are neglected, then* $\overline{d}_k=0$, $k=1,...,TR$, in which case $\mathbf{d}$ would be a random vector with zero mean and covariance $\mathbf{D}$. Therefore, neglecting the 2nd-order sensitivities would imply that the computed model responses would be consistent with the measurements only if $\overline{\mathbf{d}}=\mathbf{0}$, in other words, $\mathbf{r}^c\left(\boldsymbol{\alpha}^0\right)\cong\mathbf{r}^e$. Although such a situation would not be impossible to conceive, it would probably not occur often in practice. These considerations further underscore the significant impact that the 2nd-order response sensitivities have on the correct interpretation of actual physical systems.

Applying the saddle-point method to the expression provided in Eq. (1.70) yields the following expression:

$$J_k = I_1\left(\boldsymbol{\alpha}^{sp}-\boldsymbol{\alpha}^0\right)^\dagger\left[\mathbf{Q}_k\left(\boldsymbol{\alpha}^0\right)\right]\left(\boldsymbol{\alpha}^{sp}-\boldsymbol{\alpha}^0\right)\left[1+O\left(C_{\alpha\alpha}\right)\right], \tag{1.81}$$

where the quantity $O(C_{\alpha\alpha})$ denotes terms of the order of the components of the parameter covariance matrix $C_{\alpha\alpha}$, and can therefore be neglected by comparison to the leading term of order unity.

Replacing the expression obtained in Eqs. (1.75) and (1.81) into Eq. (1.68) and neglecting the higher-order terms yields the following expression for the partition function:

$$Z = I_1 \left\{ 1 - \frac{1}{2} \sum_{k=1}^{TR} \theta_k \left(\boldsymbol{\alpha}^{sp} - \boldsymbol{\alpha}^0 \right)^\dagger \left[\mathbf{Q}_k \left(\boldsymbol{\alpha}^0 \right) \right] \left(\boldsymbol{\alpha}^{sp} - \boldsymbol{\alpha}^0 \right) \right\}. \tag{1.82}$$

Taking the logarithm of Eq. (1.82) yields the following expression for the "free energy":

$$-\ln Z = \frac{(TR + TP)}{2} \ln(2\pi) + \frac{1}{2} \ln |C_m| - \boldsymbol{\theta}^\dagger \mathbf{d} + h_1(\boldsymbol{\theta}) + h_2(\boldsymbol{\theta}), \tag{1.83}$$

$$h_1(\boldsymbol{\theta}) \triangleq \frac{1}{2} \left[\boldsymbol{\theta}^\dagger \mathbf{C}_{rr}^e \boldsymbol{\theta} - \boldsymbol{\theta}^\dagger \left(\mathbf{C}_{r\alpha}^e \mathbf{S}^\dagger + \mathbf{S} \mathbf{C}_{\alpha r}^e \right) \boldsymbol{\theta} + \boldsymbol{\theta}^\dagger \mathbf{S} \mathbf{C}_{\alpha\alpha} \mathbf{S}^\dagger \boldsymbol{\theta} \right], \tag{1.84}$$

$$\begin{aligned} h_2(\boldsymbol{\theta}) &\triangleq \ln \left\{ 1 - \frac{1}{2} \sum_{k=1}^{TR} \theta_k \left(\boldsymbol{\alpha}^{sp} - \boldsymbol{\alpha}^0 \right)^\dagger \left[\mathbf{Q}_k \left(\boldsymbol{\alpha}^0 \right) \right] \left(\boldsymbol{\alpha}^{sp} - \boldsymbol{\alpha}^0 \right) \right\} \\ &= \ln \left\{ 1 - \frac{1}{2} \sum_{k=1}^{TR} \theta_k \left[\boldsymbol{\theta}^\dagger \mathbf{M}^{(k)} \boldsymbol{\theta} \right] \right\}, \end{aligned} \tag{1.85}$$

$$\mathbf{M}^{(k)} \triangleq \left[M_{ij}^{(k)} \right]_{TR \times TR} \triangleq \left(\mathbf{C}_{\alpha r}^e - \mathbf{C}_{\alpha\alpha} \mathbf{S}^\dagger \right)^\dagger \left[\mathbf{Q}_k \left(\boldsymbol{\alpha}^0 \right) \right] \left(\mathbf{C}_{\alpha r}^e - \mathbf{C}_{\alpha\alpha} \mathbf{S}^\dagger \right) = \left[\mathbf{M}^{(k)} \right]^\dagger. \tag{1.86}$$

The expression for $h_2(\boldsymbol{\theta})$ presented in Eq. (1.85) has been obtained by using Eq. (1.74). As indicated in Eqs. (1.84) and (1.85), respectively, the quantity $h_1(\boldsymbol{\theta})$ contains only 1st-order response sensitivities (with respect to the model's parameters) and 2nd-order bilinear terms in the components of $\boldsymbol{\theta}$, while the quantity $h_2(\boldsymbol{\theta})$ comprises all 2nd-order response sensitivities and terms higher than 2nd-order in the components of $\boldsymbol{\theta}$.

In view of Eqs. (1.83)–(1.85), the derivatives of the free energy with respect to the Lagrange multipliers θ_k are provided by the expression:

$$\frac{\partial[-\ln Z]}{\partial \theta_k} = -d_k + \frac{\partial h_1(\boldsymbol{\theta})}{\partial \theta_k} + \frac{\partial h_2(\boldsymbol{\theta})}{\partial \theta_k}, \quad k = 1, \ldots, TR. \tag{1.87}$$

It follows from Eqs. (1.87) and (1.61) that the vector of Lagrange multipliers must be the solution of the following system of coupled nonlinear equations:

$$w_k(\boldsymbol{\theta}) \triangleq \frac{\partial h_1(\boldsymbol{\theta})}{\partial \theta_k} + \frac{\partial h_2(\boldsymbol{\theta})}{\partial \theta_k} - d_k = 0, \quad k = 1, \ldots, TR \tag{1.88}$$

where:

$$\frac{\partial h_1(\boldsymbol{\theta})}{\partial \theta_k} = \sum_{j=1}^{TR} D_{kj}\theta_j ; \quad \left[D_{kj}\right]_{TR\times TR} \triangleq \mathbf{D} \triangleq \mathbf{C}_{rr}^{e} + \mathbf{S}\mathbf{C}_{\alpha\alpha}\mathbf{S}^{\dagger} - \mathbf{C}_{r\alpha}^{e}\mathbf{S}^{\dagger} - \mathbf{S}\mathbf{C}_{\alpha r}^{e}; \tag{1.89}$$

$$\begin{aligned} \frac{\partial h_2(\boldsymbol{\theta})}{\partial \theta_k} &= -\frac{1}{2}\frac{\dfrac{\partial}{\partial \theta_k}\displaystyle\sum_{i=1}^{TR}\theta_i\left[\boldsymbol{\theta}^{\dagger}\mathbf{M}^{(i)}\boldsymbol{\theta}\right]}{1-\dfrac{1}{2}\displaystyle\sum_{i=1}^{TR}\theta_i\left[\boldsymbol{\theta}^{\dagger}\mathbf{M}^{(i)}\boldsymbol{\theta}\right]} \\ &\cong -\frac{1}{2}\boldsymbol{\theta}^{\dagger}\mathbf{M}^{(k)}\boldsymbol{\theta} - \sum_{i=1}^{TR}\theta_i\sum_{j=1}^{TR}M_{kj}^{(i)}\theta_j \triangleq u_k(\boldsymbol{\theta}). \end{aligned} \tag{1.90}$$

The approximation $1-\frac{1}{2}\sum_{j=1}^{TR}\theta_j\left[\boldsymbol{\theta}^{\dagger}\mathbf{M}^{(j)}\boldsymbol{\theta}\right]\cong 1$ has been used to obtain the approximate expression on the right side of Eq. (1.90), which thus comprises just the bilinear terms in $\boldsymbol{\theta}$, neglecting the high-order terms involving the components of $\boldsymbol{\theta}$, which are expected to be comparatively small.

Introducing the vectors $\mathbf{w}(\boldsymbol{\theta}) \triangleq \left[w_1(\boldsymbol{\theta}),\ldots,w_k(\boldsymbol{\theta}),\ldots,w_{TR}(\boldsymbol{\theta})\right]^{\dagger}$ and $\mathbf{u}(\boldsymbol{\theta}) \triangleq \left[u_1(\boldsymbol{\theta}),\ldots,u_k(\boldsymbol{\theta}),\ldots,u_{TR}(\boldsymbol{\theta})\right]^{\dagger}$ enables the compact writing, in matrix vector form, of Eq. (1.88), as follows:

$$\mathbf{w}(\boldsymbol{\theta}) \triangleq -\mathbf{d} + \mathbf{D}\boldsymbol{\theta} + \mathbf{u}(\boldsymbol{\theta}) = \mathbf{0}. \tag{1.91}$$

If only the 1st-order sensitivities of the response with respect to the model parameters are considered (and thus the 2nd- and higher-order sensitivities are neglected or are unavailable), the term $\mathbf{u}(\boldsymbol{\theta})$ is neglected and the remaining terms in Eq. (1.91) yield the following 1st-order solution, denoted as $\boldsymbol{\theta}^{(1)}$, for the vector of Lagrange multipliers:

$$\boldsymbol{\theta}^{(1)} = \mathbf{D}^{-1}\mathbf{d}. \tag{1.92}$$

Notably, the matrix $\mathbf{D}$ has dimensions $TR \times TR$. Since $TR \ll TP$ in most practical situations, it follows that the computation of the coordinates of the saddle-point $\mathbf{z}^{sp} \triangleq \left(\mathbf{r}^{sp}, \boldsymbol{\alpha}^{sp}\right)^{\dagger}$ involves the inversion of only the smallest possible matrix (i.e., $\mathbf{D}$) that arises in the present "predictive modeling" methodology. Replacing the result obtained for $\boldsymbol{\theta}^{(1)}$ in Eqs. (1.73) and (1.74) provide the corresponding expressions for the coordinates $\mathbf{r}^{sp}$ and $\boldsymbol{\alpha}^{sp}$ of the saddle-point $\mathbf{z}^{sp} \triangleq \left(\mathbf{r}^{sp}, \boldsymbol{\alpha}^{sp}\right)^{\dagger}$.

Using the results obtained for $\boldsymbol{\theta}^{(1)}$ in Eq. (1.92) leads to the expressions already presented in the BERRU-PM methodology (Cacuci, 2019) for the saddle-point $\mathbf{z}^{sp} \triangleq \left(\mathbf{r}^{sp}, \boldsymbol{\alpha}^{sp}\right)^{\dagger}$ and, subsequently for the best-estimate (calibrated) mean values for the parameters and responses, as well as for the best-estimate covariance matrices for the corresponding best-estimate parameters and responses. These expressions will not be reproduced here since they will also be obtained as particular cases of the 2nd-BERRU-PM methodology to be presented in the remainder of this section.

The goal of predictive modeling is to reduce, as much as possible, the discrepancies between the measured response values and the corresponding values predicted by the computational model. However, if the nominal values of the measured responses happen – by nature, design, or accident – to coincide with the corresponding computed values, then the vector of deviations vanishes, in other words, $\mathbf{d}=\mathbf{0}$, in which case Eqs. (1.92), (1.73), (1.74), and (1.59) indicate that:

$$\mathbf{d}=\mathbf{0};\ \Rightarrow \boldsymbol{\theta}^{(1)}=\mathbf{0};\ \Rightarrow \mathbf{r}^{sp}=\mathbf{r}^{e},\ \ \boldsymbol{\alpha}^{sp}=\boldsymbol{\alpha}^{0};\Rightarrow p(\mathbf{z})=p_e\left(\mathbf{z}\middle|\mathbf{z}^{e},\mathbf{C}_{e}\right); \quad (1.93)$$

The results shown in Eq. (1.93) indicate that if only 1st-order sensitivities are used and if the nominal values of the measured responses happen to coincide with the corresponding computed values, the model would have no impact on the initial (measured) distribution of responses and parameters, so there would be no calibration of either the predicted responses or parameters. Evidently, this conclusion reveals a fundamental limitation of using just the 1st-order sensitivities for predictive modeling (and data assimilation and/or data adjustment methods). But, this conclusion also provides the fundamental motivation for using 2nd-order sensitivities, in addition to the natural expectation that using the 2nd-order sensitivities will produce more accurate results, in general.

It is also important to note that if the response measurements were perfect and if they were uncorrelated to the model parameters, in other words, if $\mathbf{C}_{rr}^{e}=\mathbf{0}$ and $\mathbf{C}_{r\alpha}^{e}=\mathbf{C}_{\alpha r}^{e}=\mathbf{0}$, then the matrix $\mathbf{D}=\mathbf{S}\mathbf{C}_{\alpha\alpha}\mathbf{S}^{\dagger}$ would nevertheless be non-singular since the matrix $\mathbf{S}\mathbf{C}_{\alpha\alpha}\mathbf{S}^{\dagger}$ is nonsingular. Thus, the solution shown in Eq. (1.92) would still exist; this is in contradistinction with the extant data assimilation schemes, which fail if the measurements are perfect. Notably, all extant data assimilation schemes consider that the response measurements and model parameters are uncorrelated.

If the 2nd-order sensitivities, which are included in the term $\mathbf{u}(\boldsymbol{\theta})$, are retained in Eq. (1.88), this equation becomes nonlinear in the components of $\boldsymbol{\theta}\triangleq(\theta_1,\ldots,\theta_k,\ldots,\theta_{TR})^{\dagger}$ and, therefore, requires an iterative method for solving. The structure of $\mathbf{u}(\boldsymbol{\theta})$ indicates that the simplest (and probably the most efficient) method for solving Eq. (1.91) would be Newton's method, as can be surmised by the similarity between this equation and a simple quadratic algebraic equation. In this vein, consider the application of Newton's method to the algebraic quadratic equation, as described here:

$$\begin{aligned} &q(x)\triangleq ax^2+bx+d=0;\ q'(x)=2ax+b;\\ &x^{(n+1)}=x^{(n)}-\left[q'(x^n)\right]^{-1}q(x^n);\ n=0,1,\ldots \end{aligned} \quad (1.94)$$

If $d\neq 0$, the usual initial guess for starting the Newton iteration is $x^{(0)}=-d/b$, which yields the first-iterate $x^{(1)}=-\left[1+ad/\left(b^2-2ad\right)\right]d/b$. The correction term included in this first-iterate solution is of first order in the product (ad), so the subsequent iterates become increasingly more accurate when this product is small; even the first-iterate becomes exact in the limit when $a \to 0$. On the other hand, if $d = 0$, the initial guess $x^{(0)} = -d/b = 0$ would yield solely the trivial solution $x = 0$. Therefore, if $d \cong 0$, the initial guess for starting the Newton iteration would be $x^{(0)} = -b/a$, which would yield the non-zero first-iterate solution $x^{(1)} = -b/a + d/b$, and so on.

Applying Newton's method to solve Eq. (1.91) yields the following iterative equation for finding its solution(s) $\boldsymbol{\theta} \triangleq (\theta_1,...,\theta_k,...,\theta_{TR})^\dagger$:

$$\boldsymbol{\theta}^{(n+1)} = \boldsymbol{\theta}^{(n)} - \left[\mathbf{A}\left(\boldsymbol{\theta}^{(n)}\right)\right]^{-1} \mathbf{w}\left(\boldsymbol{\theta}^{(n)}\right); \; n = 0,1,... \tag{1.95}$$

where the Jacobian matrix $\mathbf{A}(\boldsymbol{\theta})$ is defined as usual, namely:

$$\begin{aligned} \mathbf{A}(\boldsymbol{\theta}) &\triangleq \frac{\partial(w_1,...,w_k,...,w_{TR})}{\partial(\theta_1,...,\theta_n,...,\theta_{TR})} \\ &\triangleq \begin{bmatrix} \partial w_1/\partial\theta_1 & \bullet & \partial w_1/\partial\theta_{TR} \\ \bullet & \bullet & \bullet \\ \partial w_{TR}/\partial\theta_1 & \bullet & \partial w_{TR}/\partial\theta_{TR} \end{bmatrix} \\ &\triangleq \begin{bmatrix} A_{1,1}(\boldsymbol{\theta}) & \bullet & A_{1,TR}(\boldsymbol{\theta}) \\ \bullet & \bullet & \bullet \\ A_{TR,1}(\boldsymbol{\theta}) & \bullet & A_{TR,TR}(\boldsymbol{\theta}) \end{bmatrix}. \end{aligned} \tag{1.96}$$

Using Eqs. (1.90) and (1.91) yields the following expression for the components/elements $A_{k,n}(\boldsymbol{\theta})$ of the Jacobian matrix $\mathbf{A}(\boldsymbol{\theta})$:

$$\begin{aligned} A_{kn}(\boldsymbol{\theta}) &= \frac{\partial w_k}{\partial\theta_n} = \frac{\partial}{\partial\theta_n}\left[\sum_{j=1}^{TR} D_{kj}\theta_j - \frac{1}{2}\boldsymbol{\theta}^\dagger\mathbf{M}^{(k)}\boldsymbol{\theta} - \sum_{i=1}^{TR}\theta_i\sum_{j=1}^{TR}M_{kj}^{(i)}\theta_j\right] \\ &= D_{kn} - \sum_{j=1}^{TR} M_{nj}^{(k)}\theta_j - \sum_{j=1}^{TR} M_{kj}^{(n)}\theta_j - \sum_{i=1}^{TR}\theta_i M_{kn}^{(i)} = D_{kn} - V_{kn}(\boldsymbol{\theta}), \end{aligned} \tag{1.97}$$

where:

$$V_{kn}(\boldsymbol{\theta}) \triangleq \sum_{j=1}^{TR}\theta_j\left[M_{nj}^{(k)} + M_{kj}^{(n)} + M_{kn}^{(j)}\right], \; \mathbf{V}(\boldsymbol{\theta}) \triangleq \left[V_{kn}(\boldsymbol{\theta})\right]_{TR\times TR}. \tag{1.98}$$

As indicated in Eq. (1.97), if the initial guess $\boldsymbol{\theta}^{(0)} = \mathbf{0}$ is used, the Jacobian matrix would take on the value $\mathbf{A}(\boldsymbol{\theta} = \mathbf{0}) = \mathbf{D}$. Consequently, the resulting solution of Eq. (1.95) would be obtained in a single iteration and would be the same as the 1st-order solution obtained in Eq. (1.92), in other words, $\boldsymbol{\theta}^{(1)} = \mathbf{D}^{-1}\mathbf{d}$. Similarly, if the 2nd-order response sensitivities to the model parameters are neglected, Eq. (1.98) indicates that $\mathbf{A}\left(\mathbf{M}^{(k)} = \mathbf{0}\right) = \mathbf{D}$.

To obtain the nontrivial solution of Eq. (1.95) when $\mathbf{d}$ is very small or actually vanishes (i.e., when $\mathbf{d} = \mathbf{0}$), a non-zero initial guess ($\boldsymbol{\theta}^{(0)} \neq \mathbf{0}$) must be used for starting the Newton iterations, as was discussed in the foregoing, by analogy with the solution of a quadratic algebraic equation using Newton's method. The nontrivial solution of Eq. (1.95) which would be valid even when $\mathbf{d} = \mathbf{0}$ will be denoted as $\boldsymbol{\theta}^{(h,n+1)}$, where the superscript "h" indicates that the expression of $\boldsymbol{\theta}^{(h,n+1)}$ will include *higher-order* terms, which will involve the 2nd-order sensitivities of computed responses with respect to model parameters. The corresponding Newton iteration indicated by Eq. (1.95) will thus take on the following form for finding the nontrivial root of Eq. (1.91) when $\mathbf{d}$ is very small or actually vanishes:

$$\boldsymbol{\theta}^{(h,n+1)} = \boldsymbol{\theta}^{(h,n)} - \left[\mathbf{D} - \mathbf{V}\left(\boldsymbol{\theta}^{(h,n)}\right)\right]^{-1} \mathbf{w}\left(\boldsymbol{\theta}^{(h,n)}\right); \; n = 0,1,... \tag{1.99}$$

As has been discussed in the foregoing in conjunction with the algebraic quadratic equation, it would be advantageous to start the Newton iteration shown in Eq. (1.99) with an initial guess $\boldsymbol{\theta}^{(h,0)}$ that would have components analogous to the initial guess $x^{(0)} = -b/a$ for the quadratic equations. It would be further advantageous if the computation of the initial guess $\boldsymbol{\theta}^{(h,0)}$ would not require the inversion of matrices. An expression for $\boldsymbol{\theta}^{(h,0)}$ which satisfies these considerations can be derived by setting $\mathbf{d} = \mathbf{0}$ in Eq. (1.91) and by considering just the diagonal terms of the matrices involved in this equation, thus obtaining:

$$\boldsymbol{\theta}^{(h,0)} \triangleq \left(\theta_1^{(h,0)},...,\theta_j^{(h,0)},...,\theta_{TR}^{(h,0)}\right)^\dagger; \; \theta_j^{(h,0)} = 2D_{jj}/3M_{jj}^{(j)}; \; j = 1,...,TR. \tag{1.100}$$

Using the initial guess provided in Eq. (1.100) in Eq. (1.98) yields the following starting expressions for components of the Jacobian matrix $\mathbf{A}(\boldsymbol{\theta})$:

$$A_{kn}\left(\boldsymbol{\theta}^{(h,0)}\right) = D_{kn} - \sum_{j=1}^{TR} \theta_j^{(h,0)}\left[M_{nj}^{(k)} + M_{kj}^{(n)} + M_{kn}^{(j)}\right]. \tag{1.101}$$

Furthermore, using the initial guess provided in Eq. (1.100) in Eq. (1.91) yields the following starting expressions for components $w_k(\boldsymbol{\theta})$ of the vector $\mathbf{w}(\boldsymbol{\theta}) \triangleq \left[w_1(\boldsymbol{\theta}),...,w_k(\boldsymbol{\theta}),...,w_{TR}(\boldsymbol{\theta})\right]^\dagger$:

$$w_k\left(\boldsymbol{\theta}^{(h,0)}\right) = u_k\left(\boldsymbol{\theta}^{(h,0)}\right) + \sum_{j=1}^{TR} D_{kj}\theta_j^{(h,0)} - d_k, \quad k = 1,...,TR; \tag{1.102}$$

$$\begin{aligned} u_k\left(\boldsymbol{\theta}^{(h,0)}\right) &= -\frac{1}{2}\sum_{i=1}^{TR}\sum_{j=1}^{TR}\theta_i^{(h,0)} M_{ij}^{(k)} \theta_j^{(h,0)} - \sum_{i=1}^{TR}\theta_i^{(h,0)}\sum_{j=1}^{TR} M_{kj}^{(i)}\theta_j^{(h,0)} \\ &\cong -\frac{4}{9}\sum_{i=1}^{TR}\sum_{j=1}^{TR}\left(\frac{D_{ii}}{M_{ii}^{(i)}}\right)\left(\frac{D_{jj}}{M_{jj}^{(j)}}\right)\left[\frac{1}{2}M_{ij}^{(k)} + M_{kj}^{(i)}\right], \quad k = 1,...,TR. \end{aligned} \tag{1.103}$$

The impact of the 2nd-order sensitivities through the components $M_{ij}^{(k)}$ of the matrices $\mathbf{M}^{(k)}$ is evident in Eqs. (1.101)–(1.103). The first-iterate $\boldsymbol{\theta}^{(h,1)}$ of the solution of Eq. (1.99) has the following expression:

$$\begin{aligned}&\boldsymbol{\theta}^{(h,1)} = \boldsymbol{\theta}^{(h,0)} - \left(\mathbf{D} - \mathbf{V}^{(h,0)}\right)^{-1}\left(\mathbf{u}^{(h,0)} + \mathbf{D}\boldsymbol{\theta}^{(h,0)} - \mathbf{d}\right);\\ &\mathbf{V}^{(h,0)} \triangleq \mathbf{V}\left(\boldsymbol{\theta}^{(h,0)}\right);\ \ \mathbf{u}^{(h,0)} \triangleq \mathbf{u}\left(\boldsymbol{\theta}^{(h,0)}\right)\end{aligned} \tag{1.104}$$

It is evident from the expressions provided in Eqs. (1.101)–(1.103) that even if $\mathbf{d} = \mathbf{0}$ (i.e., if the vector of deviations between the nominally measured and computed values of the responses vanishes), the starting values of the components $w_k\left(\boldsymbol{\theta}^{(h,0)}\right) \neq 0$ are non-zero because of the impact of the 2nd-order response sensitivities. In addition, the initial Jacobian matrix $\mathbf{A}\left(\boldsymbol{\theta}^{(h,0)}\right)$ is nonsingular (also because of the impact of the 2nd-order response sensitivities), so the first-iterate $\boldsymbol{\theta}^{(h,1)}$ has a finite non-zero value. All subsequent iterates will also have non-zero values. Thus, using the 2nd-order response sensitivities overcomes the limitation of the 1st-order predictive modeling methodology BERRU-PM as well as the limitation of extant data adjustment and data assimilation procedures which all fail when the measured response values coincide with the computed response values.

Replacing the solution of Eq. (1.99) with Eqs. (1.73) and (1.74) yields the following expressions for the coordinates of the saddle-point $\mathbf{z}^{sp} \triangleq \left(\mathbf{r}^{sp}, \boldsymbol{\alpha}^{sp}\right)^{\dagger}$ of the function $h(\mathbf{r}, \boldsymbol{\alpha}; \boldsymbol{\theta})$:

$$\mathbf{r}^{sp} = \mathbf{r}^{e} + \left(\mathbf{C}_{rr}^{e} - \mathbf{C}_{r\alpha}^{e}\mathbf{S}^{\dagger}\right)\boldsymbol{\theta}^{(h,n+1)}, \tag{1.105}$$

$$\boldsymbol{\alpha}^{sp} = \boldsymbol{\alpha}^{0} + \left(\mathbf{C}_{\alpha r}^{e} - \mathbf{C}_{\alpha\alpha}\mathbf{S}^{\dagger}\right)\boldsymbol{\theta}^{(h,n+1)}. \tag{1.106}$$

The best-estimate (optimal) mean values (1st-order moments) and correlations (2nd-order moments) of the posterior distribution of the combined computational model with measured responses and parameters are obtained by using the saddle-point method to evaluate the integrals that define these 1st- and 2nd-order moments. These best-estimate values will be denoted using the superscript "*bed*" to indicate that these "best-estimate" values arise from *deterministically* incorporated/assimilated computed responses.

Thus, the predicted "best-estimated values with deterministically incorporated computed responses" for the predicted mean values of responses and parameters, respectively, have the following expressions:

$$\mathbf{r}^{bed} \triangleq \int_{Dz} \mathbf{r}\, p(\mathbf{z})\, d\mathbf{z} = \mathbf{r}^{sp} = \mathbf{r}^{e} + \left(\mathbf{C}_{rr}^{e} - \mathbf{C}_{r\alpha}^{e}\mathbf{S}^{\dagger}\right)\boldsymbol{\theta}^{(h,n+1)}; \tag{1.107}$$

$$\boldsymbol{\alpha}^{bed} \triangleq \int_{Dz} \boldsymbol{\alpha} p(\mathbf{z})\, d\mathbf{z} = \boldsymbol{\alpha}^{sp} = \boldsymbol{\alpha}^{0} + \left(\mathbf{C}_{\alpha r}^{e} - \mathbf{C}_{\alpha\alpha}\mathbf{S}^{\dagger}\right)\boldsymbol{\theta}^{(h,n+1)}. \tag{1.108}$$

The expression of the "best-estimate with deterministically incorporated computed responses" covariance matrix for the responses, denoted as $\mathbf{C}_{rr}^{bed}$, is obtained from its definition in conjunction with Eq. (1.107), which yields:

$$\begin{aligned}\mathbf{C}_{rr}^{bed} &\triangleq \int_{Dz} \left(\mathbf{r}-\mathbf{r}^{bed}\right)\left(\mathbf{r}-\mathbf{r}^{bed}\right)^{\dagger} p(\mathbf{z})\,d\mathbf{z} \\ &= \int_{Dz} \left[\mathbf{r}-\mathbf{r}^{e}-\left(\mathbf{C}_{rr}^{e}-\mathbf{C}_{r\alpha}^{e}\mathbf{S}^{\dagger}\right)\boldsymbol{\theta}^{(h,n+1)}\right] \\ &\quad \left[\mathbf{r}-\mathbf{r}^{e}-\left(\mathbf{C}_{rr}^{e}-\mathbf{C}_{r\alpha}^{e}\mathbf{S}^{\dagger}\right)\boldsymbol{\theta}^{(h,n+1)}\right]^{\dagger} p(\mathbf{z})\,d\mathbf{z}.\end{aligned} \tag{1.109}$$

The expression of the "best-estimate with deterministically incorporated computed responses" covariance matrix for the parameters, denoted as $\mathbf{C}_{\alpha\alpha}^{bed}$, is obtained from its definition in conjunction with Eq. (1.108), which yields:

$$\begin{aligned}\mathbf{C}_{\alpha\alpha}^{bed} &\triangleq \int_{Dz} \left(\boldsymbol{\alpha}-\boldsymbol{\alpha}^{bed}\right)\left(\boldsymbol{\alpha}-\boldsymbol{\alpha}^{bed}\right)^{\dagger} p(\mathbf{z})\,d\mathbf{z} \\ &= \int_{Dz} \left[\boldsymbol{\alpha}-\boldsymbol{\alpha}^{0}-\left(\mathbf{C}_{\alpha r}^{e}-\mathbf{C}_{\alpha\alpha}\mathbf{S}^{\dagger}\right)\boldsymbol{\theta}^{(h,n+1)}\right] \\ &\quad \left[\boldsymbol{\alpha}-\boldsymbol{\alpha}^{0}-\left(\mathbf{C}_{\alpha r}^{e}-\mathbf{C}_{\alpha\alpha}\mathbf{S}^{\dagger}\right)\boldsymbol{\theta}^{(h,n+1)}\right]^{\dagger} p(\mathbf{z})\,d\mathbf{z}\end{aligned} \tag{1.110}$$

The expression of the "best-estimate with deterministically incorporated computed responses" correlation matrix among parameters and responses, denoted as $\mathbf{C}_{\alpha r}^{bed}$, is obtained by using its definition in conjunction with Eqs. (1.107) and (1.108), which yields:

$$\begin{aligned}\mathbf{C}_{\alpha r}^{bed} &\triangleq \int_{Dz} \left(\boldsymbol{\alpha}-\boldsymbol{\alpha}^{bed}\right)\left(\mathbf{r}-\mathbf{r}^{bed}\right)^{\dagger} p(\mathbf{z})\,d\mathbf{z} \\ &= \int_{Dz} \left[\boldsymbol{\alpha}-\boldsymbol{\alpha}^{0}-\left(\mathbf{C}_{\alpha r}^{e}-\mathbf{C}_{\alpha\alpha}\mathbf{S}^{\dagger}\right)\boldsymbol{\theta}^{(h,n+1)}\right] \\ &\quad \left[\mathbf{r}-\mathbf{r}^{e}-\left(\mathbf{C}_{rr}^{e}-\mathbf{C}_{r\alpha}^{e}\mathbf{S}^{\dagger}\right)\boldsymbol{\theta}^{(h,n+1)}\right]^{\dagger} p(\mathbf{z})\,d\mathbf{z}\end{aligned} \tag{1.111}$$

The impact of the 2nd-order response sensitivities on the mean values and correlation/covariance matrices can be assessed already by using the first-iterate solution, $\boldsymbol{\theta}^{(h,1)}$, in Eqs. (1.107)–(1.111). Using the first-iterate, $\boldsymbol{\theta}^{(h,1)}$, in Eqs. (1.107) and (1.108), respectively, yield the following expressions for the "first-iterate best-estimate" values for the responses and parameters, respectively:

$$\mathbf{r}^{(1)bed} = \mathbf{r}^{e} + \left(\mathbf{C}_{rr}^{e}-\mathbf{C}_{r\alpha}^{e}\mathbf{S}^{\dagger}\right)\left[\boldsymbol{\theta}^{(h,0)}-\left(\mathbf{D}-\mathbf{V}^{(h,0)}\right)^{-1}\left(\mathbf{u}^{(h,0)}+\mathbf{D}\boldsymbol{\theta}^{(h,0)}-\mathbf{d}\right)\right]; \tag{1.112}$$

$$\boldsymbol{\alpha}^{(1)bed} = \boldsymbol{\alpha}^{0} + \left(\mathbf{C}_{\alpha r}^{e}-\mathbf{C}_{\alpha\alpha}\mathbf{S}^{\dagger}\right)\left[\boldsymbol{\theta}^{(h,0)}-\left(\mathbf{D}-\mathbf{V}^{(h,0)}\right)^{-1}\left(\mathbf{u}^{(h,0)}+\mathbf{D}\boldsymbol{\theta}^{(h,0)}-\mathbf{d}\right)\right]. \tag{1.113}$$

As the expressions obtained in Eqs. (1.112) and (1.113) indicate combining experimental and computational information modifies the correlations among the parameters through couplings introduced by the sensitivities of the participating responses. The expressions obtained in Eqs. (1.112) and (1.113) possess the following new features by comparisons to their counterparts provided by extant data assimilation and data adjustment procedures:

1. If the 2ond-order sensitivities are neglected, then the expressions in Eqs. (1.112) and (1.113) take on the following simplified forms:

$$\textit{no } 2^{nd} - \textit{order sensitivities}: \ \mathbf{r}^{(1)bed} = \mathbf{r}^{e} + \left(\mathbf{C}^{e}_{rr} - \mathbf{C}^{e}_{r\alpha}\mathbf{S}^{\dagger}\right)\mathbf{D}^{-1}\mathbf{d};; \tag{1.114}$$

$$\textit{no } 2^{nd} - \textit{order sensitivities}: \ \boldsymbol{\alpha}^{(1)bed} = \boldsymbol{\alpha}^{0} + \left(\mathbf{C}^{e}_{\alpha r} - \mathbf{C}_{\alpha\alpha}\mathbf{S}^{\dagger}\right)\mathbf{D}^{-1}\mathbf{d}. \tag{1.115}$$

As expected, the expressions in Eqs. (1.114) and (1.115) are identical to the expressions provided by the BERRU-PM methodology (Cacuci, 2019) and therefore possess the same strengths (easy to implement) and weaknesses (no parameter calibration if $\mathbf{d} = \mathbf{0}$) as the results produced by the 1st-BERRU-PM methodology.

2. If the vector of deviations between the nominally measured and computed values of the responses vanishes, in other words, if $\mathbf{d} = \mathbf{0}$, the expressions provided in Eqs. (1.112) and (1.113) take on the following simplified forms:

$$\begin{aligned} &\textit{if } \mathbf{d} = \mathbf{0}: \mathbf{r}^{(1)bed}\left(\mathbf{d} = \mathbf{0}\right) = \mathbf{r}^{e} + \left(\mathbf{C}^{e}_{rr} - \mathbf{C}^{e}_{r\alpha}\mathbf{S}^{\dagger}\right) \\ &\left[\boldsymbol{\theta}^{(h,0)} - \left(\mathbf{D} - \mathbf{V}^{(h,0)}\right)^{-1}\left(\mathbf{u}^{(h,0)} + \mathbf{D}\boldsymbol{\theta}^{(h,0)}\right)\right] \neq \mathbf{0}; \end{aligned} \tag{1.116}$$

$$\begin{aligned} &\textit{if } \mathbf{d} = \mathbf{0}: \boldsymbol{\alpha}^{(1)bed}\left(\mathbf{d} = \mathbf{0}\right) = \boldsymbol{\alpha}^{0} + \left(\mathbf{C}^{e}_{\alpha r} - \mathbf{C}_{\alpha\alpha}\mathbf{S}^{\dagger}\right) \\ &\left[\boldsymbol{\theta}^{(h,0)} - \left(\mathbf{D} - \mathbf{V}^{(h,0)}\right)^{-1}\left(\mathbf{u}^{(h,0)} + \mathbf{D}\boldsymbol{\theta}^{(h,0)}\right)\right] \neq \mathbf{0}. \end{aligned} \tag{1.117}$$

As evidenced by comparing Eqs. (1.116) and (1.117) with Eqs. (1.114) and (1.115), respectively, the contributions of the 2nd-order sensitivities provide the fundamental difference when $\mathbf{d} = \mathbf{0}$ between *non-calibration* (i.e., $\mathbf{r}^{(1)bed}\left(\mathbf{d} = \mathbf{0}\right) = \mathbf{r}^{e}$ and $\boldsymbol{\alpha}^{(1)bed}\left(\mathbf{d} = \mathbf{0}\right) = \boldsymbol{\alpha}^{0}$) *in the case for data adjustment, data assimilation, and 1st-BERRU-PM methodologies*, compare Eqs. (1.114) and (1.115), respectively, versus the *calibration* $\mathbf{r}^{(1)bed}\left(\mathbf{d} = \mathbf{0}\right) \neq \mathbf{r}^{e}$ and $\boldsymbol{\alpha}^{(1)bed}\left(\mathbf{d} = \mathbf{0}\right) \neq \boldsymbol{\alpha}^{0}$ provided by the 2nd-BERRU-PMD expressions obtained in Eqs. (1.116) and (1.117).

The components of the matrix $\mathbf{V}^{(h,0)}$ are usually smaller (since they contain higher-order quantities) than the corresponding components of the matrix $\mathbf{D}$, so the contributions provided by the term $\mathbf{V}^{(h,0)}$ can be considered to be "corrections" to the components of the matrix $\mathbf{D}$. If the contributions of the matrix $\mathbf{V}^{(h,0)}$ are neglected by comparison to the corresponding components of the matrix $\mathbf{D}$, the expressions provided in Eqs. (1.112) and (1.113) reduce to the following expressions:

$$if\ \mathbf{V}^{(h,0)} = \mathbf{0}: \ \mathbf{r}^{(1)bed}\left(\mathbf{V}^{(h,0)} = \mathbf{0}\right) = \mathbf{r}^e + \left(\mathbf{C}_{rr}^e - \mathbf{C}_{r\alpha}^e \mathbf{S}^\dagger\right)\mathbf{D}^{-1}\left(\mathbf{d} - \mathbf{u}^{(h,0)}\right); \tag{1.118}$$

$$if\ \mathbf{V}^{(h,0)} = \mathbf{0}: \ \boldsymbol{\alpha}^{(1)bed}\left(\mathbf{V}^{(h,0)} = \mathbf{0}\right) = \boldsymbol{\alpha}^0 + \left(\mathbf{C}_{\alpha r}^e - \mathbf{C}_{\alpha\alpha} \mathbf{S}^\dagger\right)\mathbf{D}^{-1}\left(\mathbf{d} - \mathbf{u}^{(h,0)}\right). \tag{1.119}$$

The previous expressions underscore, through the term $\mathbf{u}^{(h,0)}$, the fundamental importance of including 2nd-order sensitivities, since they provide the basic contribution to maintaining the effects of calibration for both responses and parameters, even when $\mathbf{d} = \mathbf{0}$, in other words, when the values of the measured responses coincide with the corresponding values of the computed responses. This essential contribution stemming from the 2nd-order response sensitivities is clearly highlighted by comparing the expressions obtained in Eqs. (1.118) and (1.119) with the corresponding 1st-order (BERRU-PM) expressions provided in Eqs. (1.114) and (1.115), respectively.

Using the first-iterate, $\boldsymbol{\theta}^{(h,1)}$, in Eq. (1.109) yields the following expression for the "first-iterate best-estimate" covariance matrix $\mathbf{C}_{rr}^{(1)bed}$ for the predicted responses:

$$\begin{aligned}
\mathbf{C}_{rr}^{(1)bed} &= \int_{Dz} \left\{ \begin{matrix} \mathbf{r} - \mathbf{r}^e - \left(\mathbf{C}_{rr}^e - \mathbf{C}_{r\alpha}^e \mathbf{S}^\dagger\right) \\ \left[\boldsymbol{\theta}^{(h,0)} - \left(\mathbf{D} - \mathbf{V}^{(h,0)}\right)^{-1}\left(\mathbf{u}^{(h,0)} + \mathbf{D}\boldsymbol{\theta}^{(h,0)} - \mathbf{d}\right)\right] \end{matrix} \right\} \\
&\times \left\{ \begin{matrix} \mathbf{r} - \mathbf{r}^e - \left(\mathbf{C}_{rr}^e - \mathbf{C}_{r\alpha}^e \mathbf{S}^\dagger\right) \\ \left[\boldsymbol{\theta}^{(h,0)} - \left(\mathbf{D} - \mathbf{V}^{(h,0)}\right)^{-1}\left(\mathbf{u}^{(h,0)} + \mathbf{D}\boldsymbol{\theta}^{(h,0)} - \mathbf{d}\right)\right] \end{matrix} \right\}^\dagger p(\mathbf{z})\,d\mathbf{z} \\
&= \left\langle \begin{matrix} \left[\mathbf{r} - \mathbf{r}^e + \mathbf{X}_r\boldsymbol{\theta}^{(h,0)} - \mathbf{Y}_r\mathbf{d} + \mathbf{Y}_r\mathbf{u}^{(h,0)}\right] \\ \left[\mathbf{r} - \mathbf{r}^e + \mathbf{X}_r\boldsymbol{\theta}^{(h,0)} - \mathbf{Y}_r\mathbf{d} + \mathbf{Y}_r\mathbf{u}^{(h,0)}\right]^\dagger \end{matrix} \right\rangle.
\end{aligned} \tag{1.120}$$

The following definitions were used in Eq. (1.120):

$$\langle \bullet \rangle \triangleq \int_{Dz} (\bullet)\, p(\mathbf{z})\,d\mathbf{z}, \tag{1.121}$$

$$\mathbf{X}_r \triangleq \mathbf{C}_{rd}^{(1)}\left[\left(\mathbf{D} - \mathbf{V}^{(h,0)}\right)^{-1}\mathbf{D} - \mathbf{I}\right]; \tag{1.122}$$

$$\mathbf{Y}_r \triangleq \mathbf{C}_{rd}^{(1)}\left(\mathbf{D} - \mathbf{V}^{(h,0)}\right)^{-1}; \tag{1.123}$$

$$\begin{aligned}
\mathbf{C}_{rd}^{(1)} &\triangleq \left\langle \left(\mathbf{r} - \mathbf{r}^e\right)\mathbf{d}^\dagger \right\rangle = \left\langle \left(\mathbf{r} - \mathbf{r}^e\right)\left[\mathbf{r} - \mathbf{r}^e - \mathbf{S}\left(\boldsymbol{\alpha} - \boldsymbol{\alpha}^0\right) - \mathbf{p}\left(\boldsymbol{\alpha};\boldsymbol{\alpha}^0\right)\right]^\dagger \right\rangle \\
&\cong \mathbf{C}_{rr}^e - \mathbf{C}_{r\alpha}^e \mathbf{S}^\dagger.
\end{aligned} \tag{1.124}$$

Expanding the integrand in the second equality in Eq. (1.120) yields the following expression for the "first-iterate best-estimate" covariance matrix $\mathbf{C}_{rr}^{(1)bed}$ for the predicted response:

$$\begin{aligned}\mathbf{C}_{rr}^{(1)bed} &= \mathbf{C}_{rr}^{e} - \mathbf{C}_{rd}^{(1)}\mathbf{Y}_r^{\dagger} + \mathbf{X}_r\boldsymbol{\theta}^{(h,0)}\left(\boldsymbol{\theta}^{(h,0)}\right)^{\dagger}\mathbf{X}_r^{\dagger} - \mathbf{X}_r\boldsymbol{\theta}^{(h,0)}\bar{\mathbf{d}}^{\dagger}\mathbf{Y}_r^{\dagger} \\ &+ \mathbf{X}_r\boldsymbol{\theta}^{(h,0)}\left(\mathbf{u}^{(h,0)}\right)^{\dagger}\mathbf{Y}_r^{\dagger} - \mathbf{Y}_r\left(\mathbf{C}_{rd}^{(1)}\right)^{\dagger} - \mathbf{Y}_r\bar{\mathbf{d}}\left(\boldsymbol{\theta}^{(h,0)}\right)^{\dagger}\mathbf{X}_r^{\dagger} \\ &+ \mathbf{Y}_r\mathbf{D}\mathbf{Y}_r^{\dagger} - \mathbf{Y}_r\bar{\mathbf{d}}\left(\mathbf{u}^{(h,0)}\right)^{\dagger}\mathbf{Y}_r^{\dagger} + \mathbf{Y}_r\mathbf{u}^{(h,0)}\left(\boldsymbol{\theta}^{(h,0)}\right)^{\dagger}\mathbf{X}_r^{\dagger} \\ &- \mathbf{Y}_r\mathbf{u}^{(h,0)}\bar{\mathbf{d}}^{\dagger}\mathbf{Y}_r^{\dagger} + \mathbf{Y}_r\mathbf{u}^{(h,0)}\left(\mathbf{u}^{(h,0)}\right)^{\dagger}\mathbf{Y}_r^{\dagger}.\end{aligned} \tag{1.125}$$

Using the first-iterate, $\boldsymbol{\theta}^{(h,1)}$, in Eq. (1.110) and evaluating the resulting integrals yields the following expression for the "first-iterate best-estimate" covariance matrix $\mathbf{C}_{\alpha\alpha}^{(1)be}$ for the predicted model parameters:

$$\begin{aligned}\mathbf{C}_{\alpha\alpha}^{(1)bed} &= \left\langle \begin{matrix}\left[\boldsymbol{\alpha} - \boldsymbol{\alpha}^{0} + \mathbf{X}_{\alpha}\boldsymbol{\theta}^{(h,0)} - \mathbf{Y}_{\alpha}\mathbf{d} + \mathbf{Y}_{\alpha}\mathbf{u}^{(h,0)}\right] \\ \left[\boldsymbol{\alpha} - \boldsymbol{\alpha}^{0} + \mathbf{X}_{\alpha}\boldsymbol{\theta}^{(h,0)} - \mathbf{Y}_{\alpha}\mathbf{d} + \mathbf{Y}_{\alpha}\mathbf{u}^{(h,0)}\right]^{\dagger}\end{matrix}\right\rangle \\ &= \mathbf{C}_{\alpha\alpha}^{e} - \mathbf{C}_{\alpha d}^{(1)}\mathbf{Y}_{\alpha}^{\dagger} + \mathbf{X}_{\alpha}\boldsymbol{\theta}^{(h,0)}\left(\boldsymbol{\theta}^{(h,0)}\right)^{\dagger}\mathbf{X}_{\alpha}^{\dagger} - \mathbf{X}_{\alpha}\boldsymbol{\theta}^{(h,0)}\bar{\mathbf{d}}^{\dagger}\mathbf{Y}_{\alpha}^{\dagger} \\ &+ \mathbf{X}_{\alpha}\boldsymbol{\theta}^{(h,0)}\left(\mathbf{u}^{(h,0)}\right)^{\dagger}\mathbf{Y}_{\alpha}^{\dagger} - \mathbf{Y}_{\alpha}\left(\mathbf{C}_{\alpha d}^{(1)}\right)^{\dagger} - \mathbf{Y}_{\alpha}\bar{\mathbf{d}}\left(\boldsymbol{\theta}^{(h,0)}\right)^{\dagger}\mathbf{X}_{\alpha}^{\dagger} \\ &+ \mathbf{Y}_{\alpha}\mathbf{D}\mathbf{Y}_{\alpha}^{\dagger} - \mathbf{Y}_{\alpha}\bar{\mathbf{d}}\left(\mathbf{u}^{(h,0)}\right)^{\dagger}\mathbf{Y}_{\alpha}^{\dagger} + \mathbf{Y}_{\alpha}\mathbf{u}^{(h,0)}\left(\boldsymbol{\theta}^{(h,0)}\right)^{\dagger}\mathbf{X}_{\alpha}^{\dagger} \\ &- \mathbf{Y}_{\alpha}\mathbf{u}^{(h,0)}\bar{\mathbf{d}}^{\dagger}\mathbf{Y}_{\alpha}^{\dagger} + \mathbf{Y}_{\alpha}\mathbf{u}^{(h,0)}\left(\mathbf{u}^{(h,0)}\right)^{\dagger}\mathbf{Y}_{\alpha}^{\dagger}.\end{aligned} \tag{1.126}$$

The following definitions were used in Eq. (1.126):

$$\begin{aligned}\mathbf{C}_{\alpha d}^{(1)} &\triangleq \left\langle\left(\boldsymbol{\alpha} - \boldsymbol{\alpha}^{0}\right)\mathbf{d}^{\dagger}\right\rangle \\ &= \left\langle\left(\boldsymbol{\alpha} - \boldsymbol{\alpha}^{0}\right)\left[\mathbf{r} - \mathbf{r}^{e} - \mathbf{S}\left(\boldsymbol{\alpha} - \boldsymbol{\alpha}^{0}\right) - \mathbf{p}\left(\boldsymbol{\alpha};\boldsymbol{\alpha}^{0}\right)\right]^{\dagger}\right\rangle \\ &\cong \mathbf{C}_{\alpha r}^{e} - \mathbf{C}_{\alpha\alpha}^{e}\mathbf{S}^{\dagger};\end{aligned} \tag{1.127}$$

$$\mathbf{X}_{\alpha} \triangleq \mathbf{C}_{\alpha d}^{(1)}\left[\left(\mathbf{D} - \mathbf{V}^{(h,0)}\right)^{-1}\mathbf{D} - \mathbf{I}\right]; \tag{1.128}$$

$$\mathbf{Y}_{\alpha} \triangleq \mathbf{C}_{\alpha d}^{(1)}\left(\mathbf{D} - \mathbf{V}^{(h,0)}\right)^{-1}. \tag{1.129}$$

Using the first-iterate, $\boldsymbol{\theta}^{(h,1)}$, in Eq. (1.111) and evaluating the resulting integrals yields the following expression for the "first-iterate best-estimate" correlation matrix $\mathbf{C}_{\alpha r}^{(1)bed}$ for the predicted parameters and responses:

$$\mathbf{C}_{\alpha r}^{(1)bed} = \left\langle \begin{matrix} \left[\boldsymbol{\alpha} - \boldsymbol{\alpha}^0 + \mathbf{X}_\alpha \boldsymbol{\theta}^{(h,0)} - \mathbf{Y}_\alpha \mathbf{d} + \mathbf{Y}_\alpha \mathbf{u}^{(h,0)}\right] \\ \left[\mathbf{r} - \mathbf{r}^e + \mathbf{X}_r \boldsymbol{\theta}^{(h,0)} - \mathbf{Y}_r \mathbf{d} + \mathbf{Y}_r \mathbf{u}^{(h,0)}\right]^\dagger \end{matrix} \right\rangle$$
$$= \mathbf{C}_{\alpha r}^e - \mathbf{C}_{\alpha d}^{(1)} \mathbf{Y}_r^\dagger + \mathbf{X}_\alpha \boldsymbol{\theta}^{(h,0)} \left(\boldsymbol{\theta}^{(h,0)}\right)^\dagger \mathbf{X}_r^\dagger - \mathbf{X}_\alpha \boldsymbol{\theta}^{(h,0)} \overline{\mathbf{d}}^\dagger \mathbf{Y}_r^\dagger$$
$$+ \mathbf{X}_\alpha \boldsymbol{\theta}^{(h,0)} \left(\mathbf{u}^{(h,0)}\right)^\dagger \mathbf{Y}_r^\dagger - \mathbf{Y}_\alpha \left(\mathbf{C}_{rd}^{(1)}\right)^\dagger - \mathbf{Y}_\alpha \overline{\mathbf{d}} \left(\boldsymbol{\theta}^{(h,0)}\right)^\dagger \mathbf{X}_r^\dagger \tag{1.130}$$
$$+ \mathbf{Y}_\alpha \mathbf{D} \mathbf{Y}_r^\dagger - \mathbf{Y}_\alpha \overline{\mathbf{d}} \left(\mathbf{u}^{(h,0)}\right)^\dagger \mathbf{Y}_r^\dagger + \mathbf{Y}_\alpha \mathbf{u}^{(h,0)} \left(\boldsymbol{\theta}^{(h,0)}\right)^\dagger \mathbf{X}_r^\dagger$$
$$- \mathbf{Y}_\alpha \mathbf{u}^{(h,0)} \overline{\mathbf{d}}^\dagger \mathbf{Y}_r^\dagger + \mathbf{Y}_\alpha \mathbf{u}^{(h,0)} \left(\mathbf{u}^{(h,0)}\right)^\dagger \mathbf{Y}_r^\dagger.$$

As the expressions obtained in Eqs. (1.125), (1.126), and (1.130) indicate that combining experimental and computational information modifies the correlations among the parameters through couplings introduced by the sensitivities of the participating responses.

The expressions obtained in Eqs. (1.125), (1.126), and (1.130) $\mathbf{C}_{\alpha r}^{(1)bed}$, respectively, for $\mathbf{C}_{rr}^{(1)bed}$, $\mathbf{C}_{\alpha\alpha}^{(1)bed}$, and $\mathbf{C}_{\alpha r}^{(1)bed}$, respectively, will be analyzed next by considering the same simplifying conditions as were considered for the best-estimate response and parameter mean values $\mathbf{r}^{(1)bed}$ and $\boldsymbol{\alpha}^{(1)bed}$, respectively, as follows:

1. If all of the 2nd-order sensitivities are neglected, then the following simplifications occur: $\mathbf{X}_r = \mathbf{X}_\alpha = \mathbf{0}$, $\mathbf{Y}_r = \mathbf{C}_{rd}^{(1)} \mathbf{D}^{-1}$, and $\mathbf{Y}_\alpha = \mathbf{C}_{\alpha d}^{(1)} \mathbf{D}^{-1}$. In this case, therefore, the expressions in Eqs. (1.125), (1.126), and (1.130) would take on the following simplified forms:

$$no\ 2^{nd} - order\ sensitivities:\ \mathbf{C}_{rr}^{(1)bed} = \mathbf{C}_{rr}^e - \mathbf{C}_{rd}^{(1)} \mathbf{D}^{-1} \mathbf{C}_{rd}^{(1)\dagger}. \tag{1.131}$$

$$no\ 2^{nd} - order\ sensitivities:\ \mathbf{C}_{\alpha\alpha}^{(1)bed} = \mathbf{C}_{\alpha\alpha}^e - \mathbf{C}_{\alpha d}^{(1)} \mathbf{D}^{-1} \mathbf{C}_{\alpha d}^{(1)\dagger}. \tag{1.132}$$

$$no\ 2^{nd} - order\ sensitivities:\ \mathbf{C}_{\alpha r}^{(1)bed} = \mathbf{C}_{\alpha r}^e - \mathbf{C}_{\alpha d}^{(1)} \mathbf{D}^{-1} \mathbf{C}_{rd}^{(1)\dagger}. \tag{1.133}$$

As expected, the expressions obtained in Eq. (1.131)–(1.133) are identical to the expressions provided by the 1st-order BERRU-PM methodology (Cacuci, 2019) and therefore possess the same strengths (easy to implement) and weaknesses (i.e., trivial results and no calibration if $\mathbf{d} = \mathbf{0}$) as the BERRU-PM methodology.
2. If the vector of deviations between the nominally measured and computed values of the responses vanishes, in other words, if $\mathbf{d} = \mathbf{0}$, the expressions obtained in Eqs. (1.125), (1.126), and (1.130) for $\mathbf{C}_{rr}^{(1)bed}$, $\mathbf{C}_{\alpha\alpha}^{(1)bed}$, and $\mathbf{C}_{\alpha r}^{(1)bed}$, respectively, would take on the following simplified forms:

$$if\ \mathbf{d} = \mathbf{0}:\ \mathbf{C}_{rr}^{(1)bed} = \mathbf{C}_{rr}^e - \mathbf{C}_{rd}^{(1)} \mathbf{Y}_r^\dagger + \mathbf{X}_r \boldsymbol{\theta}^{(h,0)} \left(\boldsymbol{\theta}^{(h,0)}\right)^\dagger$$
$$\mathbf{X}_r^\dagger + \mathbf{X}_r \boldsymbol{\theta}^{(h,0)} \left(\mathbf{u}^{(h,0)}\right)^\dagger \mathbf{Y}_r^\dagger - \mathbf{Y}_r \left(\mathbf{C}_{rd}^{(1)}\right)^\dagger + \mathbf{Y}_r \mathbf{D} \mathbf{Y}_r^\dagger + \mathbf{Y}_r \mathbf{u}^{(h,0)} \tag{1.134}$$
$$\left(\boldsymbol{\theta}^{(h,0)}\right)^\dagger \mathbf{X}_r^\dagger + \mathbf{Y}_r \mathbf{u}^{(h,0)} \left(\mathbf{u}^{(h,0)}\right)^\dagger \mathbf{Y}_r^\dagger.$$

$$if\ \mathbf{d}=\mathbf{0}:\ \mathbf{C}_{\alpha\alpha}^{(1)bed}=\mathbf{C}_{\alpha\alpha}^{e}-\mathbf{C}_{\alpha d}^{(1)}\mathbf{Y}_{\alpha}^{\dagger}+\mathbf{X}_{\alpha}\boldsymbol{\theta}^{(h,0)}\left(\boldsymbol{\theta}^{(h,0)}\right)^{\dagger}$$
$$\mathbf{X}_{\alpha}^{\dagger}+\mathbf{X}_{\alpha}\boldsymbol{\theta}^{(h,0)}\left(\mathbf{u}^{(h,0)}\right)^{\dagger}\mathbf{Y}_{\alpha}^{\dagger}-\mathbf{Y}_{\alpha}\left(\mathbf{C}_{\alpha d}^{(1)}\right)^{\dagger}+\mathbf{Y}_{\alpha}\mathbf{D}\mathbf{Y}_{\alpha}^{\dagger}+\mathbf{Y}_{\alpha}\mathbf{u}^{(h,0)} \tag{1.135}$$
$$\left(\boldsymbol{\theta}^{(h,0)}\right)^{\dagger}\mathbf{X}_{\alpha}^{\dagger}+\mathbf{Y}_{\alpha}\mathbf{u}^{(h,0)}\left(\mathbf{u}^{(h,0)}\right)^{\dagger}\mathbf{Y}_{\alpha}^{\dagger}.$$

$$if\ \mathbf{d}=\mathbf{0}:\ \mathbf{C}_{\alpha r}^{(1)bed}=\mathbf{C}_{\alpha r}^{e}-\mathbf{C}_{\alpha d}^{(1)}\mathbf{Y}_{r}^{\dagger}+\mathbf{X}_{\alpha}\boldsymbol{\theta}^{(h,0)}\left(\boldsymbol{\theta}^{(h,0)}\right)^{\dagger}$$
$$\mathbf{X}_{r}^{\dagger}+\mathbf{X}_{\alpha}\boldsymbol{\theta}^{(h,0)}\left(\mathbf{u}^{(h,0)}\right)^{\dagger}\mathbf{Y}_{r}^{\dagger}-\mathbf{Y}_{\alpha}\left(\mathbf{C}_{rd}^{(1)}\right)^{\dagger}+\mathbf{Y}_{\alpha}\mathbf{D}\mathbf{Y}_{r}^{\dagger}+\mathbf{Y}_{\alpha}\mathbf{u}^{(h,0)} \tag{1.136}$$
$$\left(\boldsymbol{\theta}^{(h,0)}\right)^{\dagger}\mathbf{X}_{r}^{\dagger}+\mathbf{Y}_{\alpha}\mathbf{u}^{(h,0)}\left(\mathbf{u}^{(h,0)}\right)^{\dagger}\mathbf{Y}_{r}^{\dagger}.$$

If the contributions of the 2nd-order sensitivities are neglected in the expressions obtained in Eqs. (1.134)–(1.136), these expressions would reduce to the 1st-order expressions shown in Eqs. (1.131)–(1.133).

3. If the contributions of the matrix $\mathbf{V}^{(h,0)}$ are neglected by comparison to the corresponding components of the matrix $\mathbf{D}$, then the following simplifications occur: $\mathbf{X}_r=\mathbf{X}_\alpha=\mathbf{0}$, $\mathbf{Y}_r=\mathbf{C}_{rd}^{(1)}\mathbf{D}^{-1}$, and $\mathbf{Y}_\alpha=\mathbf{C}_{\alpha d}^{(1)}\mathbf{D}^{-1}$. Consequently, the expressions provided in Eqs. (1.125), (1.126), and (1.130) would reduce to the following forms:

$$if\ \mathbf{V}^{(h,0)}=\mathbf{0}:\ \mathbf{C}_{rr}^{(1)bed}\left(\mathbf{V}^{(h,0)}=\mathbf{0}\right)=\mathbf{C}_{rr}^{e}-\mathbf{C}_{rd}^{(1)}\mathbf{D}^{-1}\mathbf{C}_{rd}^{(1)\dagger}$$
$$-\mathbf{Y}_{r}\bar{\mathbf{d}}\left(\mathbf{u}^{(h,0)}\right)^{\dagger}\mathbf{Y}_{r}^{\dagger}-\mathbf{Y}_{r}\mathbf{u}^{(h,0)}\bar{\mathbf{d}}^{\dagger}\mathbf{Y}_{r}^{\dagger}+\mathbf{Y}_{r}\mathbf{u}^{(h,0)}\left(\mathbf{u}^{(h,0)}\right)^{\dagger}\mathbf{Y}_{r}^{\dagger}; \tag{1.137}$$

$$if\ \mathbf{V}^{(h,0)}=\mathbf{0}:\ \mathbf{C}_{\alpha\alpha}^{(1)bed}\left(\mathbf{V}^{(h,0)}=\mathbf{0}\right)=\mathbf{C}_{\alpha\alpha}^{e}-\mathbf{C}_{\alpha d}^{(1)}\mathbf{D}^{-1}\left(\mathbf{C}_{\alpha d}^{(1)}\right)^{\dagger}$$
$$-\mathbf{Y}_{\alpha}\bar{\mathbf{d}}\left(\mathbf{u}^{(h,0)}\right)^{\dagger}\mathbf{Y}_{\alpha}^{\dagger}-\mathbf{Y}_{\alpha}\mathbf{u}^{(h,0)}\bar{\mathbf{d}}^{\dagger}\mathbf{Y}_{\alpha}^{\dagger}+\mathbf{Y}_{\alpha}\mathbf{u}^{(h,0)}\left(\mathbf{u}^{(h,0)}\right)^{\dagger}\mathbf{Y}_{\alpha}^{\dagger}; \tag{1.138}$$

$$if\ \mathbf{V}^{(h,0)}=\mathbf{0}:\ \mathbf{C}_{\alpha r}^{(1)bed}\left(\mathbf{V}^{(h,0)}=\mathbf{0}\right)=\mathbf{C}_{\alpha r}^{e}-\mathbf{C}_{\alpha d}^{(1)}\mathbf{D}^{-1}\left(\mathbf{C}_{rd}^{(1)}\right)^{\dagger}$$
$$-\mathbf{Y}_{\alpha}\bar{\mathbf{d}}\left(\mathbf{u}^{(h,0)}\right)^{\dagger}\mathbf{Y}_{r}^{\dagger}-\mathbf{Y}_{\alpha}\mathbf{u}^{(h,0)}\bar{\mathbf{d}}^{\dagger}\mathbf{Y}_{r}^{\dagger}+\mathbf{Y}_{\alpha}\mathbf{u}^{(h,0)}\left(\mathbf{u}^{(h,0)}\right)^{\dagger}\mathbf{Y}_{r}^{\dagger}. \tag{1.139}$$

The previous expressions underscore, through the terms involving the vector $\mathbf{u}^{(h,0)}$, the fundamental importance of including 2nd-order sensitivities, which provide non-zero contributions even when $\bar{\mathbf{d}}=\mathbf{0}$.

It is important to emphasize that the posterior distribution of responses and parameters is non-Gaussian since the "best-estimate" triple correlations among the best-estimate responses and parameters are non-zero even if the experimentally measured responses and parameters were normally distributed initially. This fact can be shown by considering Eqs. (1.112) and (1.113), which can be written as follows:

$$\mathbf{r}^{(1)bed} = \mathbf{r}^e + \mathbf{r}^{(0)}; \ \mathbf{r}^{(0)} \triangleq \left(\mathbf{C}_{rr}^e - \mathbf{C}_{r\alpha}^e \mathbf{S}^\dagger\right) \left[\boldsymbol{\theta}^{(h,0)} - \left(\mathbf{D} - \mathbf{V}^{(h,0)}\right)^{-1} \left(\mathbf{u}^{(h,0)} + \mathbf{D}\boldsymbol{\theta}^{(h,0)} - \mathbf{d}\right)\right]; \tag{1.140}$$

$$\boldsymbol{\alpha}^{(1)bed} = \boldsymbol{\alpha}^0 + \boldsymbol{\beta}^{(0)}; \ \boldsymbol{\beta}^{(0)} \triangleq \left(\mathbf{C}_{\alpha r}^e - \mathbf{C}_{\alpha\alpha} \mathbf{S}^\dagger\right) \left[\boldsymbol{\theta}^{(h,0)} - \left(\mathbf{D} - \mathbf{V}^{(h,0)}\right)^{-1} \left(\mathbf{u}^{(h,0)} + \mathbf{D}\boldsymbol{\theta}^{(h,0)} - \mathbf{d}\right)\right]. \tag{1.141}$$

Using Eqs. (1.140) and (1.141) in conjunction with the definition of triple correlations for multivariate random quantities yields the following "lowest-order estimate" for the best-estimate triple correlations among responses and parameters, which are denoted as $T_{rr,ijk}^{(1)bed}$ and, respectively, $T_{\alpha\alpha,ijk}^{(1)bed}$:

$$\begin{aligned} T_{rr,ijk}^{(1)bed} &\triangleq \left\langle \left(r_i - r_i^e - r_i^{(0)}\right)\left(r_j - r_j^e - r_j^{(0)}\right)\left(r_k - r_k^e - r_k^{(0)}\right) \right\rangle \\ &= E\left[\left(r_i - r_i^e\right)\left(r_j - r_j^e\right)\left(r_k - r_k^e\right)\right] - r_i^{(0)} \operatorname{cov}\left(r_j, r_k\right) - r_j^{(0)} \operatorname{cov}\left(r_i, r_k\right) \\ &-r_k^{(0)} \operatorname{cov}\left(r_i, r_j\right) - r_i^{(0)} r_j^{(0)} r_k^{(0)}; \quad i,j,k = 1,\dots,TR; \end{aligned} \tag{1.142}$$

$$\begin{aligned} T_{\alpha\alpha,ijk}^{(1)bed} &\triangleq \left\langle \left(\alpha_i - \alpha_i^0 - \beta_i^{(0)}\right)\left(\alpha_j - \alpha_j^0 - \beta_j^{(0)}\right)\left(\alpha_k - \alpha_k^e - \beta_k^{(0)}\right) \right\rangle \\ &= E\left[\left(\alpha_i - \alpha_i^0\right)\left(\alpha_j - \alpha_j^0\right)\left(\alpha_j - \alpha_j^0\right)\right] - \beta_i^{(0)} \operatorname{cov}\left(\alpha_j, \alpha_k\right) \\ &-\beta_j^{(0)} \operatorname{cov}\left(\alpha_i, \alpha_k\right) - \beta_k^{(0)} \operatorname{cov}\left(\alpha_i, \alpha_j\right) - \beta_i^{(0)}\beta_j^{(0)}\beta_k^{(0)}; \quad i,j,k = 1,\dots,TP. \end{aligned} \tag{1.143}$$

If the responses are normally distributed initially, then the initial triple correlations are all zero, in other words, $E\left[\left(\alpha_i - \alpha_i^0\right)\left(\alpha_j - \alpha_j^0\right)\left(\alpha_j - \alpha_j^0\right)\right]$ for all $i,j,k = 1,\dots,TR$. Nevertheless, as indicated in Eq. (1.142), the "best-estimate response triple correlations" $T_{rr,ijk}^{(1)bed} \neq 0$ are non-zero even in such a case, unless there is "no calibration," in other words, if $r_i^{(0)} = r_j^{(0)} = r_k^{(0)} = 0$ for all $i,j,k = 1,\dots,TR$. Similarly, if the parameters are normally distributed initially, then the initial triple correlations are all zero, in other words, $E\left[\left(r_i - r_i^e\right)\left(r_j - r_j^e\right)\left(r_k - r_k^e\right)\right] = 0$ for all $i,j,k = 1,\dots,TP$. Nevertheless, as indicated in Eq. (1.143), the "best-estimate parameter triple correlations" $T_{\alpha\alpha,ijk}^{(1)bed} \neq 0$ are non-zero even in such a case, unless there is "no calibration," in other words, if $\alpha_i^{(0)} = \alpha_j^{(0)} = \alpha_k^{(0)} = 0$ for all $i,j,k = 1,\dots,TP$.

In summary, the *2nd-BERRU-PMD* methodology is fundamentally based on the maximum entropy (MaxEnt) principle; the attribute "2nd" indicates that this methodology incorporates 2nd-order uncertainties and 2nd-order sensitivities of computed model responses to model parameters; the acronym BERRU stands for "best-estimate results with reduced uncertainties;" and the letter "*D*" indicates "deterministic," referring to the deterministic inclusion of the computational model responses. The 2nd-BERRU-PMD methodology includes all of the previous results produced by the 1st-order BERRU-PM methodology, generalizing and extending current data assimilation and/or data adjustment procedures while overcoming the fundamental limitations of these procedures. Since the framework of the 2nd-BERRU-PMD methodology comprises the combined phase-space of parameters and responses, the expressions obtained for the predicted best-estimate responses and calibrated parameters can be used for solving both forward/direct and inverse problems.

1.6 2ND-BERRU-PMP: 2ND-ORDER MAXENT PREDICTIVE MODELING METHODOLOGY WITH PROBABILISTICALLY INCLUDED COMPUTED RESPONSES

This section presents the "*2nd-BERRU-PMP*" methodology, which is the alternative to the 2nd-BERRU-PMD methodology presented in Section 1.5. The *2nd-BERRU-PMP* methodology is also based on the MaxEnt principle but includes the computational model responses *probabilistically* – through a MaxEnt representation – in contradistinction to its deterministic inclusion within the 2nd-BERRU-PMD methodology. This (contra)distinction is indicated by the last letter ("*P*") in the acronym, which refers to the "probabilistic" inclusion of the computed model responses. Although both the 2nd-BERRU-PMP and the 2nd-BERRU-PMD methodologies yield expressions that include second (and higher) order sensitivities of responses to model parameters, the respective expressions for the predicted responses, calibrated predicted parameters and their predicted uncertainties (covariances) are not identical to each other. Both the 2nd-BERRU-PMP and the 2nd-BERRU-PMD methodologies encompass, as particular cases, the results produced by the extant data assimilation and data adjustment procedures, as well as the results produced by the 1st-order BERRU-PM methodology (Cacuci, 2019). The intercomparison between the 2nd-BERRU-PMP and the 2nd-BERRU-PMD methodologies, and a discussion of the additional considerations that make the corresponding expressions predicted by the 2nd-BERRU-PMD and the 2nd-BERRU-PMP become identical, will be presented further in Section 1.7.

1.6.1 2ND-ORDER MAXENT PROBABILISTIC REPRESENTATION OF THE COMPUTATIONAL MODEL

The generic mathematical modeling of a physical system has been presented in Section 1.2. As shown in Eq. (1.11), each of the model's computed responses is formally a function (implicit and/or explicit) of the model parameters $\boldsymbol{\alpha}$, in other words, $r_k = r_k(\boldsymbol{\alpha})$, $k = 1,\ldots,TR$. The uncertainties affecting the model parameters $\boldsymbol{\alpha}$ will "propagate" both directly and indirectly (through the model's dependent variables) to induce uncertainties in the computed model responses $r_k(\boldsymbol{\alpha})$. It is therefore reasonable to consider that the *computed model responses* are quasi-random quantities distributed according to an unknown distribution, which will be denoted as $p_c(\mathbf{r})$. Furthermore, the unknown joint probability distribution of model parameters and responses will be denoted as $p_c(\boldsymbol{\alpha},\mathbf{r}) \triangleq p_\alpha(\boldsymbol{\alpha})\, p_c(\mathbf{r})$. The approximate moments of the unknown distribution of $r_k(\boldsymbol{\alpha})$ can be computed by using the so-called "propagation of errors" methodology, which entails the formal integration over $p_c(\boldsymbol{\alpha},\mathbf{r}) \triangleq p_\alpha(\boldsymbol{\alpha})\,p_c(\mathbf{r})$ of various expressions involving the truncated Taylor series expansion of the response in terms of parameter variations. This procedure was first used by Tukey (1957). Tukey's results were generalized to the 6th-order by Cacuci (2022).

Each computed model response can be formally expanded in a multivariate Taylor series around the parameters' mean (or nominal) values. In particular, the 4th-order Taylor series of a system response, denoted as $r_k(\boldsymbol{\alpha})$, around the expected (or nominal) parameter values $\boldsymbol{\alpha}^0$ has the following formal expression:

$$
\begin{aligned}
r_k(\boldsymbol{\alpha}) = {} & r_k(\boldsymbol{\alpha}^0) + \sum_{j_1=1}^{TP}\left\{\frac{\partial r_k(\boldsymbol{\alpha})}{\partial \alpha_{j_1}}\right\}_{\alpha^0}\delta\alpha_{j_1} + \frac{1}{2}\sum_{j_1=1}^{TP}\sum_{j_2=1}^{TP}\left\{\frac{\partial^2 r_k(\boldsymbol{\alpha})}{\partial \alpha_{j_1}\partial \alpha_{j_2}}\right\}_{\alpha^0}\delta\alpha_{j_1}\delta\alpha_{j_2} \\
& + \frac{1}{3!}\sum_{j_1=1}^{TP}\sum_{j_2=1}^{TP}\sum_{j_3=1}^{TP}\left\{\frac{\partial^3 r_k(\boldsymbol{\alpha})}{\partial \alpha_{j_1}\partial \alpha_{j_2}\partial \alpha_{j_3}}\right\}_{\alpha^0}\delta\alpha_{j_1}\delta\alpha_{j_2}\delta\alpha_{j_3} \\
& + \frac{1}{4!}\sum_{j_1=1}^{TP}\sum_{j_2=1}^{TP}\sum_{j_3=1}^{TP}\sum_{j_4=1}^{TP}\left\{\frac{\partial^4 r_k(\boldsymbol{\alpha})}{\partial \alpha_{j_1}\partial \alpha_{j_2}\partial \alpha_{j_3}\partial \alpha_{j_4}}\right\}_{\alpha^0}\delta\alpha_{j_1}\delta\alpha_{j_2}\delta\alpha_{j_3}\delta\alpha_{j_4} + \varepsilon_k .
\end{aligned}
\tag{1.144}
$$

In Eq. (1.144), the quantity $r_k(\boldsymbol{\alpha}^0)$ indicates the computed value of the response using the expected/nominal parameter values $\boldsymbol{\alpha}^0 \triangleq (\alpha_1^0,...,\alpha_{TP}^0)^\dagger$. The notation $\{\;\}_{\alpha^0}$ indicates that the quantities within the braces are also computed using the expected/nominal parameter values. The quantity ε_k in Eq. (1.144) comprises all quantifiable errors in the representation of the computed response as a function of the model parameters, including the truncation errors $O(\delta\alpha_j)^5$ of the Taylor series expansion, possible bias errors due to incompletely modeled physical phenomena, and possible random errors due to numerical approximations. The radius/domain of convergence of the series in Eq. (1.144) determines the largest values of the parameter variations $\delta\alpha_j$ which are admissible before the respective series becomes divergent. In turn, these maximum admissible parameter variations limit, through Eq. (1.31), the largest parameter covariances/standard deviations that can be considered for using the Taylor expansion for the subsequent purposes of computing moments of the distribution of computed responses.

As is well-known, and as indicated by Eq. (1.144), the Taylor series of a function of TP-variables (e.g., $r_k(\boldsymbol{\alpha})$) comprises TP distinct 1st-order derivatives, $TP(TP+1)/2$ distinct 2nd-order derivatives, and so on. The computation by conventional methods of the nth-order functional derivatives (called "sensitivities" in the field of sensitivity analysis) of a response with respect to the TP-parameters (on which it depends) would require at least $O(TP^n)$ large-scale computations. The exponential increase – with the order of response sensitivities – of the number of large-scale computations needed to determine higher-order sensitivities is the manifestation of the "curse of dimensionality in sensitivity analysis," by analogy to the expression coined by Bellman (1957) to express the difficulty of using "brute force" grid search when optimizing a function with many input variables. The "nth-order Comprehensive

Adjoint Sensitivity Analysis Methodology for Response-Coupled Forward/Adjoint Linear Systems" (n^{th}-CASAM-L) conceived by Cacuci (2022) and the "n^{th}-order Comprehensive Adjoint Sensitivity Analysis Methodology for Nonlinear Systems" (n^{th}-CASAM-N) conceived by Cacuci (2023a) are currently the only methodologies that enable the exact and efficient computation of arbitrarily high-order sensitivities while overcoming the curse of dimensionality.

The expectation value, $E_c(r_k) \triangleq \int r_k(\boldsymbol{\alpha}) p_c(\boldsymbol{\alpha},\mathbf{r}) d\boldsymbol{\alpha} d\mathbf{r}$, of a computed response $r_k(\boldsymbol{\alpha})$ is obtained by integrating formally Eq. (1.144) over the unknown distribution $p_c(\boldsymbol{\alpha},\mathbf{r})$, which yields the following expression up to and including the contributions from the 4th-order sensitivities:

$$
\begin{aligned}
E_c(r_k) = {} & r_k(\boldsymbol{\alpha}^0) + \frac{1}{2}\sum_{j_1=1}^{TP}\sum_{j_2=1}^{TP}\left\{\frac{\partial^2 r_k(\boldsymbol{\alpha})}{\partial\alpha_{j_1}\partial\alpha_{j_2}}\right\}_{\boldsymbol{\alpha}^0} \operatorname{cov}(\alpha_{j_1},\alpha_{j_2}) \\
& + \frac{1}{6}\sum_{j_1=1}^{TP}\sum_{j_2=1}^{TP}\sum_{j_3=1}^{TP}\left\{\frac{\partial^3 r_k(\boldsymbol{\alpha})}{\partial\alpha_{j_1}\partial\alpha_{j_2}\partial\alpha_{j_3}}\right\}_{\boldsymbol{\alpha}^0} \\
& \times t_{j_1 j_2 j_3}\sigma_{j_1}\sigma_{j_2}\sigma_{j_3} \\
& + \frac{1}{24}\sum_{j_1=1}^{TP}\sum_{j_2=1}^{TP}\sum_{j_3=1}^{TP}\sum_{j_4=1}^{TP}\left\{\frac{\partial^4 r_k(\boldsymbol{\alpha})}{\partial\alpha_{j_1}\partial\alpha_{j_2}\partial\alpha_{j_3}\partial\alpha_{j_4}}\right\}_{\boldsymbol{\alpha}^0} q_{j_1 j_2 j_3 j_4}\sigma_{j_1}\sigma_{j_2}\sigma_{j_3}\sigma_{j_4} + \ldots
\end{aligned}
\tag{1.145}
$$

In Eq. (1.145), the 3rd-order correlation, t_{ijk}, of three parameters α_i, α_j,α_k, is formally defined as follows:

$$
t_{ijk}\sigma_i\sigma_j\sigma_k \triangleq \int_{D_\alpha} \delta\alpha_i\delta\alpha_j\delta\alpha_k p_\alpha(\boldsymbol{\alpha}) d\boldsymbol{\alpha}; \quad i,j,k = 1,\ldots,TP \tag{1.146}
$$

Also, in Eq. (1.145), the 4th-order correlation, $q_{ijk\ell}$, among four parameters is formally defined as follows:

$$
q_{ijk\ell}\sigma_i\sigma_j\sigma_k\sigma_\ell \triangleq \int_{D_\alpha} \delta\alpha_i\delta\alpha_j\delta\alpha_k\,\delta\alpha_\ell p_\alpha(\boldsymbol{\alpha}) d\boldsymbol{\alpha}; \quad i,j,k,\ell = 1,\ldots,TP. \tag{1.147}
$$

For subsequent mathematical derivations, it is convenient to consider that the expectation values $E_c(r_k)$, $k = 1,\ldots,TR$, are the components of a vector defined as follows: $\mathbf{E}_c(\mathbf{r}) \triangleq [E_c(r_1),\ldots,E_c(r_k),\ldots,E_c(r_{TR})]^\dagger$.

The expression of the correlation between a parameter and a computed response, which will be denoted as $\operatorname{cor}(\alpha_i, r_k)$, is presented in what follows, up to and including the contributions from the 3rd-order sensitivities:

$$
\begin{aligned}
\operatorname{cor}(\alpha_i, r_k) & \triangleq \int \delta\alpha_i \left[r_k(\boldsymbol{\alpha}) - E_c(r_k)\right] p_c(\boldsymbol{\alpha},\mathbf{r}) d\boldsymbol{\alpha} d\mathbf{r} \\
& = \sum_{j=1}^{TP}\left\{\frac{\partial r_k(\boldsymbol{\alpha})}{\partial\alpha_j}\right\}_{\boldsymbol{\alpha}^0} \operatorname{cov}(\alpha_i,\alpha_j)
\end{aligned}
\tag{1.148}
$$

$$+\frac{1}{2}\sum_{j_1=1}^{TP}\sum_{j_2=1}^{TP}\left\{\frac{\partial^2 r_k(\boldsymbol{\alpha})}{\partial\alpha_{j_1}\partial\alpha_{j_2}}\right\}_{\boldsymbol{\alpha}^0} t_{i,j_1j_2}\sigma_i\sigma_{j_1}\sigma_{j_2}$$
$$+\frac{1}{6}\sum_{j_1=1}^{TP}\sum_{j_2=1}^{TP}\sum_{j_3=1}^{TP}\left\{\frac{\partial^3 r_k(\boldsymbol{\alpha})}{\partial\alpha_{j_1}\partial\alpha_{j_2}\partial\alpha_{j_3}}\right\}_{\boldsymbol{\alpha}^0} q_{i,j_1j_2j_3}\sigma_{j_1}\sigma_{j_2}\sigma_{j_3}\sigma_{j_4}$$
$$+O\left(\sigma_{j_1}\sigma_{j_2}\sigma_{j_3}\sigma_{j_4}\sigma_{j_5}\right).$$

The correlations $\mathrm{cor}(\alpha_i, r_k)$, $i=1,...,TP$, $k=1,...,TR$, are considered to be the components of a "parameter-response computed correlation matrix" denoted as $\mathbf{C}^c_{\alpha r}$ and defined as follows:

$$\mathbf{C}^c_{\alpha r} \triangleq \begin{pmatrix} \mathrm{cor}(\alpha_1, r_1) & \cdots & \mathrm{cor}(\alpha_1, r_{TR}) \\ \vdots & \ddots & \vdots \\ \mathrm{cor}(\alpha_{TP}, r_1) & \cdots & \mathrm{cor}(\alpha_{TP}, r_{TR}) \end{pmatrix}. \quad (1.149)$$

The expression of the covariance between two responses r_k and r_ℓ will be denoted as $\mathrm{cov}(r_k, r_\ell)$ and is presented in Eq. (1.150), up to and including the contributions from the 4th-order sensitivities:

$$\mathrm{cov}(r_k, r_\ell) \triangleq \int\left[r_k(\boldsymbol{\alpha})-E_c(r_k)\right]\left[r_\ell(\boldsymbol{\alpha})-E_c(r_\ell)\right]p_c(\boldsymbol{\alpha},\mathbf{r})d\boldsymbol{\alpha}d\mathbf{r}$$
$$=\sum_{i=1}^{TP}\sum_{j=1}^{TP}\left\{\frac{\partial r_k(\boldsymbol{\alpha})}{\partial\alpha_i}\frac{\partial r_\ell(\boldsymbol{\alpha})}{\partial\alpha_j}\right\}_{\boldsymbol{\alpha}^0}\mathrm{cov}(\alpha_i,\alpha_j)$$
$$+\frac{1}{2}\sum_{j_1=1}^{TP}\sum_{j_2=1}^{TP}\sum_{j_3=1}^{TP}\left\{\frac{\partial^2 r_k(\boldsymbol{\alpha})}{\partial\alpha_{j_1}\partial\alpha_{j_2}}\frac{\partial r_\ell(\boldsymbol{\alpha})}{\partial\alpha_{j_3}}+\frac{\partial r_k(\boldsymbol{\alpha})}{\partial\alpha_{j_1}}\frac{\partial^2 r_\ell(\boldsymbol{\alpha})}{\partial\alpha_{j_2}\partial\alpha_{j_3}}\right\}_{\boldsymbol{\alpha}^0} t_{j_1j_2j_3}\sigma_{j_1}\sigma_{j_2}\sigma_{j_3}$$
$$+\frac{1}{4}\sum_{j_1=1}^{TP}\sum_{j_2=1}^{TP}\sum_{j_3=1}^{TP}\sum_{j_4=1}^{TP}\left\{\frac{\partial^2 r_k(\boldsymbol{\alpha})}{\partial\alpha_{j_1}\partial\alpha_{j_2}}\frac{\partial^2 r_\ell(\boldsymbol{\alpha})}{\partial\alpha_{j_2}\partial\alpha_{j_3}}\right\}_{\alpha^0}\left(q_{j_1j_2j_3j_4}-\rho_{j_1j_2}\rho_{j_3j_4}\right)\sigma_{j_1}\sigma_{j_2}\sigma_{j_3}\sigma_{j_4}$$
$$+\frac{1}{6}\sum_{j_1=1}^{TP}\sum_{j_2=1}^{TP}\sum_{j_3=1}^{TP}\sum_{j_4=1}^{TP}\left\{\frac{\partial^3 r_k(\boldsymbol{\alpha})}{\partial\alpha_{j_1}\partial\alpha_{j_2}\partial\alpha_{j_3}}\frac{\partial r_\ell(\boldsymbol{\alpha})}{\partial\alpha_{j_4}}+\frac{\partial r_k(\boldsymbol{\alpha})}{\partial\alpha_{j_1}}\frac{\partial^3 r_k(\boldsymbol{\alpha})}{\partial\alpha_{j_2}\partial\alpha_{j_3}\partial\alpha_{j_4}}\right\}_{\boldsymbol{\alpha}^0} q_{j_1j_2j_3j_4}\sigma_{j_1}\sigma_{j_2}\sigma_{j_3}\sigma_{j_4}$$
$$+\frac{1}{12}\sum_{j1=1}^{TP}\sum_{j2=1}^{TP}\sum_{j3=1}^{TP}\sum_{j4=1}^{TP}\sum_{j5=1}^{TP}\left[\frac{\partial^3 R_k(\boldsymbol{\alpha}^0)}{\partial\alpha_{j1}\partial\alpha_{j2}\partial\alpha_{j3}}\frac{\partial^2 R_\ell(\boldsymbol{\alpha}^0)}{\partial\alpha_{j4}\partial\alpha_{j5}}\left(p_{j1j2j3j4j5}-t_{j1j2j3}\rho_{j4j5}\right)\right.$$
$$\left.+\frac{\partial^2 R_k(\boldsymbol{\alpha}^0)}{\partial\alpha_{j1}\partial\alpha_{j2}}\frac{\partial^3 R_\ell(\boldsymbol{\alpha}^0)}{\partial\alpha_{j3}\partial\alpha_{j4}\partial\alpha_{j5}}\left(p_{j1j2j3j4j5}-\rho_{j1j2}t_{j3j4j5}\right)\right]\sigma_{j1}\sigma_{j2}\sigma_{j3}\sigma_{j4}\sigma_{j5}$$
$$+\frac{1}{24}\sum_{j1=1}^{TP}\sum_{j2=1}^{TP}\sum_{j3=1}^{TP}\sum_{j4=1}^{TP}\sum_{j5=1}^{TP}\left[\frac{\partial^4 R_k(\boldsymbol{\alpha}^0)}{\partial\alpha_{j1}\partial\alpha_{j2}\partial\alpha_{j3}\partial\alpha_{j4}}\frac{\partial R_\ell(\boldsymbol{\alpha}^0)}{\partial\alpha_{j5}}\right.$$

$$
\begin{aligned}
&+\frac{\partial R_k\left(\boldsymbol{\alpha}^0\right)}{\partial \alpha_{j1}} \frac{\partial^4 R_\ell\left(\boldsymbol{\alpha}^0\right)}{\partial \alpha_{j2}\partial \alpha_{j3}\partial \alpha_{j4}\partial \alpha_{j5}}\Bigg] \times p_{j1j2j3j4j5}\sigma_{j1}\sigma_{j2}\sigma_{j3}\sigma_{j4}\sigma_{j5} \\
&+\frac{1}{36}\sum_{j1=1}^{J_\alpha}\sum_{j2=1}^{J_\alpha}\sum_{j3=1}^{J_\alpha}\sum_{j4=1}^{N_\alpha}\sum_{j5=1}^{N_\alpha}\sum_{j6=1}^{N_\alpha} \frac{\partial^3 R_k\left(\boldsymbol{\alpha}^0\right)}{\partial \alpha_{j1}\partial \alpha_{j2}\partial \alpha_{j3}} \frac{\partial^3 R_l\left(\boldsymbol{\alpha}^0\right)}{\partial \alpha_{j4}\partial \alpha_{j5}\partial \alpha_{j6}} \\
&\times\left(s_{j1j2j3j4j5j6} - t_{j1j2j3}t_{j4j5j6}\right)\sigma_{j1}\sigma_{j2}\sigma_{j3}\sigma_{j4}\sigma_{j5}\sigma_{j6} \\
&+\frac{1}{48}\sum_{j1=1}^{N_\alpha}\sum_{j2=1}^{N_\alpha}\sum_{j3=1}^{N_\alpha}\sum_{j4=1}^{N_\alpha}\sum_{j5=1}^{N_\alpha}\sum_{j6=1}^{N_\alpha} \frac{\partial^4 R_k\left(\boldsymbol{\alpha}^0\right)}{\partial \alpha_{j1}\partial \alpha_{j2}\partial \alpha_{j3}\partial \alpha_{j4}} \frac{\partial^2 R_l\left(\boldsymbol{\alpha}^0\right)}{\partial \alpha_{j5}\partial \alpha_{j6}} \\
&\times\left(s_{j1j2j3j4j5j6} - q_{j1j2j3j4}p_{j5j6}\right)\sigma_{j1}\sigma_{j2}\sigma_{j3}\sigma_{j4}\sigma_{j5}\sigma_{j6} \\
&+\frac{1}{48}\sum_{j1=1}^{J_\alpha}\sum_{j2=1}^{J_\alpha}\sum_{j3=1}^{N_\alpha}\sum_{j4=1}^{N_\alpha}\sum_{j5=1}^{N_\alpha}\sum_{j6=1}^{N_\alpha} \frac{\partial^2 R_k\left(\boldsymbol{\alpha}^0\right)}{\partial \alpha_{j1}\partial \alpha_{j2}} \frac{\partial^4 R_l\left(\boldsymbol{\alpha}^0\right)}{\partial \alpha_{j3}\partial \alpha_{j4}\partial \alpha_{j5}\partial \alpha_{j6}} \\
&\times\left(s_{j1j2j3j4j5j6} - p_{j1j2}q_{j3j4j5j6}\right)\sigma_{j1}\sigma_{j2}\sigma_{j3}\sigma_{j4}\sigma_{j5}\sigma_{j6} + \ldots
\end{aligned} \tag{1.150}
$$

The covariances $\operatorname{cov}\left(r_k, r_\ell\right)$, $k,\ell = 1,\ldots,TR$, are considered to be the components of a "computed-responses covariance matrix" denoted as $\mathbf{C}_{rr}^c$ and defined as follows:

$$
\mathbf{C}_{rr}^c \triangleq \begin{pmatrix} \operatorname{cor}\left(r_1, r_1\right) & \cdots & \operatorname{cor}\left(r_1, r_{TR}\right) \\ \vdots & \ddots & \vdots \\ \operatorname{cor}\left(r_{TP}, r_1\right) & \cdots & \operatorname{cor}\left(r_{TP}, r_{TR}\right) \end{pmatrix}. \tag{1.151}
$$

The information provided in Eqs. (1.145)–(1.151) regarding the mean values and correlations for the model parameters and computed model responses will be used to construct a 2nd-order MaxEnt approximation, denoted as $p_c\left(\mathbf{z}|\mathbf{z}_c, \mathbf{C}_c\right)$, to represent the distribution $p_c\left(\boldsymbol{\alpha}, \mathbf{r}\right) \triangleq p_\alpha\left(\boldsymbol{\alpha}\right) p_c\left(\mathbf{r}\right)$ of computed model responses and parameters in the joint phase-space of model parameters and responses. The distribution $p_c\left(\mathbf{z}|\mathbf{z}_c, \mathbf{C}_c\right)$ is constructed by following the procedure outlined in Section 1.4, to obtain the following expression:

$$
p_c\left(\boldsymbol{\alpha}, \mathbf{r}\right) \cong p_c\left(\mathbf{z}|\mathbf{z}_c, \mathbf{C}_c\right) \triangleq \frac{\left(2\pi\right)^{-(TR+TP)/2}}{\sqrt{Det\left(\mathbf{C}_c\right)}} \exp\left[-\frac{1}{2}\left(\mathbf{z}-\mathbf{z}_c\right)^\dagger \mathbf{C}_c^{-1}\left(\mathbf{z}-\mathbf{z}_c\right)\right], \tag{1.152}
$$

where:

$$
\mathbf{C}_c \triangleq \begin{pmatrix} \mathbf{C}_{rr}^c & \mathbf{C}_{r\alpha}^c \\ \mathbf{C}_{\alpha r}^c & \mathbf{C}_{\alpha\alpha} \end{pmatrix}; \quad \mathbf{C}_{r\alpha}^c \triangleq \left(\mathbf{C}_{\alpha r}^c\right)^\dagger; \quad \mathbf{z} = \begin{pmatrix} \mathbf{r} \\ \boldsymbol{\alpha} \end{pmatrix}; \quad \mathbf{z}_c = \begin{pmatrix} \mathbf{E}_c\left(\mathbf{r}\right) \\ \boldsymbol{\alpha}^0 \end{pmatrix}. \tag{1.153}
$$

It is important to note that even though the MaxEnt distribution represented in Eq. (1.152) is a multivariate normal distribution characterized by just the 1st-order moments (mean values) and 2nd-order moments (variance/covariances) of the full but unknown "joint distribution $p_c\left(\alpha, \mathbf{r}\right)$ of *computed* responses and parameters," the components of the vector $\mathbf{z}_c$ of mean values and the components of the covariance matrix $\mathbf{C}_c$ may comprise terms involving 3rd- and higher-order response sensitivities and parameter correlations (e.g., triple and quadruple correlations), if available, as indicated in Eqs. (1.145), (1.148), and (1.150).

1.6.2 General Case 2nd-BERRU-PMP: Inclusion of Additional External Measurements for Both Responses and Parameters

In the general case, measurements outside of (and in addition to) the computational model could be available, not only for the model responses (results) of interest but also for the parameters used in the model to compute the respective results. In such a case, the 2nd-order MaxEnt distribution which will represent probabilistically the *computed* model responses and model parameters is to be combined with a 2nd-order MaxEnt distribution which is to represent probabilistically the measured model responses and the additionally measured model parameters (i.e., model parameters measured in addition to, and independently of, the computational model). Since only 1st- and 2nd-order moments are considered to be available, each of these MaxEnt distributions will be multivariate normal distributions, so their combination will also yield a multivariate posterior MaxEnt distribution (for all of the 1st- and 2nd-order available moments).

Consider that the total number of experimentally measured responses is TR. The information usually available regarding the distribution of such measured responses comprises the 1st-order moments (mean values), which will be denoted as r_i^e, $i = 1,\dots,TR$, and the 2nd-order moments (variances/covariances), which will be denoted as $\mathrm{cov}\left(r_i, r_j\right)_e$, $i, j = 1,\dots,TR$, for the measured responses. Occasionally, measured correlations between the model parameters and the measured responses could also be available. The letter "e" will be used either as a superscript or a superscript to indicate *experimentally* measured quantities. The expected values of the measured responses will be considered to constitute the components of a vector denoted as $\mathbf{r}^e \triangleq \left(r_1^e,\dots,r_{TR}^e\right)^\dagger$. The covariances of the measured responses are considered to be components of the $TR \times TR$-dimensional covariance matrix of measured responses, which will be denoted as $\mathbf{C}_{rr}^e \triangleq \left[\mathrm{cov}\left(r_i, r_j\right)_e\right]_{TR\times TR}$. In principle, it is also possible to obtain correlations between some measured responses and some model parameters. When such correlations between measured responses and measured model parameters are available, they are denoted as $\mathrm{cor}\left(\alpha_i, r_j\right)_e$, $i = 1,\dots,TP;\ j = 1,\dots,TR$, and can formally be considered to be elements of a rectangular correlation matrix which will be denoted as $\mathbf{C}_{\alpha r}^e \triangleq \left[\mathrm{cov}\left(\alpha_i, r_j\right)_e\right]_{TP\times TR}$. In the most general situation, it is possible to have not only measured responses but also new measurements for the parameters, in addition to and independently of, the information regarding parameters used in the mathematical/computational model (i.e., where the parameters are considered to have nominal values $\boldsymbol{\alpha}^0$ and covariance matrix $\mathbf{C}_{\alpha\alpha}$). The independently obtained additional information measurement parameters will be considered to have a vector of mean values denoted as $\boldsymbol{\alpha}^e \triangleq \left(\alpha_1^e,\dots,\alpha_{TP}^e\right)$, and a covariance matrix denoted as $\mathbf{C}_{\alpha\alpha}^e \triangleq \left[\mathrm{cov}\left(\alpha_i, \alpha_j\right)_e\right]_{TP\times TR}$.

The MaxEnt principle can now be applied, as described in Section 1.3, to construct the least informative (and hence, most conservative) distribution using the available experimental information from external measurements of 1st- and 2nd-order moments for responses and model parameters, to obtain the following expression:

$$p_e\left(\mathbf{z}\middle|\mathbf{z}_e,\mathbf{C}_e\right)=\frac{(2\pi)^{-(TR+TP)/2}}{\sqrt{Det\left(\mathbf{C}_e\right)}}\exp\left[-\frac{1}{2}\left(\mathbf{z}-\mathbf{z}_e\right)^{\dagger}\mathbf{C}_e^{-1}\left(\mathbf{z}-\mathbf{z}_e\right)\right], \tag{1.154}$$

where:

$$\mathbf{C}_e\triangleq\begin{pmatrix}\mathbf{C}_{rr}^{e} & \mathbf{C}_{r\alpha}^{e}\\ \mathbf{C}_{\alpha r}^{e} & \mathbf{C}_{\alpha\alpha}^{e}\end{pmatrix};\ \mathbf{z}=\begin{pmatrix}\mathbf{r}\\ \boldsymbol{\alpha}\end{pmatrix};\ \mathbf{z}_e=\begin{pmatrix}\mathbf{r}^{e}\\ \boldsymbol{\alpha}^{e}\end{pmatrix}. \tag{1.155}$$

The joint posterior probability distribution of all computed and experimentally measured means and covariances will be denoted as $p_p\left(\mathbf{z}\middle|\mathbf{z}_p,\mathbf{C}_p\right)$, where the subscript "p" indicates "posterior," is obtained as the properly normalized product of the distributions $p_e\left(\mathbf{z}\middle|\mathbf{z}_e,\mathbf{C}_e\right)$ and $p_c\left(\mathbf{z}\middle|\mathbf{z}_c,\mathbf{C}_c\right)$. Since both $p_e\left(\mathbf{z}\middle|\mathbf{z}_e,\mathbf{C}_e\right)$ and $p_c\left(\mathbf{z}\middle|\mathbf{z}_c,\mathbf{C}_c\right)$ are multivariate normal distributions, it follows that the posterior probability $p_p\left(\mathbf{z}\middle|\mathbf{z}_p,\mathbf{C}_p\right)$ will also be a multivariate normal distribution, given by the following expression:

$$p_p\left(\mathbf{z}\middle|\mathbf{z}_p,\mathbf{C}_p\right)\triangleq p_c\left(\mathbf{z}\middle|\mathbf{z}_c,\mathbf{C}_c\right)p_e\left(\mathbf{z}\middle|\mathbf{z}_e,\mathbf{C}_e\right)=K\exp\left[-\frac{1}{2}Q\left(\mathbf{z}\middle|\mathbf{z}_p,\mathbf{C}_p\right)\right], \tag{1.156}$$

where the normalization constant K and the quadratic form $Q\left(\mathbf{z}_p,\mathbf{C}_p\right)$ have the following expressions, respectively:

$$K\triangleq(2\pi)^{-N/2}\left[Det\left(\mathbf{C}_c+\mathbf{C}_e\right)\right]^{-1/2}\exp\left[-\frac{1}{2}\left(\mathbf{z}_c-\mathbf{z}_e\right)^{\dagger}\left(\mathbf{C}_c+\mathbf{C}_e\right)^{-1}\left(\mathbf{z}_c-\mathbf{z}_e\right)\right]; \tag{1.157}$$

$$Q\left(\mathbf{z}\middle|\mathbf{z}_p,\mathbf{C}_p\right)\triangleq\left(\mathbf{z}-\mathbf{z}_p\right)^{\dagger}\mathbf{C}_p^{-1}\left(\mathbf{z}-\mathbf{z}_p\right); \tag{1.158}$$

$$\mathbf{C}_p\triangleq\left(\mathbf{C}_c^{-1}+\mathbf{C}_e^{-1}\right)^{-1}=\mathbf{C}_c-\mathbf{C}_c\left(\mathbf{C}_c+\mathbf{C}_e\right)^{-1}\mathbf{C}_c=\mathbf{C}_e-\mathbf{C}_e\left(\mathbf{C}_c+\mathbf{C}_e\right)^{-1}\mathbf{C}_e; \tag{1.159}$$

$$\mathbf{z}_p\triangleq\mathbf{C}_p\left(\mathbf{C}_c^{-1}\mathbf{z}_c+\mathbf{C}_e^{-1}\mathbf{z}_e\right). \tag{1.160}$$

The expression obtained in Eq. (1.156) provides the exact 1st-order (mean values) and 2nd-order (variances and co-variances) of the most comprehensive combined distribution of computations and measurements of responses and parameters. In practice, however, such a comprehensive amount of experimental information is highly unlikely to be available. Furthermore, even if such a massive amount of experimental information was available, the inversion of the matrix $\left(\mathbf{C}_c+\mathbf{C}_e\right)^{-1}$ may be impractical since it would require massive computational resources when the physical systems comprise many (thousands of) parameters.

1.6.3 Practical Case 2nd-BERRU-PMP: Inclusion of Response Measurements

In practice, the information (i.e., mean values and covariances) about the model parameters indicated in Eqs. (1.30) and (1.31) are obtained and assessed to construct the components of the vector $\boldsymbol{\alpha}^0 \triangleq \left(\alpha_1^0,...,\alpha_{TP}^0\right)^\dagger$ of parameter nominal values and the components of the parameter covariance matrix $\mathbf{C}_{\alpha\alpha}$ prior to using this information in the mathematical/computational model used for computing responses. Therefore, only information about measured responses would become additionally available in practice, in other words, only the vector $\mathbf{r}^e \triangleq \left(r_1^e,...,r_{TR}^e\right)^\dagger$ and the covariance matrix $\mathbf{C}_{rr}^e \triangleq \left[\text{cov}\left(r_i,r_j\right)_e\right]_{TR\times TR}$ would become available for subsequent combination with the computational information. In such a case, the MaxEnt distribution corresponding to Eq. (1.55) reduces to the following expression:

$$p_e\left(\mathbf{r}\middle|\mathbf{r}^e,\mathbf{C}_{rr}^e\right)=\left(2\pi\right)^{-TR/2}\left[Det\left(\mathbf{C}_{rr}^e\right)\right]^{-1/2}\exp\left[-\frac{1}{2}\left(\mathbf{r}-\mathbf{r}^e\right)^\dagger\left(\mathbf{C}_{rr}^e\right)^{-1}\left(\mathbf{r}-\mathbf{r}^e\right)\right]. \quad (1.161)$$

Furthermore, when only the experimental information represented by the distribution $p_e\left(\mathbf{r}\middle|\mathbf{r}^e,\mathbf{C}_{rr}^e\right)$ is available, the posterior joint probability distribution of the computed and measured quantities, which will be denoted as $p_p\left(\mathbf{r},\boldsymbol{\alpha}\right)$, will have the following form:

$$p_p\left(\mathbf{r},\boldsymbol{\alpha}\right)=N\exp\left[-\frac{1}{2}Q\left(\mathbf{r},\boldsymbol{\alpha}\right)\right];\; N\triangleq\left[\int\exp\left[-\frac{1}{2}Q\left(\mathbf{r},\boldsymbol{\alpha}\right)\right]d\boldsymbol{\alpha}\,d\mathbf{r}\right]^{-1};$$
$$Q\left(\mathbf{r},\boldsymbol{\alpha}\right)\triangleq\left(\mathbf{z}-\mathbf{z}_c\right)^\dagger\mathbf{C}_c^{-1}\left(\mathbf{z}-\mathbf{z}_c\right)+\left(\mathbf{r}-\mathbf{r}^e\right)^\dagger\left(\mathbf{C}_{rr}^e\right)^{-1}\left(\mathbf{r}-\mathbf{r}^e\right). \quad (1.162)$$

The evaluation of the moments of the posterior distribution represented by Eq. (1.162) will involve the evaluation of ratios of integrals having the following form:

$$I=\left[\int\exp\left[-g\left(\mathbf{z}\right)\right]d\mathbf{z}\right]^{-1}\int f\left(\mathbf{z}\right)\exp\left[-g\left(\mathbf{z}\right)\right]d\mathbf{z}. \quad (1.163)$$

The evaluations of expressions such as shown in Eq. (1.163) can be performed to a high degree of accuracy, with a controlled error, by employing the saddle-point (Laplace) method (see, e.g., Shun and McCullagh, 1995; Evangelou et al., 2011; Tang and Reid, 2023), which yields the following result:

$$I=\hat{f}-\frac{1}{2}\hat{f}_{i_1}\hat{g}_{i_2i_3i_4}\hat{g}^{i_1i_2}\hat{g}^{i_3i_4}+\frac{1}{2}\hat{f}_{j_1j_2}\hat{g}^{j_1j_2}+O\left(\hat{f}_{j_1j_2j_3}\right), \quad (1.164)$$

where:

1. The derivative of a function with respect to a component of $\mathbf{z}$ is denoted using a subscript, for example, $f_i \triangleq \partial f/\partial z_i$, $f_{ij}\triangleq\partial^2 f/\partial z_i\partial z_j$, $i,j=1,...,TI$, where TI denotes the total number of independent variables;

2. The superscripts denote the respective component of the inverse Hessian of the respective function, e.g., f^{ij} denotes the (i,j)-element of the inverse Hessian matrix $\left[f_{ij}^{-1}\right]$;
3. An index that appears as a subscript and a superscript implies a summation over all possible values of that index;
4. The "hat" denotes that the respective quantity is to be evaluated at the saddle-point of $\exp\left[-g(\mathbf{z})\right]$, which is defined as the point at which the gradient $g_{\mathbf{z}}(\mathbf{z}) \triangleq (\partial g/\partial z_1,\ldots,\partial g/\partial z_{TI})$ of $g(\mathbf{z})$ vanishes, in other words, $g_{\mathbf{z}}(\mathbf{z}) = \mathbf{0}$.

The saddle-point of the normalization integral N will be denoted as $(\mathbf{r}_s, \boldsymbol{\alpha}_s)$ and is defined by the following relations:

$$\frac{\partial Q(\mathbf{r},\boldsymbol{\alpha})}{\partial \mathbf{r}} = \mathbf{0}, \quad \frac{\partial Q(\mathbf{r},\boldsymbol{\alpha})}{\partial \boldsymbol{\alpha}} = \mathbf{0}, \; at \; (\mathbf{r},\boldsymbol{\alpha}) = (\mathbf{r}_s,\boldsymbol{\alpha}_s). \tag{1.165}$$

To obtain the partial gradients (differentials) shown in Eq. (1.165), it is convenient to write the matrix $\mathbf{C}_c^{-1}$ in the form $\mathbf{C}_c^{-1} = \begin{pmatrix} \mathbf{C}_{11} & \mathbf{C}_{12} \\ \mathbf{C}_{21} & \mathbf{C}_{22} \end{pmatrix}$, and use this form together with the definitions provided in Eqs. (1.153) and (1.155) in Eq. (1.162) to expand the functional $Q(\mathbf{r},\boldsymbol{\alpha})$ into the following form:

$$\begin{aligned} Q(\mathbf{r},\boldsymbol{\alpha}) &= \left[\mathbf{r} - \mathbf{E}_c(\mathbf{r})\right]^{\dagger} \mathbf{C}_{11}\left[\mathbf{r} - \mathbf{E}_c(\mathbf{r})\right] + \left[\mathbf{r} - \mathbf{E}_c(\mathbf{r})\right]^{\dagger} \mathbf{C}_{12}\left(\boldsymbol{\alpha} - \boldsymbol{\alpha}^0\right) \\ &+ \left(\boldsymbol{\alpha} - \boldsymbol{\alpha}^0\right)^{\dagger} \mathbf{C}_{21}\left[\mathbf{r} - \mathbf{E}_c(\mathbf{r})\right] + \left(\boldsymbol{\alpha} - \boldsymbol{\alpha}^0\right)^{\dagger} \mathbf{C}_{22}\left(\boldsymbol{\alpha} - \boldsymbol{\alpha}^0\right) \\ &+ \left(\mathbf{r} - \mathbf{r}^e\right)^{\dagger}\left(\mathbf{C}_{rr}^e\right)^{-1}\left(\mathbf{r} - \mathbf{r}^e\right). \end{aligned} \tag{1.166}$$

Taking the partial differentials of the expression in Eq. (1.166) yields the following equation for determining the coordinates of the saddle-point $(\mathbf{r},\boldsymbol{\alpha}) = (\mathbf{r}_s,\boldsymbol{\alpha}_s)$:

$$\begin{pmatrix} \mathbf{r}_s - \mathbf{E}_c(\mathbf{r}) \\ \boldsymbol{\alpha}_s - \boldsymbol{\alpha}^0 \end{pmatrix} = -\begin{pmatrix} \mathbf{C}_{rr}^c & \mathbf{C}_{r\alpha}^c \\ \mathbf{C}_{\alpha r}^c & \mathbf{C}_{\alpha\alpha} \end{pmatrix} \begin{pmatrix} \left(\mathbf{C}_{rr}^e\right)^{-1}\left(\mathbf{r}_s - \mathbf{r}^e\right) \\ \mathbf{0} \end{pmatrix}. \tag{1.167}$$

Solving Eq. (1.167) leads to the following expressions for the coordinates $\mathbf{r}_s$ and $\boldsymbol{\alpha}_s$ of the saddle-point:

$$\mathbf{r}_s = \mathbf{r}^e + \mathbf{C}_{rr}^e\left(\mathbf{C}_{rr}^e + \mathbf{C}_{rr}^c\right)^{-1}\left[\mathbf{E}_c(\mathbf{r}) - \mathbf{r}^e\right], \tag{1.168}$$

$$\alpha_s = \alpha^0 - \mathbf{C}_{\alpha r}^c\left(\mathbf{C}_{rr}^e + \mathbf{C}_{rr}^c\right)^{-1}\left[\mathbf{E}_c(\mathbf{r}) - \mathbf{r}^e\right]. \tag{1.169}$$

The best-estimate predicted mean values for the responses and parameters are defined:

$$\mathbf{r}^{bep} \triangleq \int_{D_p} \mathbf{r} p_p(\boldsymbol{\alpha},\mathbf{r})\, d\boldsymbol{\alpha}\, d\mathbf{r}, \tag{1.170}$$

$$\boldsymbol{\alpha}^{bep} \triangleq \int_{D_p} \boldsymbol{\alpha} p_p(\boldsymbol{\alpha},\mathbf{r})\, d\boldsymbol{\alpha}\, d\mathbf{r}. \tag{1.171}$$

The superscript "*bep*" has been used in Eqs. (1.170) and (1.171), respectively, to indicate that the expressions obtained for the respective predicted responses and parameters are "*best-estimates* with *probabilistic* computational model."

The saddle-point method is applied to evaluate the integrals represented by Eqs. (1.170) and (1.171), respectively, to obtain the following expressions for the optimally predicted best-estimate values for the responses and calibrated parameters:

$$\mathbf{r}^{bep} = \mathbf{r}_s = \mathbf{r}^e + \mathbf{C}_{rr}^e \left(\mathbf{C}_{rr}^e + \mathbf{C}_{rr}^c\right)^{-1} \left[\mathbf{E}_c(\mathbf{r}) - \mathbf{r}^e\right], \tag{1.172}$$

$$\boldsymbol{\alpha}^{bep} = \boldsymbol{\alpha}_s = \boldsymbol{\alpha}^0 - \mathbf{C}_{\alpha r}^c \left(\mathbf{C}_{rr}^e + \mathbf{C}_{rr}^c\right)^{-1} \left[\mathbf{E}_c(\mathbf{r}) - \mathbf{r}^e\right]. \tag{1.173}$$

Since the components of the vector $\mathbf{E}_c(\mathbf{r})$, and the components of the matrices $\mathbf{C}_{rr}^c$ and $\mathbf{C}_{\alpha r}^c$ can contain arbitrarily high-order response sensitivities to model parameters, the expressions presented in Eqs. (1.172) and (1.173) generalize the previous formulas of this type found in data adjustment/assimilation procedures published to date (which contain at most 2nd-order sensitivities). The best-estimate parameter values are the "calibrated model parameters" which can be used for subsequent computations with the "calibrated model."

The 2nd-order moments of the posterior distribution $p_p(\mathbf{r},\boldsymbol{\alpha})$ comprise the covariances between the best-estimated response, which will be denoted as $\mathbf{C}_{rr}^{bep}$, the covariances between the best-estimate parameters, which will be denoted as $\mathbf{C}_{\alpha\alpha}^{bep}$, and the correlations between the best-estimate parameters and responses, which will be denoted as $\mathbf{C}_{\alpha r}^{bep}$. The expression of the "best-estimate probabilistic model" posterior parameter covariance matrix, $\mathbf{C}_{rr}^{bep}$, for the best-estimate responses $\mathbf{r}^{bep}$ is derived by using the result given in Eq. (1.172) to obtain the following expression:

$$\begin{aligned} \mathbf{C}_{rr}^{bep} &\triangleq \int_{D_p} \left(\mathbf{r} - \mathbf{r}^{bep}\right)\left(\mathbf{r} - \mathbf{r}^{bep}\right)^{\dagger} p_p(\boldsymbol{\alpha},\mathbf{r})\, d\boldsymbol{\alpha}\, d\mathbf{r} \\ &= \mathbf{C}_{rr}^e - \mathbf{C}_{rr}^e \left(\mathbf{C}_{rr}^e + \mathbf{C}_{rr}^c\right)^{-1} \mathbf{C}_{rr}^e. \end{aligned} \tag{1.174}$$

The following important result has been used to obtain the expression in Eq. (1.174):

$$\begin{aligned} \left[\mathbf{E}_c(\mathbf{r}) - \mathbf{r}^e\right]\left[\mathbf{E}_c(\mathbf{r}) - \mathbf{r}^e\right]^{\dagger} &= \left[\mathbf{r} - \mathbf{r}^e - \mathbf{r} + \mathbf{E}_c(\mathbf{r})\right]\left[\mathbf{r} - \mathbf{r}^e - \mathbf{r} + \mathbf{E}_c(\mathbf{r})\right]^{\dagger} \\ &= \mathbf{C}_{rr}^e + \mathbf{C}_{rr}^c. \end{aligned} \tag{1.175}$$

As indicated in Eq. (1.174), the initial covariance matrix $\mathbf{C}_{rr}^e$ for the experimentally measured responses is multiplied by the matrix $\left[\mathbf{I} - \left(\mathbf{C}_{rr}^e + \mathbf{C}_{rr}^c\right)^{-1} \mathbf{C}_{rr}^e\right]$, which means that the variances contained on the diagonal of the best-estimate matrix $\mathbf{C}_{rr}^{bep}$ will be smaller than the experimentally measured variances contained in $\mathbf{C}_{rr}^e$. Hence, the incorporation of experimental information reduces the predicted best-estimate response variances in $\mathbf{C}_{rr}^{bep}$ by comparison to the measured variances contained a priori in $\mathbf{C}_{rr}^{e}$. Since the components of the matrix $\mathbf{C}_{rr}^{bep}$ contain high-order sensitivities, the formula presented in Eq. (1.174) generalizes the previous formulas of this type found in data adjustment/assimilation procedures published to date.

The expression of the "best-estimate" posterior parameter covariance matrix $\mathbf{C}_{\alpha\alpha}^{bep}$ for the best-estimate parameters $\boldsymbol{\alpha}^{bep}$ is derived by using the result given in Eq. (1.173) to obtain:

$$\begin{aligned}\mathbf{C}_{\alpha\alpha}^{bep} &\triangleq \int_{D_p} \left(\boldsymbol{\alpha}-\boldsymbol{\alpha}^{bep}\right)\left(\boldsymbol{\alpha}-\boldsymbol{\alpha}^{bep}\right)^{\dagger} p_p\left(\boldsymbol{\alpha},\mathbf{r}\right) d\boldsymbol{\alpha}\, d\mathbf{r} \\ &= \mathbf{C}_{\alpha\alpha} - \mathbf{C}_{\alpha r}^{c}\left(\mathbf{C}_{rr}^{e}+\mathbf{C}_{rr}^{c}\right)^{-1}\mathbf{C}_{\alpha r}^{c}.\end{aligned} \tag{1.176}$$

The matrices $\mathbf{C}_{\alpha\alpha}$ and $\mathbf{C}_{\alpha r}^{c}\left(\mathbf{C}_{rr}^{e}+\mathbf{C}_{rr}^{c}\right)^{-1}\mathbf{C}_{\alpha r}^{c}$ are symmetric and positive definite. Therefore, the subtraction indicated in Eq. (1.176) implies that the components of the main diagonal of $\mathbf{C}_{\alpha\alpha}^{bep}$ must have smaller values than the corresponding elements of the main diagonal of $\mathbf{C}_{\alpha\alpha}$. In this sense, the combination of computational and experimental information has reduced the best-estimate parameter variances on the diagonal of $\mathbf{C}_{\alpha\alpha}^{bep}$. Since the components of the matrices $\mathbf{C}_{\alpha\alpha}$, $\mathbf{C}_{\alpha r}^{c}$, and $\mathbf{C}_{rr}^{c}$ contain high-order response sensitivities, the formula presented in Eq. (1.176) generalizes the previous formulas of this type found in data adjustment/assimilation procedures published to date.

The expressions of the "best-estimate" posterior parameter correlation matrix $\mathbf{C}_{\alpha r}^{bep}$ and/or its transpose $\mathbf{C}_{r\alpha}^{bep}$, for the best-estimate parameters $\boldsymbol{\alpha}^{bep}$ and best-estimate responses $\mathbf{r}^{bep}$, are derived by using the results given in Eq. (1.172) and Eq. (1.173) to obtain the following expressions:

$$\begin{aligned}\mathbf{C}_{\alpha r}^{bep} &\triangleq \int_{D_p} \left(\boldsymbol{\alpha}-\boldsymbol{\alpha}^{bep}\right)\left(\mathbf{r}-\mathbf{r}^{bep}\right)^{\dagger} p_p\left(\boldsymbol{\alpha},\mathbf{r}\right) d\boldsymbol{\alpha}\, d\mathbf{r} \\ &= \mathbf{C}_{\alpha r}^{e} - \mathbf{C}_{\alpha r}^{c}\left(\mathbf{C}_{rr}^{e}+\mathbf{C}_{rr}^{c}\right)^{-1}\mathbf{C}_{rr}^{e};\end{aligned} \tag{1.177}$$

$$\begin{aligned}\mathbf{C}_{r\alpha}^{bep} &\triangleq \int_{D_p} \left(\mathbf{r}-\mathbf{r}^{bep}\right)\left(\boldsymbol{\alpha}-\boldsymbol{\alpha}^{bep}\right)^{\dagger} p_p\left(\boldsymbol{\alpha},\mathbf{r}\right) d\boldsymbol{\alpha}\, d\mathbf{r} \\ &= \mathbf{C}_{r\alpha}^{e} - \mathbf{C}_{rr}^{e}\left(\mathbf{C}_{rr}^{e}+\mathbf{C}_{rr}^{c}\right)^{-1}\mathbf{C}_{r\alpha}^{c} = \left(\mathbf{C}_{\alpha r}^{bep}\right)^{\dagger}.\end{aligned} \tag{1.178}$$

Since the components of the matrices $\mathbf{C}_{c}^{\alpha r}$ and $\mathbf{C}_{c}^{rr}$ contain high-order sensitivities, the formulas presented in Eqs. (1.177) and (1.178) generalize the previous formulas of this type found in data adjustment/assimilation procedures published to date.

It is important to note from the results shown in Eqs. (1.172)–(1.178) that the computation of the best-estimate parameter and response values, together with their corresponding best-estimate covariance matrices, only involves a single matrix inversion when computing $\left(\mathbf{C}_{rr}^{e}+\mathbf{C}_{rr}^{c}\right)^{-1}$, which entails the inversion of a matrix of size $TR \times TR$. This is computationally very advantageous, since $TR \ll TP$, in other words, the number of responses is much less than the number of model parameters in the overwhelming majority of practical situations.

Using Eqs. (1.167), (1.172), and (1.173) in Eq. (1.166) yield the following expression for the minimum value, $Q_{\min} = Q\left(\boldsymbol{\alpha}^{bep}, \mathbf{r}^{bep}\right)$, of the quadratic form $Q(\boldsymbol{\alpha}, \mathbf{r})$:

$$
\begin{aligned}
Q_{\min} &= \begin{pmatrix} \mathbf{r}^{bep} - \mathbf{E}_c(\mathbf{r}) \\ \boldsymbol{\alpha}^{bep} - \boldsymbol{\alpha}^0 \end{pmatrix}^{\dagger} \begin{pmatrix} \mathbf{C}_{rr}^{c} & \mathbf{C}_{r\alpha}^{c} \\ \mathbf{C}_{\alpha r}^{c} & \mathbf{C}_{\alpha\alpha} \end{pmatrix}^{-1} \begin{pmatrix} \mathbf{r}^{bep} - \mathbf{E}_c(\mathbf{r}) \\ \boldsymbol{\alpha}^{bep} - \boldsymbol{\alpha}^0 \end{pmatrix} \\
&\quad + \left(\mathbf{r}^{bep} - \mathbf{r}^{e}\right)^{\dagger} \left(\mathbf{C}_{rr}^{e}\right)^{-1} \left(\mathbf{r}^{bep} - \mathbf{r}^{e}\right) \\
&= \left[\mathbf{E}_c(\mathbf{r}) - \mathbf{r}^{e}\right]^{\dagger} \left(\mathbf{C}_{rr}^{e} + \mathbf{C}_{rr}^{c}\right)^{-1} \left[\mathbf{E}_c(\mathbf{r}) - \mathbf{r}^{e}\right].
\end{aligned}
\tag{1.179}
$$

As the expression obtained in Eq. (1.179) indicates, the quantity $Q_{\min}$ represents the square of the length of the vector $\left[\mathbf{E}_c(\mathbf{r}) - \mathbf{r}^{e}\right]$, measuring (in the corresponding metric) the deviations between the experimental and nominally computed responses. The quantity $Q_{\min}$ is independent of calibrating (or adjusting) the original data, so it can be evaluated directly from the given data (i.e., model parameters and computed and measured responses, together with their original uncertainties) after having computed the matrix $\left(\mathbf{C}_{rr}^{e} + \mathbf{C}_{rr}^{c}\right)^{-1}$. As the dimension of the vector $\left[\mathbf{E}_c(\mathbf{r}) - \mathbf{r}^{e}\right]$ indicates, the number of degrees of freedom characteristic of the calibration under consideration is equal to the number TR of experimental responses.

1.7 INTER-COMPARISON: 2ND-BERRU-PMP VS. 2ND-BERRU-PMD

In this section, the results obtained for the best-estimate mean values for the responses and parameters, together with their corresponding best-estimate covariances/correlations will be compared to the corresponding results produced by the 2nd-BERRU-PMD methodology, as well as to the corresponding results produced by the 1st-order BERRU-PM methodology (Cacuci, 2019).

1.7.1 Inter-Comparison: Best-Estimate Predicted Mean Values for Responses

Recall that the 2nd-BERRU-PMD (best-estimated values with deterministically incorporated computed responses) methodology incorporates deterministically (into the 2nd-order MaxEnt representation of the experimentally measured responses) the following 2nd-order Taylor expansion of the computed responses:

$$
\begin{aligned}
&\mathbf{r} = \mathbf{r}^{c}\left(\boldsymbol{\alpha}^0\right) + \left[\mathbf{S}\left(\boldsymbol{\alpha}^0\right)\right]\left(\boldsymbol{\alpha} - \boldsymbol{\alpha}^0\right) + \mathbf{p}\left(\boldsymbol{\alpha};\boldsymbol{\alpha}^0\right); \\
&\mathbf{p}\left(\boldsymbol{\alpha};\boldsymbol{\alpha}^0\right) \triangleq \left[p_1\left(\boldsymbol{\alpha};\boldsymbol{\alpha}^0\right), \ldots, p_{TR}\left(\boldsymbol{\alpha};\boldsymbol{\alpha}^0\right)\right]^{\dagger}; \\
&p_k\left(\boldsymbol{\alpha};\boldsymbol{\alpha}^0\right) \triangleq \frac{1}{2}\left(\boldsymbol{\alpha} - \boldsymbol{\alpha}^0\right)^{\dagger}\left[\mathbf{Q}_k\left(\boldsymbol{\alpha}^0\right)\right]\left(\boldsymbol{\alpha} - \boldsymbol{\alpha}^0\right).
\end{aligned}
\tag{1.180}
$$

In Eq. (1.180), the components of the vector $\mathbf{r}^c(\boldsymbol{\alpha}^0) \triangleq [r_1^c(\boldsymbol{\alpha}^0),...,r_{TR}^c(\boldsymbol{\alpha}^0)]^\dagger$ represent the values of the model responses computed at the nominal parameter value $\boldsymbol{\alpha}^0$, and the other vectors and matrices contain 1st-order and 2nd-order sensitivities of responses with respect to the model parameters, evaluated at the nominal parameter value $\boldsymbol{\alpha}^0$, as defined:

$$\mathbf{S}(\boldsymbol{\alpha}) \triangleq \begin{pmatrix} s_{11} & \cdots & s_{1,TR} \\ \vdots & \ddots & \vdots \\ s_{TR,1} & \cdots & s_{TR,TR} \end{pmatrix} \triangleq \begin{pmatrix} \frac{\partial r_1(\boldsymbol{\alpha})}{\partial \alpha_1} & \cdots & \frac{\partial r_1(\boldsymbol{\alpha})}{\partial \alpha_{TP}} \\ \vdots & \ddots & \vdots \\ \frac{\partial r_{TR}(\boldsymbol{\alpha})}{\partial \alpha_1} & \cdots & \frac{\partial r_{TR}(\boldsymbol{\alpha})}{\partial \alpha_{TP}} \end{pmatrix};$$

$$\mathbf{Q}_k(\boldsymbol{\alpha}) \triangleq \begin{pmatrix} q_{11}^{(k)} & \cdots & q_{1,TP}^{(k)} \\ \vdots & \ddots & \vdots \\ q_{TP,1}^{(k)} & \cdots & q_{TP,TP}^{(k)} \end{pmatrix} \triangleq \begin{pmatrix} \frac{\partial^2 r_k^c(\boldsymbol{\alpha})}{\partial \alpha_1 \partial \alpha_1} & \cdots & \frac{\partial^2 r_k^c(\boldsymbol{\alpha})}{\partial \alpha_1 \partial \alpha_{TP}} \\ \vdots & \ddots & \vdots \\ \frac{\partial^2 r_k^c(\boldsymbol{\alpha})}{\partial \alpha_{TP} \partial \alpha_1} & \cdots & \frac{\partial^2 r_k^c(\boldsymbol{\alpha})}{\partial \alpha_{TP} \partial \alpha_{TP}} \end{pmatrix}. \quad (1.181)$$

The expressions of the end results produced the "best-estimated values with deterministically incorporated computed responses" (2nd-BERRU-PMD) methodology will be designated using the superscript "*bed*." The expression obtained using the 2nd-BERRU-PMD methodology for the best-estimate responses is as follows:

$$\mathbf{r}^{bed} = \mathbf{r}^e + (\mathbf{C}_{rr}^e - \mathbf{C}_{r\alpha}^e \mathbf{S}^\dagger)\boldsymbol{\theta}^{(h,n+1)}, \quad (1.182)$$

where the vector $\boldsymbol{\theta}^{(h,n+1)} \triangleq \left(\theta_1^{(h,n+1)},...,\theta_k^{(h,n+1)},...,\theta_{TR}^{(h,n+1)}\right)^\dagger$ is the solution of the following iterative (Newton's method) equation

$$\boldsymbol{\theta}^{(n+1)} = \boldsymbol{\theta}^{(n)} - \left[\mathbf{A}\left(\boldsymbol{\theta}^{(n)}\right)\right]^{-1} \mathbf{w}\left(\boldsymbol{\theta}^{(n)}\right); \; n = 0,1,... \quad (1.183)$$

In Eq. (1.183), the components/elements $A_{k,n}(\boldsymbol{\theta})$ of the Jacobian matrix $\mathbf{A}(\boldsymbol{\theta})$ have the following expressions:

$$A_{kn}(\boldsymbol{\theta}) = D_{kn} - V_{kn}(\boldsymbol{\theta}), \quad (1.184)$$

where:

$$D_{kj} \triangleq \left(\mathbf{C}_{rr}^e + \mathbf{S}\mathbf{C}_{\alpha\alpha}\mathbf{S}^\dagger - \mathbf{C}_{r\alpha}^e\mathbf{S}^\dagger - \mathbf{S}\mathbf{C}_{\alpha r}^e\right)_{kj}, \; \mathbf{D} \triangleq [D_{kj}]_{TR \times TR}; \quad (1.185)$$

$$V_{kn}(\boldsymbol{\theta}) \triangleq \sum_{j=1}^{TR} \theta_j \left[M_{nj}^{(k)} + M_{kj}^{(n)} + M_{kn}^{(j)}\right], \; \mathbf{V}(\boldsymbol{\theta}) \triangleq [V_{kn}(\boldsymbol{\theta})]_{TR \times TR} \quad (1.186)$$

$$\mathbf{M}^{(k)} \triangleq \left[M_{ij}^{(k)}\right]_{TR\times TR} \triangleq \left(\mathbf{C}_{\alpha r}^{e} - \mathbf{C}_{\alpha\alpha}\mathbf{S}^{\dagger}\right)^{\dagger}\left[\mathbf{Q}_{k}\left(\boldsymbol{\alpha}^{0}\right)\right]\left(\mathbf{C}_{\alpha r}^{e} - \mathbf{C}_{\alpha\alpha}\mathbf{S}^{\dagger}\right) = \left[\mathbf{M}^{(k)}\right]^{\dagger}. \quad (1.187)$$

$$\mathbf{w}(\boldsymbol{\theta}) \triangleq \left[w_{1}(\boldsymbol{\theta}),...,w_{k}(\boldsymbol{\theta}),...,w_{TR}(\boldsymbol{\theta})\right]^{\dagger} \triangleq -\mathbf{d} + \mathbf{D}\boldsymbol{\theta} + \mathbf{u}(\boldsymbol{\theta}). \quad (1.188)$$

$$\mathbf{d} \triangleq \mathbf{r}^{c}\left(\boldsymbol{\alpha}^{0}\right) - \mathbf{r}^{e};\ \ \mathbf{d} \triangleq \left(d_{1},...,d_{TR}\right)^{\dagger};\ \ d_{i} \triangleq r_{i}^{c}\left(\boldsymbol{\alpha}^{0}\right) - r_{i}^{e};\ \ i = 1,...,TR. \quad (1.189)$$

$$\mathbf{u}(\boldsymbol{\theta}) \triangleq \left[u_{1}(\boldsymbol{\theta}),..,u_{TR}(\boldsymbol{\theta})\right]^{\dagger};\ u_{k}(\boldsymbol{\theta}) \triangleq -\frac{1}{2}\boldsymbol{\theta}^{\dagger}\mathbf{M}^{(k)}\boldsymbol{\theta} - \sum_{i=1}^{TR}\theta_{i}\sum_{j=1}^{TR}M_{kj}^{(i)}\theta_{j}. \quad (1.190)$$

The initial iterate for starting the iterations for obtaining the solution of Eq. (1.183) can be chosen as follows:

$$\boldsymbol{\theta}^{(h,0)} \triangleq \left(\theta_{1}^{(h,0)},...,\theta_{j}^{(h,0)},...,\theta_{TR}^{(h,0)}\right)^{\dagger};\ \ \theta_{j}^{(h,0)} = 2D_{jj}/3M_{jj}^{(j)};\ \ j = 1,...,TR. \quad (1.191)$$

$$A_{kn}\left(\boldsymbol{\theta}^{(h,0)}\right) = D_{kn} - \sum_{j=1}^{TR}\theta_{j}^{(h,0)}\left[M_{nj}^{(k)} + M_{kj}^{(n)} + M_{kn}^{(j)}\right]. \quad (1.192)$$

$$w_{k}\left(\boldsymbol{\theta}^{(h,0)}\right) = u_{k}\left(\boldsymbol{\theta}^{(h,0)}\right) + \sum_{j=1}^{TR}D_{kj}\theta_{j}^{(h,0)} - d_{k},\quad k = 1,...,TR; \quad (1.193)$$

$$\begin{aligned} u_{k}\left(\boldsymbol{\theta}^{(h,0)}\right) &= -\frac{1}{2}\sum_{i=1}^{TR}\sum_{j=1}^{TR}\theta_{i}^{(h,0)}M_{ij}^{(k)}\theta_{j}^{(h,0)} - \sum_{i=1}^{TR}\theta_{i}^{(h,0)}\sum_{j=1}^{TR}M_{kj}^{(i)}\theta_{j}^{(h,0)} \\ &\cong -\frac{4}{9}\sum_{i=1}^{TR}\sum_{j=1}^{TR}\left(\frac{D_{ii}}{M_{ii}^{(i)}}\right)\left(\frac{D_{jj}}{M_{jj}^{(j)}}\right)\left[\frac{1}{2}M_{ij}^{(k)} + M_{kj}^{(i)}\right],\quad k = 1,...,TR. \end{aligned} \quad (1.194)$$

The first-iterate $\boldsymbol{\theta}^{(h,1)}$ of the solution of Eq. (1.183) has the following expression:

$$\begin{aligned} &\boldsymbol{\theta}^{(h,1)} = \boldsymbol{\theta}^{(h,0)} - \left(\mathbf{D} - \mathbf{V}^{(h,0)}\right)^{-1}\left(\mathbf{u}^{(h,0)} + \mathbf{D}\boldsymbol{\theta}^{(h,0)} - \mathbf{d}\right); \\ &\mathbf{V}^{(h,0)} \triangleq \mathbf{V}\left(\boldsymbol{\theta}^{(h,0)}\right);\ \ \mathbf{u}^{(h,0)} \triangleq \mathbf{u}\left(\boldsymbol{\theta}^{(h,0)}\right). \end{aligned} \quad (1.195)$$

Using the first-iterate, $\boldsymbol{\theta}^{(h,1)}$, in Eq. (1.182) yields the following expressions for the "first-iterate best-estimate" values for the responses:

$$\begin{aligned} &\mathbf{r}^{(1)bed} = \mathbf{r}^{e} + \mathbf{r}^{(0)}; \\ &\mathbf{r}^{(0)} \triangleq \left(\mathbf{C}_{rr}^{e} - \mathbf{C}_{r\alpha}^{e}\mathbf{S}^{\dagger}\right)\left[\boldsymbol{\theta}^{(h,0)} - \left(\mathbf{D} - \mathbf{V}^{(h,0)}\right)^{-1}\left(\mathbf{u}^{(h,0)} + \mathbf{D}\boldsymbol{\theta}^{(h,0)} - \mathbf{d}\right)\right]; \end{aligned} \quad (1.196)$$

As has been shown in Section 1.5, the "vector of deviations" $\mathbf{d}$ can be considered to be a (quasi-) random variable; the "vector of mean values of deviations" $\bar{\mathbf{d}} \triangleq \left[\bar{d}_{1},...,\bar{d}_{TR}\right]^{\dagger}$ and the covariance matrix $\mathbf{D} \triangleq \left[D_{ij}\right]_{TR\times TR}$ have the following expressions (up to 2nd-order sensitivities):

$$\bar{\mathbf{d}} = \left[\mathbf{Q}_1\left(\boldsymbol{\alpha}^0\right)\mathbf{C}_{\alpha\alpha},...,\mathbf{Q}_{TR}\left(\boldsymbol{\alpha}^0\right)\mathbf{C}_{\alpha\alpha}\right]^\dagger; \tag{1.197}$$

$$\mathbf{D} = \mathbf{C}_{rr}^e + \mathbf{S}\mathbf{C}_{\alpha\alpha}\mathbf{S}^\dagger - \mathbf{C}_{r\alpha}^e\mathbf{S}^\dagger - \mathbf{S}\mathbf{C}_{\alpha r}^e. \tag{1.198}$$

Comparing the expression of the first-iterate, compare Eq. (1.140) produced by the 2nd-BERRU-PMD methodology with the corresponding expression produced by the 2nd-BERRU-PMP methodology, compare Eq. (1.172), indicates that both expressions include the 2nd-order sensitivities of the responses with respect to the parameters, but in distinct ways. Furthermore, *the expression produced by the 2nd-BERRU-PMP methodology, compare Eq. (1.172), can include all of the response sensitivities and parameter correlations of order higher than two*, if available. Such an inclusion of higher-order sensitivities and parameter correlations is not provided by the result presented in Eq. (1.140), which was obtained using the 2nd-BERRU-PMD methodology.

The expression provided in Eq. (1.172) coincides with the expression provided in Eq. (1.140) only if the following relationship holds:

$$\begin{aligned}\mathbf{r}^{bep} = \mathbf{r}^{(1)bed} \;\; iff: \quad & \mathbf{C}_{rr}^e\left(\mathbf{C}_{rr}^e + \mathbf{C}_{rr}^c\right)^{-1}\left[\mathbf{E}_c(\mathbf{r}) - \mathbf{r}^e\right] \\ = \left(\mathbf{C}_{rr}^e - \mathbf{C}_{r\alpha}^e\mathbf{S}^\dagger\right)&\left[\boldsymbol{\theta}^{(h,0)} - \left(\mathbf{D} - \mathbf{V}^{(h,0)}\right)^{-1}\left(\mathbf{u}^{(h,0)} + \mathbf{D}\boldsymbol{\theta}^{(h,0)} - \mathbf{d}\right)\right].\end{aligned} \tag{1.199}$$

The equality in Eq. (1.199) can hold if the following conditions are simultaneously satisfied: $\mathbf{C}_{r\alpha}^e = \mathbf{0}$, $\boldsymbol{\theta}^{(h,0)} = \mathbf{0}$, $\mathbf{V}^{(h,0)} = \mathbf{0}$, $\mathbf{u}^{(h,0)} = \mathbf{0}$, $\mathbf{C}_{rr}^c = \mathbf{S}\mathbf{C}_{\alpha\alpha}\mathbf{S}^\dagger$, $\mathbf{d} = \left[\mathbf{E}_c(\mathbf{r}) - \mathbf{r}^e\right]$, in which case we also have the relation $\mathbf{D} = \mathbf{C}_{rr}^e + \mathbf{S}\mathbf{C}_{\alpha\alpha}\mathbf{S}^\dagger = \mathbf{C}_{rr}^e + \mathbf{C}_{rr}^c$. These conditions are satisfied if all of the sensitivities higher than 1st-order are ignored, and if the experimental responses are uncorrelated to the model parameters. In this particular case, the results produced by the 2nd-BERRU-PMP and 2nd-BERRU-PMD methodologies also coincide with the results produced by the 1st-order 1st-BERRU-PM methodology.

1.7.2 Inter-Comparison: Best-Estimate Predicted Mean Values for Parameters

The expression obtained using the 2nd-BERRU-PMD methodology for the best-estimate parameters is as follows:

$$\boldsymbol{\alpha}^{bed} = \boldsymbol{\alpha}^0 + \left(\mathbf{C}_{\alpha r}^e - \mathbf{C}_{\alpha\alpha}\mathbf{S}^\dagger\right)\boldsymbol{\theta}^{(h,n+1)}. \tag{1.200}$$

Using the first-iterate $\boldsymbol{\theta}^{(h,1)}$ in Eq. (1.200) yields the following first-iterate expression for the best-estimate parameters:

$$\begin{aligned}&\boldsymbol{\alpha}^{(1)bed} = \boldsymbol{\alpha}^0 + \boldsymbol{\beta}^{(0)}; \\ &\boldsymbol{\beta}^{(0)} \triangleq \left(\mathbf{C}_{\alpha r}^e - \mathbf{C}_{\alpha\alpha}\mathbf{S}^\dagger\right)\left[\boldsymbol{\theta}^{(h,0)} - \left(\mathbf{D} - \mathbf{V}^{(h,0)}\right)^{-1}\left(\mathbf{u}^{(h,0)} + \mathbf{D}\boldsymbol{\theta}^{(h,0)} - \mathbf{d}\right)\right].\end{aligned} \tag{1.201}$$

Comparing the expression provided in Eq. (1.141), namely the first-iterate produced by the 2nd-BERRU-PMD methodology, with the corresponding expression produced by the 2nd-BERRU-PMP methodology, compare Eq. (1.173), indicates that both expressions include the 2nd-order sensitivities of the responses with respect to the parameters, but in distinct ways. Furthermore, *the expression produced by the 2nd-BERRU-PMP methodology, compare Eq.* (1.173), *can include all of the response sensitivities and parameter correlations of order higher than two*, if available. Such an inclusion of higher-order sensitivities and parameter correlations is not possible with the result obtained using the 2nd-BERRU-PMD methodology, compare Eq. (1.141).

The expression provided in Eq. (1.173) coincides with the expression provided in Eq. (1.141) only if the following relationship holds:

$$\begin{aligned}\boldsymbol{\alpha}^{bep} = \boldsymbol{\alpha}^{(1)bed} \quad \textit{iff}: \quad & -\mathbf{C}^{c}_{\alpha r}\left(\mathbf{C}^{e}_{rr}+\mathbf{C}^{c}_{rr}\right)^{-1}\left[\mathbf{E}_{c}(\mathbf{r})-\mathbf{r}^{e}\right] \\ = \left(\mathbf{C}^{e}_{\alpha r}-\mathbf{C}_{\alpha\alpha}\mathbf{S}^{\dagger}\right) & \left[\boldsymbol{\theta}^{(h,0)}-\left(\mathbf{D}-\mathbf{V}^{(h,0)}\right)^{-1}\left(\mathbf{u}^{(h,0)}+\mathbf{D}\boldsymbol{\theta}^{(h,0)}-\mathbf{d}\right)\right].\end{aligned} \tag{1.202}$$

The equality in Eq. (1.202) can hold if the following conditions are simultaneously satisfied: $\mathbf{C}^{e}_{\alpha r}=\mathbf{0}$, $\theta^{(h,0)}=\mathbf{0}$, $\mathbf{V}^{(h,0)}=\mathbf{0}$, $\mathbf{u}^{(h,0)}=\mathbf{0}$, $\mathbf{C}^{c}_{rr}=\mathbf{S}\mathbf{C}_{\alpha\alpha}\mathbf{S}^{\dagger}$, $\mathbf{d}=\left[\mathbf{E}_{c}(\mathbf{r})-\mathbf{r}^{e}\right]$, in which case we also have the relation $\mathbf{D}=\mathbf{C}^{e}_{rr}+\mathbf{S}\mathbf{C}_{\alpha\alpha}\mathbf{S}^{\dagger}=\mathbf{C}^{e}_{rr}+\mathbf{C}^{c}_{rr}$. These conditions are satisfied if all of the sensitivities higher than 1st-order are ignored. In this particular case, the results produced by the 2nd-BERRU-PMP and 2nd-BERRU-PMD methodologies also coincide with the results produced by the 1st-order 1st-BERRU-PM methodology.

1.7.3 Inter-Comparison: Best-Estimate Predicted Response Covariances

The expression obtained in Eq. (1.125) using the 2nd-BERRU-PMD methodology for the "first-iterate best-estimate" covariance matrix $\mathbf{C}^{(1)bed}_{rr}$ for the responses are reproduced for convenient comparison:

$$\begin{aligned}\mathbf{C}^{(1)bed}_{rr} = & \mathbf{C}^{e}_{rr}-\mathbf{C}^{(1)}_{rd}\mathbf{Y}^{\dagger}_{r}+\mathbf{X}_{r}\boldsymbol{\theta}^{(h,0)}\left(\boldsymbol{\theta}^{(h,0)}\right)^{\dagger}\mathbf{X}^{\dagger}_{r}-\mathbf{X}_{r}\boldsymbol{\theta}^{(h,0)}\overline{\mathbf{d}}^{\dagger}\mathbf{Y}^{\dagger}_{r} \\ & +\mathbf{X}_{r}\boldsymbol{\theta}^{(h,0)}\left(\mathbf{u}^{(h,0)}\right)^{\dagger}\mathbf{Y}^{\dagger}_{r}-\mathbf{Y}_{r}\left(\mathbf{C}^{(1)}_{rd}\right)^{\dagger}-\mathbf{Y}_{r}\overline{\mathbf{d}}\left(\boldsymbol{\theta}^{(h,0)}\right)^{\dagger}\mathbf{X}^{\dagger}_{r} \\ & +\mathbf{Y}_{r}\mathbf{D}\mathbf{Y}^{\dagger}_{r}-\mathbf{Y}_{r}\overline{\mathbf{d}}\left(\mathbf{u}^{(h,0)}\right)^{\dagger}\mathbf{Y}^{\dagger}_{r}+\mathbf{Y}_{r}\mathbf{u}^{(h,0)}\left(\boldsymbol{\theta}^{(h,0)}\right)^{\dagger}\mathbf{X}^{\dagger}_{r} \\ & -\mathbf{Y}_{r}\mathbf{u}^{(h,0)}\overline{\mathbf{d}}^{\dagger}\mathbf{Y}^{\dagger}_{r}+\mathbf{Y}_{r}\mathbf{u}^{(h,0)}\left(\mathbf{u}^{(h,0)}\right)^{\dagger}\mathbf{Y}^{\dagger}_{r}.\end{aligned} \tag{1.203}$$

where:

$$\mathbf{C}^{(1)}_{rd}=\mathbf{C}^{e}_{rr}-\mathbf{C}^{e}_{r\alpha}\mathbf{S}^{\dagger};\ \mathbf{X}_{r}\triangleq\mathbf{C}^{(1)}_{rd}\left[\left(\mathbf{D}-\mathbf{V}^{(h,0)}\right)^{-1}\mathbf{D}-\mathbf{I}\right];\ \mathbf{Y}_{r}\triangleq\mathbf{C}^{(1)}_{rd}\left(\mathbf{D}-\mathbf{V}^{(h,0)}\right)^{-1}. \tag{1.204}$$

Comparing the 2nd-BERRU-PMD-expression provided in Eqs. (1.203) and (1.204) with the corresponding expression obtained using the 2nd-BERRU-PMP methodology, compare Eq. (1.174), indicates that both expressions include the 2nd-order sensitivities of the responses with respect to the parameters, but in distinct ways. Furthermore, *the expression produced by the 2nd-BERRU-PMP methodology, compare Eq.* (1.174), *can include all of the response sensitivities and parameter correlations of order higher than two*, if available. Such an inclusion of higher-order sensitivities and parameter correlations is not possible with the result obtained using the 2nd-BERRU-PMD methodology, compare Eq. (1.125).

The expression provided in Eq. (1.174) coincides with the expression provided in Eq. (1.125) only if the following relationship holds:

$$\begin{aligned}\mathbf{C}_{rr}^{bep} = \mathbf{C}_{rr}^{(1)bed} \; iff: \; -\mathbf{C}_{rr}^{e}\left(\mathbf{C}_{rr}^{e}+\mathbf{C}_{rr}^{c}\right)^{-1}\mathbf{C}_{rr}^{e} = -\mathbf{C}_{rd}^{(1)}\mathbf{Y}_r^{\dagger}+\mathbf{X}_r\boldsymbol{\theta}^{(h,0)}\left(\boldsymbol{\theta}^{(h,0)}\right)^{\dagger}\mathbf{X}_r^{\dagger}\\ -\mathbf{X}_r\boldsymbol{\theta}^{(h,0)}\bar{\mathbf{d}}^{\dagger}\mathbf{Y}_r^{\dagger}+\mathbf{X}_r\boldsymbol{\theta}^{(h,0)}\left(\mathbf{u}^{(h,0)}\right)^{\dagger}\mathbf{Y}_r^{\dagger}-\mathbf{Y}_r\left(\mathbf{C}_{rd}^{(1)}\right)^{\dagger}-\mathbf{Y}_r\bar{\mathbf{d}}\left(\boldsymbol{\theta}^{(h,0)}\right)^{\dagger}\mathbf{X}_r^{\dagger}+\mathbf{Y}_r\mathbf{D}\mathbf{Y}_r^{\dagger}. \quad (1.205)\\ -\mathbf{Y}_r\bar{\mathbf{d}}\left(\mathbf{u}^{(h,0)}\right)^{\dagger}\mathbf{Y}_r^{\dagger}+\mathbf{Y}_r\mathbf{u}^{(h,0)}\left(\boldsymbol{\theta}^{(h,0)}\right)^{\dagger}\mathbf{X}_r^{\dagger}-\mathbf{Y}_r\mathbf{u}^{(h,0)}\bar{\mathbf{d}}^{\dagger}\mathbf{Y}_r^{\dagger}+\mathbf{Y}_r\mathbf{u}^{(h,0)}\left(\mathbf{u}^{(h,0)}\right)^{\dagger}\mathbf{Y}_r^{\dagger}.\end{aligned}$$

The equality shown in Eq. (1.205) can hold only if all of the 2nd-order sensitivities are neglected, and if the experimental responses are uncorrelated to the model parameters, in which case the following simplifications occur: $\mathbf{X}_r=\mathbf{0}$, $\mathbf{Y}_r=\mathbf{C}_{rd}^{(1)}\mathbf{D}^{-1}$, and $\mathbf{C}_{rd}^{(1)}=\mathbf{C}_{rr}^{e}$, in which case we also have the relation $\mathbf{D}=\mathbf{C}_{rr}^{e}+\mathbf{S}\mathbf{C}_{\alpha\alpha}\mathbf{S}^{\dagger}=\mathbf{C}_{rr}^{e}+\mathbf{C}_{rr}^{c}$. In this particular case, the results produced by the 2nd-BERRU-PMP and 2nd-BERRU-PMD methodologies also coincide with the results produced by the 1st-order 1st-BERRU-PM methodology.

1.7.4 Inter-Comparison: Best-Estimate Predicted Parameter Covariances

The expression obtained in Eq. (1.126) using the 2nd-BERRU-PMD methodology for the "first-iterate best-estimate" covariance matrix $\mathbf{C}_{\alpha\alpha}^{(1)bed}$ for the model parameters are reproduced in Eqs. (1.206) and (1.207) for convenient comparison:

$$\begin{aligned}\mathbf{C}_{\alpha\alpha}^{(1)bed} = \mathbf{C}_{\alpha\alpha}^{e}-\mathbf{C}_{\alpha d}^{(1)}\mathbf{Y}_\alpha^{\dagger}+\mathbf{X}_\alpha\boldsymbol{\theta}^{(h,0)}\left(\boldsymbol{\theta}^{(h,0)}\right)^{\dagger}\mathbf{X}_\alpha^{\dagger}-\mathbf{X}_\alpha\boldsymbol{\theta}^{(h,0)}\bar{\mathbf{d}}^{\dagger}\mathbf{Y}_\alpha^{\dagger}+\mathbf{X}_\alpha\boldsymbol{\theta}^{(h,0)}\left(\mathbf{u}^{(h,0)}\right)^{\dagger}\mathbf{Y}_\alpha^{\dagger}\\ -\mathbf{Y}_\alpha\left(\mathbf{C}_{\alpha d}^{(1)}\right)^{\dagger}-\mathbf{Y}_\alpha\bar{\mathbf{d}}\left(\boldsymbol{\theta}^{(h,0)}\right)^{\dagger}\mathbf{X}_\alpha^{\dagger}+\mathbf{Y}_\alpha\mathbf{D}\mathbf{Y}_\alpha^{\dagger}-\mathbf{Y}_\alpha\bar{\mathbf{d}}\left(\mathbf{u}^{(h,0)}\right)^{\dagger}\mathbf{Y}_\alpha^{\dagger} \quad (1.206)\\ +\mathbf{Y}_\alpha\mathbf{u}^{(h,0)}\left(\boldsymbol{\theta}^{(h,0)}\right)^{\dagger}\mathbf{X}_\alpha^{\dagger}-\mathbf{Y}_\alpha\mathbf{u}^{(h,0)}\bar{\mathbf{d}}^{\dagger}\mathbf{Y}_\alpha^{\dagger}+\mathbf{Y}_\alpha\mathbf{u}^{(h,0)}\left(\mathbf{u}^{(h,0)}\right)^{\dagger}\mathbf{Y}_\alpha^{\dagger}.\end{aligned}$$

where:

$$\begin{aligned}&\mathbf{C}_{\alpha d}^{(1)}=\mathbf{C}_{\alpha r}^{e}-\mathbf{C}_{\alpha\alpha}^{e}\mathbf{S}^{\dagger};\quad \mathbf{X}_\alpha \triangleq \mathbf{C}_{\alpha d}^{(1)}\left[\left(\mathbf{D}-\mathbf{V}^{(h,0)}\right)^{-1}\mathbf{D}-\mathbf{I}\right];\\ &\mathbf{Y}_\alpha \triangleq \mathbf{C}_{\alpha d}^{(1)}\left(\mathbf{D}-\mathbf{V}^{(h,0)}\right)^{-1}.\end{aligned} \quad (1.207)$$

Comparing the 2nd-BERRU-PMD-expression previously provided with the corresponding expression obtained using the 2nd-BERRU-PMP methodology, compare Eq. (1.176), indicates that both expressions include the 2nd-order sensitivities of the responses with respect to the parameters, but in distinct ways. Furthermore, *the expression produced by the 2nd-BERRU-PMP methodology, compare Eq. (1.176), can include all of the response sensitivities and parameter correlations of order higher than two*, if available. Such an inclusion of higher-order sensitivities and parameter correlations is not possible with the result obtained using the 2nd-BERRU-PMD methodology, compare Eq. (1.126).

The expression provided in Eq. (1.176) coincides with the expression provided in Eq. (1.126) only if the following relationship holds:

$$\begin{aligned}
&\mathbf{C}_{\alpha\alpha}^{bep} = \mathbf{C}_{\alpha\alpha}^{(1)bed} \;\; \textit{iff}: \; -\mathbf{C}_{\alpha r}^{c}\left(\mathbf{C}_{rr}^{e}+\mathbf{C}_{rr}^{c}\right)^{-1}\mathbf{C}_{\alpha r}^{c} = -\mathbf{C}_{\alpha d}^{(1)}\mathbf{Y}_{\alpha}^{\dagger} \\
&+\mathbf{X}_{\alpha}\boldsymbol{\theta}^{(h,0)}\left(\boldsymbol{\theta}^{(h,0)}\right)^{\dagger}\mathbf{X}_{\alpha}^{\dagger} - \mathbf{X}_{\alpha}\boldsymbol{\theta}^{(h,0)}\overline{\mathbf{d}}^{\dagger}\mathbf{Y}_{\alpha}^{\dagger} + \mathbf{X}_{\alpha}\boldsymbol{\theta}^{(h,0)}\left(\mathbf{u}^{(h,0)}\right)^{\dagger}\mathbf{Y}_{\alpha}^{\dagger} \\
&-\mathbf{Y}_{\alpha}\left(\mathbf{C}_{\alpha d}^{(1)}\right)^{\dagger} - \mathbf{Y}_{\alpha}\overline{\mathbf{d}}\left(\boldsymbol{\theta}^{(h,0)}\right)^{\dagger}\mathbf{X}_{\alpha}^{\dagger} + \mathbf{Y}_{\alpha}\mathbf{D}\mathbf{Y}_{\alpha}^{\dagger} - \mathbf{Y}_{\alpha}\overline{\mathbf{d}}\left(\mathbf{u}^{(h,0)}\right)^{\dagger}\mathbf{Y}_{\alpha}^{\dagger} \\
&+\mathbf{Y}_{\alpha}\mathbf{u}^{(h,0)}\left(\boldsymbol{\theta}^{(h,0)}\right)^{\dagger}\mathbf{X}_{\alpha}^{\dagger} - \mathbf{Y}_{\alpha}\mathbf{u}^{(h,0)}\overline{\mathbf{d}}^{\dagger}\mathbf{Y}_{\alpha}^{\dagger} + \mathbf{Y}_{\alpha}\mathbf{u}^{(h,0)}\left(\mathbf{u}^{(h,0)}\right)^{\dagger}\mathbf{Y}_{\alpha}^{\dagger}.
\end{aligned} \tag{1.208}$$

The equality shown in Eq. (1.208) can hold only if all of the 2nd-order sensitivities are neglected, and if the experimental responses are uncorrelated to the model parameters, in which case the following simplifications occur: $\mathbf{X}_{\alpha} = \mathbf{0}$, $\mathbf{Y}_{\alpha} = \mathbf{C}_{\alpha d}^{(1)}\mathbf{D}^{-1}$, in which case we also have the relation $\mathbf{D} = \mathbf{C}_{rr}^{e} + \mathbf{S}\mathbf{C}_{\alpha\alpha}\mathbf{S}^{\dagger} = \mathbf{C}_{rr}^{e} + \mathbf{C}_{rr}^{c}$. In this particular case, the results produced by the 2nd-BERRU-PMP and 2nd-BERRU-PMD methodologies also coincide with the results produced by the 1st-order 1st-BERRU-PM methodology.

1.7.5 Inter-Comparison: Best-Estimate Predicted Correlations Between Parameters and Responses

The expression obtained using the 2nd-BERRU-PMD methodology for the "first-iterate best-estimate" correlation matrix $\mathbf{C}_{\alpha r}^{(1)bed}$ for the parameters and responses are as follows:

$$\begin{aligned}
&\mathbf{C}_{\alpha r}^{(1)bed} = \mathbf{C}_{\alpha r}^{e} - \mathbf{C}_{\alpha d}^{(1)}\mathbf{Y}_{r}^{\dagger} + \mathbf{X}_{\alpha}\boldsymbol{\theta}^{(h,0)}\left(\boldsymbol{\theta}^{(h,0)}\right)^{\dagger}\mathbf{X}_{r}^{\dagger} - \mathbf{X}_{\alpha}\boldsymbol{\theta}^{(h,0)}\overline{\mathbf{d}}^{\dagger}\mathbf{Y}_{r}^{\dagger} \\
&+\mathbf{X}_{\alpha}\boldsymbol{\theta}^{(h,0)}\left(\mathbf{u}^{(h,0)}\right)^{\dagger}\mathbf{Y}_{r}^{\dagger} - \mathbf{Y}_{\alpha}\left(\mathbf{C}_{rd}^{(1)}\right)^{\dagger} - \mathbf{Y}_{\alpha}\overline{\mathbf{d}}\left(\boldsymbol{\theta}^{(h,0)}\right)^{\dagger}\mathbf{X}_{r}^{\dagger} \\
&+\mathbf{Y}_{\alpha}\mathbf{D}\mathbf{Y}_{r}^{\dagger} - \mathbf{Y}_{\alpha}\overline{\mathbf{d}}\left(\mathbf{u}^{(h,0)}\right)^{\dagger}\mathbf{Y}_{r}^{\dagger} + \mathbf{Y}_{\alpha}\mathbf{u}^{(h,0)}\left(\boldsymbol{\theta}^{(h,0)}\right)^{\dagger}\mathbf{X}_{r}^{\dagger} \\
&-\mathbf{Y}_{\alpha}\mathbf{u}^{(h,0)}\overline{\mathbf{d}}^{\dagger}\mathbf{Y}_{r}^{\dagger} + \mathbf{Y}_{\alpha}\mathbf{u}^{(h,0)}\left(\mathbf{u}^{(h,0)}\right)^{\dagger}\mathbf{Y}_{r}^{\dagger}.
\end{aligned} \tag{1.209}$$

Comparing the 2nd-BERRU-PMD-expression provided in Eq. (1.130) with the corresponding expression obtained using the 2nd-BERRU-PMP methodology, compare

Eq. (1.177), indicates that both expressions include the 2nd-order sensitivities of the responses with respect to the parameters, but in distinct ways. Furthermore, *the expression produced by the 2nd-BERRU-PMP methodology, compare Eq.* (1.177), *can include all of the response sensitivities and parameter correlations of order higher than two*, if available. Such an inclusion of higher-order sensitivities and parameter correlations is not possible with the result obtained using the 2nd-BERRU-PMD methodology, compare Eq. (1.130).

The expression provided in Eq. (1.177) coincides with the expression provided in Eq. (1.130) only if the following relationship holds:

$$\begin{aligned}
&\mathbf{C}_{\alpha r}^{bep} = \mathbf{C}_{\alpha r}^{(1)bed} \; \textit{iff}: \; -\mathbf{C}_{\alpha r}^{c}\left(\mathbf{C}_{rr}^{e} + \mathbf{C}_{rr}^{c}\right)^{-1}\mathbf{C}_{rr}^{e} \\
&= -\mathbf{C}_{\alpha d}^{(1)}\mathbf{Y}_{r}^{\dagger} + \mathbf{X}_{\alpha}\boldsymbol{\theta}^{(h,0)}\left(\boldsymbol{\theta}^{(h,0)}\right)^{\dagger}\mathbf{X}_{r}^{\dagger} - \mathbf{X}_{\alpha}\boldsymbol{\theta}^{(h,0)}\overline{\mathbf{d}}^{\dagger}\mathbf{Y}_{r}^{\dagger} \\
&+ \mathbf{X}_{\alpha}\boldsymbol{\theta}^{(h,0)}\left(\mathbf{u}^{(h,0)}\right)^{\dagger}\mathbf{Y}_{r}^{\dagger} - \mathbf{Y}_{\alpha}\left(\mathbf{C}_{rd}^{(1)}\right)^{\dagger} - \mathbf{Y}_{\alpha}\overline{\mathbf{d}}\left(\boldsymbol{\theta}^{(h,0)}\right)^{\dagger}\mathbf{X}_{r}^{\dagger} \\
&+ \mathbf{Y}_{\alpha}\mathbf{D}\mathbf{Y}_{r}^{\dagger} - \mathbf{Y}_{\alpha}\overline{\mathbf{d}}\left(\mathbf{u}^{(h,0)}\right)^{\dagger}\mathbf{Y}_{r}^{\dagger} + \mathbf{Y}_{\alpha}\mathbf{u}^{(h,0)}\left(\boldsymbol{\theta}^{(h,0)}\right)^{\dagger}\mathbf{X}_{r}^{\dagger} \\
&- \mathbf{Y}_{\alpha}\mathbf{u}^{(h,0)}\overline{\mathbf{d}}^{\dagger}\mathbf{Y}_{r}^{\dagger} + \mathbf{Y}_{\alpha}\mathbf{u}^{(h,0)}\left(\mathbf{u}^{(h,0)}\right)^{\dagger}\mathbf{Y}_{r}^{\dagger}.
\end{aligned} \tag{1.210}$$

The equality shown in Eq. (1.210) can hold only if all of the 2nd-order sensitivities are neglected, and if the experimental responses are uncorrelated to the model parameters. In this particular case, the results produced by the 2nd-BERRU-PMP and 2nd-BERRU-PMD methodologies also coincide with the results produced by the 1st-order 1st-BERRU-PM methodology.

1.8 REVIEW OF PRINCIPLES UNDERLYING THE DATA ADJUSTMENT AND DATA ASSIMILATION PROCEDURES

This section briefly summarizes the principles underlying the data adjustment and data assimilation procedures in order to compare these with the principles underlying the 2nd-BERRU-PMP and 2nd-BERRU-PMD methodologies.

1.8.1 Principles Underlying the Data Adjustment Procedure

The earliest systematic activities aimed at extracting best-estimate values for model parameters appear to have been initiated in the 1960s (see, e.g., Humi et al., 1964; Cecchini et al., 1964; Usachhev, 1964), in the course of evaluating neutron cross-sections by using time-independent reactor physics experiments for measuring "integral quantities" (also called "system responses") such as reaction rates and multiplication factors. A decade later, these activities had reached conceptual maturity under the name of "cross-section adjustment" (see, e.g., Rowlands et al., 1973; Gandini and Petilli, 1973), a methodology which essentially amounted to using a weighted least-square procedure (with response sensitivities used as weighting functions) for combining uncertainties in the model parameters with uncertainties in the experimental data, subject to the constraint imposed by the linearized model. The resulting "adjusted" neutron cross-sections (model parameters) and their "adjusted" uncertainties were subsequently employed in the respective reactor core model to predict improved model responses (e.g., improved reaction rates and reaction rate ratios, reactor multiplication factors, Doppler coefficients) in an extended application domain (e.g., a new or improved reactor core design). By the late 1970s, adjoint neutron fluxes were introduced and used (see, e.g., Kuroi and Mitani, 1975; Dragt et al., 1977; Weisbin et al., 1978) to compute efficiently the 1st-order response sensitivities, which appeared as weighting functions in the least squares adjustment procedure. It is important to note that all of these works dealt with the time-independent linear neutron transport or diffusion equation, as encountered in reactor physics and shielding, for which the corresponding adjoint equations were already known and readily available. For *nonlinear* systems, Cacuci (1981a, 1981b) pioneered the adjoint method for computing efficiently response sensitivities to model parameters and introduced it to applications in the earth sciences (see, e.g., Práger and Kelemen, 2013) and many other fields (see, e.g.), also including this method in the formulation of a data adjustment methodology for time-dependent nonlinear system (Barhen et al., 1982).

The latest version of the "data adjustment" methodology for time-independent systems is described in the leading nuclear engineering software packages SCALE (2009) and CRISTAL (Venard et al., 2009). This data adjustment methodology is intended for assimilating experimental information in order to "adjust" (calibrate) nuclear cross-sections ("model parameters") in linear reactor physics problems modeled by the time-independent linear Boltzmann equation. The data adjustment methodology relies on a so-called "Generalized Linear Least Squares Methodology (GLLSM)" which involves minimizing the following *user-defined* Lagrangian functional, denoted as $L(\mathbf{y},\mathbf{z})$:

$$\min_{\mathbf{y},\mathbf{z}} L(\mathbf{y},\mathbf{z}); \; L(\mathbf{y},\mathbf{z}) \triangleq (\mathbf{y},\mathbf{z})^{\dagger} \begin{pmatrix} \mathbf{C}_{mm} & \mathbf{C}_{m\alpha} \\ \mathbf{C}_{\alpha m} & \mathbf{C}_{\alpha\alpha} \end{pmatrix} \begin{pmatrix} \mathbf{y} \\ \mathbf{z} \end{pmatrix} + 2\boldsymbol{\lambda}^{\dagger}(\mathbf{S}_k \mathbf{z} - \mathbf{y}). \tag{1.211}$$

The covariance matrices appearing in functional $L(\mathbf{y},\mathbf{z})$ defined in Eq. (1.211) have the following meanings/definitions:

1. The (symmetric) relative correlation matrix for the model parameters (nuclear cross-sections), $\mathbf{C}_{\alpha\alpha}$, is defined as follows:

$$\mathbf{C}_{\alpha\alpha} \triangleq \left[\frac{\operatorname{cov}(\alpha_n, \alpha_p)}{\alpha_n \alpha_p} \right]; \; n, p = 1, \ldots, M, \tag{1.212}$$

   where α_n denotes the nth model parameter (cross sections); $\operatorname{cov}(\alpha_n, \alpha_p)$ denotes the covariance between the two respective model parameters; M denotes the total number of model parameters.
2. The (symmetric) relative correlation matrix for the measured model responses (effective multiplication factors), $\mathbf{C}_{mm}$, is defined as follows:

$$\mathbf{C}_{mm} \triangleq \left(\frac{m_i}{k_i}\right)\left[\frac{\operatorname{cov}\left(m_i, m_j\right)}{m_i m_j}\right]\left(\frac{m_j}{k_j}\right); \; i, j = 1, \ldots, I, \tag{1.213}$$

where k_i denotes the nominal value of ith-response, computed using the nominal values of the model parameters (cross sections); m_i denotes the nominal values of the measured i^{th}-response; $\operatorname{cov}\left(m_i,\, m_j\right)$ denotes the covariance between two measured effective multiplication factors; I denotes the total number of model responses (measured or computed effective multiplication factors).

3. The rectangular matrix $\mathbf{C}_{\alpha m}$ contains as elements the relative covariances (correlations) between the measured responses and model parameters, and is defined as follows:

$$\mathbf{C}_{\alpha m} \triangleq \left[\frac{\operatorname{cov}\left(\alpha_n, m_i\right)}{\alpha_n m_i}\right]\left(\frac{m_i}{k_i}\right); \; n = 1, \ldots, M; \; i = 1, \ldots, I; \; \mathbf{C}_{m\alpha} \triangleq \mathbf{C}_{\alpha m}^{\dagger}. \tag{1.214}$$

4. Each model response (effective multiplication factor), denoted as $k_i\left(\boldsymbol{\alpha}\right)$, is considered to be a linear function of the model parameters, denoted as $\boldsymbol{\alpha} \triangleq \left(\alpha_1, \ldots, \alpha_M\right)^{\dagger}$, having the following form:

$$k_i\left(\boldsymbol{\alpha} + \delta\boldsymbol{\alpha}\right) = k_i\left(\boldsymbol{\alpha}\right)\left[1 + \sum_{n=1}^{M} S_n^i \frac{\delta\alpha_n}{\alpha_n}\right]; \; i = 1, \ldots, I; \; \delta\boldsymbol{\alpha} \triangleq \left(\delta\alpha_1, \ldots, \delta\alpha_M\right)^{\dagger}, \tag{1.215}$$

where: (i) the quantity $\delta\alpha_n$ denotes an arbitrary perturbation in the n-th model parameter, and (ii) the quantity S_n^i represents the 1st-order relative sensitivity of the i-th response to the n-th parameter, computed at the nominal parameter values, and is defined as follows:

$$S_n^i \triangleq \frac{\partial k_i}{\partial \alpha_n}\frac{\alpha_n}{k_i}; \; i = 1, \ldots, I; \; n = 1, \ldots, M. \tag{1.216}$$

The linear relationship representing the model response as a function of the model parameters is re-written in the following form:

$$\mathbf{y} = \mathbf{d} + \mathbf{S}_k \mathbf{z}, \tag{1.217}$$

where:

$$\mathbf{S}_k \triangleq \left[S_n^i\right] \triangleq \left[\frac{\partial k_i}{\partial \alpha_n}\frac{\alpha_n}{k_i}\right]; \; i = 1, \ldots, I; \; n = 1, \ldots, M; \tag{1.218}$$

$$\mathbf{z} \triangleq \left(z_1, \ldots, z_M\right)^{\dagger}; \; z_n \triangleq \frac{\delta\alpha_n}{\alpha_n}; \tag{1.219}$$

$$\mathbf{d} \triangleq (d_1,\ldots,d_I)^\dagger;\ d_i \triangleq \frac{k_i(\boldsymbol{\alpha}) - m_i}{k_i(\boldsymbol{\alpha})}; \tag{1.220}$$

$$\mathbf{y} \triangleq (y_1,\ldots,y_I)^\dagger;\ y_i \triangleq \frac{k_i(\boldsymbol{\alpha} + \delta\boldsymbol{\alpha}) - m_i}{k_i(\boldsymbol{\alpha})}. \tag{1.221}$$

The (column) vector of Lagrange multipliers $\boldsymbol{\lambda} \triangleq (\lambda_1,\ldots,\lambda_I)^\dagger$, which appears in the functional $R(\mathbf{y},\mathbf{z})$ defined in Eq. (1.211), enforces the linear model defined by Eq. (1.217) as a "hard constraint."

The "adjusted" values for parameters and responses are obtained by solving the following equations, which are to be satisfied at the minimum of the functional $R(\mathbf{y},\mathbf{z})$:

$$\frac{\partial R(\mathbf{y},\mathbf{z})}{\partial \mathbf{z}} = \frac{\partial R(\mathbf{y},\mathbf{z})}{\partial \mathbf{y}} = 0. \tag{1.222}$$

The solution of Eq. (1.222) yields the following relations for the "adjusted" values for parameters and responses:

$$\mathbf{z} = \left(\mathbf{C}_{\alpha m} - \mathbf{C}_{\alpha\alpha}\mathbf{S}_k^\dagger\right)\mathbf{C}_{dd}^{-1}\mathbf{d}, \tag{1.223}$$

$$\mathbf{y} = \left(\mathbf{C}_{mm} - \mathbf{C}_{m\alpha}\mathbf{S}_k^\dagger\right)\mathbf{C}_{dd}^{-1}\mathbf{d}, \tag{1.224}$$

$$\mathbf{C}_{dd} \triangleq \mathbf{S}_k\mathbf{C}_{\alpha\alpha}\mathbf{S}_k^\dagger + \mathbf{C}_{mm} - \mathbf{S}_k\mathbf{C}_{\alpha m} - \mathbf{C}_{m\alpha}\mathbf{S}_k^\dagger. \tag{1.225}$$

The "adjusted covariance matrix," $\mathbf{C}_{mm}^{adj}$, for the "adjusted" responses (measured or computed) and "the adjusted covariance matrix," $\mathbf{C}_{\alpha\alpha}^{adj}$, for the "adjusted" parameters, respectively, are obtained by using Eqs. (1.223) and (1.224), to obtain the following expressions:

$$\mathbf{C}_{mm}^{adj} = \mathbf{C}_{mm} - \left(\mathbf{C}_{mm} - \mathbf{C}_{m\alpha}\mathbf{S}_k^\dagger\right)\mathbf{C}_{dd}^{-1}\left(\mathbf{C}_{mm} - \mathbf{S}_k\mathbf{C}_{\alpha m}\right), \tag{1.226}$$

$$\mathbf{C}_{\alpha\alpha}^{adj} = \mathbf{C}_{\alpha\alpha} - \left(\mathbf{C}_{\alpha m} - \mathbf{C}_{\alpha\alpha}\mathbf{S}_k^\dagger\right)\mathbf{C}_{dd}^{-1}\left(\mathbf{C}_{m\alpha} - \mathbf{S}_k\mathbf{C}_{\alpha\alpha}\right). \tag{1.227}$$

Care must be exercised, however, since the indiscriminate incorporation of all seemingly relevant experimental response data could produce a set of calibrated (adjusted) parameter values that might differ unreasonably much from the corresponding original nominal values and might even fail to improve the agreement between the calculated and measured values of some of the very responses that were used to calibrate the model parameters. When calibrating (adjusting) a library of model parameters, it is tacitly assumed that the given parameters are basically "correct," except that they are not sufficiently accurate for the objective at hand. The calibration procedure uses additional data (e.g., experimentally measured responses) for improving the parameter values while reducing their uncertainties. Although such additional information induces modifications of the original parameter values, the adjusted parameters are still generally expected to remain consistent with their original nominal values, within the range of their original uncertainties. However, calibration of a library of model parameters by experimental responses that significantly deviate from their respective computed values would significantly modify the resulting adjusted parameters, perhaps even violating the restriction of linearity expressed by Eq. (1.217). Such adjustments would most probably lead to the failure of even reproducing the original experimental responses.

On the other hand, calibrating the model parameter by using measured responses that are very close to their respective computed values would cause minimal parameter modifications and a nearly perfect reproduction of the given responses by the calibrated parameters, as would be expected. In such a case, the given responses would be considered as being consistent with the parameter library, in contradistinction to adjustment by inconsistent experimental information, in which case the adjustment (calibration) could fail because of inconsistencies. These considerations clearly underscore the need for using a quantitative indicator to measure the mutual and joint consistency of the information available for model calibration.

1.8.2 Principles Underlying the Data Assimilation Procedure

The predecessors to data assimilation methods were called "objective analyses" (Gandin, 1963), to distinguish them from the "subjective analyses" in which data was manipulated according to the opinion of experts. Subsequently, the so-called "nudging" or "Newtonian relaxation" methods were introduced (Anthes, 1974) to drive the model variables towards the observations by adding empirical forcing terms. The variational principle introduced by Sasaki (1955, 1958) provided the basis for the development of "optimal interpolation," "three-dimensional variational data assimilation," and "physical space statistical analysis" methods (see, e.g., the books by Lewis et al., 2006; Lahoz et al., 2010; Faragó et al., 2013) which were developed independently but were shown by Lorenc (1986) to be equivalent.

The introduction (Hall et al., 1982; Hall and Cacuci, 1983; see also: Práger and Kelemen, 2013) of Cacuci's (1981a, 1981b) adjoint sensitivity analysis method (for nonlinear systems) to the atmospheric sciences has enabled the subsequent development of the modern techniques nowadays called "variational data assimilation" methodology (also called "4D-VAR," where "4D" refers to a framework involving the four independent variables "time plus three spatial dimensions"). Variational data assimilation aims at determining the model's evolution "trajectory" which fits "best" – in a "least-square" sense – the observational data over an assimilation time interval by adjusting the initial conditions supplied for advancing (integrating) the forecast model forward in time. The earliest work on developing the 4D-VAR methodology using adjoint functions was by Derber (1985, 1987) and Hoffmann (1986), while Navon (1986, 1998) introduced augmented Lagrangian methods for enforcing the conservation of integral invariants. Most of the meteorological centers around the world have since implemented 4D-VAR methods in conjunction with comprehensive numerical weather prediction models (Derber et al., 1991; Mahfouf et al., 1996; Rabier, 2005).

The so-called *strong constraint* "variational data assimilation (VDA)" or *classical* VDA framework assumes that the forecast model perfectly represents the evolution of the actual physical system (atmosphere), and the "best fit" model trajectory is obtained by adjusting only the initial conditions via minimization of a user-defined cost functional, subject to the model equations as strong constraints. However, like any other models of large-scale processes, numerical weather prediction (NWP) models are affected by errors arising from the omission of sub-grid processes; also, the discretization of continuous processes introduces dissipative and dispersion errors. Furthermore, the mathematical modeling of the boundary conditions and forcing terms is imperfect, and most of the physical processes and their interactions in the atmosphere are parameterized. In general, these modeling approximations/errors comprise both systematic (so-called "bias") and stochastic components, which vary in space and time. Customarily, the *model error* (ME) term is formally introduced as a *correction* to the time derivatives of model variables, and the resulting methodology is called the *weak constraint VDA*.

Using the notation of Cacuci et al. (2014), the state of the physical system (e.g., atmosphere) can be described by the following evolution equation:

$$\frac{d\mathbf{x}(t)}{dt} = \mathbf{M}[\mathbf{x}(t)] + \mathbf{T}[\eta(t)], \; \mathbf{x}(t_0) = \mathbf{x}_0, \tag{1.228}$$

where: (i) the vector $\mathbf{x}(t)$ represents the state of the atmosphere, with an initial condition $\mathbf{x}_0$ at the initial time t_0; (ii) the operator $\mathbf{M}[\bullet]$ denotes all the mathematical operations involved in the model; (iii) $\eta(t)$ represents the model errors; (iv) $\mathbf{T}[\bullet]$ represents an operator that maps the space of model errors (ME) to the space of the model state $\mathbf{x}(t)$. If the model state has an associated model error at every grid point, then the operator $\mathbf{T}[\bullet]$ is identically equal to the unit matrix and the dimension of $\eta(t)$ is the same as the dimension of the model state $\mathbf{x}(t)$. On the other hand, if only some components of the state vector (e.g., the atmosphere in the polar regions) are affected by modeling errors, then the operator $\mathbf{T}[\bullet]$ can be specified in such a way that only those model grid points (in the polar regions) have modeling errors, while (by comparison) the rest of the model states are free of MEs. Some authors formally/artificially write $\eta(t)$ as the sum of a stochastic error ("noise") component and a "systematic error" component but such a separation is superfluous in practice since both components are anyway estimated based on "expert opinion" rather than based on exact modeling.

The model error is considered to evolve according to the following evolution model:

$$\frac{d\eta}{dt} = \mathbf{\Phi}[\eta(t), \mathbf{x}(t)], \;\; \eta(t_0) = \eta_0, \tag{1.229}$$

where the rate of growth or decay of $\eta(t)$ in time is governed by the particular form of the mapping $\mathbf{\Phi}$.

Data assimilation schemes determine the analyzed atmospheric state as an *optimal* combination of a-priori background information and observational information, by minimizing, in the least-squares sense, the following quadratic *user-defined* functional:

$$J[\mathbf{x}(t_0), \boldsymbol{\eta}(t_0)] \triangleq J_o + J_b + J_\eta, \tag{1.230}$$

where:

$$J_b \triangleq \frac{1}{2}[\mathbf{x}_0 - \mathbf{x}^b]^T \mathbf{B}^{-1}[\mathbf{x}_0 - \mathbf{x}^b], \tag{1.231}$$

$$J_0 \triangleq \frac{1}{2}\sum_{i=0}^{n}[\mathbf{H}(\mathbf{x}_i) - \mathbf{y}^0(t_i)]^T \mathbf{R}^{-1}[\mathbf{H}(\mathbf{x}_i) - \mathbf{y}^0(t_i)], \tag{1.232}$$

$$J_\eta \triangleq \frac{1}{2}[\eta_0 - \eta^b]^T \mathbf{Q}^{-1}[\eta_0 - \eta^b]. \tag{1.233}$$

The components of the functionals J_b, J_o, and J_η are defined as follows:

1. The functional J_b quantifies, in a least-square sense, the squared differences between the initial state $\mathbf{x}_0$ and the background state $\mathbf{x}^b$; $\mathbf{B}$ denotes the estimated background covariance matrix. The background state $\mathbf{x}^b$ provides an initial guess for minimization of J.
2. The functional J_0 quantifies, in a least-square sense, the squared differences between the observed state and initial state at the time instance t_i; $\mathbf{R}$ denotes the estimated observation covariance matrix and $\mathbf{H}(\mathbf{x}_i)$ denotes the observation operator at the time instance t_i.
3. The functional J_η quantifies, in a least-square sense, the model errors; $\mathbf{Q}$ denotes the covariance matrix of model errors.

The constrained minimization of the functional $J(\mathbf{x}_0, \boldsymbol{\eta}_0)$ subject to the "hard constraints" represented by Eqs. (1.228) and (1.229) is performed by minimizing the unconstrained augmented Lagrangian functional defined here:

$$\begin{aligned} L(\mathbf{x}, \boldsymbol{\eta}, \mathbf{x}^*, \boldsymbol{\eta}^*) = J(\mathbf{x}_0, \boldsymbol{\eta}_0) &+ \int_{t_0}^{t_n} \left\langle \mathbf{x}^*(t), \left\{ \frac{d\mathbf{x}(t)}{dt} - \mathbf{M}[\mathbf{x}(t)] - \boldsymbol{\eta}(t) \right\} \right\rangle dt \\ &+ \int_{t_0}^{t_n} \left\langle \boldsymbol{\eta}^*(t), \left\{ \frac{d\boldsymbol{\eta}}{dt} - \boldsymbol{\Phi}[\boldsymbol{\eta}(t), \mathbf{x}(t)] \right\} \right\rangle dt, \end{aligned} \tag{1.234}$$

where $\mathbf{x}^*(t)$ and $\boldsymbol{\eta}^*(t)$ are the Lagrange multiplier vectors corresponding to $\mathbf{x}(t)$ and $\boldsymbol{\eta}(t)$, respectively, while $\langle \bullet, \bullet \rangle$ denotes Euclidean inner product in the three-dimensional physical space.

The extrema of $L(\mathbf{x}, \boldsymbol{\eta}, \mathbf{x}^*, \boldsymbol{\eta}^*)$ are the solutions of the following Euler-Lagrange equations:

$$\frac{\partial L}{\partial \mathbf{x}} = \mathbf{0}, \quad \frac{\partial L}{\partial \boldsymbol{\eta}} = \mathbf{0}, \tag{1.235}$$

$$\frac{\partial L}{\partial \mathbf{x}^*} = \mathbf{0}, \quad \frac{\partial L}{\partial \eta^*} = \mathbf{0}. \tag{1.236}$$

Imposing Eq. (1.236) ensures that the equations describing the evolution of the model state and ME are shown in Eqs. (1.228) and (1.229) are satisfied. Solving Eq. (1.235) yields the following adjoint equations which describe the evolution of the adjoint variables $\mathbf{x}^*(t)$ and $\eta^*(t)$:

$$\begin{aligned} -\frac{d\mathbf{x}^*(t)}{dt} &= \left[\frac{\partial \mathbf{M}}{\partial \mathbf{x}}\right]^T \mathbf{x}^*(t) + \left[\frac{\partial \mathbf{\Phi}}{\partial \mathbf{x}}\right]^T \eta^*(t) \\ &+ \delta(t-t_i)\sum_{i=0}^{n}\left[\frac{\partial H}{\partial \mathbf{x}}\right]^T \mathbf{R}^{-1}[H(\mathbf{x}_i) - \mathbf{y}^0(t_i)]; \ \mathbf{x}^*(t_n) = \mathbf{0}, \end{aligned} \tag{1.237}$$

$$-\frac{d\eta^*(t)}{dt} = \left[\frac{\partial \mathbf{\Phi}}{\partial \eta}\right]^T \eta^*(t) + \mathbf{x}^*(t); \quad \eta^*(t_n) = \mathbf{0}. \tag{1.238}$$

As indicated by Eqs. (1.237) and (1.238), the evolution of $\mathbf{x}^*(t)$is coupled to the evolution of $\eta^*(t)$ via the operator $\mathbf{\Phi}$. Backward integration of the adjoint models represented by Eqs. (1.237) and (1.238) from time $t_n \to t_0$ provides the values of initial adjoint states: $\mathbf{x}^*(t_0)$ and $\eta^*(t_0)$. Furthermore, the optimality conditions expressed by Eqs. (1.235) and (1.236) also provide the following expressions for the partial gradients of the functional $L(\mathbf{x}, \eta, \mathbf{x}^*, \eta^*)$ or, equivalently, the functional $J(\mathbf{x}_0, \eta_0)$, with respect to the initial model state, $\mathbf{x}_0$, and the initial ME state, η_0:

$$\nabla_{\mathbf{x}_0} J = \nabla_{\mathbf{x}_0} J_b + \nabla_{\mathbf{x}_0} J_o = \mathbf{B}^{-1}[\mathbf{x}_0 - \mathbf{x}^b] + \mathbf{x}^*(t_0); \tag{1.239}$$

$$\nabla_{\eta_0} J = \nabla_{\eta_0} J_\eta + \nabla_{\eta_0} J_o = \mathbf{Q}^{-1}[\eta_0 - \eta^b] + \eta^*(t_0). \tag{1.240}$$

If the model's errors had been neglected, Eq. (1.240) would not exist, so the amount of computations would be just half of the computations needed when modeling errors are explicitly considered.

Although they are not identical, the data adjustment and data assimilation methodologies share several fundamental common features which stem from their common root in least-square minimization procedures, so the expressions of the end results produced by these methodologies look similar. Neither of these methodologies uses the complete 1st-order information, however. The *data adjustment* methodology is incomplete in 1st-order because it has no provisions for incorporating the 1st-order correlations between the model parameters and the *computed responses*, which are defined by Eq. (1.148) and which can be readily computed. The applicability of the data adjustment methodology is also restricted by the domain of applicability of the "generalized linear least-squares methodology" (GLLSM), which requires that: (i) the uncertainty of the calculated response (e.g., neutron multiplication factor) be dominated by uncertainties in the data (e.g., cross-sections, neutron multiplicities, fission spectra), having negligible systematic errors; and (ii) the data uncertainties be sufficiently small for approximating the response (e.g., neutron multiplication factor) by a 1st-order Taylor series expansion about the values of the given data. The analysis performed by Hoefer et al. (2011a, 2011b) indicated that neither condition is fulfilled in general, since (i) the linear perturbation assumption underlying the GLLSM predictions breaks down for relative standard deviations (in the data) larger than ca. 1%; and (ii) systematic errors larger than ca. 1% also cause the brake-down of the GLLSM predictions, since such systematic errors are not explicitly considered within the GLLSM formalism. Consequently, for real application cases, which are always characterized by systematic computational errors, the work reported by Hoefer et al. (2011a, 2011b) indicates that true coverage probabilities of GLLSM confidence intervals for the computational bias may strongly differ from their expected confidence levels, particularly if a chi-square filter is used. The user-defined "cost functional" represented by Eq. (1.211), which is minimized in the least-square sense, cannot be extended in an obvious manner to incorporate higher-order correlations and sensitivities. A fundamentally different type of "cost functional" would need to be introduced for incorporating higher-order correlations and sensitivities, the form of which has not been produced yet in the literature.

The *data assimilation* methodology is also incomplete in 1st-order because it has no provisions for incorporating the 1st-order correlations between the model parameters and responses, such as represented by Eq. (1.148). On the other hand, the currently most advanced 4D-VAR data assimilation methodologies (including the extended/ensemble Kalman filter, 2nd-order filter, and variants thereof; see, for example, the books by Lewis et al., 2006; Lahoz et al., 2010; Faragó et al., 2013; Cacuci et al., 2014) allow the incorporation of some partial 2nd-order information through the so-called "Hessian-vector product" which appears as an additional term in the model dynamics/forecast step and the data assimilation step. Thus, although the most advanced data assimilation methods can be considered to be "more inclusive" (in the sense of incorporating some partial 2nd-order information) than the most advanced data adjustment methods, neither of these methodologies are complete regarding the incorporation of *all extant 1st- and/or 2nd-order information*, and neither methodology provides a mechanism (even a theoretical one) for incorporating response sensitivities (to model parameters) higher than 1st-order and/or incorporating correlations among responses and/or parameters higher than 2nd-order. Of course, the predictions of the data adjustment or assimilation methodologies are also limited to 1st-order mean values and 2nd-order covariances/correlations for parameters or responses.

1.9 DISCUSSION AND CONCLUSIONS

This chapter has presented the two variants of the "2nd-order best-estimate results with reduced uncertainties predictive modeling (2nd-BERRU-PM)" methodology, namely: (i) the *2nd-BERRU-PMP* methodology with *probabilistically* included computed responses, as underscored by the letter "*P*" in the acronym, and (ii) the alternative *2nd-BERRU-PMD* methodology, in which the computational model is included *deterministically*, as underscored by the letter "*D*" in this acronym.

Both the *2nd-BERRU-PMP* and the *2nd-BERRU-PMD* methodologies are fundamentally based on the MaxEnt principle and yield expressions that include all 2nd-order (and possibly higher-order) sensitivities of responses to model parameters. Although the *2nd-BERRU-PMP* and the *2nd-BERRU-PMD* yield expressions for the predicted responses, calibrated predicted parameters, and predicted covariances that are consistent within 2nd-order (sensitivities and covariances), the respective predicted quantities are not identical to each other, except in special circumstances that were analyzed in this chapter. The *2nd-BERRU-PMP* and the *2nd-BERRU-PMD* methodologies encompass, as particular cases, the results produced by the extant data assimilation, data adjustment procedures, and the 1st-order BERRU-PM methodology.

Notably, the 2nd-BERRU-PMP methodology enables the use of sensitivities and parameter correlations beyond 2nd-order, if available; this opportunity is not available within the 2nd-BERRU-PMD methodology. If it is necessary to combine external experimental information for both the model responses and parameters with the computational information produced by the model, then the complete Gaussian expression of the 2nd-BERRU-PMP should be used, if it is possible to accommodate the massive computational requirements underlying the inversion of the matrices that arise within this methodology. On the other hand, the simplest methodology to implement computationally is the simplified 2nd-BERRU-PMP, which should be used when additional experimental information about the model parameters (i.e., in addition to the information used in the computational model) is not important. The 2nd-BERRU-PMD methodology offers an intermediary alternative, being easier to implement computationally than the complete version of the 2nd-BERRU-PMP methodology but involving more computations than the simplified 2nd-BERRU-PMP methodology. Whenever the particular respective characteristics need not be highlighted, the 2nd-BERRU-PMD and 2nd-BERRU-PMP methodologies will be collectively called the "2nd-BERRU-PM methodology."

Since the framework of the 2nd-BERRU-PM methodology comprises the combined phase space of parameters and responses, it can be used for solving both forward/direct and inverse problems. The "forward" or "direct problem" solves the "parameter-to-output" mapping that describes the "cause-to-effect" relationship in the physical process being modeled. The "inverse problem" attempts to solve the "output-to-parameters" mapping. In particular, "measurement problems" are "inverse" to the direct problem, in that they seek to determine (from measurements) the properties of the host medium (e.g., composition, geometry, including internal interfaces) or the properties of the source (e.g., strength, location, direction) and/or the size of the medium on its boundaries. Such inverse problems are encountered in fields as diverse as astrophysics (in which one measures the intensity and spectral distribution of light in order to infer properties of starts), nuclear medicine (where radioisotopes are injected into patients and the radiation emitted is used in diagnostics to reconstruct body properties, e.g., tumors), non-destructive fault detection in materials, underground (oil, water) logging, and detection of sensitive materials. Some authors further group inverse problems into "invasive" (when the interior particle distribution is accessible for measurements) and "non-invasive" (when only particle distributions on the boundaries of, or exterior to, the medium can be measured). The solution for a forward/direct problem is provided by the 2nd-BERRU-PM expressions for the predicted best-estimate responses, together with the corresponding reduced predicted standard deviations. Conversely, the solution for an "inverse problem" is provided by the 2nd-BERRU-PM expressions for the best-estimate predicted parameters and their reduced predicted uncertainties. The 2nd-BERRU-PM expressions for the parameter-response correlations and the "goodness of fit indicator" are additional features provided by this methodology.

Notably, the 2nd-BERRU-PMP methodology explicitly includes the 3rd- and 4th-order response sensitivities to model parameters, thereby indicating the path for incorporating explicitly higher-order sensitivities, if available. Illustrative applications of the *2nd-BERRU-PMP* and the *2nd-BERRU-PMD* methodologies to paradigm "large-scale" models are presented in Chapter 2.

2 Application of the 2nd-BERRU-PM Methodology to the PERP Reactor Physics Benchmark

2.1 INTRODUCTION

This chapter presents illustrative results obtained by applying the 2nd-BERRU-PM methodology which was introduced in Chapter 1, to a *p*oly*e*thylene-*r*eflected *p*lutonium (acronym: PERP) OECD/NEA reactor physics benchmark. The response of interest for this benchmark is the leakage of neutrons through the benchmark's outer surface. As has been amply described in the book by Cacuci and Fang (2023), this benchmark is modeled using the neutron transport Boltzmann equation, involving 21976 imprecisely known parameters, the solution of which is representative of "large-scale computations." The computational model of the PERP benchmark is summarized in Section 2.2. It has been shown by Cacuci and Fang (2023) that the largest sensitivities are the benchmark's response sensitivities to the 180 microscopic group-averaged total cross-sections. Since there is no information regarding the actual distribution obeyed by these cross-sections, they are assumed to be uncorrelated and normally distributed, since this assumption is the least biased and most conservative possible among all distributions. Under these circumstances, the mathematical expressions for computing the mean and variance of the PERP benchmark's *computed* leakage response are presented in Section 2.3, while Section 2.4 presents the corresponding mathematical expressions for computing the *best-estimate predicted* mean and variance for the leakage response.

Section 2.5 illustrates the results predicted by the 2nd-BERRU-PM methodology when the measured response, having a "typical precision experimental standard deviation of 5%" is consistent with the computed response, within the 1st-order computed standard deviation. The effects of the precision (standard deviations) of the parameters (total cross-sections) on the 2nd-BERRU-PMP predicted results are illustrated by considering "high precision" parameters (with relative standard deviations of 3%), "typical precision" parameters (relative standard deviations of 5%), and "low precision" parameters (relative standard deviations of 10%). Section 2.6 analyzes a "low precision response measurement" (relative standard deviations of 10%) but "high precision" parameters (relative SD = 3%),

DOI: 10.1201/9781003478119-2

thus enabling a direct quantification of the impact of the precision (low vs. typical) of the measured response.

Section 2.7 considers a typical precision (relative SD = 5%) measured response, which appears initially to be inconsistent with the computed value of the response. Again, the impact of the parameter standard deviations and sensitivities is illustrated by considering "high precision" (2%), "typical precision" (5%), and "low precision" (10%) parameters, respectively. Section 2.8 analyzes the situation of a high precision measured response which is, apparently, inconsistent with the computed response, contrasting this situation with the "typical precision" measured response in Section 2.7.

Section 2.9 presents the results produced by using the 2nd-BERRU-PM in conjunction with an experimental response measured with high precision, but with rather imprecisely known model parameters, all having relative standard deviations of 10%, for which the Taylor series expansion of the model's leakage response in terms of the total cross-section parameters is divergent. This illustrative example, which involves precise measurements but rather imprecise parameters, has been deliberately chosen in order to underscore the decisive impact of higher-order sensitivities and the ability of the 2nd-BERRU-PM methodology to yield best-estimate results with reduced uncertainties despite the large uncertainties considered here for the imprecisely known model parameters.

Section 2.10 presents the results produced by using the 2nd-BERRU-PM methodology in conjunction with an experimental response measured with low precision, and relatively imprecisely known model parameters, all having relative standard deviations of 5%, which often occurs in practice, and for which the Taylor series expansion of the model's leakage response in terms of the total cross-section parameters is slowly divergent. This example underscores the impact of the higher-order response sensitivities when the parameter uncertainties are representative of the uncertainties usually encountered in practice while also being "borderline" in terms of the convergence of the Taylor series that underlies the determination of the statistics (expected values, variance, etc.) of the distribution of the computed response in the phase-space of imprecisely known model parameters.

Section 2.11 analyzes the unusual but possible situation when the measured value coincides with the computed value of the leakage response, both having the value $L^e = 1.765 \times 10^6$ neutrons/sec. As discussed in Chapter 1, *the conventional data adjustment and/or assimilation methods fail when the measured and computed response values coincide*. The measurement is considered to have a "moderate" precision, with $SD^{(e)} = 5\%$ and that the parameters are also known with moderate precision, all having relative standard deviations of 5%, which are typical measurement relative standard deviations encountered in practice. Even though the Taylor series is divergent for 5% parameter standard deviations, the 2nd-BERRU-PM methodology is shown not only to "work" (as opposed to failing, like the data adjustment and assimilation procedures) but actually predicts the best-estimate response value to be very closed to the measured one, having a predicted standard deviation also close to (albeit marginally smaller, as expected) the measured standard deviation, since the measured standard deviation is smaller than the computed one. This chapter is concluded with summarizing remarks in Section 2.12.

2.2 MATHEMATICAL MODELING OF THE OECD/NEA POLYETHYLENE-REFLECTED PLUTONIUM METAL SPHERE (PERP) REACTOR PHYSICS BENCHMARK

The OECD/NEA *P*oly*e*thylene-*R*eflected *P*lutonium Metal Sphere (which will be designated in this book using the acronym "PERP") reactor physics benchmark (Valentine, 2006) is a one-dimensional spherical subcritical nuclear system driven by a source of spontaneous fission neutrons. The PERP benchmark comprises an inner sphere (designated as "material 1") which is surrounded by a spherical shell (designated as "material 2"). The inner sphere of the PERP benchmark contains α-phase plutonium which acts as the source of particles; it has a radius $r_1 = 3.794\ cm$. This inner sphere is surrounded by a spherical shell reflector made of polyethylene of thickness 3.81 cm; the radius of the outer shell containing polyethylene is $r_s = 7.064\ cm$. Table 2.1 specifies the constitutive materials of the PERP benchmark.

The distribution of neutrons within the benchmark is modeled by the linear time-independent inhomogeneous neutron transport (Boltzmann) equation in six independent variables, which has the following standard form:

$$\begin{aligned}
&\boldsymbol{\Omega} \bullet \nabla \varphi(r,E,\boldsymbol{\Omega}) + \Sigma_t(r,E)\varphi(r,E,\boldsymbol{\Omega}) \\
&= \int_0^{E_f} dE' \int_{4\pi} d\boldsymbol{\Omega}' \Sigma_s(r,E' \to E, \boldsymbol{\Omega}' \bullet \boldsymbol{\Omega})\varphi(r,E',\boldsymbol{\Omega}') \\
&+ \int_0^{E_f} dE' \int_{4\pi} \chi(r,E' \to E)\nu\Sigma(r,E')\varphi(r,E',\boldsymbol{\Omega}')d\boldsymbol{\Omega}' + Q(r,E,\boldsymbol{\Omega}).
\end{aligned} \tag{2.1}$$

The boundary condition appropriate for this physical system is that of "no incoming flux" which means that:

$$\varphi(r_s,E,\boldsymbol{\Omega}) = 0,\ r_s \in S_b,\ \boldsymbol{\Omega}\cdot\mathbf{n} < 0, \tag{2.2}$$

TABLE 2.1
Dimensions and Composition of the PERP Benchmark

Materials	Isotopes	Weight Fraction	Density (g/cm^3)	Zones
Material 1 (plutonium metal)	Isotope 1 (^{239}Pu)	9.3804×10^{-1}	19.6	Material 1 is assigned to zone 1, which has a radius of 3.794 cm.
	Isotope 2 (^{240}Pu)	5.9411×10^{-2}		
	Isotope 3 (^{69}Ga)	1.5152×10^{-3}		
	Isotope 4 (^{71}Ga)	1.0346×10^{-3}		
Material 2 (polyethylene)	Isotope 5 (^{12}C)	8.5630×10^{-1}	0.95	Material 2 is assigned to zone 2, which has an inner radius of 3.794 cm and an outer radius of 7.604 cm.
	Isotope 6 (^{1}H)	1.4370×10^{-1}		

where r_s denotes the external radius of the PERP benchmark and where the vector **n** denotes the outward unit normal vector at each point on the sphere's outer surface (boundary), denoted as S_b. The physical quantities which appear in the standard notation used in Eq. (2.1) are as follows:

1. $\varphi(r,E,\mathbf{\Omega})$ denotes the flux of particles (i.e., particle number density multiplied by the particle speed) in the energy range dE about E, in the volume element dr about r, with directions of motion in the solid angle element $d\mathbf{\Omega}$ about $\mathbf{\Omega}$.
2. $Q(r,E,\mathbf{\Omega})$ denotes the rate at which particles are produced in the same element of phase-space from sources that are independent of the flux.
3. $\Sigma_t(r,E)$ denotes the macroscopic total cross-section.
4. Assuming isotropic media, $\Sigma_s(r,E' \rightarrow E,\mathbf{\Omega}' \bullet \mathbf{\Omega})$ denotes the macroscopic scattering transfer cross-section from energy E' to energy E through a scattering angle $\mathbf{\Omega}' \bullet \mathbf{\Omega}$.
5. ν denotes the number of particles emitted isotropically $(1/4\pi)$ per fission.
6. $\Sigma_f(r,E)$ denotes the macroscopic fission cross-section.
7. $\chi(r,E' \rightarrow E)$ denotes the fraction of fission particles appearing in energy dE about E from fissions in dE' about E'.

The neutron flux distribution within the PERP benchmark has been computed by using the deterministic software package PARTISN (Alcouffe et al. 2008), which solves the multigroup approximation of the transport equation. This approximation is obtained by partitioning the domain of the energy variable into G intervals, called "energy groups," and integrating Eqs. (2.1) and (2.2) formally over a generic energy group, labeled as "g," to obtain the following "multigroup transport equation" to be solved for the group-fluxes $\varphi^g(r,\mathbf{\Omega})$:

$$B^g(\boldsymbol{\alpha})\varphi^g(r,\mathbf{\Omega}) = Q^g(\boldsymbol{\alpha};r), \quad g = 1,...,G, \tag{2.3}$$

$$\varphi^g(r_s,\mathbf{\Omega}) = 0, \; r_s \in S_b, \; \mathbf{\Omega}\cdot\mathbf{n} < 0, \quad g = 1,...,G, \tag{2.4}$$

where:

$$\begin{aligned} B^g(\boldsymbol{\alpha})\varphi^g(r,\mathbf{\Omega}) &\triangleq \mathbf{\Omega}\bullet\nabla\varphi^g(r,\mathbf{\Omega}) + \Sigma_t^g(r)\varphi^g(r,\mathbf{\Omega}) \\ &-\sum_{g'=1}^{G}\int_{4\pi}\Sigma_s^{g'\rightarrow g}(r,\mathbf{\Omega}'\bullet\mathbf{\Omega})\varphi^{g'}(r,\mathbf{\Omega}')d\mathbf{\Omega}' - \chi^g(r) \\ &\sum_{g'=1}^{G}\int_{4\pi}(\nu\Sigma)_f^{g'}(r)\varphi^{g'}(r,\mathbf{\Omega}')d\mathbf{\Omega}', \end{aligned} \tag{2.5}$$

$$Q^g(\boldsymbol{\alpha};r) \triangleq \sum_{i=1}^{N_f}\lambda_i N_{i,1} F_i^{SF}\nu_i^{SF}\frac{1}{I_0}\int_{E^{g+1}}^{E^g} dE\, e^{-E/a_i}\sinh\sqrt{b_i E}, g = 1,...,G; \tag{2.6}$$

with

$$I_0 \triangleq \frac{\sqrt{\pi a_i^3 b_i}}{2} e^{\frac{a_i b_i}{4}}. \tag{2.7}$$

In Eqs. (2.6) and (2.7), the subscript "i" denotes the number of nuclides within the spontaneous fission source.

The PARTISN (Alcouffe et al., 2008) computations used the MENDF71X library (Conlin et al., 2013) which comprises 618 group cross-sections. These cross-sections were collapsed to $G = 30$ energy groups, with group boundaries, E^g, as presented in Table 2.2. The MENDF71X library (Conlin et al., 2013) uses ENDF/B-VII.1 nuclear data (Chadwick et al., 2011). The group boundaries, E^g, are user-defined and are therefore considered to be perfectly well-known parameters.

The source of neutrons in the PERP benchmark is provided by the spontaneous fissions stemming from ^{239}Pu (Isotope 1) and ^{240}Pu (Isotope 2); there are no delayed neutron or (α, n) sources. The spontaneous fission source has been computed using the code SOURCES4C (Wilson et al., 2002). For an actinide nuclide k, the spontaneous source depends on the following six model parameters: the decay constant λ_k, the atom density $N_{k,m}$, the average number of neutrons per spontaneous fission ν_k^{SF}, the spontaneous fission branching ratio F_k^{SF}, and the two parameters a_k and b_k used in a Watt's fission spectrum to approximate the spontaneous fission neutron spectrum. The PERP benchmark contains two actinide nuclides, so that $k = 1,2$ for the PERP benchmark. The nominal values of these parameters (except for $N_{k,m}$) are available

TABLE 2.2
Energy Group Structure, in [MeV], for PERP Benchmark Computations

g	1	2	3	4	5	6
E^g	1.50×10^{1}	1.35×10^{1}	1.20×10^{1}	1.00×10^{1}	7.79×10^{0}	6.07×10^{0}
E^{g-1}	1.70×10^{1}	1.50×10^{1}	1.35×10^{1}	1.20×10^{1}	1.00×10^{1}	7.79×10^{0}
g	7	8	9	10	11	12
E^g	3.68×10^{0}	2.87×10^{0}	2.23×10^{0}	1.74×10^{0}	1.35×10^{0}	8.23×10^{-1}
E^{g-1}	6.07×10^{0}	3.68×10^{0}	2.87×10^{0}	2.23×10^{0}	1.74×10^{0}	1.35×10^{0}
g	13	14	15	16	17	18
E^g	5.00×10^{-1}	3.03×10^{-1}	1.84×10^{-1}	6.76×10^{-2}	2.48×10^{-2}	9.12×10^{-3}
E^{g-1}	8.23×10^{-1}	5.00×10^{-1}	3.03×10^{-1}	1.84×10^{-1}	6.76×10^{-2}	2.48×10^{-2}
g	19	20	21	22	23	24
E^g	3.35×10^{-3}	1.24×10^{-3}	4.54×10^{-4}	1.67×10^{-4}	6.14×10^{-5}	2.26×10^{-5}
E^{g-1}	9.12×10^{-3}	3.35×10^{-3}	1.24×10^{-3}	4.54×10^{-4}	1.67×10^{-4}	6.14×10^{-5}
g	25	26	27	28	29	30
E^g	8.32×10^{-6}	3.06×10^{-6}	1.13×10^{-6}	4.14×10^{-7}	1.52×10^{-7}	1.39×10^{-10}
E^{g-1}	2.26×10^{-5}	8.32×10^{-6}	3.06×10^{-6}	1.13×10^{-6}	4.14×10^{-7}	1.52×10^{-7}

from a library file contained in SOURCES4C (Wilson et al., 2002), while the nominal values for $N_{k,m}$ are specified from the PERP benchmark. These imprecisely known source parameters also affect the accuracy of the neutron transport calculation.

PARTISN (Alcouffe et al., 2008) uses the discrete ordinates approximation to discretize the angular variable in the first and second terms on the right side of Eq. (2.5), and it uses a finite-moments expansion in spherical harmonics to approximate the angular variable in the third and fourth terms on the right side Eq. (2.5). The specific computations in this work were performed using a P_3 Legendre expansion of the scattering cross-section, an angular quadrature of S_{256}, and a fine-mesh spacing of 0.005 cm (comprising 759 meshes for the plutonium sphere of radius of 3.794 cm, and 762 meshes for the polyethylene shell of thickness of 3.81 cm). It is convenient to retain the continuous representation in the angular and radial variables since the spatial and angular discretization parameters are considered to be perfectly well-known. The various quantities in Eqs. (2.3)–(2.7) have their usual meanings for the standard form of the multigroup neutron transport equation, as follows:

1. Using the notation employed in PARTISN (Alcouffe et al., 2008), the quantity $\varphi^g(r,\boldsymbol{\Omega}) \triangleq \int_{E^{g-1/2}}^{E^{g+1/2}} \varphi(r,E,\boldsymbol{\Omega})\,dE$ denotes the "group-flux" for group g, and is the unknown state function obtained by solving Eqs. (2.3) and (2.4).
2. The spontaneous-fission isotopes in the PERP benchmark are "isotope 1" (^{239}Pu) and "isotope 2" (^{240}Pu). The quantity N_f denotes the total number of spontaneous fission isotopes; for the PERP benchmark, $N_f = 2$. The spontaneous fission neutron spectra of ^{239}Pu and, respectively, ^{240}Pu, are approximated by Watt's fission spectra, each spectrum using two evaluated parameters, denoted as a_k and b_k, respectively. The decay constant for actinide nuclide k is denoted as λ_k, while F_k^{SF} denotes the fraction of decays that are spontaneous fission (the "spontaneous fission branching fraction").
3. The quantity $N_{i,m}$ denotes the atom density of isotope i in material m; $i = 1,...,I$, $m = 1,...,M$, where I denotes the total number of isotopes, and M denotes the total number of materials. The computation of $N_{i,m}$ uses the following well-known expression:

$$N_{i,m} \triangleq \frac{\rho_m w_{i,m} N_A}{A_i}, \tag{2.8}$$

where ρ_m denotes the mass density of material m, $m = 1,...,M$; $w_{i,m}$ denotes the weight fraction of isotope i in material m; A_i denotes the atomic weight of isotope i, $i = 1,...,I$; N_A denotes the Avogadro's number. For the PERP benchmark, $I = 6$ and $M = 2$, but since the respective isotopes are all distinct (i.e., are not repeated) in the PERP benchmark's distinct materials, as specified in Table 2.1, it follows that only the following isotopic number densities exist for this benchmark: $N_{1,1}, N_{2,1}, N_{3,1}, N_{4,1}, N_{5,2}, N_{6,2}$.
4. The quantity $\Sigma_s^{g' \to g}(r, \boldsymbol{\Omega}' \bullet \boldsymbol{\Omega})$ represents the scattering transfer cross-section from energy group g', $g' = 1,...,G$ into energy group g, $g = 1,...,G$. The transfer cross-sections are computed in terms of the lth-order Legendre

coefficients $\sigma_{s,l,i}^{g'\to g}$ (of the Legendre-expanded microscopic scattering cross-section from energy group g' into energy group g for isotope i) which are tabulated parameters using the following finite-order expansion:

$$\begin{aligned} \Sigma_s^{g'\to g}\left(r,\boldsymbol{\Omega}'\bullet\boldsymbol{\Omega}\right) &= \sum_{m=1}^{M=2}\Sigma_{s,m}^{g'\to g}\left(r,\boldsymbol{\Omega}'\bullet\boldsymbol{\Omega}\right), \\ \Sigma_{s,m}^{g'\to g}\left(r,\boldsymbol{\Omega}'\bullet\boldsymbol{\Omega}\right) &\cong \sum_{i=1}^{I=6}N_{i,m}\sum_{l=0}^{ISCT=3}(2l+1)\sigma_{s,l,i}^{g'\to g}(r)P_l\left(\boldsymbol{\Omega}'\cdot\boldsymbol{\Omega}\right),\; m=1,2, \end{aligned} \tag{2.9}$$

where $ISCT=3$ denotes the order of the respective finite expansion in Legendre polynomial. The variable r will, henceforth, no longer appear in the arguments of the various cross-sections since the cross-sections for every material are treated in the PARTISN (Alcouffe et al., 2008) computations as being space-independent within the respective material.

5. The total cross-section Σ_t^g for energy group g, $g=1,...,G$, and material m, is computed for the PERP benchmark using the following expression:

$$\begin{aligned} \Sigma_t^g &= \sum_{m=1}^{M=2}\Sigma_{t,m}^g;\; \Sigma_{t,m}^g = \sum_{i}^{I}N_{i,m}\sigma_{t,i}^g \\ &= \sum_{i}^{I}N_{i,m}\left[\sigma_{f,i}^g+\sigma_{c,i}^g+\sum_{g'=1}^{G}\sigma_{s,l=0,i}^{g\to g'}\right], m=1,2, \end{aligned} \tag{2.10}$$

where $\sigma_{f,i}^g$ and $\sigma_{c,i}^g$ denote, respectively, the tabulated group microscopic fission and neutron capture cross-sections for group g, $g=1,...,G$. Other nuclear reactions including (n,2n) and (n,3n) reactions, are not present in this benchmark. The expressions in Eqs. (2.9) and (2.10) indicate that the 0th-order (i.e., $l=0$) scattering cross-sections must be separately considered from the higher order (i.e., $l\geq 1$) scattering cross-sections, since the $l=0$ term contributes to the total cross-sections, while the $l\geq 1$ terms do not contribute.

6. PARTISN (Alcouffe et al., 2008) computes the quantity $\left(\nu\Sigma_f\right)^g$ using the quantities $\left(\nu\sigma\right)_{f,i}^g$, which are provided in data files for each isotope i and energy group g, as follows:

$$\left(\nu\Sigma_f\right)^g = \sum_{m=1}^{M=2}\left(\nu\Sigma_f\right)_m^g;\; \left(\nu\Sigma_f\right)_m^g = \sum_{i=1}^{I=6}N_{i,m}\left(\nu\sigma_f\right)_i^g,\; m=1,2. \tag{2.11}$$

For the purposes of sensitivity analysis, the quantity ν_i^g, which denotes the number of neutrons that are produced per fission by isotope i and energy group g, can be obtained by using the relation $\nu_i^g=\left(\nu\sigma_f\right)_i^g/\sigma_{f,i}^g$, where the isotopic fission cross-sections $\sigma_{f,i}^g$ are available in data files for computing reaction rates.

7. The quantity χ^g denotes the fission spectrum in the energy group g; χ^g is defined in PARTISN (Alcouffe et al., 2008) as a space-independent quantity, as follows:

$$\chi^g \triangleq \frac{\sum_{i=1}^{N_f} \chi_i^g N_{i,m} \sum_{g'=1}^{G} \left(\nu\sigma_f\right)_i^{g'} f_i^{g'}}{\sum_{i=1}^{N_f} N_{i,m} \sum_{g'=1}^{G} \left(\nu\sigma_f\right)_i^{g'} f_i^{g'}}, \quad with \sum_{g=1}^{G} \chi_i^g = 1, \tag{2.12}$$

where χ_i^g denotes the isotopic fission spectrum in group g, while f_i^g denotes the corresponding spectrum weighting function.

8. The vector $\boldsymbol{\alpha}$, which appears in the argument of the Boltzmann operator $B^g(\boldsymbol{\alpha})$, represents the "vector of imprecisely known model parameters," comprising 21,976 components, which are presented in Table 2.3.

In view of Eq. (2.10), the total cross-section $\Sigma_t^g \rightarrow \Sigma_t^g(\mathbf{t})$ is characterized by the vector of parameters $\mathbf{t}$, which is defined, as follows:

$$\mathbf{t} \triangleq \left[t_1,\ldots,t_{J_t}\right]^\dagger \triangleq \left[t_1,\ldots,t_{J_{\sigma t}};n_1,\ldots,n_{J_n}\right]^\dagger \triangleq \left[\boldsymbol{\sigma}_t;\mathbf{N}\right]^\dagger, \; J_t \triangleq J_{\sigma t} + J_n, \tag{2.13}$$

where:

$$\mathbf{N} \triangleq \left[n_1,\ldots,n_{J_n}\right]^\dagger \triangleq \left[N_{1,1},N_{2,1},N_{3,1},N_{4,1},N_{5,2},N_{6,2}\right]^\dagger, \; J_n = 6, \tag{2.14}$$

$$\begin{aligned} &\boldsymbol{\sigma}_t \triangleq \left[t_1,\ldots,t_{J_{\sigma t}}\right]^\dagger \triangleq \left[\sigma_{t,i=1}^1,\sigma_{t,i=1}^2,\ldots,\sigma_{t,i=1}^G,\ldots,\sigma_{t,i}^g,\ldots,\sigma_{t,i=I}^1,\ldots,\sigma_{t,i=I}^G\right]^\dagger, \\ &i = 1,\ldots,I; \; g = 1,\ldots,G; \; J_{\sigma t} = I \times G. \end{aligned} \tag{2.15}$$

In Eqs. (2.13)–(2.15), the dagger denotes "transposition," $\sigma_{t,i}^g$ denotes the microscopic total cross-section for isotope i and energy group g, $N_{i,m}$ denotes the respective isotopic number density, and J_n denotes the total number of isotopic number densities in the model. Thus, the vector $\mathbf{t}$ comprises a total of $J_t = 30 \times 6 + 6 = 186$ imprecisely known components ("model parameters").

In view of Eq. (2.9), the scattering cross-section $\Sigma_s^{g' \rightarrow g}(\boldsymbol{\Omega}' \bullet \boldsymbol{\Omega}) \rightarrow \Sigma_s^{g' \rightarrow g}(\mathbf{s};\boldsymbol{\Omega}' \bullet \boldsymbol{\Omega})$ is characterized by the vector of parameters $\mathbf{s}$ which is defined as follows:

$$\mathbf{s} \triangleq \left[s_1,\ldots,s_{J_s}\right]^\dagger \triangleq \left[s_1,\ldots,s_{J_{\sigma s}};n_1,\ldots,n_{J_n}\right]^\dagger \triangleq \left[\boldsymbol{\sigma}_s;\mathbf{N}\right]^\dagger, \; J_s \triangleq J_{\sigma s} + J_n, \tag{2.16}$$

where:

$$\begin{aligned} &\boldsymbol{\sigma}_s \triangleq \left[s_1,\ldots,s_{J_{\sigma s}}\right]^\dagger \triangleq \left[\sigma_{s,l=0,i=1}^{g'=1\rightarrow g=1},\sigma_{s,l=0,i=1}^{g'=2\rightarrow g=1},\ldots,\sigma_{s,l=0,i=1}^{g'=G\rightarrow g=1},\sigma_{s,l=0,i=1}^{g'=1\rightarrow g=2},\sigma_{s,l=0,i=1}^{g'=2\rightarrow g=2},\right. \\ &\left.\ldots,\sigma_{s,l,i}^{g'\rightarrow g},\ldots,\sigma_{s,ISCT,i=I}^{G\rightarrow G}\right]^\dagger, \; l = 0,\ldots,ISCT; \; i = 1,\ldots,I; \; g,g' = 1,\ldots,G; \\ &J_{\sigma s} = (G \times G) \times I \times (ISCT + 1). \end{aligned} \tag{2.17}$$

In view of Eq. (2.11), the quantity $(\nu\Sigma_f)^g \to [\nu\Sigma_f(\mathbf{f})]^g$ in the fission integral $\int_{4\pi} (\nu\Sigma)_f^{g'} \varphi^{g'}(r,\mathbf{\Omega}')d\mathbf{\Omega}'$ depends on the vector of parameters $\mathbf{f}$ which are defined as follows:

$$\begin{aligned}\mathbf{f} &\triangleq \left[f_1,...,f_{J_{\sigma f}};f_{J_{\sigma f}+1},...,f_{J_{\sigma f}+J_\nu};f_{J_{\sigma f}+J_\nu+1},...,f_{J_f}\right]^\dagger \\ &\triangleq \left[\boldsymbol{\sigma}_f;\boldsymbol{\nu};\mathbf{N}\right]^\dagger,\ J_f = J_{\sigma f}+J_\nu+J_n,\end{aligned} \tag{2.18}$$

where:

$$\begin{aligned}\boldsymbol{\sigma}_f &\triangleq \left[\sigma_{f,i=1}^1,\sigma_{f,i=1}^2,...,\sigma_{f,i=1}^G,...,\sigma_{f,i}^g,...,\sigma_{f,i=N_f}^1,...,\sigma_{f,i=N_f}^G\right]^\dagger \\ &\triangleq \left[f_1,...,f_{J_{\sigma f}}\right]^\dagger,\ i=1,...,N_f;\ g=1,...,G;\ J_{\sigma f}=G\times N_f,\end{aligned} \tag{2.19}$$

$$\begin{aligned}\boldsymbol{\nu} &\triangleq \left[\nu_{i=1}^1,\nu_{i=1}^2,...,\nu_{i=1}^G,...,\nu_i^g,...,\nu_{i=N_f}^1,...,\nu_{i=N_f}^G\right]^\dagger \\ &\triangleq \left[f_{J_{\sigma f}+1},...,f_{J_{\sigma f}+J_\nu}\right]^\dagger,\ i=1,...,N_f;\ g=1,...,G;\ J_\nu=G\times N_f,\end{aligned} \tag{2.20}$$

and where $\sigma_{f,i}^g$ denotes the microscopic fission cross-section for isotope i and energy group g, ν_i^g denotes the average number of neutrons per fission for isotope i and energy group g, and N_f denotes the total number of fissionable isotopes.

The fission spectrum is considered to depend on the vector of parameters $\mathbf{p}$ which is defined as follows:

$$\begin{aligned}&\mathbf{p} \triangleq \left[p_1,...,p_{J_p}\right]^\dagger \triangleq \left[\chi_{i=1}^{g=1},\chi_{i=1}^{g=2},...,\chi_{i=1}^G,...,\chi_i^g,...,\chi_{N_f}^G\right]^\dagger, \\ &i=1,...,N_f;\ g=1,...,G;\ J_p=G\times N_f.\end{aligned} \tag{2.21}$$

In view of Eq. (2.12), the quantities χ^g further depend on the parameters χ_i^g, $N_{i,m}$, f_i^g, $(\nu\sigma_f)_i^g$, but these latter dependences can be taken into account by applying the chain rule to the 1st-order sensitivities $\partial L/\partial\chi^g$, once these sensitivities have been obtained.

In view of Eq. (2.6), the source $Q^g(r) \to Q^g(\mathbf{q};\mathbf{N})$ depends on the vector of model parameters $\mathbf{q}$ which are defined as follows:

$$\mathbf{q} \triangleq \left[q_1,...,q_{J_q}\right]^\dagger \triangleq \left[\lambda_1,\lambda_2;F_1^{SF},F_2^{SF};a_1,a_2;b_1,b_2;\nu_1^{SF},\nu_2^{SF}\right]^\dagger,\ J_q=10. \tag{2.22}$$

In view of Eqs. (2.13)–(2.22), the model parameters characterizing the PERP benchmark can all be considered to be the components of the "vector of model parameters" $\boldsymbol{\alpha}$ which are defined here:

$$\begin{aligned}&\boldsymbol{\alpha} \triangleq \left[\alpha_1,...,\alpha_{J_\alpha}\right]^\dagger \triangleq \left[\boldsymbol{\sigma}_t;\boldsymbol{\sigma}_s;\boldsymbol{\sigma}_f;\boldsymbol{\nu};\mathbf{p};\mathbf{q};\mathbf{N}\right]^\dagger, \\ &J_\alpha = J_{\sigma t}+J_{\sigma s}+J_{\sigma f}+J_\nu+J_p+J_q+J_n.\end{aligned} \tag{2.23}$$

TABLE 2.3
Summary of Imprecisely Known Parameters for the PERP Benchmark

Symbol	Parameter Name	Number of Parameters
$\sigma_{t,i}^{g}$	Multigroup microscopic total cross-section for isotope i and energy group g	180; $for\ i=1,...,6;\ g=1,...,30$
$\sigma_{s,l,i}^{g'\to g}$	Multigroup microscopic scattering cross-section for l-th order Legendre expansion, from energy group g' into energy group g for isotope i	21,600; $for\ l=0,...,3; i=1,...,6;$ $g,g'=1,...,30$
$\sigma_{f,i}^{g}$	Multigroup microscopic fission cross-section i and energy group g	60; $for\ i=1,2;\ g=1,...,30$
ν_i^g	Average number of neutrons per fission for isotope i and energy group g	60; $for\ i=1,2;\ g=1,...,30$
χ_i^g	Fission spectrum for isotope i and energy group g	60; $for\ i=1,2;\ g=1,...,30$
q_j	Source parameters $\lambda_1,\lambda_2;F_1^{SF},F_2^{SF};a_1,a_2;b_1,b_2;\nu_1^{SF},\nu_2^{SF}$	10
$N_{i,m}$	Isotopic number density for isotope $i$and material m	6; $N_{1,1},N_{2,1},N_{3,1},N_{4,1},N_{5,2},N_{6,2}$
J_α	Total number of parameters:	21,976

Thus, the PERP benchmark comprises a total of $J_\alpha=(I\times G)+(G\times G)\times I\times(ISCT+1)+2(G\times N_f)+G\times N_f+10+6=21{,}976.$ imprecisely known (i.e., uncertain) model parameters, as summarized in Table 2.3.

Although the numerical model of the PERP benchmark comprises 21,976 uncertain parameters, only 7,477 parameters have non-zero nominal values, as follows: 180 group-averaged total microscopic cross-sections, 7,101 non-zero group-averaged scattering microscopic cross-sections (the other scattering cross-sections, of which there are 21,600 in total, are zero due to scattering laws of neutrons with matter); 120 fission process parameters; 60 fission spectrum parameters; ten parameters describing the experiment's nuclear sources; and six isotopic number densities.

The fundamental quantities of interest (i.e., "system responses") for *subcritical* benchmarks (such as the PERP benchmark) are single-counting rates, double-counting rates, the leakage multiplication, and the total leakage. Many systems are characterized for practical applications by their total leakage rather than by the count rate that a particular detector would see at a particular distance. This is because the total leakage does not depend on the detector configuration (as opposed to count rates) and is, therefore, more meaningful physically than the count rates. For this reason, the *total leakage* of neutrons out of the PERP benchmark will be considered to be the paradigm *model response* of interest for sensitivity analysis, uncertainty quantification, and predictive modeling.

Mathematically, the total neutron leakage from the PERP sphere, which is denoted as $L(\alpha)$, will depend on all model parameters (indirectly, through the neutron flux) and is defined as follows:

$$L(\boldsymbol{\alpha}) \triangleq \int_V dV \int_0^\infty dE \int_{\boldsymbol{\Omega}\cdot\mathbf{n}>0} d\boldsymbol{\Omega}\,\boldsymbol{\Omega}\cdot\mathbf{n}\,\delta(r-r_s)\,\varphi(r,E,\boldsymbol{\Omega})$$
$$= \int_{S_b} dS \sum_{g=1}^{G} \int_{\boldsymbol{\Omega}\cdot\mathbf{n}>0} d\boldsymbol{\Omega}\,\boldsymbol{\Omega}\cdot\mathbf{n}\,\varphi^g(r,\boldsymbol{\Omega}). \tag{2.24}$$

The nominal value of total leakage, computed by using Eq. (2.24) at the nominal parameter values (which are denoted using the usual notation $\boldsymbol{\alpha}^0$) is $L(\boldsymbol{\alpha}^0) = 1.7648 \times 10^6$ neutrons/sec. Figure 2.1, depicts the histogram plot of the leakage for each energy group for the PERP benchmark.

The 1st-order through 4th-order sensitivities of the PERP leakage response, with respect to the PERP model parameters, were computed exactly and efficiently using the nth-CASAM-L (n^{th}-Order Comprehensive Adjoint Sensitivity Analysis Methodology for Linear Systems) conceived by Cacuci (2022), for n = 1,2,3,4. These sensitivities are conveniently summarized in the book by Cacuci and Fang (2023), where it is shown that a significant number of 2nd-order sensitivities are larger than the 1st-order sensitivities, several 3rd-order sensitivities are much larger than the 2nd-order ones, and several 4th-order sensitivities are much larger than the 3rd-order ones. The largest unmixed sensitivities occur for isotope 6 (^{1}H). Table 2.4, presents an illustrative comparison of the values of the unmixed relative sensitivities, from 1st-order through 4th-order, for isotope 6 (^{1}H). Sensitivities that have absolute values larger than unity are presented in bold characters.

As shown in Table 2.4, the largest absolute values for the 1st-, 2nd-, 3rd-, and 4th-order unmixed relative sensitivities all occur for the lowest energy group (g = 30; thermal neutrons), which are significantly larger than the values of the sensitivities in other energy groups. Notably, the largest 4th-order unmixed relative

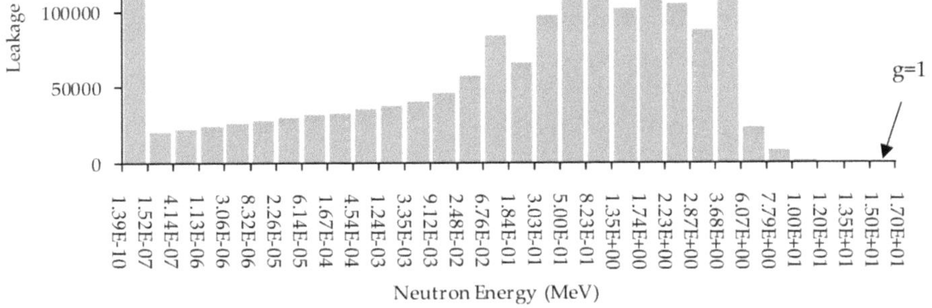

FIGURE 2.1 Histogram of neutron leakage for each energy group (PERP benchmark).

TABLE 2.4

Comparison of the Unmixed Relative Sensitivities, $S^{(1)}\left(\sigma_{t,6}^{g}\right)$, $S^{(2)}\left(\sigma_{t,6}^{g},\sigma_{t,6}^{g}\right)$, $S^{(3)}\left(\sigma_{t,6}^{g},\sigma_{t,6}^{g},\sigma_{t,6}^{g}\right)$, and $S^{(4)}\left(\sigma_{t,6}^{g},\sigma_{t,6}^{g},\sigma_{t,6}^{g},\sigma_{t,6}^{g}\right), g = 1,\ldots,30$, for Isotope 6 ($^{1}H$)

g	1st-order	2nd-order	3rd-order	4th-order
1	-8.471×10^{-6}	7.636×10^{-7}	6.322×10^{-8}	1.460×10^{-7}
2	-2.060×10^{-5}	2.280×10^{-6}	4.516×10^{-8}	4.956×10^{-7}
3	-6.810×10^{-5}	9.021×10^{-6}	-4.677×10^{-7}	2.245×10^{-6}
4	-3.932×10^{-4}	6.673×10^{-5}	-8.758×10^{-6}	2.039×10^{-5}
5	-2.449×10^{-3}	5.549×10^{-4}	-1.216×10^{-4}	2.142×10^{-4}
6	-9.342×10^{-3}	2.935×10^{-3}	-1.123×10^{-3}	1.553×10^{-3}
7	-7.589×10^{-2}	3.949×10^{-2}	-2.690×10^{-2}	3.513×10^{-2}
8	-9.115×10^{-2}	5.604×10^{-2}	-4.380×10^{-2}	5.536×10^{-2}
9	-1.358×10^{-1}	1.014×10^{-1}	-9.758×10^{-2}	1.416×10^{-1}
10	-1.659×10^{-1}	1.428×10^{-1}	-1.604×10^{-1}	2.582×10^{-1}
11	-1.899×10^{-1}	1.849×10^{-1}	-2.385×10^{-1}	4.233×10^{-1}
12	-4.446×10^{-1}	6.620×10^{-1}	$\mathbf{-1.373 \times 10^{0}}$	$\mathbf{3.815 \times 10^{0}}$
13	-5.266×10^{-1}	9.782×10^{-1}	$\mathbf{-2.590 \times 10^{0}}$	$\mathbf{9.015 \times 10^{0}}$
14	-5.772×10^{-1}	$\mathbf{1.262 \times 10^{0}}$	$\mathbf{-3.991 \times 10^{0}}$	$\mathbf{1.650 \times 10^{1}}$
15	-5.820×10^{-1}	$\mathbf{1.391 \times 10^{0}}$	$\mathbf{-4.581 \times 10^{0}}$	$\mathbf{2.208 \times 10^{1}}$
16	$\mathbf{-1.164 \times 10^{0}}$	$\mathbf{4.460 \times 10^{0}}$	$\mathbf{-2.530 \times 10^{1}}$	$\mathbf{1.890 \times 10^{2}}$
17	$\mathbf{-1.173 \times 10^{0}}$	$\mathbf{4.853 \times 10^{0}}$	$\mathbf{-2.991 \times 10^{1}}$	$\mathbf{2.432 \times 10^{2}}$
18	$\mathbf{-1.141 \times 10^{0}}$	$\mathbf{4.828 \times 10^{0}}$	$\mathbf{-3.049 \times 10^{1}}$	$\mathbf{2.543 \times 10^{2}}$
19	$\mathbf{-1.094 \times 10^{0}}$	$\mathbf{4.619 \times 10^{0}}$	$\mathbf{-2.913 \times 10^{1}}$	$\mathbf{2.428 \times 10^{2}}$
20	$\mathbf{-1.033 \times 10^{0}}$	$\mathbf{4.284 \times 10^{0}}$	$\mathbf{-2.655 \times 10^{1}}$	$\mathbf{2.175 \times 10^{2}}$
21	-9.692×10^{-1}	$\mathbf{3.937 \times 10^{0}}$	$\mathbf{-2.388 \times 10^{1}}$	$\mathbf{1.915 \times 10^{2}}$
22	-8.917×10^{-1}	$\mathbf{3.515 \times 10^{0}}$	$\mathbf{-2.069 \times 10^{1}}$	$\mathbf{1.609 \times 10^{2}}$
23	-8.262×10^{-1}	$\mathbf{3.177 \times 10^{0}}$	$\mathbf{-1.823 \times 10^{1}}$	$\mathbf{1.382 \times 10^{2}}$
24	-7.495×10^{-1}	$\mathbf{2.792 \times 10^{0}}$	$\mathbf{-1.552 \times 10^{1}}$	$\mathbf{1.140 \times 10^{2}}$
25	-7.087×10^{-1}	$\mathbf{2.604 \times 10^{0}}$	$\mathbf{-1.427 \times 10^{1}}$	$\mathbf{1.033 \times 10^{2}}$
26	-6.529×10^{-1}	$\mathbf{2.349 \times 10^{0}}$	$\mathbf{-1.260 \times 10^{1}}$	$\mathbf{8.932 \times 10^{1}}$
27	-5.845×10^{-1}	$\mathbf{2.039 \times 10^{0}}$	$\mathbf{-1.061 \times 10^{1}}$	$\mathbf{7.288 \times 10^{1}}$
28	-5.474×10^{-1}	$\mathbf{1.885 \times 10^{0}}$	$\mathbf{-9.678 \times 10^{0}}$	$\mathbf{6.565 \times 10^{1}}$
29	-5.439×10^{-1}	$\mathbf{1.891 \times 10^{0}}$	$\mathbf{-9.800 \times 10^{0}}$	$\mathbf{6.705 \times 10^{1}}$
30	$\mathbf{-9.366 \times 10^{0}}$	$\mathbf{4.296 \times 10^{2}}$	$\mathbf{-2.966 \times 10^{4}}$	$\mathbf{2.720 \times 10^{6}}$

sensitivity attains a very large value: $S^{(4)}\left(\sigma_{t,6}^{g=30},\sigma_{t,6}^{g=30},\sigma_{t,6}^{g=30},\sigma_{t,6}^{g=30}\right)=2.720\times10^{6}$. By comparison, the largest values for the 1st-, 2nd-, and 3rd-order unmixed relative sensitivities are: $S^{(1)}\left(\sigma_{t,6}^{g=30}\right)=-9.366$, $S^{(2)}\left(\sigma_{t,6}^{g=30},\sigma_{t,6}^{g=30}\right)=4.296\times10^{2}$, and $S^{(3)}\left(\sigma_{t,6}^{g=30},\sigma_{t,6}^{g=30},\sigma_{t,6}^{g=30}\right)=-2.966\times10^{4}$, respectively. The remainder of the parameters have very small sensitivities. Their combined impact on the expectation value and variance of the computed leakage response is of the order of 5%. Therefore, these parameters can be safely neglected, thus greatly reducing the number of computations which would otherwise be required, without affecting the conclusions that will be reached by considering just the microscopic total cross-sections. Hence, only the 180 group-averaged microscopic total cross-sections will be considered as the benchmark's imprecisely known parameters for the illustrative application of the 2nd-BERRU-PM methodology to be presented in the remainder of this chapter. The mathematical expressions for computing the 1st-, 2nd-, 3rd-, and 4th-order sensitivities of the PERP benchmark's leakage response with respect to these 180 total cross-sections are presented in Chapter 7, for convenient referencing.

The value of Keff computed using the PARTISN model for the PERP benchmark is 0.8986. The ratio of neutrons born from spontaneous fission versus neutron-induced fission is 0.113. This number is obtained from the PARTISN results for the PERP benchmark, where the neutrons born from spontaneous fission source is 2.7839×10^{5} n/s, and the neutron-induced fission is 2.4584×10^{6} n/s. As has been mentioned in Table 2.1, the largest absolute values for the 1st-, 2nd-, 3rd-, and 4th-order unmixed relative sensitivities all occur for the lowest energy group ($g = 30$; thermal neutrons), for isotope 6 (^{1}H), which is bound in the polyethylene that constitutes the outer shell ("material 2") of the PERP Benchmark. It is, therefore, of interest to present here the basic expressions that underly the computations of the thermal cross-sections of polyethylene. As has also been mentioned in the description of the computational model of the PERP benchmark, the PARTISN (Alcouffe et al., 2008) neutron transport equation solver uses the (collapsed) MENDF71X library (Conlin et al., 2013), which was created at Los Alamos National Laboratory (LANL) by processing ENDF/B-VII.1 nuclear data (Chadwick et al., 2011) using the LANL-code NJOY (MacFarlane et al., 2017). Within NJOY, the thermal cross-sections for heterogeneous materials like polyethylene are processed using the THERMR module, which models the following scattering processes for partially ordered systems such as polyethylene:

1. *Thermal inelastic scattering*, using the following expression for the respective differential cross-section: $\sigma\left(E\rightarrow E',\mu\right)=\sigma_{b}/4\pi T\sqrt{E'/E}e^{-\beta/2}S\left(\alpha,\beta\right)$, $\alpha\triangleq\left(E'+E-2\mu\sqrt{E'E}\right)/\left(AkT\right)\beta=\left(E'-E\right)/\left(kT\right)$, where E' and E are the incident and outgoing neutron energies, μ is the scattering cosine, T is the absolute temperature, A is the mass ratio to the neutron of the scatterer, k is Boltzmann's constant, $S\left(\alpha,\beta\right)$ is the evaluated scattering law data from an ENDF-format, and σ_{b} is the characteristic bound cross-section.

2. *Incoherent elastic scattering*, using the following expression for the respective differential cross-section: $\sigma(E,\mu)=\frac{\sigma_b}{2}\exp\left[-2EW(1-\mu)/A\right]$, where W is the Debye-Waller integral.

2.3 MEAN AND VARIANCE OF THE PERP BENCHMARK'S COMPUTED LEAKAGE RESPONSE

The correlations for the 180 total cross-sections used in the computational modeling of the PERP benchmark are not available. Therefore, these cross-sections will be considered to be *uncorrelated* and *normally* distributed, since such a distribution would be the least informative (i.e., least biased) according to the MaxEnt principle (Jaynes, 1957). Under these circumstances, the expected (mean) value of the leakage computed response, which will be denoted as $E(L^c)$, will take on the following particular form of Eq. (1.128) of Chapter 1:

$$E(L^c)=L^c\left(\boldsymbol{\alpha}^0\right)+\frac{1}{2}\sum_{i=1}^{TP}\frac{\partial^2 L\left(\boldsymbol{\alpha}^0\right)}{\left(\partial\alpha_i\right)^2}\sigma_i^2+\frac{1}{8}\sum_{i=1}^{TP}\frac{\partial^4 L\left(\boldsymbol{\alpha}^0\right)}{\left(\partial\alpha_i\right)^4}\sigma_i^4 \tag{2.25}$$

The following notations were used in Eq. (2.25): (i) the value of the computed leakage response at the nominal parameter values is denoted as $L^c\left(\boldsymbol{\alpha}^0\right)=1.7648\times10^6$ neutrons/sec; (ii) the "total number" of microscopic total cross-sections is denoted as $TP=180$; (iii) the "i^{th}" total microscopic cross-section is denoted as α_i; (iv) the standard deviation of the "i^{th}" total microscopic cross-section is denoted as σ_i.

For the 180 uncorrelated and normally distributed total microscopic cross-sections, the variance of the computed leakage response, L^c, will be denoted as "$\mathrm{var}(L^c)$" and will take on the following simplified form of the corresponding expression provided in Eq. (1. 133) of Chapter 1:

$$\begin{aligned}\mathrm{var}(L^c)=&\sum_{i=1}^{TP}\left[\frac{\partial L\left(\boldsymbol{\alpha}^0\right)}{\partial\alpha_i}\right]^2\sigma_i^2+\frac{1}{2}\sum_{i=1}^{TP}\left[\frac{\partial^2 L\left(\boldsymbol{\alpha}^0\right)}{\left(\partial\alpha_i\right)^2}\right]^2\sigma_i^4\\&+\sum_{i=1}^{TP}\left[\frac{\partial^3 L\left(\boldsymbol{\alpha}^0\right)}{\left(\partial\alpha_i\right)^3}\times\frac{\partial L\left(\boldsymbol{\alpha}^0\right)}{\partial\alpha_i}\right]\sigma_i^4\\&+\frac{15}{36}\sum_{i=1}^{TP}\left[\frac{\partial^3 L\left(\boldsymbol{\alpha}^0\right)}{\left(\partial\alpha_i\right)^3}\right]^2\sigma_i^6\\&+\frac{1}{2}\sum_{i=1}^{TP}\left[\frac{\partial^4 L\left(\boldsymbol{\alpha}^0\right)}{\left(\partial\alpha_i\right)^4}\times\frac{\partial^2 L\left(\boldsymbol{\alpha}^0\right)}{\left(\partial\alpha_i\right)^2}\right]\sigma_i^6.\end{aligned} \tag{2.26}$$

The expressions provided in Eqs. (2.25) and (2.26) are valid only if the Taylor series expansion of the leakage response in terms of the cross-section variations converges.

The largest admissible cross-section variations $\delta\alpha_j$, which are still within the radius of convergence of this Taylor series provide the largest parameter covariances/standard deviations which can be considered to ensure that the expressions provided in Eqs. (2.25) and (2.26) are valid.

The effects of various values for the standard deviations for the PERP parameters (i.e., the 180 group-averaged microscopic total cross-sections), which are considered to be uncorrelated and normally distributed, will be illustrated in Subsections 2.3.1–2.3.3 for parameters that are known with "high precision" (having uniform standard deviations of 2%); parameters known with "typical precision" (having uniform standard deviations of 5%); and parameters known with "low precision" (having uniform standard deviations of 10%), respectively.

2.3.1 "High Precision" Parameters; Uniform Relative Standard Deviations *SD* = 2%

As has been discussed by Cacuci and Fang (2023), when the uncorrelated and normally distributed parameters are assumed to have uniform relative standard deviations of 3%, the convergence "ratio test" of the 3rd-order term with respect to the 2nd-order term of the Taylor series (of the response in terms of parameter variations) is 0.58, while the ratio of the 4th-order term with respect to the 3rd-order term of this Taylor series is 0.68. Both of these results are below 1.00, which implies that for uniform relative standard deviations of 3% or less, the Taylor series expansion which underlies the computation of the expressions shown in Eqs. (2.25) and (2.26) are expected to be convergent. Figure 2.2 depicts the magnitudes of the higher-order contributions to the expected value, $E(L^c)$, of the computed leakage response, while Figure 2.3 depicts the magnitudes of the higher-order contributions to the variance,

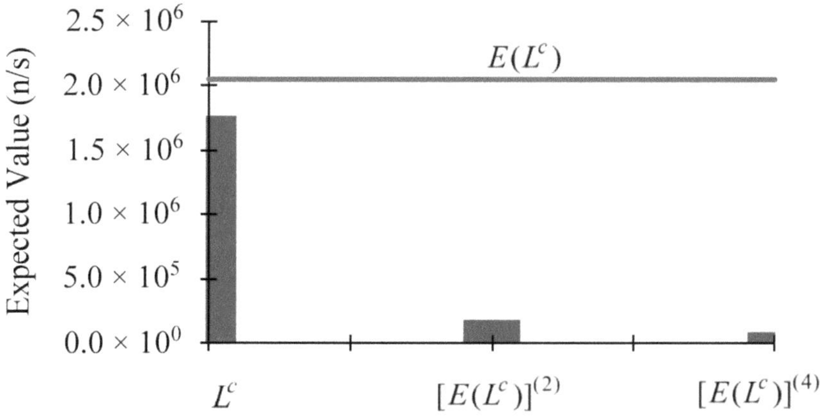

FIGURE 2.2 Contributions to the expected value, $E(L^c)$, of the computed leakage response from parameters having uniform relative standard deviations $SD = 2\%$: (i) 0th-order: L^c; (ii) 2nd-order: $[E(L^c)]^{(2)}$; (iii) 4th-order: $[E(L^c)]^{(4)}$; (iv) the odd-order contributions are null.

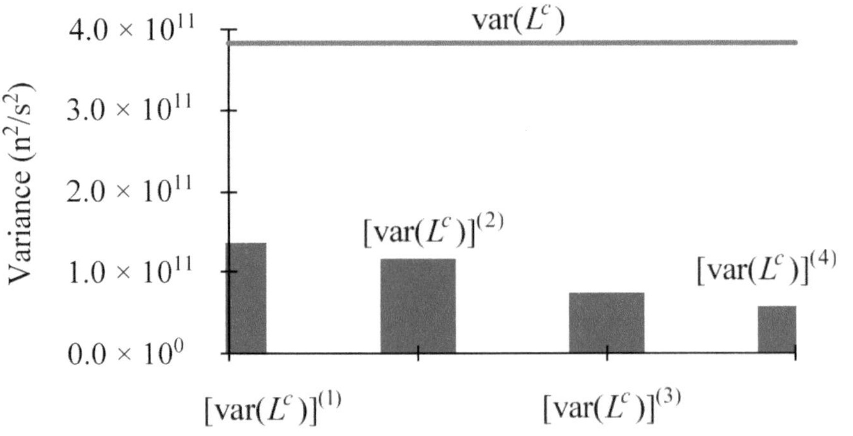

FIGURE 2.3 Contributions to the variance, var(L^c), of the computed leakage response from parameters having uniform relative standard deviations $SD = 2\%$: (i) 1st-order: $[\text{var}(L^c)]^{(1)}$; (ii) 2nd-order: $[\text{var}(L^c)]^{(2)}$; (iii) 3rd-order: $[\text{var}(L^c)]^{(3)}$; (iv) 4th-order: $[\text{var}(L^c)]^{(4)}$.

var(L^c), of the computed leakage response, for parameters that are assumed to have uniform relative standard deviations of 2% (which is smaller than 3%). The numerical results depicted in Figures 2.2 and 2.3 indicate that the contributions of the increasingly higher-order terms to the expected value, $E(L^c)$, and the variance, var(L^c), of the computed leakage response become increasingly smaller (as the order of the respective terms increases), thus confirming the expectation that the underlying Taylor series is convergent. For practical purposes, therefore, the contributions from terms involving sensitivities of order five and higher become negligible by comparison to the contributions from the terms comprising the sensitivities of 1st- through 4rth-order to $E(L^c)$ and var(L^c), respectively, when the uncorrelated and normally-distributed parameters have uniform relative standard deviations $SD = 2\%$.

2.3.2 "Typical Precision" Parameters; Uniform Relative Standard Deviations $SD = 5\%$

For uniform relative standard deviations of 5% for the uncorrelated and normally distributed model parameters, it has been shown by Cacuci and Fang (2023) that the ratio of the 3rd-order term with respect to the 2nd-order term of the Taylor series (of the response in terms of parameter variations) is $0.97 < 1.00$, but the ratio of the 4th-order term with respect to the 3rd-order term of this Taylor series is $1.13 > 1.00$. These ratios indicate that relative standard deviations of 5% for the model parameters are outside of the radius of convergence of the Taylor series which underlies the computation of the expressions shown in Eqs. (2.25) and (2.26). These indications are confirmed by the results depicted in Figures 2.4 and 2.5.

The results depicted in Figure 2.4 indicate that the contributions to $E(L^c)$ stemming from 2nd-order sensitivities are smaller than those stemming from the

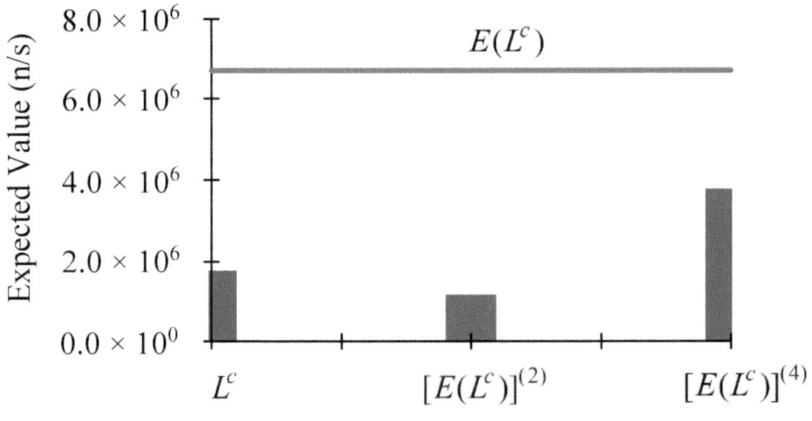

FIGURE 2.4 Contributions to the expected value, $E(L^c)$, of the computed leakage response from parameters having uniform relative standard deviations $SD = 5\%$: (i) zero-order: L^c; (ii) 2nd-order: $[E(L^c)]^{(2)}$; (iii) 4th-order: $[E(L^c)]^{(4)}$; (iv) the odd-order contributions are null.

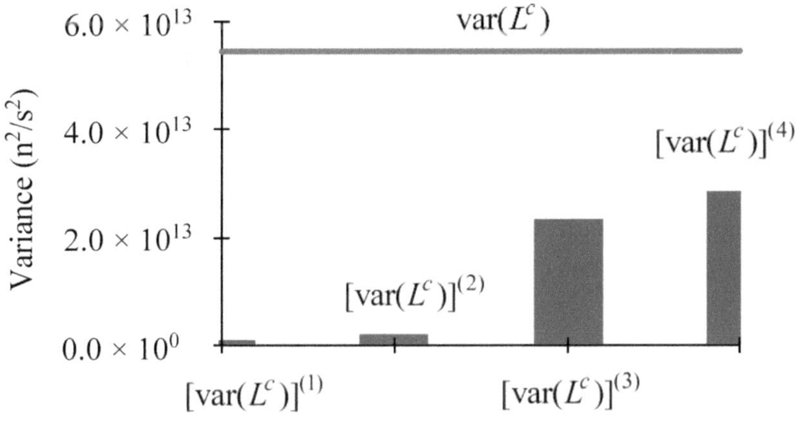

FIGURE 2.5 Contributions to the variance, $\text{var}(L^c)$, of the computed leakage response from parameters having uniform relative standard deviations $SD = 5\%$: (i) 1st-order: $[\text{var}(L^c)]^{(1)}$; (ii) 2nd-order: $[\text{var}(L^c)]^{(2)}$; (iii) 3rd-order: $[\text{var}(L^c)]^{(3)}$; (iv) 4th-order: $[\text{var}(L^c)]^{(4)}$.

zero-order term, L^c, but the contributions to $E(L^c)$ stemming from 4th-order sensitivities are larger than those stemming from the zero- and 2nd-order terms. This oscillatory behavior with increasing amplitudes is indicative of the divergence of the Taylor series underlying the computation of the expected value, $E(L^c)$.

The results depicted in Figure 2.5 for the variance, $\text{var}(L^c)$, of the computed leakage response indicate that the contributions to $\text{var}(L^c)$ increase as the order of the contributing terms increases, thus underscoring the divergent nature of the underlying Taylor series when the parameters have uniform relative standard deviations of

5%. On the other hand, the relative standard deviation of 5% is often encountered in measurements of total cross-sections, which highlights the need for computing the 2nd- and higher-order response sensitivities in order to investigate the convergence properties of the Taylor series that underlies the determination of the statistics (expected values, variance, etc.) of the distribution of computed responses in the phase-space of imprecisely known model parameters.

2.3.3 "Low Precision" Parameters; Uniform Relative Standard Deviations $SD = 10\%$

When considering uniform relative standard deviations of 10% for the uncorrelated and normally distributed parameters (total cross-sections), it has been shown by Cacuci and Fang (2023) that the ratio of the 3rd-order term with respect to the 2nd-order term of the Taylor series (of the response in terms of parameter variations) is 1.93; the ratio of the 4th-order term with respect to the 3rd-order term of this Taylor series is 2.26. Both of these results are larger than 1.00, indicating that the Taylor series which underlies the computation of the expressions shown in Eqs. (2.25) and (2.26) would be divergent if used for parameters having standard deviations of 10%. The divergence of the Taylor series for such parameter standard deviations is underscored by the corresponding results depicted in Figure 2.6 for the expected value, $E(L^c)$, of the computed leakage response, which clearly indicates the massive increase of the contributions to $E(L^c)$ as the order of the retained terms increases. The conclusion that the Taylor series expansion is divergent and should therefore not be used for parameters with uniform relative standard deviations $SD = 10\%$ is reinforced by the corresponding results depicted in Figure 2.7 for the variance, $\text{var}(L^c)$, of the computed leakage response. Figure 2.7 also highlights that the contributions to $\text{var}(L^c)$ increase massively as the order of the retained terms increases.

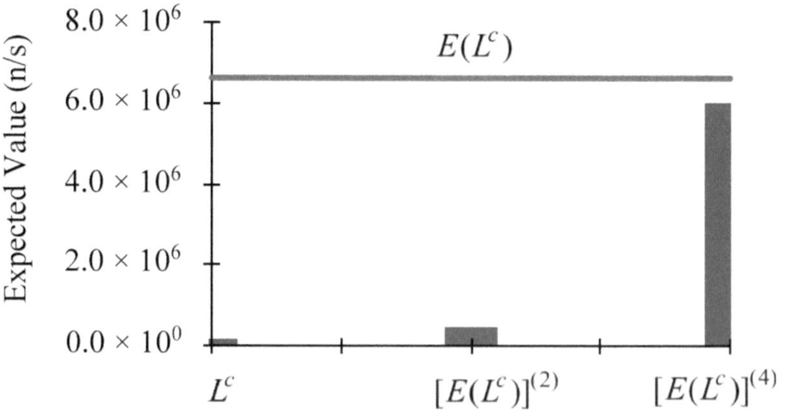

FIGURE 2.6 Contributions to the expected value, $E(L^c)$, of the computed leakage response from parameters having uniform relative standard deviations $SD = 10\%$: (i) zero-order: L^c; (ii) 2nd-order: $[E(L^c)]^{(2)}$; (iii) 4th-order: $[E(L^c)]^{(4)}$.

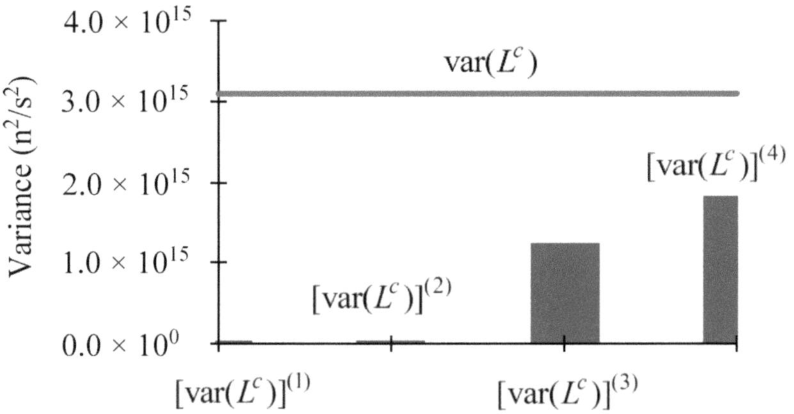

FIGURE 2.7 Contributions to the variance, $\mathrm{var}(L^c)$, of the computed leakage response from parameters having uniform relative standard deviations $SD = 10\%$: (i) 1st-order: $[\mathrm{var}(L^c)]^{(1)}$; (ii) 2nd-order: $[\mathrm{var}(L^c)]^{(2)}$; (iii) 3rd-order: $[\mathrm{var}(L^c)]^{(3)}$; (iv) 4th-order: $[\mathrm{var}(L^c)]^{(4)}$.

2.4 ILLUSTRATIVE APPLICATION OF THE 2ND-BERRU-PM METHODOLOGY TO THE PERP BENCHMARK: MATHEMATICAL EXPRESSIONS FOR THE BEST-ESTIMATE PREDICTED MEAN AND VARIANCE FOR THE PERP LEAKAGE RESPONSE

It is considered that multiple measurements of the leakage response, if available, have been averaged correspondingly so that the 2nd-BERRU-PM methodology will incorporate the moments of a single response, with known measured nominal value, denoted as L^e, and a known standard deviation, denoted as $SD^{(e)}$. The expressions for the best-estimate predicted mean value and predicted standard deviation for the PERP leakage response will take on the simplified expressions obtained from Chapter 1, Eqs. (1.155) and (1.157), respectively, are presented in Subsections 2.4.1 and 2.4.2.

2.4.1 Best-Estimate Predicted Mean Value, $L^{(be,i)}$, $i = 1,2,3,4$, for the PERP Leakage Response

Generically, the particular expression of the best-estimate predicted mean value, L^{be}, for the PERP benchmark takes on the following simplified form of Eq. (1.155), from Chapter 1:

$$L^{be} = L^e + \mathrm{var}(L^e)\frac{E(L^c) - L^e}{\mathrm{var}(L^e) + \mathrm{var}(L^c)} = \frac{L^e\,\mathrm{var}(L^c) + E(L^c)\,\mathrm{var}(L^e)}{\mathrm{var}(L^e) + \mathrm{var}(L^c)}, \quad (2.27)$$

where: L^e denotes the experimentally measured mean value of the leakage response; $\mathrm{var}(L^e)$ denotes the experimentally measured variance of the leakage response;

$E(L^c)$ denotes the computed mean value of the leakage response; and $\mathrm{var}(L^c)$ denotes the computed variance of the leakage response. The relation provided in Eq. (2.27) implies the following inequalities:

$$\mathit{If} \quad 0 < L^e < E(L^c), \quad \mathit{then} \;\; L^e < L^{be} < E(L^c), \tag{2.28}$$

$$\mathit{If} \quad 0 < E(L^c) < L^e, \quad \mathit{then} \;\; E(L^c) < L^{be} < L^e. \tag{2.29}$$

The inequalities shown in Eqs. (2.28) and (2.29) indicate that the best-estimate predicted value L^{be} of the leakage response will fall in between the initially measured value, L^e, and the computed value, $E(L^c)$, of the leakage response. Using the index $i = 1,2,3,4$, to denote the highest order of the retained sensitivities in Eq. (2.27), the expression of the best-estimate predicted mean value, $L^{(be,i)}$, will have the following form:

$$L^{(be,i)} = L^e + \left[SD^{(e)}\right]^2 \left\{\left[SD^{(e)}\right]^2 + \left[\mathrm{var}(L)\right]^{(i,U,N)}\right\}^{-1} \left[E^{(i)}(L^c) - L^e\right], \; i = 1,2,3,4, \tag{2.30}$$

where the superscript "U" denotes "uncorrelated parameters," the superscript "N" denotes "normally distributed parameters" and where the following definitions are used:

1. Only 1st-order sensitivities are retained $(i = 1)$:

$$\left[\mathrm{var}(L)\right]^{(1,U,N)} \triangleq \sum_{i=1}^{TP} \left[\frac{\partial L(\boldsymbol{\alpha})}{\partial \alpha_i}\right]^2 \sigma_i^2; \;\; SD^{(1)} \triangleq \sqrt{\left[\mathrm{var}(L)\right]^{(1,U,N)}}; \tag{2.31}$$

$$E^{(1)}(L^c) = L(\boldsymbol{\alpha}^0). \tag{2.32}$$

2. Ther 1st- and 2nd-order sensitivities are retained $(i = 2)$:

$$\begin{aligned} &\left[\mathrm{var}(L)\right]^{(2,U,N)} \triangleq \left[\mathrm{var}(L)\right]^{(1,U,N)} + \frac{1}{2}\sum_{i=1}^{TP} \left[\frac{\partial^2 L(\boldsymbol{\alpha})}{\partial \alpha_i \partial \alpha_i}\right]^2 \sigma_i^4; \\ &SD^{(2)} \triangleq \sqrt{\left[\mathrm{var}(L)\right]^{(2,U,N)}}; \end{aligned} \tag{2.33}$$

$$E^{(2)}(L^c) = L(\boldsymbol{\alpha}^0) + \frac{1}{2}\sum_{i=1}^{TP} \frac{\partial^2 L(\boldsymbol{\alpha})}{(\partial \alpha_i)^2} \sigma_i^2. \tag{2.34}$$

3. The 1st-, 2nd-, and 3rd-order sensitivities are retained $(i = 3)$:

$$\begin{aligned} &\left[\mathrm{var}(L)\right]^{(3,U,N)} \triangleq \left[\mathrm{var}(L)\right]^{(2,U,N)} + \sum_{i=1}^{TP} \left[\frac{\partial^3 L(\boldsymbol{\alpha})}{(\partial \alpha_i)^3} \frac{\partial L(\boldsymbol{\alpha})}{\partial \alpha_i}\right] \sigma_i^4 \\ &+ \frac{15}{36}\sum_{i=1}^{TP} \left[\frac{\partial^3 L(\boldsymbol{\alpha})}{(\partial \alpha_i)^3_i}\right]^2 \sigma_i^6; \;\; SD^{(3)} \triangleq \sqrt{\left[\mathrm{var}(L)\right]^{(3,U,N)}}; \end{aligned} \tag{2.35}$$

$$E^{(3)}(L^c) = L(\boldsymbol{\alpha}^0) + \frac{1}{2}\sum_{i=1}^{TP} \frac{\partial^2 L(\boldsymbol{\alpha})}{(\partial \alpha_i)^2} \sigma_i^2 . \tag{2.36}$$

4. The 1st-, 2nd-, 3rd-, and 4th-order sensitivities are retained $(i = 4)$:

$$\left[\text{var}(L)\right]^{(4,U,N)} \triangleq \left[\text{var}(L)\right]^{(3,U,N)} + \frac{1}{2}\sum_{i=1}^{TP}\left[\frac{\partial^4 L(\boldsymbol{\alpha})}{(\partial \alpha_i)^4}\frac{\partial^2 L(\boldsymbol{\alpha})}{(\partial \alpha_i)^2}\right]\sigma_i^6; \qquad SD^{(4)} \triangleq \sqrt{\left[\text{var}(L)\right]^{(4,U,N)}}; \tag{2.37}$$

$$E^{(4)}(L^c) = L(\boldsymbol{\alpha}^0) + \frac{1}{2}\sum_{i=1}^{TP} \frac{\partial^2 L(\boldsymbol{\alpha})}{(\partial \alpha_i)^2} \sigma_i^2 + \frac{1}{8}\sum_{i=1}^{TP} \frac{\partial^4 L(\boldsymbol{\alpha})}{(\partial \alpha_i)^4} \sigma_i^4 . \tag{2.38}$$

2.4.2 Best-Estimate Predicted Standard Deviation for PERP Leakage Response

Generically, the particular expression of the predicted variance, $\text{var}(L^{be})$, of the best-estimate (leakage) response L^{be} for the PERP benchmark takes on the following simplified form of Eq. (1.157), from Chapter 1:

$$\begin{aligned}\text{var}(L^{be}) &= \text{var}(L^e)\left\{1 - \text{var}(L^e)\left[\text{var}(L^e) + \text{var}(L^c)\right]^{-1}\right\} \\ &= \frac{\text{var}(L^e)\,\text{var}(L^c)}{\text{var}(L^e) + \text{var}(L^c)}.\end{aligned} \tag{2.39}$$

The expression in Eq. (2.39) implies the following inequalities:

$$\text{var}(L^{be}) < \text{var}(L^e); \quad \text{var}(L^{be}) < \text{var}(L^c). \tag{2.40}$$

The aforementioned inequalities indicate a reduction of the predicted variance $\text{var}(L^{be})$ by comparison to either the variance $\text{var}(L^e)$ of the experimentally measured response or the variance $\text{var}(L^c)$ of the computed response.

Using the index $i = 1,2,3,4$, to denote the highest order of the retained sensitivities in Eq. (2.39), the expression of the predicted best-estimate variance of the predicted best-estimate leakage response, which will be denoted as $SD^{(be,i)}$, will have the following expression:

$$\begin{aligned}Var^{(be,i)} &= \left[SD^{(e)}\right]^2 - \left[SD^{(e)}\right]^2\left\{\left[SD^{(e)}\right]^2 + \left[\text{var}(L)\right]^{(i,U,N)}\right\}^{-1}\left[SD^{(e)}\right]^2, \\ SD^{(be,i)} &\triangleq \sqrt{Var^{(be,i)}}, \quad i = 1,2,3,4.\end{aligned} \tag{2.41}$$

2.5 TYPICAL PRECISION CONSISTENT MEASURED RESPONSE ($L^e = 1.6 \times 10^6$ NEUTRONS/SEC; $SD^{(e)} = 5\%$)

The results presented in this section consider that the *experimentally measured* value of the leakage response is $L^e = 1.6 \times 10^6$ neutrons/sec and that the relative standard deviation of the measured value is $SD^{(e)} = 5\%$ which is a usual (typical) value for this type of measurement. Recall that the nominal value of the computed leakage response is $L^c\left(\alpha^0\right) = 1.7648 \times 10^6$ neutrons/sec. The results to be presented next will include the following quantities:

1. The numerical values for the computed quantity $E^{(i)}(L^c) \pm SD^{(i)}$, where $i = 1,...,4$; for $i = 1$, only the contributions stemming from the 1st-order sensitivities are included; for $i = 2$, the contributions stemming from the 1st- and 2nd-order sensitivities are included; for $i = 3$, the contributions stemming from the 1st-, 2nd-, and 3rd-order sensitivities are included; for $i = 4$, all contributions stemming from the 1st- through 4th-order sensitivities are included.
2. The numerical values for the quantity $L^{be} \pm SD^{(be,i)}$, where $i = 1,...,4$, and where L^{be} denotes the value of the corresponding best-estimate predicted response and where $SD^{(be,i)}$ denotes the accompanying best-estimate (reduced) predicted standard deviation. As has been described above, only the contributions stemming from the 1st-order sensitivities are included when $i = 1$; for $i = 2$, the contributions stemming from both the 1st- and 2nd-order sensitivities are included; for $i = 3$, the contributions stemming from the 1st-, 2nd-, and 3rd-order sensitivities are included; all contributions stemming from the 1st- through 4th-order sensitivities are included when $i = 4$.

Three sub-cases will be considered to illustrate the effects of the precision (standard deviations) of the parameters (total cross-sections) on the results predicted by the 2nd-BERRU-PM methodology, as follows: (i) Subsection 2.5.1: high precision parameters, relative standard deviations of 3%; (ii) Subsection 2.5.2: typical precision parameters, relative standard deviations of 5%; (iii) Subsection 2.5.3: low precision parameters, relative standard deviations of 10%.

2.5.1 High Precision (3% Relative Standard Deviations) Parameters

It has been shown by Cacuci and Fang (2023) that for uniform relative standard deviations of 3% for the model parameters (total cross-sections), the Taylor series expansion of the computed response in terms of the variations in the model parameters [see Chapter 1, Eq. (1.127)] is expected to be convergent since the convergence "ratio test" of the 3rd-order term with respect to the 2nd-order term of the Taylor series is 0.58 and the ratio of the 4th-order term with respect to the 3rd-order term of the Taylor series is 0.68. Both of these results are below 1.00; therefore, even though it is not possible to conduct a bona fide convergence test involving the ratio of the nth-term and the $(n+1)^{th}$-term, this Taylor series is expected to be convergent.

The numerical values for the quantities $E^{(i)}(L^c) \pm SD^{(i)}$ and $L^{(be,i)} \pm SD^{(be,i)}$ for $i = 1,...,4$ are presented in Table 2.5, for the case when the model parameters (which are 180 groups averaged total microscopic cross-sections, assumed to be uncorrelated and normally distributed) are known with relatively high precision, all having relative standard deviations of 3%. The numerical results presented in Table 2.5 are depicted in Figure 2.8.

The following conclusions follow from the results presented in Table 2.5 and Figure 2.8:

1. Since the standard deviation $SD^{(e)}$ of the measured response overlaps with the standard deviation $SD^{(1)}$ of the computed result already when considering just the 1st-order sensitivities, the measured response value can be considered to be consistent (as opposed to "inconsistent" or "discrepant") with the computed nominal value of the response.
2. As expected from Equations (2.25) and (2.26), the values of both $E^{(i)}(L^c)$ and $SD^{(i)}$ increase progressively as the contributions from the higher-order sensitivities are progressively taken into account, for $i = 1,...,4$.
3. The best-estimate predicted response value, $L^{(be,i)}$, is expected to fall between the measured and the computed values, being closer to the value which is more precisely known (i.e., least uncertain). These expectations are confirmed already when considering just the 1st-order sensitivities. The predicted value is very close to the experimentally measured value of the response, the standard deviation of which is smaller than the standard deviation of the computed response.
4. As expected, the best-estimate predicted standard deviation, $SD^{(be,i)}$, is smaller than either the computed or measured standard deviations. The

TABLE 2.5

Results for $E^{(i)}(L^c) - SD^{(i)}$ and $L^{(be,i)} \pm SD^{(be,i)}$, $i = 1,...,4$, for Response Measurement Relative $SD^{(e)} = 5\%$, Parameters Relative $SD = 3\%$

Order of Sensitivities Considered	Values of $L^e \pm SD^{(e)}$ (Neutrons/sec)	Values of $E^{(i)}(L^c) - SD^{(i)}$ (Neutrons/sec)	Values of $L^{(be,i)} \pm SD^{(be,i)}$ (Neutrons/sec)
1st-order ($i = 1$)	1.600×10^6 $\pm 8.000 \times 10^4$	$1.765 \times 10^6 \pm 5.548 \times 10^5$	1.603×10^6 $\pm 7.918 \times 10^4$
1st+2nd-order ($i = 2$)		$2.179 \times 10^6 \pm 7.355 \times 10^5$	1.607×10^6 $\pm 7.953 \times 10^4$
1st+2nd+3rd-order ($i = 3$)		$2.179 \times 10^6 \pm 1.443 \times 10^6$	1.602×10^6 $\pm 7.988 \times 10^4$
1st+2nd+3rd+4th-order ($i = 4$)		$2.667 \times 10^6 \pm 1.847 \times 10^6$	1.602×10^6 $\pm 7.993 \times 10^4$

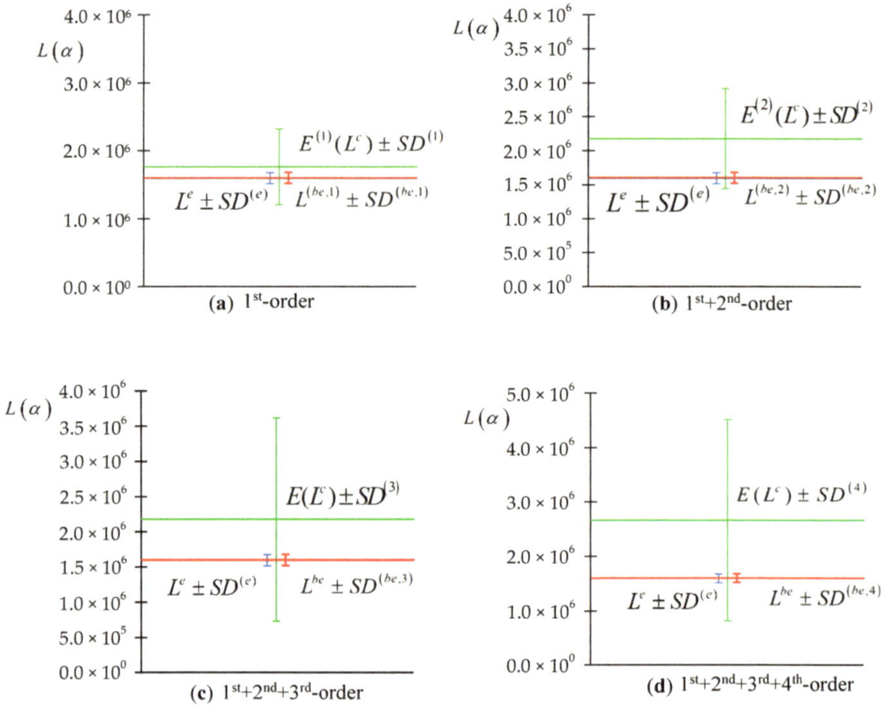

FIGURE 2.8 Comparison of $E^{(i)}(L^c) \pm SD^{(i)}$ (in green), $L^{(be,i)} \pm SD^{(be,i)}$ (in red), and $L^e \pm SD^{(e)}$ (in blue), for response measurement $SD^{(e)} = 5\%$, parameters $SD = 3\%$. (i) 1st-order $(i = 1)$ sensitivities are considered. (ii) 1st+2nd-order $(i = 2)$ sensitivities are included. (iii) 1st+2nd+3rd-order $(i = 3)$ sensitivities are included. (iv) 1st+2nd+3rd+4th-order $(i = 4)$ sensitivities are included.

actual predicted value for the best-estimate predicted response is very close to (albeit still smaller than) the standard deviation of the measured response while being much smaller than the standard deviation of the computed leakage response.

2.5.2 Typical Precision (5% Relative Standard Deviations) Parameters

It has been shown by Cacuci and Fang (2023) that, for relative standard deviations of 5% for the model parameters, the ratio of the 3rd-order term with respect to the 2nd-order term of the Taylor series expansion of the computed response in terms of the model parameters is $0.97 < 1.00$ but the ratio of the 4th-order term with respect to the 3rd-order term of the Taylor series is $1.13 > 1.00$. These ratios indicate that relative standard deviations of 5% for the model parameters are "borderline" values and the Taylor series expansion of the computed response in terms of the model parameters is expected to be divergent. On the other hand, relative standard deviations of 5% are representative of the "usual uncertainties encountered in practice" and are therefore often encountered in measurements of total cross-sections. This situation, therefore, underscores the impact of the higher-order response sensitivities when the parameter uncertainties are representative of uncertainties usually encountered in practice while also being "borderline" in terms of the convergence of the Taylor series expansion of the computed response in terms of the model parameters, which underlies the determination of the statistics (expected values, variance, etc.) of the distribution of the computed response in the phase-space of imprecisely known model parameters. The numerical values for the quantities $E(L^c) \pm SD^{(i)}$ and $L^{be} \pm SD^{(be,i)}$, for $i = 1,...,4$, are presented in Table 2.6 for the case when the model parameters (which are the uncorrelated and normally distributed total microscopic cross-sections) are known with typical/average precision, all having relative standard deviations of 5%. The numerical results presented in Table 2.6 are depicted in Figure 2.9.

The following conclusions follow from the results presented in Table 2.6 and Figure 2.9:

1. The results in Table 2.6 and Figure 2.9 display the same trends as already displayed by the results presented in Table 2.5 and Figure 2.8, respectively, but these trends are significantly more accentuated regarding the progressive increase of the values of $E^{(i)}(L^c)$ and $SD^{(i)}$, as the contributions from the higher-order sensitivities are progressively taken into account ($i = 1,...,4$). This massive increase (leading to non-physical results for $SD^{(3)}$ and $SD^{(4)}$) indicates the non-convergence of the Taylor series expansion of the computed response in terms of the model parameters that underlie the computations of $E^{(i)}(L^c)$ and $SD^{(i)}$.
2. However, despite the massive increase in the computed values $E^{(i)}(L^c) \pm SD^{(i)}$, $i = 1,...,4$, both the best-estimate predicted response value, $L^{(be,i)}$, and the

TABLE 2.6

Results for $E^{(i)}(L^c) - SD^{(i)}$ and $L^{(be,i)} \pm SD^{(be,i)}$, $i = 1,...,4$, for Response Measurement Relative $SD^{(e)} = 5\%$, Parameters Relative $SD = 5\%$

Order of Sensitivities Considered	Values of $L^e \pm SD^{(e)}$ (Neutrons/sec)	Values of $E^{(i)}(L^c) - SD^{(i)}$ (Neutrons/sec)	Values of $L^{(be,i)} \pm SD^{(be,i)}$ (Neutrons/sec)
1st-order ($i = 1$)	1.600×10^6 $\pm 8.000 \times 10^4$	$1.765 \times 10^6 \pm 9.246 \times 10^5$	1.601×10^6 $\pm 7.970 \times 10^4$
1st+2nd-order ($i = 2$)		$2.914 \times 10^6 \pm 1.629 \times 10^6$	1.603×10^6 $\pm 7.990 \times 10^4$
1st+2nd+3rd-order ($i = 3$)		$2.914 \times 10^6 \pm 5.103 \times 10^6$	1.601×10^6 $\pm 7.999 \times 10^4$
1st+2nd+3rd+4th-order ($i = 4$)		$6.681 \times 10^6 \pm 7.386 \times 10^6$	1.601×10^6 $\pm 7.999 \times 10^4$

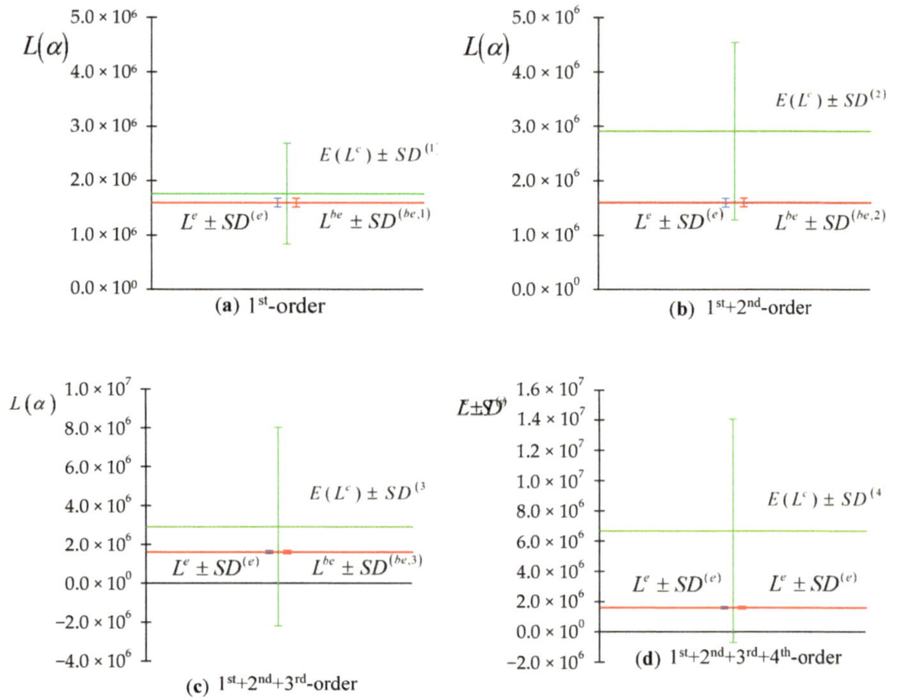

FIGURE 2.9 Comparison of $E^{(i)}(L^c) \pm SD^{(i)}$ (in green), $L^{(be,i)} \pm SD^{(be,i)}$ (in red), and $L^e \pm SD^{(e)}$ (in blue), for response measurement $SD^{(e)} = 5\%$, parameters $SD = 5\%$. (i) 1st-order $(i = 1)$ sensitivities are considered. (ii) 1st+2nd-order $(i = 2)$ sensitivities are included. (iii) 1st+2nd+3rd-order $(i = 3)$ sensitivities are included. (iv) 1st+2nd+3rd+4th-order $(i = 4)$ sensitivities are included.

best-estimate predicted standard deviation, $SD^{(be,i)}$, have values that are very close to the corresponding experimentally measured values, thus proving that the 2nd-BERRU-PM methodology correctly predicts best-estimate values even when the quantities that characterize the computed responses may be unreliable because of non-convergence of the respective mathematical expressions.

2.5.3 Low Precision (10% Relative Standard Deviations) Parameters

It has been shown by Cacuci and Fang (2023) that, when considering uniform relative standard deviation of 10% for all of the parameters (total cross-sections), the ratio of the 3rd-order term with respect to the 2nd-order term of the Taylor series expansion of the computed response in terms of the model parameters is 1.93; the ratio of the 4th-order term with respect to the 3rd-order term of this Taylor series is 2.26. Both of these results are larger than 1.00, which indicates that this Taylor series is being used outside of its radius of convergence to compute the response's expected value and variance. The numerical values for the quantities $E^{(i)}(L^c) \pm SD^{(i)}$ and $L^{(be,i)} \pm SD^{(be,i)}$, $i = 1,...,4$, are presented in Table 2.7 and depicted in Figure 2.10, for the case when the model parameters (which are the 180 group-averaged total microscopic cross-sections, assumed to be uncorrelated and normally distributed) are known with low precision, all having relative standard deviations of 10%.

The following conclusions follow from the results presented in Table 2.7 and Figure 2.10:

1. The results in Table 2.7 and Figure 2.10 display the same trends as already displayed by the results presented in Tables 2.6 and 2.5 (and Figures 2.9 and 2.8), respectively, but the trends displayed in Table 2.7 and Figure 2.10 are substantially accentuated regarding the progressive increase of the values of $E(L^c)$ and $SD^{(i)}$, as the contributions from the higher-order sensitivities are progressively taken into account ($i = 1,...,4$). This massive increase confirms the non-convergence of the Taylor series expansion of the computed response in terms of the model parameters, which underlies the computations of $E(L^c)$ and $SD^{(i)}$.
2. Nevertheless, despite the massive increase in the computed values $E(L^c) \pm SD^{(i)}$, $i = 1,...,4$, both the best-estimate predicted response value, L^{be}, and the best-estimate predicted standard deviation, $SD^{(be,i)}$, have values that are very close to the corresponding experimentally measured values, thus proving that the 2nd-BERRU-PM methodology correctly predicts best-estimate values even when the quantities that characterize the computed responses may be unreliable because of non-convergence of the respective mathematical expressions.

TABLE 2.7

Results for $E^{(i)}(L^c) \pm SD^{(i)}$ and $L^{(be,i)} \pm SD^{(be,i)}$, $i = 1,...,4$, for Response Measurement Relative $SD^{(e)}$ = 5%, Parameters Relative SD = 10%

Order of Sensitivities Considered	Values of $L^e \pm SD^{(e)}$ (Neutrons/sec)	Values of $E^{(i)}(L^c) - SD^{(i)}$ (Neutrons/sec)	Values of $L^{(be,i)} \pm SD^{(be,i)}$ (Neutrons/sec)
1st-order ($i = 1$)	1.600×10^6 $\pm 8.000 \times 10^4$	$1.765 \times 10^6 \pm 1.849 \times 10^6$	1.601×10^6 $\pm 7.993 \times 10^4$
1st+2nd-order ($i = 2$)		$6.363 \times 10^6 \pm 5.675 \times 10^6$	1.601×10^6 $\pm 7.999 \times 10^4$
1st+2nd+3rd-order ($i = 3$)		$6.363 \times 10^6 \pm 3.561 \times 10^7$	1.601×10^6 $\pm 7.999 \times 10^4$
1st+2nd+3rd+4th-order ($i = 4$)		$6.662 \times 10^7 \pm 5.562 \times 10^7$	1.601×10^6 $\pm 7.999 \times 10^4$

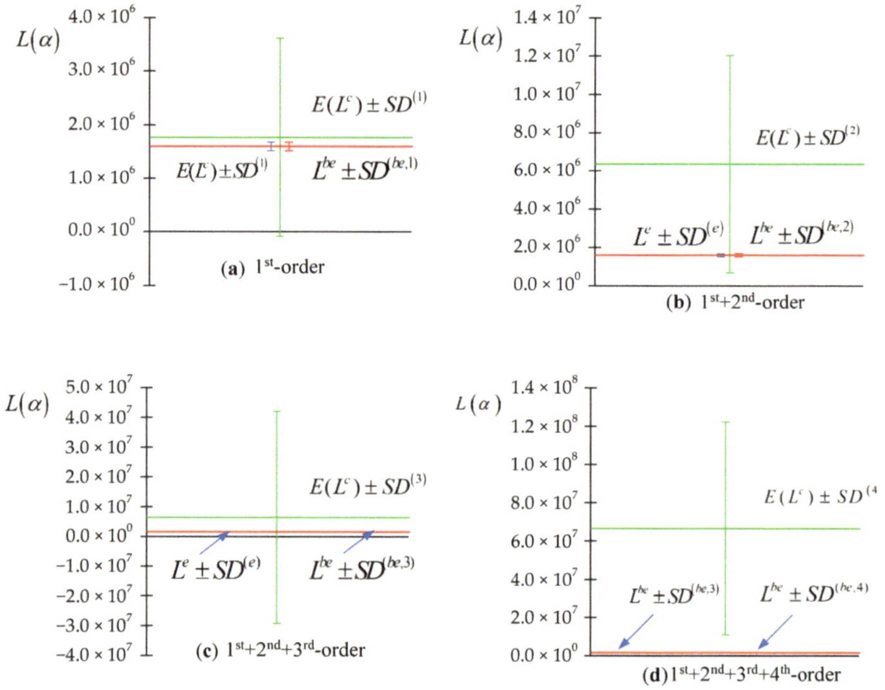

FIGURE 2.10 Comparison of $E(L^c) \pm SD^{(i)}$ (in green), $L^{be} \pm SD^{(be,i)}$ (in red), and $L^e \pm SD^{(e)}$ (in blue), for response measurement $SD^{(e)} = 5\%$, parameters $SD = 10\%$. (i) 1st-order ($i = 1$) sensitivities are considered. (ii) 1st+2nd-order ($i = 2$) sensitivities are included. (iii) 1st+2nd+3rd-order ($i = 3$) sensitivities are included. (iv) 1st+2nd+3rd+4th-order ($i = 4$) sensitivities are included.

2.6 LOW PRECISION CONSISTENT MEASURED RESPONSE ($L^e = 1.6 \times 10^6$ NEUTRONS/SEC; $SD^{(e)} = 10\%$); HIGH PRECISION PARAMETERS (RELATIVE SD=3%)

The situation analyzed in this section differs from the situation analyzed in Subsection 2.5.1, only regarding the precision of the measured leakage response, in that the relative standard deviation of the experimentally measured response is $SD^{(e)} = 10\%$, which is twice as large (i.e., 10% vs. 5%) as the relative standard deviation of the experimentally measured response considered in Subsection 2.5.1. In other words, the situation analyzed in this section involves a low precision response measurement but "high precision" parameters (relative SD = 3%), thus enabling a direct quantification of the impact of the precision (low vs. typical) of the measured response. The numerical values obtained for this case are presented in Table 2.8 and depicted in Figure 2.11.

These results presented in Table 2.8 and depicted in Figure 2.11 can be directly compared to the corresponding values presented in Table 2.5 and depicted in Figure 2.8, respectively. This comparison shows that the values of the computed quantities $E(L^c) \pm SD^{(i)}$ are identical in the two tables, as expected. In both

TABLE 2.8

Results for $E^{(i)}(L^c) \pm SD^{(i)}$ and $L^{(be,i)} \pm SD^{(be,i)}$, $i = 1,\ldots,4$, for Response Measurement Relative $SD^{(e)} = 10\%$, Parameters Relative $SD = 3\%$

Order of Sensitivities Considered	Values of $L^e \pm SD^{(e)}$ (Neutrons/sec)	Values of $E^{(i)}(L^c) \pm SD^{(i)}$ (Neutrons/sec)	Values of $L^{(be,i)} \pm SD^{(be,i)}$ (Neutrons/sec)
1st-order ($i = 1$)	$1.600 \times 10^6 \pm 1.600 \times 10^5$	$1.765 \times 10^6 \pm 5.548 \times 10^5$	$1.613 \times 10^6 \pm 1.537 \times 10^5$
1st+2nd-order ($i = 2$)		$2.179 \times 10^6 \pm 7.355 \times 10^5$	$1.626 \times 10^6 \pm 1.563 \times 10^5$
1st+2nd+3rd-order ($i = 3$)		$2.179 \times 10^6 \pm 1.443 \times 10^6$	$1.607 \times 10^6 \pm 1.590 \times 10^5$
1st+2nd+3rd+4th-order ($i = 4$)		$2.667 \times 10^6 \pm 1.847 \times 10^6$	$1.608 \times 10^6 \pm 1.594 \times 10^5$

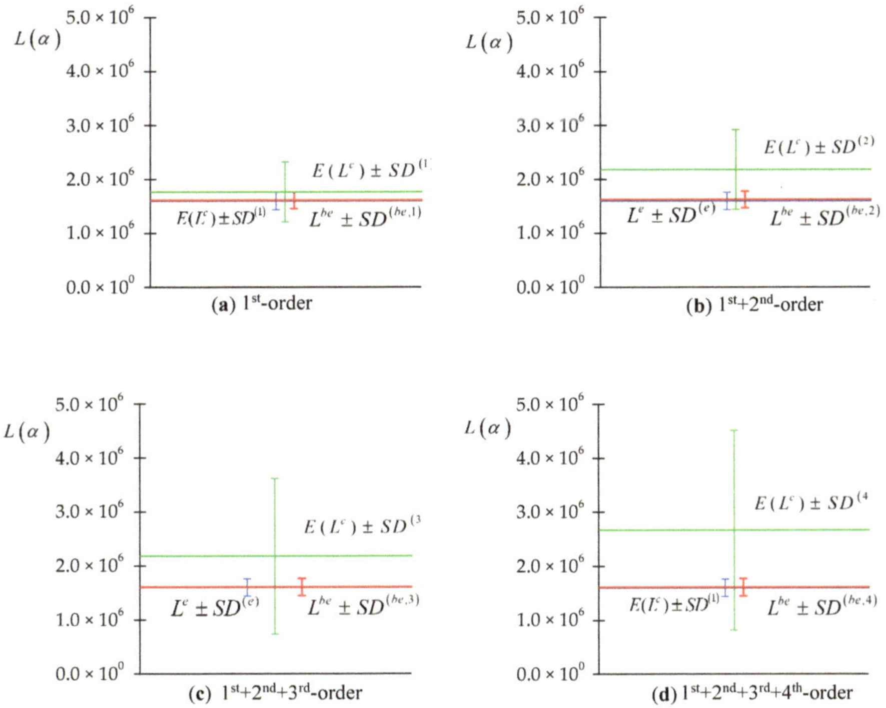

FIGURE 2.11 Comparison of $E^{(i)}(L^c) \pm SD^{(i)}$ (in green), $L^{(be,i)} \pm SD^{(be,i)}$ (in red), and $L^e \pm SD^{(e)}$ (in blue), for response measurement $SD^{(e)} = 10\%$, parameters $SD = 3\%$. (i) 1st-order ($i = 1$) sensitivities are considered. (ii) 1st+2nd-order ($i = 2$) sensitivities are included. (iii) 1st+2nd+3rd-order ($i = 3$) sensitivities are included. (iv) 1st+2nd+3rd+4th-order ($i = 4$) sensitivities are included.

Tables 2.8 and 2.5, the best-estimate predicted values, L^{be}, of the leakage response are all very close to the experimentally measured value indicating that the precision (low of typical) of the respective measurement insignificantly affects the prediction of the 2nd-BERRU-PM methodology. Only the predicted standard deviation of the best-estimate predicted leakage response is affected, as can be seen by comparing the results presented in the rightmost column of Table 2.8 with the corresponding results reported in the rightmost column of Table 2.56. This comparison indicates that, for the low precision measurement, the results presented in the rightmost column of Table 2.8 indicate that the standard deviation is approximately 1.563×10^5 neutrons/sec, which is smaller than the corresponding measured SD (1.600×10^5) and is much smaller than the computed ones (ranging from 5.548×10^5 to 1.847×10^6). On the other hand, for the typical precision measurement, the results presented in the rightmost column of Table 2.5 indicate that the standard deviation is approximately 7.918×10^4 neutrons/sec which is smaller than the corresponding measured SD (8.00×10^4) and is much smaller than the computed ones (ranging from 5.548×10^5 to 1.847×10^6, just as in Table 2.8). Thus, the reduction in the respective predicted standard deviation of the leakage response is larger when the precision of the measured response is higher. In all cases, however, the predicted standard deviation of the response is smaller than the respective initially computed and/or measured values (for the response standard deviation).

The situations analyzed in Sections 2.5 and 2.6 pertain to values of the measured responses which are *consistent* with the computed response values, in that the standard deviations of the measured responses overlap with the standard deviations of the computed responses. The results obtained for these situations have indicated that the nominal value of the computed response, L^c, coincides with the expected value of the computed response, $E^{(i)}(L^c)$, *only if* all sensitivities higher than 1st-order are ignored. Otherwise, the incorporation of the 2nd- and higher-order sensitivities cause the expected value of the computed response, $E^{(i)}(L^c)$, to become increasingly larger than the nominal computed value L^c, in other words, $E^{(i)}(L^c) \geq L^c$, when $i>1$. Similarly, the standard deviation of the computed response increases as sensitivities of increasingly higher-order are incorporated, as would logically be expected. However, this fact has no adverse consequences after applying the 2nd-BERRU-PM methodology to combine the computational results with the experimental results, since the 2nd-BERRU-PM methodology reduces the predicted best-estimate standard deviations to values that are smaller than both the computed and the experimentally measured values of the initial standard deviations. The 2nd-BERRU-PM methodology also predicts that the best-estimate response value, L^{be}, will fall in between the "expected value of the computed response," $E^{(i)}(L^c)$, and the experimentally measured value L^e, being closer to the value that is more precisely known (i.e., has a lower accompanying standard deviation). These properties of the 2nd-BERRU-PM methodology (i.e., prediction of best-estimate results that fall in between the corresponding computed and measured values, while reducing the predicted standard deviations of the predicted results to values smaller than either the experimentally measured or the computed values of the respective standard deviations) stem from the information-theoretical foundation of the 2nd-BERRU-PM, which ensures that the incorporation of additional (consistent) information reduces the predicted uncertainties.

Section 2.7 presents situations that can arise when the measured response values appear to be inconsistent with the computed response values.

2.7 TYPICAL PRECISION INCONSISTENT MEASURED RESPONSE ($L^e = 3.0\times10^6$ NEUTRONS/SEC; $SD^{(e)} = 5\%$)

Recall that the nominal value of the computed total leakage response for the PERP benchmark is $L^c = 1.7648\times10^6$ neutrons/sec. The situations to be analyzed in this section will consider a measured response having a nominal value $L^e = 3.0\times10^6$ neutrons/sec, with a relative standard deviation having a "moderate" value of $SD^{(e)} = 5\%$. The impact of the precision of the uncertain parameters will be quantified by considering the following values for their relative standard deviations: "high precision" (2%), presented in Subsection 2.7.1; "typical precision" (5%), presented in Subsection 2.7.2; "low precision" (10%), presented in Subsection 2.7.3, respectively.

2.7.1 High Precision (2% Relative Standard Deviations) Parameters

The results presented in this Subsection consider that the experimentally measured value of the leakage response is $L^e = 3.0\times10^6$ neutrons/sec, with a relative standard deviation of $SD^{(e)} = 5\%$, which is a usual (typical) value for this type of measurement. The model parameters are the microscopic total cross-sections, considered to be uncorrelated and normally distributed, and having relative standard deviations of 2%. As has been previously discussed in Section 2.6, the Taylor series expansion of the computed leakage response in terms of the model's total cross-sections converges for uniform relative standard deviations for these cross-sections of 3% or less, so this Taylor series is expected to be convergent for relative standard deviations of 2%. The numerical values obtained for the quantities $E(L^c)\pm SD^{(i)}$ and $L^{be}\pm SD^{(be,i)}$ for $i=1,...,4$ are presented in Table 2.9 and depicted in Figure 2.12.

TABLE 2.9

Numerical Values of $E(L^c)\pm SD^{(i)}$ and $L^{be}\pm SD^{(be,i)}$, $i=1,...,4$, for Measured Response $L^e = 3.0\times10^6$ Neutrons/Sec with $SD^{(e)} = 5\%$; Parameters Relative Standard Deviations of 2%

Order of Sensitivities Considered	Values of $L^e \pm SD^{(e)}$ (Neutrons/sec)	Values of $E(L^c)\pm SD^{(i)}$ (Neutrons/sec)	Values of $L^{be}\pm SD^{(be,i)}$ (Neutrons/sec)
1st-order ($i=1$)	3.000×10^6 $\pm 1.500\times10^5$	$1.765\times10^6 \pm 3.698\times10^5$	2.826×10^6 $\pm 1.390\times10^5$
1st+2nd-order ($i=2$)		$1.949\times10^6 \pm 4.276\times10^5$	2.885×10^6 $\pm 1.415\times10^5$
1st+2nd+3rd-order ($i=3$)		$1.949\times10^6 \pm 6.288\times10^5$	2.943×10^6 $\pm 1.459\times10^5$
1st+2nd+3rd+4th-order ($i=4$)		$2.045\times10^6 \pm 7.157\times10^5$	2.960×10^6 $\pm 1.468\times10^5$

The following conclusions follow from the results presented in Table 2.9 and Figure 2.12:

1. The computed results appear to be inconsistent with the measured results because the small standard deviations considered for the parameters yield small values for the standard deviation of the computed response. This fact is particularly evident when only the 1st-order sensitivities are considered.
2. As higher-order sensitivities are progressively considered, for $i = 1,...,4$, the agreement between the measured and the computed quantities is progressively improved, indicating a possible convergence if sensitivities of order higher than four were included.
3. In all cases, though, the predicted best-estimate values $L^{be} \pm SD^{(be,4)}$ remain consistent with the measured values. The best-estimate predicted response value, L^{be}, moves progressively closer to the measured value, as progressively higher-order sensitivities are considered, while the best-estimate standard deviation, $SD^{(be,i)}$ is in all cases close to, albeit smaller, than the experimentally measured standard deviation $SD^{(e)}$.

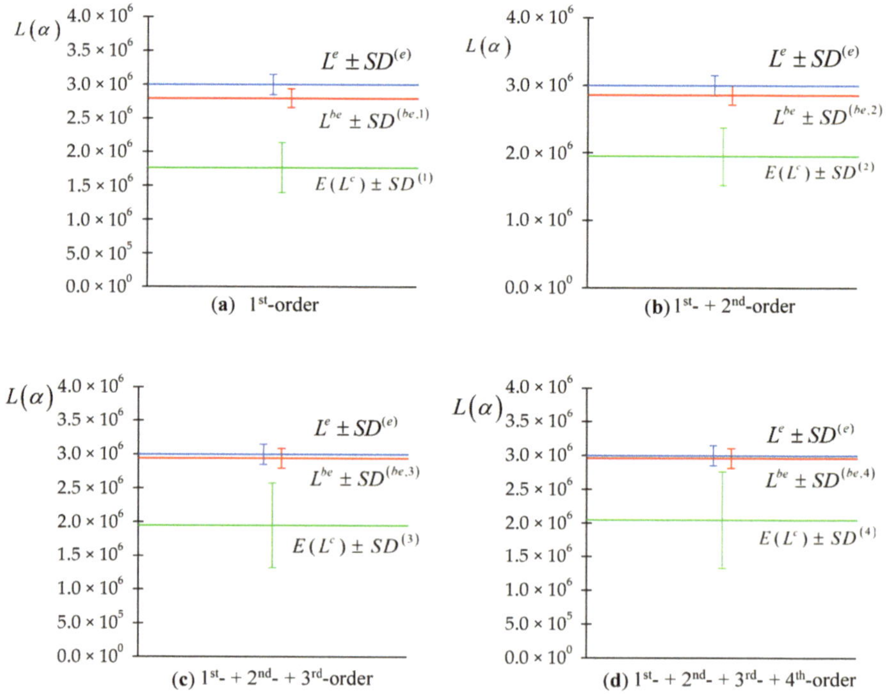

FIGURE 2.12 Comparison of $E(L^c) \pm SD^{(i)}$ (in green), $L^{be} \pm SD^{(be,i)}$ (in red), and $L^e \pm SD^{(e)}$ (in blue), for measured response $L^e = 3.0 \times 10^6$ neutrons/sec with $SD^{(e)} = 5\%$; parameters relative standard deviations = 2%. (i) 1st-order ($i = 1$) sensitivities are considered. (ii) 1st+2nd-order ($i = 2$) sensitivities are included. (iii) 1st+2nd+3rd-order ($i = 3$) sensitivities are included. (iv) 1st+2nd+3rd+4th-order ($i = 4$) sensitivities are included.

2.7.2 Typical Precision (5% Relative Standard Deviations) Parameters

As in the previous Subsection, the results presented in this Subsection also consider that the experimentally measured value of the leakage response is $L^e = 3.0\times 10^6$ neutrons/sec with a relative standard deviation of $SD^{(e)} = 5\%$. The model parameters (uncorrelated and normally distributed microscopic total cross-sections) are considered to be known with "typical precision" having relative standard deviations of 5%. As has been previously discussed in Section 2.6, relative standard deviations of 5% for the model parameters are outside the radius of convergence of the Taylor series expansion of the leakage response in terms of the model parameters. The numerical values for the computed quantities $E(L^c)\pm SD^{(i)}$ (i.e., computed response mean value plus/minus computed standard deviation) and $L^{be} \pm SD^{(be,i)}$ (i.e., best-estimate predicted response value plus/minus predicted standard deviation) for $i = 1,...,4$ are presented in Table 2.10. As before, only the contributions stemming from the 1st-order sensitivities are included for $i=1$; for $i=2$, the contributions stemming from the 1st- and 2nd-order sensitivities are included; for $i=3$, the contributions stemming from the 1st-, 2nd-, and 3rd-order sensitivities are included; contributions stemming from the 1st- through 4th-order sensitivities are included when $i=4$. The numerical results presented in Table 2.10 are depicted in Figure 2.13.

The following conclusions follow from the results presented in Table 2.10 and Figure 2.13:

1. When considering only the 1st-order sensitivities, the standard deviation $SD^{(e)}$ of the measured response does *not* overlap with the standard deviation $SD^{(1)}$ of the computed response. Thus, when considering just the 1st-order sensitivities, the measured response value appears to be "inconsistent" ("discrepant") with the computed nominal value of the response.

TABLE 2.10

Numerical Values of $E(L^c)\pm SD^{(i)}$ and $L^{be} \pm SD^{(be,i)}$, $i = 1,...,4$, for Measured Response $L^e = 3.0\times 10^6$ Neutrons/Sec with $SD^{(e)} = 5\%$; Parameters Relative Standard Deviations of 5%

Order of Sensitivities Considered	Values of $L^e \pm SD^{(e)}$ (Neutrons/sec)	Values of $E(L^c) \pm SD^{(i)}$ (Neutrons/sec)	Values of $L^{be} \pm SD^{(be,i)}$ (Neutrons/sec)
1st-order ($i=1$)	$3.000\times 10^6 \pm 1.500\times 10^5$	1.765×10^6 $\pm 9.246\times 10^5$	2.968×10^6 $\pm 1.481\times 10^5$
1st+2nd-order ($i=2$)		2.914×10^6 $\pm 1.629\times 10^6$	2.969×10^6 $\pm 1.494\times 10^5$
1st+2nd+3rd-order ($i=3$)		2.914×10^6 $\pm 5.103\times 10^6$	2.999×10^6 $\pm 1.499\times 10^5$
1st+2nd+3rd+4th-order ($i=4$)		6.681×10^6 $\pm 7.386\times 10^6$	3.002×10^6 $\pm 1.499\times 10^5$

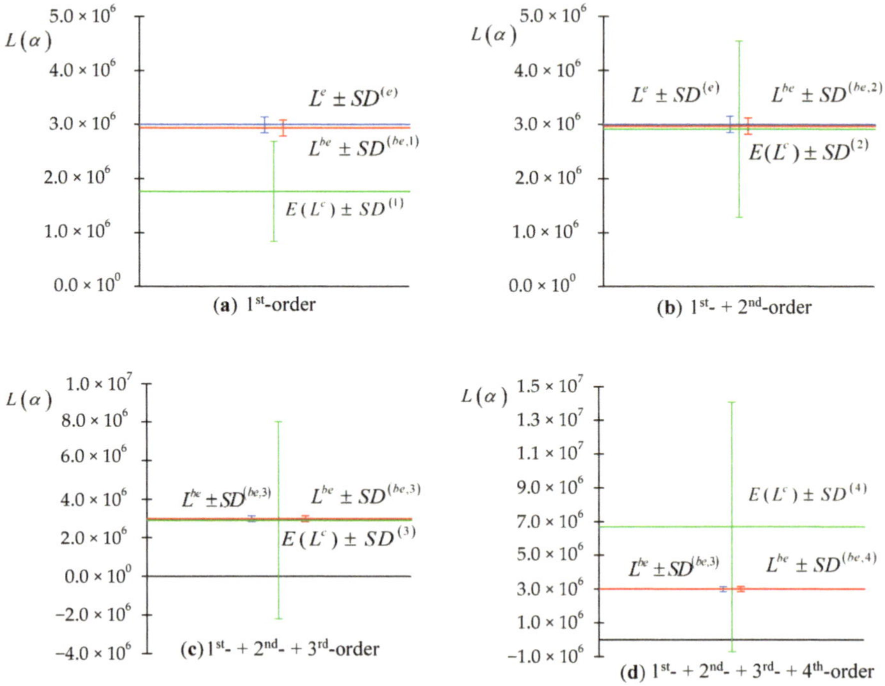

FIGURE 2.13 Comparison of $E(L^c) \pm SD^{(i)}$ (in green), $L^{be} \pm SD^{(be,i)}$ (in red), and $L^e \pm SD^{(e)}$ (in blue), for measured response $L^e = 3.0 \times 10^6$ neutrons/sec with $SD^{(e)} = 5\%$; uniform parameters relative standard deviations of 5%. (i) 1st-order $(i = 1)$sensitivities are considered; (ii) 1st+2nd-order $(i = 2)$ sensitivities are included; (iii) 1st+2nd+3rd-order $(i = 3)$ sensitivities are included; (iv) 1st+2nd+3rd+4th-order $(i = 4)$ sensitivities are included.

Nevertheless, the best-estimate predicted response value, L^{be}, is very close to the measured value, with a standard deviation that is also close to (and smaller than) the measured value.

2. As expected, the values of both $E(L^c)$ and $SD^{(i)}$ increase progressively as the contributions from the higher-order sensitivities are progressively taken into account for $i = 1,...,4$. Already when the 2nd-order sensitivities are considered, the computational results appear to become consistent with the measured values, since the respective measured and computed standard deviation overlap as soon as the 2nd-order sensitivities are considered.
3. The best-estimate predicted response value, L^{be}, remains close to the measured value, even though the expected response value continues to increase divergently as higher-order sensitivities are considered.

2.7.3 Low Precision (10% Relative Standard Deviations) Parameters

The results presented in this Subsection also consider that the experimentally measured value of the leakage response is $L^e = 3.0 \times 10^6$ neutrons/sec, with a relative

TABLE 2.11
Numerical Values of $E(L^c) \pm SD^{(i)}$ and $L^{be} \pm SD^{(be,i)}$, $i = 1,...,4$, for Measured Response $L^e = 3.0 \times 10^6$ Neutrons/Sec with $SD^{(e)} = 5\%$; Parameters Relative Standard Deviations of 10%

Order of Sensitivities Considered	Values of $L^e \pm SD^{(e)}$ (Neutrons/sec)	Values of $E(L^c) \pm SD^{(i)}$ (Neutrons/sec)	Values of $L^{be} \pm SD^{(be,i)}$ (Neutrons/sec)
1st-order ($i = 1$)	3.000×10^6 $\pm 1.500 \times 10^5$	1.765×10^6 $\pm 1.849 \times 10^6$	2.992×10^6 $\pm 1.495 \times 10^5$
1st+2nd-order ($i = 2$)		6.363×10^6 $\pm 5.675 \times 10^6$	3.002×10^6 $\pm 1.499 \times 10^5$
1st+2nd+3rd-order ($i = 3$)		6.363×10^6 $\pm 3.561 \times 10^7$	3.001×10^6 $\pm 1.499 \times 10^5$
1st+2nd+3rd+4th-order ($i = 4$)		6.662×10^7 $\pm 5.562 \times 10^7$	3.001×10^6 $\pm 1.499 \times 10^5$

standard deviation of $SD^{(e)} = 5\%$ but the uncorrelated and normally distributed model parameters have relative standard deviations of 10%. As has been previously discussed in Section 2.6, the Taylor series expansion of the model's leakage response in terms of the total cross-section parameters diverges for relative standard deviations of 10% for these parameters. Thus, the numerical values for $E(L^c) \pm SD^{(i)}$ and $L^{be} \pm SD^{(be,i)}$ for $i = 1,...,4$, which are presented in Table 2.11 and depicted in Figure 2.14, are computed based on a truncated non-convergent Taylor series expansion.

The following conclusions follow from the results presented in Table 2.11 and Figure 2.14:

1. The computed results appear to be consistent with the measured results.
2. The divergent nature of the computations for the values of $E(L^c)$ and $SD^{(i)}$ is apparent, as these values increase massively when the contributions from the higher-order sensitivities are progressively taken into account for $i = 1,...,4$. Nevertheless, the best-estimate predicted response value, L^{be}, remains close to the measured value even when only the 1st-order sensitivities are considered.

2.8 HIGH PRECISION APPARENTLY INCONSISTENT MEASURED RESPONSE ($L^e = 3.0 \times 10^6$ NEUTRONS/SEC; $SD^{(e)} = 2\%$) AND HIGH PRECISION PARAMETERS (SD = 3%)

This section presents the results produced by using the 2nd-BERRU-PM methodology in conjunction with an experimental response measured with high precision, and model parameters which are also known fairly precisely, all having relative standard deviations of 3%, for which the Taylor series expansion of the model's leakage response

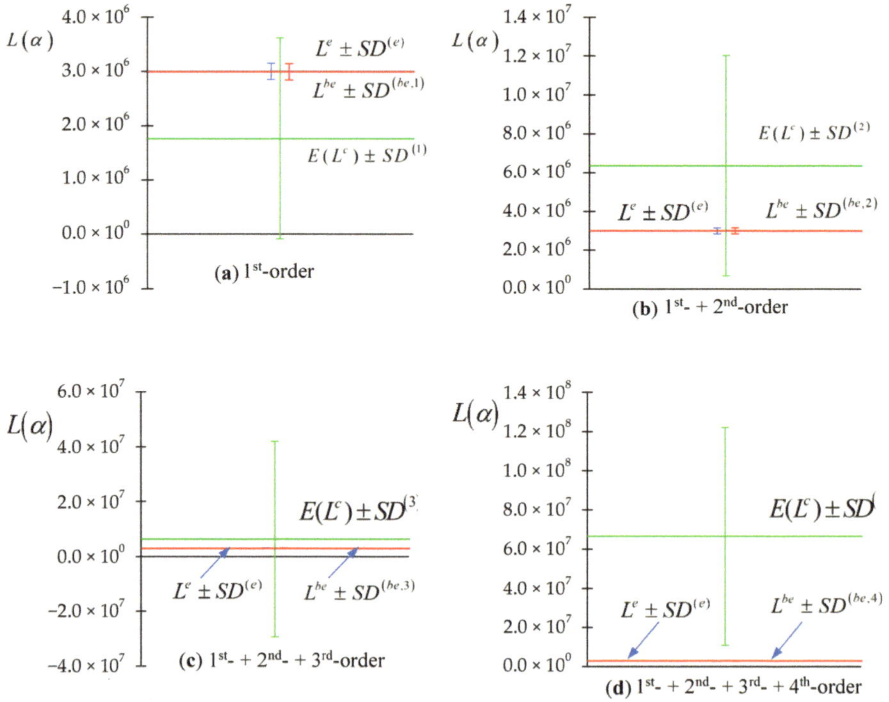

FIGURE 2.14 Comparison of $E(L^c) \pm SD^{(i)}$ (in green), $L^{be} \pm SD^{(be,i)}$ (in red), and $L^e \pm SD^{(e)}$ (in blue), for measured response $L^e = 3.0 \times 10^6$ neutrons/sec with $SD^{(e)} = 5\%$; parameters relative standard deviations of 10%. (i) 1st-order $(i=1)$ sensitivities are considered; (ii) 1st+2nd-order $(i=2)$ sensitivities are included; (iii) 1st+2nd+3rd-order $(i=3)$ sensitivities are included; (iv) 1st+2nd+3rd+4th-order $(i=4)$ sensitivities are included.

in terms of the total cross-section parameters are expected to be convergent, as has been previously discussed (the ratio of the 3rd-order term with respect to the 2nd-order term of the Taylor series is 0.58, and the ratio of the 4th-order term with respect to the 3rd-order term of the Taylor series is 0.68). Subsection 2.8.1, below, presents results obtained when considering just the contributions from the 1st-order sensitivities of the leakage response to the total cross-sections; Subsection 2.8.2 presents results obtained when considering the contributions from the 1st- and 2nd-order sensitivities; Subsection 2.8.3 presents results obtained when considering the contributions from 1st-, 2nd-, and 3rd-order sensitivities; Subsection 2.8.4 presents results obtained when considering the contributions from 1st-, 2nd-, 3rd-, and 4th-order sensitivities.

2.8.1 Including Only Contributions from the 1st-Order Sensitivities of the Leakage Response to the Total Cross-Sections

When only the 1st-order sensitivities, with respect to the total cross-sections, are considered, the 2nd-BERRU-PMD expressions shown in Eqs. (2.30) and (2.41), respectively, yield the results presented in Table 2.13 and depicted in Figure 2.15.

The results presented in Table 2.12 and Figure 2.15 indicate that:

1. The inequality $E(L^c) < L^{(be,1)} < L^e$ is fulfilled, as predicted by Eq. (2.29).
2. The reduction of the predicted standard deviation of the predicted response has been accomplished as predicted by Eq. (2.40); the numerical results given in Table 2.13 indicate that $SD^{(be,1)} < SD^{(e)} < SD^{(1)}$.
3. However, the computed results, which only include the contributions stemming from the 1st-order sensitivities, appear to be inconsistent with the experimental results. This apparent inconsistency occurs despite the fact that the Taylor series expansion of the model's leakage response in terms of the total cross-section parameters is expected to be convergent. Thus, this apparent "inconsistency between computation and experiment" indicates that, at the very least, the contributions stemming from the 2nd-order sensitivities must also be included.

TABLE 2.12

Values of $E(L^c) \pm SD^{(1)}$, $L^{be} \pm SD^{(be,1)}$, $L^e \pm SD^{(e)}$, Due to 3% Standard Deviations for Uncorrelated and Normally Distributed Total Microscopic Cross-Sections σ_t, when Only the 1st-Order Sensitivities Are Considered

Responses	Numerical Values (only σ_t)
$E(L^c) \pm SD^{(1)}$	$1.765 \times 10^6 \pm 5.548 \times 10^5$
$L^{(be,1)} \pm SD^{(be,1)}$	$2.956 \times 10^6 \pm 5.965 \times 10^4$
$L^e \pm SD^{(e)}$	$3.000 \times 10^6 \pm 6.000 \times 10^4$

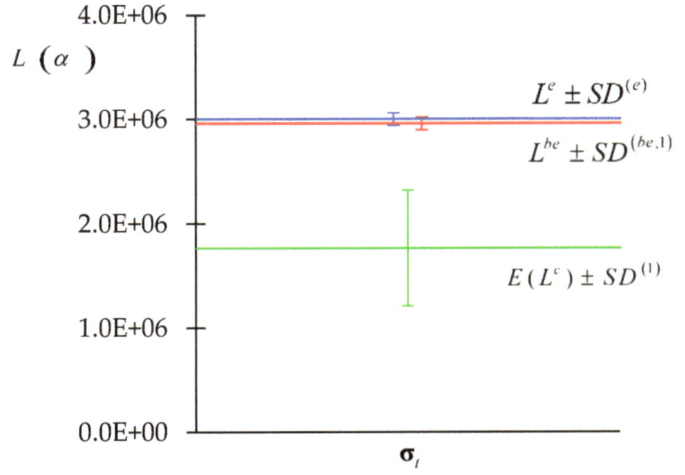

FIGURE 2.15 Comparison of $E(L^c) \pm SD^{(1)}$ (in green), $L^{(be,1)} \pm SD^{(be,1)}$ (in red), $L^e \pm SD^{(e)}$ (in blue), due to 3% standard deviations for uncorrelated and normally distributed total microscopic cross-sections σ_t, when only the 1st-order sensitivities are considered.

2.8.2 Including Contributions from the 1st+2nd-Order Sensitivities of the Leakage Response to the Total Cross-Sections

The numerical results obtained by using Eqs. (2.30) and (2.41) when including the contributions stemming from the 1st+2nd-order sensitivities of the leakage response to the total cross-sections, are presented in Table 2.13 and depicted in Figure 2.16.

The results presented in Table 2.14 and Figure 2.16 indicate that:

1. The inequality $E(L^c) < L^{(be,2)} < L^e$ is fulfilled as predicted by Eq. (2.29).
2. The expected value $E(L^c)$ of the computed response approaches from below the values of both the best-estimate value $L^{(be,2)}$ and the experimentally measured mean/nominal value L^e.

TABLE 2.13
Values of $E(L^c) \pm SD^{(2)}$, $L^{(be,2)} \pm SD^{(be,2)}$, $L^e \pm SD^{(e)}$, Due to 3% Standard Deviations for Uncorrelated and Normally Distributed Total Microscopic Cross-Sections σ_t, when the 1st+2nd-Order Sensitivities Are Included

Responses	Numerical Values (only σ_t)
$E(L^c) \pm SD^{(2)}$	$2.179 \times 10^6 \pm 7.355 \times 10^5$
$L^{(be,2)} \pm SD^{(be,2)}$	$2.995 \times 10^6 \pm 5.980 \times 10^4$
$L^e \pm SD^{(e)}$	$3.000 \times 10^6 \pm 6.000 \times 10^4$

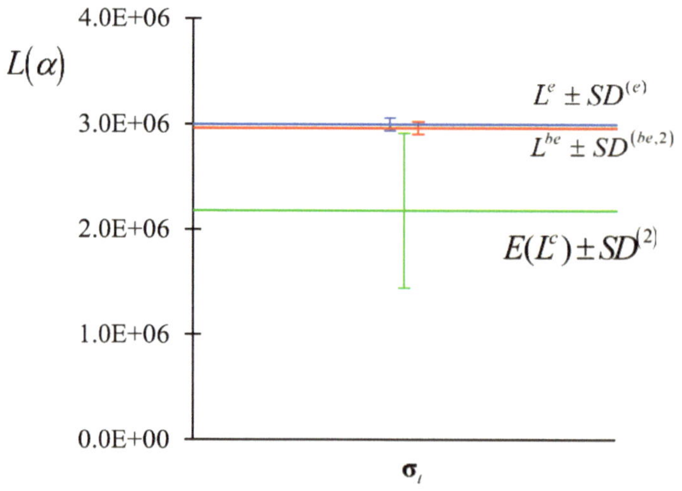

FIGURE 2.16 Comparison of $E(L^c) \pm SD^{(2)}$ (in green), $L^{be} \pm SD^{(be,2)}$ (in red), $L^e \pm SD^{(e)}$ (in blue), due to 3% standard deviations for uncorrelated and normally distributed total microscopic cross-sections σ_t, when the 1st+2nd-order sensitivities are included.

3. The inequality $SD^{(be,2)} < SD^{(e)} < SD^{(2)}$ is preserved as predicted by Eq. (2.40). By comparison to the results shown in Table 2.12, the inclusion of the contributions stemming from the 2nd-order sensitivities (in addition to the contributions stemming from the 1st-order sensitivities) brings the best-estimate predicted value $L^{(be,2)}$ closer to the precisely measured value L^{e}.
4. However, the "computed results" for the response are still slightly inconsistent with the "experimental results." This inconsistency, albeit small, indicates that the contributions from the 3rd-order sensitivities need to be included as well in addition to the contributions stemming from the 1st-order and the 2nd-order sensitivities.

2.8.3 Including Contributions from the 1st+2nd+3rd-Order Sensitivities of the Leakage Response to the Total Cross-Sections

The numerical results obtained by using Eqs. (2.30) and (2.41) when including the contributions stemming from the 1st+2nd+3rd-order sensitivities of the leakage response to the total cross-sections, are presented in Table 2.14 and depicted in Figure 2.17.

The results presented in Table 2.14 and Figure 2.17 indicate that:

1. The inequality $E(L^{c}) < L^{(be,3)} < L^{e}$ is fulfilled as predicted by Eq. (2.29).
2. The expected value $E(L^{c})$ of the computed response continues to approach monotonically from below the values of both the best-estimate value L^{be} and the experimentally measured mean/nominal value L^{e}.
3. The inequality $SD^{(be,3)} < SD^{(e)} < SD^{(3)}$, is preserved as predicted by Eq. (2.40). By comparison to the results shown in Table 2.13, the inclusion of the contributions stemming from the 3rd-order sensitivities (in addition to the contributions stemming from the 1st- and 2nd-order sensitivities) brings the best-estimate predicted value L^{be} even closer to the precisely measured value L^{e} .
4. The "computed results" for the response are now consistent with the "experimental results."

TABLE 2.14

Values of $E(L^{c}) \pm SD^{(3)}$, $L^{(be,3)} \pm SD^{(be,3)}$, $L^{e} \pm SD^{(e)}$, Due to 3% Standard Deviations for Uncorrelated and Normally Distributed Total Microscopic Cross-Sections σ_t, when the 1st+2nd+3rd-Order Sensitivities Are Included

Responses	Numerical Values (only σ_t)
$E(L^{c}) \pm SD^{(3)}$	$2.179 \times 10^{6} \pm 1.443 \times 10^{6}$
$L^{(be,3)} \pm SD^{(be,3)}$	$2.999 \times 10^{6} \pm 5.995 \times 10^{4}$
$L^{e} \pm SD^{(e)}$	$3.000 \times 10^{6} \pm 6.000 \times 10^{4}$

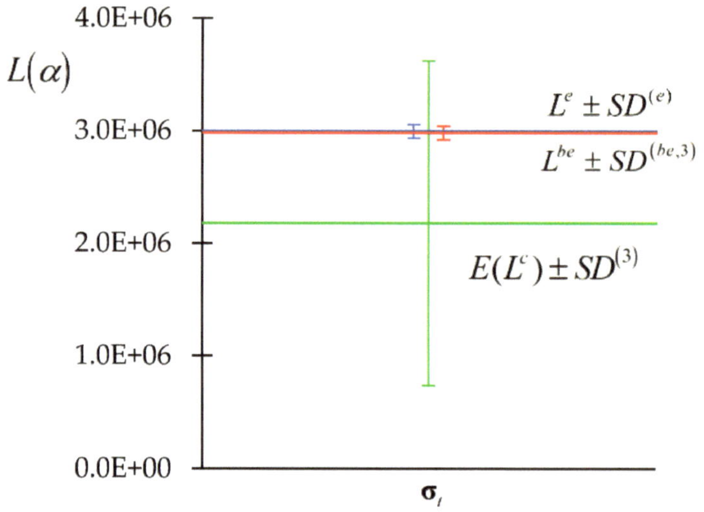

FIGURE 2.17 Comparison of $E(L^c) \pm SD^{(3)}$ (in green), $L^{(be,3)} \pm SD^{(be,3)}$ (in red), $L^e \pm SD^{(e)}$ (in blue), due to 3% standard deviations for uncorrelated and normally distributed total microscopic cross-sections σ_t, when the 1st+2nd+3rd-order sensitivities are included.

2.8.4 Including Contributions from the 1st+2nd+3rd+4th-Order Sensitivities of the Leakage Response to the Total Cross-Sections

The numerical results obtained by using Eqs. (2.30) and (2.41) when including the contributions stemming from the 1st+2nd+3rd+4th-order sensitivities of the leakage response to the total cross-sections, are presented in Table 2.15 and depicted in Figure 2.18.

The results presented in Table 2.15 and in Figure 2.18 indicate that:

1. The inequality $E(L^c) < L^{(be,4)} < L^e$ is fulfilled as predicted by Eq. (2.29).
2. The expected value $E(L^c)$ of the computed response continued to approach monotonically from below the values of both the best-estimate value L^{be} and the experimentally measured mean/nominal value L^e .
3. The inequality $SD^{(be,4)} < SD^{(e)} < SD^{(4)}$ is preserved as predicted by Eq. (2.40). By comparison to the results shown in Table 2.14, the inclusion of the contributions stemming from the 4th-order sensitivities (in addition to the contributions stemming from the 1st-, 2nd-, and 3rd-order sensitivities) practically renders the best-estimate predicted value $L^{(be,4)}$ to be the same as the precisely measured value L^e .
4. The "computed results" for the response are consistent with the "experimental results."
5. The additional contributions from the 4th-order sensitivities have only insignificantly changed the results obtained by having included only the sensitivities up to the 3rd-order. This result indicates that it is not necessary to

TABLE 2.15

Values of $E(L^c) \pm SD^{(4)}$, $L^{(be,4)} \pm SD^{(be,4)}$, $L^e \pm SD^{(e)}$, Due to 3% Standard Deviations for Uncorrelated and Normally Distributed Total Microscopic Cross-Sections σ_t, when the 1st+2nd+3rd+4th-Order Sensitivities Are Included

Responses	Numerical Values (only σ_t)
$E(L^c) \pm SD^{(4)}$	$2.667 \times 10^6 \pm 1.847 \times 10^6$
$L^{(be,4)} \pm SD^{(be,4)}$	$2.999 \times 10^6 \pm 5.997 \times 10^4$
$L^e \pm SD^{(e)}$	$3.000 \times 10^6 \pm 6.000 \times 10^4$

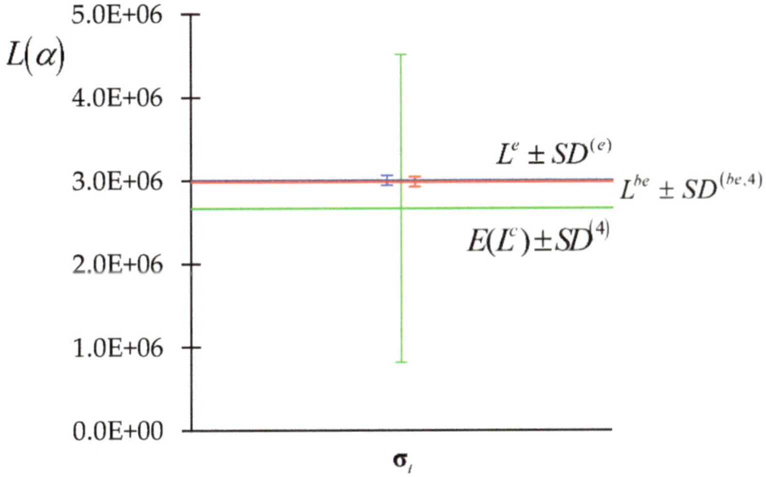

FIGURE 2.18 Comparison of $E(L^c) \pm SD^{(4)}$ (in green), $L^{(be,4)} \pm SD^{(be,4)}$ (in red), $L^e \pm SD^{(e)}$ (in blue), due to 3% standard deviations for uncorrelated and normally distributed total microscopic cross-sections σ_t, when the 1st+2nd+3rd+4th-order sensitivities are included.

include contributions from 5th- and higher-order, since their contributions are expected to be increasingly less significant as their order increases, as predicted by the convergent Taylor series expansion of the model's leakage response in terms of the total cross-section parameters, for the parameter standard deviations (3%) considered in this illustrative example.

2.9 HIGH PRECISION POSSIBLY INCONSISTENT MEASURED RESPONSE ($L^e = 3.0 \times 10^6$ NEUTRONS/SEC; $SD^{(e)} = 2\%$) AND LOW PRECISION PARAMETERS (SD = 10%)

This section presents the results produced by using the 2nd-BERRU-PM methodology in conjunction with an experimental response measured with high precision, but with rather imprecisely known model parameters, all having relative standard deviations of 10%. Such standard deviations are not prevalent in practice but may nevertheless be encountered for total cross-sections. As has been discussed in previous Sections, the ratio of the 3rd-order term with respect to the 2nd-order term of the Taylor series is 1.93; the ratio of the 4th-order term with respect to the 3rd-order term of the Taylor series is 2.26. Both of these results are larger than 1.00, which indicates that the Taylor series expansion of the model's leakage response in terms of the total cross-section parameters is being used outside its radius of convergence to compute the response's expected value and variance. This illustrative example, which involves precise measurements but rather imprecise parameters, has been deliberately chosen in order to underscore the decisive impact of higher-order sensitivities and the ability of the 2nd-BERRU-PM methodology to yield best-estimate results with reduced uncertainties despite the large uncertainties considered here for the imprecisely known model parameters. Subsection 2.9.1 presents results obtained when including just the 1st-order sensitivities while Subsection 2.9.2 presents results obtained when including the 1st- and the 2nd-order sensitivities of the leakage response to the total cross-sections.

2.9.1 Including Only Contributions from the 1st-Order Sensitivities of the Leakage Response to the Total Cross-Sections

When only the 1st-order sensitivities with respect to the total cross-sections are considered, the expressions provided in Eqs. (2.30) and (2.41), respectively, yield the results presented in Table 2.16 and depicted in Figure 2.19.

The results presented in Table 2.16 and Figure 2.19 indicate that:

1. The inequality $E(L^c) < L^{(be,1)} < L^e$ is fulfilled as predicted by Eq. (2.29); the best-estimate response value $L^{(be,1)}$ practically coincides with the nominal value L^e of the experimental response.
2. The predicted standard deviation of the predicted response is reduced; the numerical results given in Table 2.16 indicate that $SD^{(be,1)} < SD^{(e)} < SD^{(1)}$; $SD^{(e)} \cong SD^{(be,1)}$.

TABLE 2.16
Values of $E(L^c) \pm SD^{(1)}$, $L^{(be,1)} \pm SD^{(be,1)}$, $L^e \pm SD^{(e)}$, Due to 10% Standard Deviations for Uncorrelated and Normally Distributed Total Microscopic Cross-Sections σ_t, when Only the 1st-Order Sensitivities Are Considered

Responses	Numerical Values (only σ_t)
$E(L^c) \pm SD^{(1)}$	$1.765 \times 10^6 \pm 1.849 \times 10^6$
$L^{(be,1)} \pm SD^{(be,1)}$	$2.999 \times 10^6 \pm 5.997 \times 10^4$
$L^e \pm SD^{(e)}$	$3.000 \times 10^6 \pm 6.000 \times 10^4$

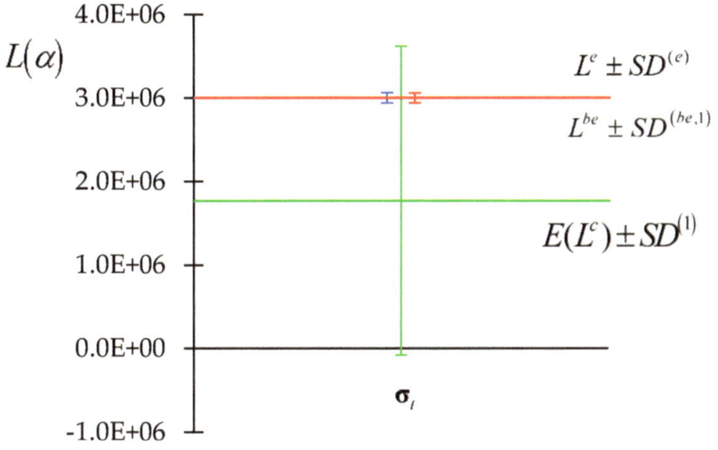

FIGURE 2.19 Comparison of $E(L^c) \pm SD^{(1)}$ (in green), $L^{(be,1)} \pm SD^{(be,1)}$ (in red), $L^e \pm SD^{(e)}$ (in blue), due to 10% standard deviations for uncorrelated and normally distributed total microscopic cross-sections σ_t, when only the 1st-order sensitivities are considered.

3. The computed results are consistent with the experimental results, even though only the contributions stemming from the 1st-order sensitivities are included. This apparent consistency occurs despite the fact that the Taylor series expansion of the model's leakage response in terms of the total cross-section parameters is expected to be divergent for the value of 10% considered (uniformly) for the relative standard deviations of parameters.

2.9.2 Including Contributions from the 1st+2nd-Order Sensitivities of the Leakage Response to the Total Cross-Sections

When both the 1st-order and 2nd-order sensitivities with respect to the total cross-sections are considered, Eqs. (2.30) and (2.41) yield the results presented in Table 2.17 and depicted in Figure 2.20.

The results presented in Table 2.17 and Figure 2.20 indicate that:

1. The inequality $E(L^c) < L^{(be,2)} < L^e$ is *not* fulfilled as predicted by Eq. (2.29), thus indicating the effects of the divergence of the Taylor series expansion of the model's leakage response in terms of the total cross-section parameters for the value of 10% considered for the relative standard deviations of parameters. The best-estimate response value $L^{(be,2)}$ practically coincides with the nominal value L^e of the experimental response.
2. The numerical results given in Table 2.17 indicate that $SD^{(e)} \cong SD^{(be,2)}$.
3. Comparing the results provided in Table 2.17 with the results provided in Table 2.13 indicates that when considering an accurate measurement but inaccurate parameters, the 2nd-BERRU-PM uses the "flexibility" provided

TABLE 2.17

Values of $E(L^c) \pm SD^{(2)}$, $L^{(be,2)} \pm SD^{(be,2)}$, $L^e \pm SD^{(e)}$, Due to 10% Standard Deviations for Uncorrelated and Normally Distributed Total Microscopic Cross-Sections σ_t, when the 1st+2nd-Order Sensitivities Are Included

Responses	Numerical Values (only σ_t)
$E(L^c) \pm SD^{(2)}$	$6.363 \times 10^6 \pm 5.675 \times 10^6$
$L^{(be,2)} \pm SD^{(be,2)}$	$3.001 \times 10^6 \pm 5.999 \times 10^4$
$L^e \pm SD^{(e)}$	$3.000 \times 10^6 \pm 6.000 \times 10^4$

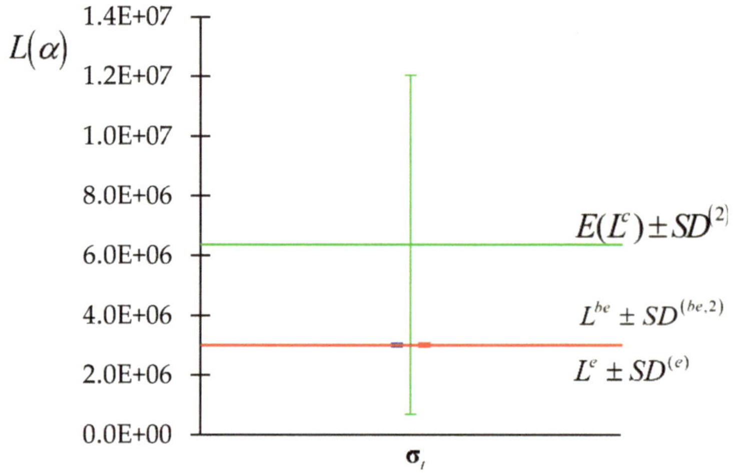

FIGURE 2.20 Comparison of $E(L^c) \pm SD^{(2)}$ (in green), $L^{(be,2)} \pm SD^{(be,2)}$ (in red), $L^e \pm SD^{(e)}$ (in blue), due to 10% standard deviations for uncorrelated and normally distributed total microscopic cross-sections σ_t, when the 1st+2nd-order sensitivities are included.

by the inaccurate parameters to yield a best-estimate nominal response value that practically coincides with the precisely measured response value, while significantly reducing the accompanying predicted best-estimate standard deviations.

4. Intercomparing the results presented in Tables 2.16 and 2.17 indicates that the inclusion of the 2nd-order sensitivities had no practical effects on either the predicted best-estimate response value or on its predicted standard deviation. In turn, this result indicates that higher-order sensitivities are not needed; the values of the best-estimate response and accompanying best-estimate standard deviation have been already provided by using just the 1st-order sensitivities (results presented in Table 2.16), as confirmed by the results presented in Table 2.17, which imply that even the 2nd-order sensitivities have insignificant impact in the case considered (high precision measurement and imprecisely known model parameters).

2.10 LOW PRECISION APPARENTLY INCONSISTENT MEASURED RESPONSE ($L^e = 7.0 \times 10^6$ NEUTRONS/SEC; $SD^{(e)} = 10\%$); TYPICAL PRECISION PARAMETERS (SD=5%)

This section presents the results produced by using the 2nd-BERRU-PM methodology in conjunction with an experimental response measured with low precision, and relatively imprecisely known model parameters, all having relative standard deviations of 5%, which often occurs in practice. As has been previously discussed, the ratio of the 3rd-order term with respect to the 2nd-order term of the Taylor series is $0.97 < 1.00$, but the ratio of the 4th-order term with respect to the 3rd-order term of the Taylor series is $1.13 > 1.00$. These ratios indicate that relative standard deviations of 5% for the model parameters are "borderline" values, just outside of the radius of convergence of the Taylor series expansion of the model's leakage response in terms of the total cross-section parameters. Thus, this illustrative example involves a low precision measurement and model parameters that are representative of the "usual uncertainties encountered in practice." This example will underscore the impact of the higher-order response sensitivities when the parameter uncertainties are representative of the uncertainties usually encountered in practice while also being "borderline" in terms of the convergence of the Taylor series that underlies the determination of the statistics (expected values, variance, etc.) of the distribution of the computed response in the phase-space of imprecisely known model parameters. Subsection 2.10.1, presents results obtained when considering just the contributions from the 1st-order sensitivities of the leakage response to the total cross-sections; Subsection 2.10.2 presents results obtained when considering the contributions from the 1st- and 2nd-order sensitivities; Subsection 2.10.3 presents results obtained when considering the contributions from 1st-, 2nd -, and 3rd-order sensitivities; Subsection 2.10.4 presents results obtained when considering the contributions from 1st-, 2nd-, 3rd-, and 4th-order sensitivities.

2.10.1 Including Only Contributions from the 1st-Order Sensitivities of the Leakage Response to All Important Parameters

Table 2.18 presents numerical results obtained for the expected value and standard deviation of the computed response, together with the best-estimate mean value and best-estimate standard deviation for the best-estimate response, when considering the 1st-order sensitivities with respect to the total cross-sections and, in parallel, considering all of the important 1st-order response sensitivities (i.e., considering the largest 1st-order sensitivities with respect to the fission cross-section, fission parameters, and isotopic atomic densities, in addition to the 1st-order sensitivities with respect to the total cross). The numerical results presented in Table 2.18 are depicted in Figure 2.21.

The results presented in Table 2.18 and Figure 2.21 indicate that:

1. The inequality $E(L^c) < L^{(be,1)} < L^e$ is fulfilled as predicted by Eq. (2.29), not only when considering just the total cross-sections but also when considering all of the important model parameters. The predicted response value,

TABLE 2.18

Values of $E(L^c) \pm SD^{(1)}$, $L^{(be,1)} \pm SD^{(be,1)}$, $L^e \pm SD^{(e)}$, Due to 5% Standard Deviations for Uncorrelated and Normally Distributed Total Microscopic Cross-Sections σ_t, and All Important Parameters $(\sigma_t + \sigma_f + \nu + N)$, when Only the 1st-Order Sensitivities Are Considered

Responses	Numerical Values For σ_t	Numerical Values For $(\sigma_t + \sigma_f + \nu + N)$
$E(L^c) \pm SD^{(1)}$	$1.765 \times 10^6 \pm 9.246 \times 10^5$	$1.765 \times 10^6 \pm 1.107 \times 10^6$
$L^{(be,1)} \pm SD^{(be,1)}$	$5.093 \times 10^6 \pm 5.581 \times 10^5$	$5.505 \times 10^6 \pm 5.916 \times 10^5$
$L^e \pm SD^{(e)}$	$7.000 \times 10^6 \pm 7.000 \times 10^5$	$7.000 \times 10^6 \pm 7.000 \times 10^5$

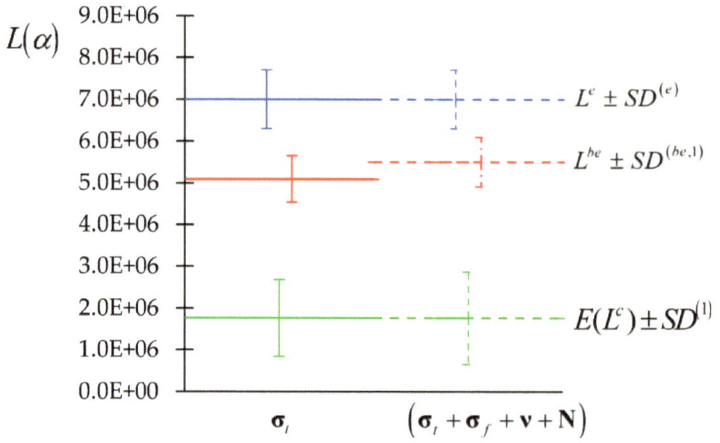

FIGURE 2.21 Comparison of $E(L^c) \pm SD^{(1)}$ (in green), $L^{(be,1)} \pm SD^{(be,1)}$ (in red), $L^e \pm SD^{(e)}$ (in blue), due to 5% standard deviations for uncorrelated and normally distributed total microscopic cross-sections σ_t, and all important parameters $(\sigma_t + \sigma_f + \nu + N)$, when the 1st-order sensitivities are considered.

$L^{(be,1)}$, is closer to the experimentally measured value when the 1st-order sensitivities to all important parameters are included.

2. The reduction of the predicted standard deviation to a value smaller than either the originally computed or the measured standard deviations, in other words, $SD^{(be,1)} < SD^{(e)}$ and $SD^{(be,1)} < SD^{(1)}$ are also apparent.
3. The computed results are apparently inconsistent with the experimental results, both when only the total cross-sections are considered and also when all-important model parameters are considered.
4. The contributions of the 1st-order sensitivities stemming from the additional parameters (i.e., the parameters in addition to the total cross-sections) are insignificant.

2.10.2 Including Contributions from the 1st+2nd-Order Sensitivities of the Leakage Response to All Important Parameters

Table 2.19 presents numerical results obtained for the expected value and standard deviation of the computed response, together with the best-estimate mean value and best-estimate standard deviation for the best-estimate response, when considering the 1st- and 2nd-order sensitivities with respect to the total cross-sections. In parallel, Table 2.19 also presents numerical results obtained for the expected value and standard deviation of the computed response, together with the best-estimate mean value and best-estimate standard deviation for the best-estimate response, when considering the 1st- and 2nd-order sensitivities with respect to all of the significant model parameters (total cross-sections, fission cross-section, fission parameters, and isotopic atomic densities). The numerical results presented in Table 2.19 are depicted in Figure 2.22.

The results presented in Table 2.19 and Figure 2.22 indicate that:

1. The inequality $E(L^c) < L^{(be,2)} < L^e$ is fulfilled both when considering only the total cross-sections and when considering all of the important model parameters. The predicted response value, $L^{(be,2)}$, is closer to the experimentally measured value when the 1st- and 2nd-order sensitivities to all important parameters are included.
2. The reduction of the predicted standard deviation to a value smaller than either the originally computed or the measured standard deviations, in other words, $SD^{(be,2)} < SD^{(e)}$ and $SD^{(be,2)} < SD^{(2)}$ are apparent.
3. The computed results are apparently still inconsistent (albeit less so than when only the 1st-order sensitivities are included) with the experimental results, both if only the total cross-sections and also if all-important model parameters are considered.
4. The contributions of the 1st- and 2nd-order sensitivities stemming from the additional parameters (i.e., the parameters in addition to the total

TABLE 2.19

Values of $E(L^c) \pm SD^{(2)}$, $L^{(be,2)} \pm SD^{(be,2)}$, $L^e \pm SD^{(e)}$, Due to 5% Standard Deviations for Uncorrelated Total Microscopic Cross-Sections σ_t and All Important Paramters $(\sigma_t + \sigma_f + \nu + N)$, when the 1st+2nd-Order Sensitivities Are Included

Responses	Numerical Values For σ_t	Numerical Values For $(\sigma_t + \sigma_f + \nu + N)$
$E(L^c) \pm SD^{(2)}$	$2.914 \times 10^6 \pm 1.629 \times 10^6$	$3.120 \times 10^6 \pm 1.754 \times 10^6$
$L^{(be,2)} \pm SD^{(be,2)}$	$6.363 \times 10^6 \pm 6.431 \times 10^5$	$6.467 \times 10^6 \pm 6.502 \times 10^5$
$L^e \pm SD^{(e)}$	$7.000 \times 10^6 \pm 7.000 \times 10^5$	$7.000 \times 10^6 \pm 7.000 \times 10^5$

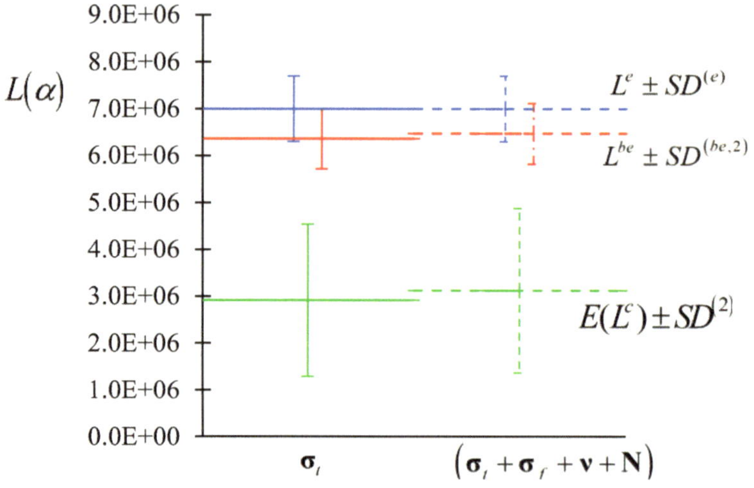

FIGURE 2.22 Comparison of $E(L^c) \pm SD^{(2)}$ (in green), $L^{(be,2)} \pm SD^{(be,2)}$ (in red), $L^e \pm SD^{(e)}$ (in blue), due to 5% standard deviations of the uncorrelated and normally distributed total microscopic cross-sections σ_t and all important parameters $(\sigma_t + \sigma_f + \nu + \mathbf{N})$, when the 1st+2nd-order sensitivities are included.

cross-sections) are insignificant. This observation indicates that the contributions of the higher-order sensitivities stemming from the model parameters other than the total cross-sections can be omitted by comparison to the contributions of the higher-order sensitivities stemming just from the total cross-sections.

2.10.3 Including Contributions from the 1st+2nd+3rd-Order Sensitivities of the Leakage Response to the Total Cross-Sections

Table 2.20 presents numerical results obtained for the expected value and standard deviation of the computed response, together with the best-estimate mean value and best-estimate standard deviation for the best-estimate response, when considering the 1st-, 2nd-, and 3rd-order sensitivities with respect to the total cross-sections. The numerical results presented in Table 2.20 are depicted in Figure 2.23.

The results presented in Table 2.20 and Figure 2.23 indicate that:

1. The inequality $E(L^c) < L^{(be,3)} < L^e$ is fulfilled. The predicted response value, $L^{(be,3)}$, is much closer to the experimentally measured value when also including the contributions stemming from the 3rd-order sensitivities (in addition to the contributions stemming from the 1st- and 2nd-order ones).
2. The computed standard deviation, $SD^{(3)}$, of the expected response value becomes unphysically large, indicating the divergence of the underlying Taylor series expansion of the response in terms of parameter variations.

TABLE 2.20
Values of $E(L^c) \pm SD^{(3)}$, $L^{(be,3)} \pm SD^{(be,3)}$, $L^e \pm SD^{(e)}$, Due to 5% Standard Deviations for Uncorrelated and Normally Distributed Total Microscopic Cross-Sections σ_t, when the 1st+2nd+3rd-Order Sensitivities Are Included

Responses	Numerical Values (only σ_t)
$E(L^c) \pm SD^{(3)}$	$2.914 \times 10^6 \pm 5.103 \times 10^6$
$L^{(be,3)} \pm SD^{(be,3)}$	$6.925 \times 10^6 \pm 6.935 \times 10^5$
$L^e \pm SD^{(e)}$	$7.000 \times 10^6 \pm 7.000 \times 10^5$

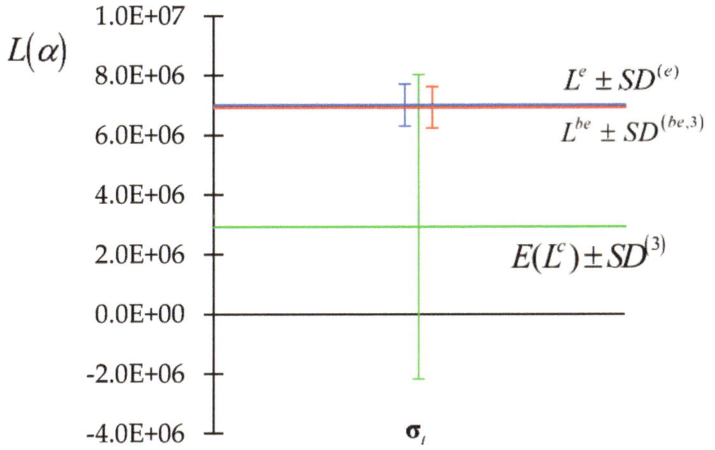

FIGURE 2.23 Comparison of $E(L^c) \pm SD^{(3)}$ (in green), $L^{(be,3)} \pm SD^{(be,3)}$ (in red), $L^e \pm SD^{(e)}$ (in blue), due to 5% standard deviations for uncorrelated and normally distributed total microscopic cross-sections σ_t, when the 1st+2nd+3rd-order sensitivities are included.

3. The reduction of the predicted standard deviation to a value smaller than either the originally computed or the measured standard deviations, in other words, $SD^{(be,3)} < SD^{(e)}$ and $SD^{(be,3)} < SD^{(3)}$ are apparent.
4. When the contributions stemming from the 3rd-order are also included, the computed results become consistent with the experimental results.

2.10.4 Including Contributions from the 1st+2nd+3rd+4th-Order Sensitivities of the Leakage Response to the Total Cross-Sections

Table 2.21 presents numerical results obtained for the best-estimate mean value and best-estimate standard deviation for the best-estimate response, together with numerical results for the expected value and standard deviation of the computed response, when considering the 1st-, 2nd-, 3rd-, and 4th-order sensitivities with respect to the total cross-sections. The numerical results presented in Table 2.21 are depicted in Figure 2.24.

The results presented in Table 2.21 and Figure 2.24 indicate that:

1. The inequality $E(L^c) < L^{(be,4)} < L^e$ is fulfilled. All three of these quantities have remarkably clustered together, with $L^{(be,4)} \cong L^e$.
2. The computed standard deviation, $SD^{(4)}$, of the expected response value becomes unphysically large, indicating the divergence of the underlying Taylor series expansion of the response in terms of parameter variations.
3. The reduction of the predicted standard deviation to a value smaller than either the originally computed or the measured standard deviations, (in other words, $SD^{(be,4)} < SD^{(e)}$ and $SD^{(be,4)} < SD^{(4)}$) is apparent.

TABLE 2.21

Values of $E(L^c) \pm SD^{(4)}$, $L^{(be,4)} \pm SD^{(be,4)}$, $L^e \pm SD^{(e)}$, Due to 5% Standard Deviations for Uncorrelated and Normally Distributed Total Microscopic Cross-Sections σ_t, when the 1st+2nd+3rd+4th-Order Sensitivities Are Included

Responses	Numerical Values (only σ_t)
$E(L^c) \pm SD^{(4)}$	$6.681 \times 10^6 \pm 7.386 \times 10^6$
$L^{(be,4)} \pm SD^{(be,4)}$	$6.997 \times 10^6 \pm 6.969 \times 10^5$
$L^e \pm SD^{(e)}$	$7.000 \times 10^6 \pm 7.000 \times 10^5$

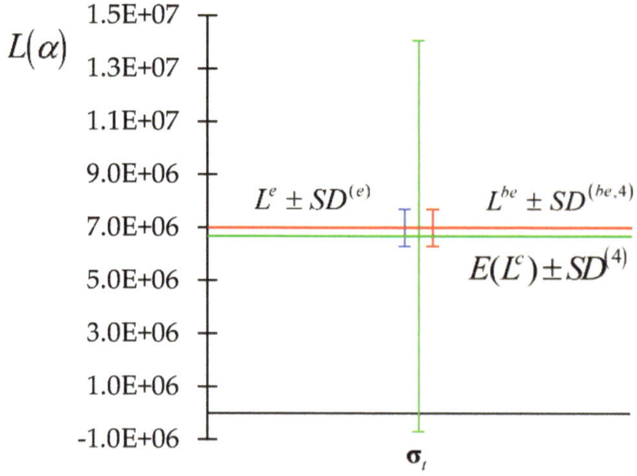

FIGURE 2.24 Comparison of $E(L^c) \pm SD^{(4)}$ (in green), $L^{(be,4)} \pm SD^{(be,4)}$ (in red), $L^e \pm SD^{(e)}$ (in blue), due to 5% standard deviations for uncorrelated and normally distributed total microscopic cross-sections σ_t, when the 1st+2nd+3rd+4th-order sensitivities are included.

4. Comparing the results presented in Table 2.20 with the corresponding results presented in Table 2.21 indicates that the 4th-order sensitivities contribute very little to the respective results; in turn, this observation indicates that higher-order sensitivities need not be considered.

2.11 MEASURED RESPONSE VALUE COINCIDES WITH NOMINALLY COMPUTED RESPONSE VALUE

Recall that the computed nominal value of the total leakage response for this benchmark is $L^c = 1.7648 \times 10^6$ neutrons/sec (where the superscript "c" denotes "computed quantities"). The situation that will be analyzed in this Subsection involves the unusual but possible situation when the measured value coincides with the computed value of the leakage response, both having the value $L^e = 1.765 \times 10^6$ neutrons/sec. As discussed in Chapter 1, *the conventional data adjustment and/or assimilation methods fail when the measured and computed response values coincide.*

The measurement is considered to have a "moderate" precision, with $SD^{(e)} = 5\%$, and that the parameters are also known with moderate precision, all having relative standard deviations of 5%. As has been already discussed, the Taylor series expansion of the computed response in terms of the model parameters is divergent for uniform relative standard deviations of 5% for the model parameters (total cross-sections).

The results for this case are presented in Table 2.22 and depicted in Figure 2.25. As expected, the Taylor series is divergent, as indicated by the unphysically large standard deviations $SD^{(3)}$ and $SD^{(4)}$ depicted in Figures 2.25c and 2.25.d. Nevertheless, the 2nd-BERRU-PM methodology not only "works" (as opposed to failing, like the data adjustment and assimilation procedures) but actually predicts the best-estimate response value to be very close to the measured one, having a

TABLE 2.22

Results for $E^{(i)}(L^c) \pm SD^{(i)}$ and $L^{(be,i)} \pm SD^{(be,i)}$, $i = 1,...,4$, for Coinciding Measured and Computed Response Values, $L^e = 1.765 \times 10^6 = L^c(\alpha^0)$, $SD^{(e)} = 5\%$; Parameters Relative $SD = 5\%$

Order of Sensitivities Considered	Values of $L^e \pm SD^{(e)}$ (Neutrons/sec)	Values of $E^{(i)}(L^c) \pm SD^{(i)}$ (Neutrons/sec)	Values of $L^{(be,i)} \pm SD^{(be,i)}$ (Neutrons/sec)
1st-order ($i = 1$)	$1.765 \times 10^6 \pm 8.824 \times 10^4$	$1.765 \times 10^6 \pm 9.246 \times 10^5$	1.765×10^6 $\pm 8.784 \times 10^4$
1st+2nd-order ($i = 2$)		$2.914 \times 10^6 \pm 1.629 \times 10^6$	1.768×10^6 $\pm 8.811 \times 10^4$
1st+2nd+3rd-order ($i = 3$)		$2.914 \times 10^6 \pm 5.103 \times 10^6$	1.766×10^6 $\pm 8.822 \times 10^4$
1st+2nd+3rd+4th-order ($i = 4$)		$6.681 \times 10^6 \pm 7.386 \times 10^6$	1.766×10^6 $\pm 8.823 \times 10^4$

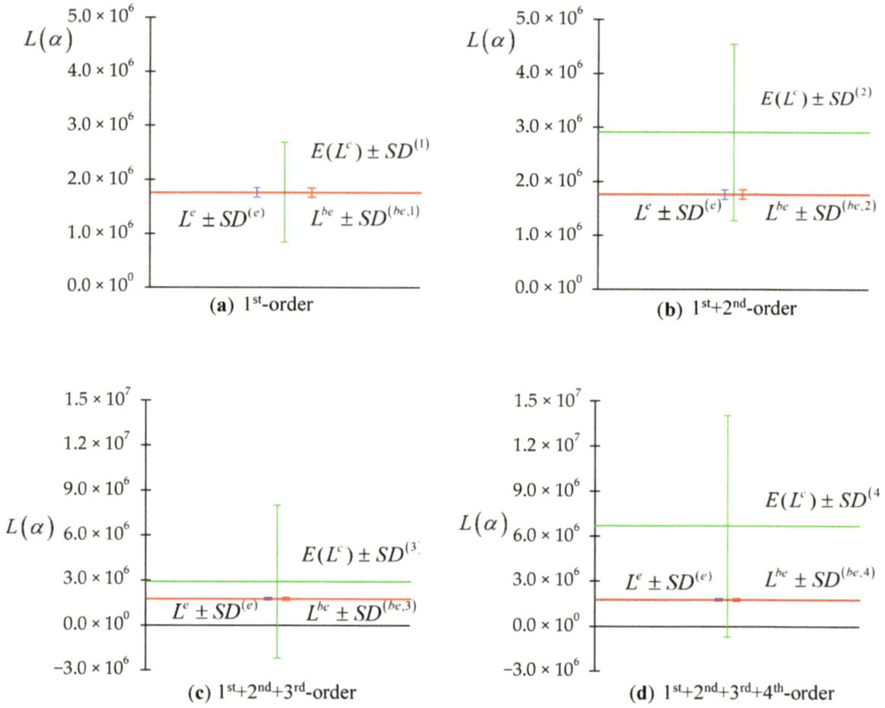

FIGURE 2.25 Comparison of $E^{(i)}(L^c) \pm SD^{(i)}$ (in green), $L^{(be,i)} \pm SD^{(be,i)}$ (in red), and $L^e \pm SD^{(e)}$ (in blue), for response measurement $L^e = 1.765 \times 10^6 = L^c\left(\boldsymbol{\alpha}^0\right)$ with $SD^{(e)} = 5\%$, parameters $SD = 5\%$. (i) 1st-order $(i = 1)$ sensitivities are considered. (ii) 1st+2nd-order $(i = 2)$ sensitivities are included. (iii) 1st+2nd+3rd-order $(i = 3)$ sensitivities are included. (iv) 1st+2nd+3rd+4th-order $\left(i = 4\right)$ sensitivities are included.

predicted standard deviation also close to (albeit marginally smaller, as expected) the measured standard deviation, since the measured standard deviation is smaller than the computed one.

2.12 CONCLUDING REMARKS

This chapter has presented applications of the 2nd-BERRU-PM methodology to the *p*oly*e*thylene-*r*eflected *p*lutonium (acronym: PERP) OECD/NEA reactor physics benchmark (Valentine, 2006) predicted best-estimate standard deviations. This benchmark is modeled using the neutron transport Boltzmann equation, the solution of which is representative of "large-scale computations" and involves a large number (21,976) of uncertain parameters. Several representative practical situations have been considered, as follows: (i) high precision measured response and high precision parameters; (ii) high precision measured response and low precision parameters that would cause the Taylor series expansion of the response in terms of parameter deviations to diverge; (iii) typical and/or low precision measured response and various parameter precisions. In all situations, the 2nd-BERRU-PM methodology predicted best-estimate response values that would fall in between the computed and measured values of the response. Also as predicted by the 2nd-BERRU-PM methodology, it has been observed that the predicted standard deviation of the predicted response is smaller than either the measured or the computed standard deviation, regardless of the order of sensitivities included in the respective computation, in other words, $SD^{(be)} < SD^{(e)}$ and $SD^{(be)} < SD^{(comp)}$. It has also been observed that when the parameters are known with high precision, the contributions of the higher-order sensitivities diminish with increasing order, so that the inclusion of the 2nd-order sensitivities (in addition to the 1st-order sensitivities), while necessary, may suffice for obtaining accurate predicted best-estimate response values and best-estimate standard deviations. On the other hand, when the parameters' standard deviations are sufficiently large to approach (or be outside of) the radius of convergence of the multivariate Taylor series which represents the response in the phase-space of model parameters, the contributions of the 3rd- and even 4th-order sensitivities are necessary to investigate the consistency (or inconsistency) between the computed and measured response, even though the contributions from these higher-order sensitivities may cause the standard deviation of the computed response to become unphysically large.

This chapter has also considered situations in which the measurements and the computations of the leakage response may appear to be inconsistent with one another in that their respective computed and measured standard deviations do not initially overlap when including just the 1st-order sensitivities. However, this inconsistency diminishes as higher-order sensitivities are progressively included, since the standard deviation of the computed response increases. In all cases, the 2nd-BERRU-PM methodology yields a predicted mean value for the response which is closer to either the measured or the computed response, depending on which of these has a smaller standard deviation response, in other words, is known more accurately. Furthermore, in all cases, the 2nd-BERRU-PM methodology yields predicted best-estimate standard deviations which are smaller than both the computed and the experimentally measured values of the standard deviations for the leakage response. This unique feature enjoyed by the 2nd-BERRU-PM methodology (to reduce the predicted standard deviations of the predicted results to values smaller than either the experimentally measured or the computed values of the respective standard deviations) stems from the information-theoretical foundation of the 2nd-BERRU-PM methodology, which ensures that the incorporation of additional (consistent) information reduces the predicted uncertainties.

The results presented in this chapter indicate the general conclusion that at least the 1st- and 2nd-order sensitivities need to be computed and included in any methodology that combines computational and experimental information. In particular, when the computed responses appear to be inconsistent with the measured responses, the use of only the 1st-order sensitivities may erroneously indicate that the respective measurement and computation are inconsistent with each other. Although the higher-order sensitivities may contribute little in special situations, for example, when the model parameters are known with high precision, at least the 2nd-order sensitivities need to be computed and possibly included, while the 3rd-order sensitivities would be needed in order to quantitatively confirm or infirm the necessity of including sensitivities of order three and/or higher-order.

3 A Novel Generic 4th-Order Moment-Constrained Maximum Entropy Distribution

3.1. INTRODUCTION

Although the 2nd-BERRU-PM methodology presented in Chapter 1 can incorporate arbitrarily high sensitivities, this methodology is limited to considering just 2nd-order moments (hence the designation "2nd-") of the experimentally measured and computed model responses. The 3rd-order moments (skewness) and 4th-order moments (kurtosis) of the distribution of measured and computed responses cannot be considered within the framework of the 2nd-BERRU-PM methodology. Furthermore, the "output" produced by the 2nd-BERRU-PM methodology is limited to yielding optimal best-estimate values for the means and covariances (i.e., the 1st- and 2nd-order moments) of the best-estimate predicted distribution of responses and parameters.

On the other hand, skewness and kurtosis play an essential role in determining the asymmetries of distributions. It is therefore paramount to generalize the 2nd-BERRU-PM methodology to enable the incorporation of 3rd- and 4th-order moments of measured and computed responses, as well as to enable the computation of skewness and kurtosis of the best-estimate predicted posterior distribution of calibrated model parameters and responses. However, such a generalization can be achieved only if the underlying moment-constrained MaxEnt distribution can be correspondingly generalized to include 3rd- and 4th-order moments of the distributions of measured and computed model responses.

Variants of the "moment-constrained maximum entropy problem" arise in areas as diverse as solid-state physics (see, e.g., Bandyopadhyay et al., 2005) econometrics (Wu, 2003), statistical description of gas flows (see, e.g., Muller and Ruggeri, 1993), weather and climate prediction (see, e.g., Roulston and Smith, 2002; Abramov et al., 2005; Haven et al., 2005). It appears that the most efficient computational algorithm currently available for solving the "multidimensional moment-constrained maximum entropy problem" is the algorithm devised by Abramov (2009), which is practically capable of solving two-dimensional problems with moments of order up to eight, three-dimensional problems with moments of order up to six, and four-dimensional problems of order up to four. But, being able to solve "four-dimensional problems of order up to four" is woefully insufficient for realistic problems, which are characterized by multivariate quantities of interest having dimensions much larger than just "four-dimensional."

 DOI: 10.1201/9781003478119-3

This chapter presents the novel, closed-form, 4th-order moment-constrained MaxEnt probabilistic representation of the distribution of an empirically (or experimentally) obtained multivariate quantity, the components of which will be called "responses." Such responses can be produced by computations, measurements, or inferences based on expert opinion, and are all subject to uncertainties. The first four moments of the unknown multivariate distribution of these responses, namely their mean values, variances/covariances, skewness, and kurtosis, are assumed to be known. The generic 4th-order moment-constrained MaxEnt distribution presented in this chapter is based on the original work by Cacuci (2023d) and will provide the foundation for constructing the mathematical framework of the 4th-BERRU-PM (4th-order) methodology, which will be presented in Chapter 4.

3.2 CONSTRUCTION OF THE 4TH-ORDER MOMENT-CONSTRAINED MAXIMUM ENTROPY (MAXENT) REPRESENTATION OF UNCERTAIN MULTIVARIATE QUANTITIES

In order to introduce the mathematical definitions of the known information (i.e., the moments of the unknown distribution of system responses), the unknown distribution of measured model system responses will be denoted as $p_e(\mathbf{r})$, $\mathbf{r} \triangleq (r_1,\ldots,r_{TR})^\dagger$, where r_i denotes the "ith-system response", $i = 1,\ldots,TR$, and where TR denotes the total number of system responses of interest. The unknown distribution $p_e(\mathbf{r})$ is formally defined on a TR-dimensional real-valued domain $D_e \subset \mathbb{R}^{TR}$. The letter "e" will be used either as a superscript or a superscript to indicate "experimentally" or "empirically" obtained (e.g., measured or computed with an inaccessible model) quantities.

For developing the mathematical framework of the *4th-order MaxEnt distribution corresponding to* $p_e(\mathbf{r})$, it is considered that the first four moments, namely: mean values, variances/covariances, triple correlations (skewness), and quadruple correlations (kurtosis), of the unknown distribution of system responses are available. The formal mathematical definitions of the first four moments, considered to be known, of the unknown distribution $p_e(\mathbf{r})$, are as follows:

1. Known expectation values, denoted as r_i^e, for the system responses r_i, where $i = 1,\ldots,TR$:

$$r_i^e \triangleq \int_{D_e} r_i p_e(\mathbf{r})\, d\mathbf{r};\ \mathbf{r}^e \triangleq \left(r_1^e,\ldots,r_i^e,\ldots,r_{TR}^e\right)^\dagger;\ i = 1,\ldots,TR. \tag{3.1}$$

2. Known covariances, denoted as c_{ij}^e, for two system responses r_i and r_j, where $i, j = 1,\ldots,TR$:

$$c_{ij}^e \triangleq \operatorname{cov}\left(r_i, r_j\right)_e \triangleq \int_{D_e} \left(r_i - r_i^e\right)\left(r_j - r_j^e\right) p_e(\mathbf{r})\, d\mathbf{r};\ i, j = 1,\ldots,TR; \tag{3.2}$$

The covariances, $\text{cov}\left(r_i, r_j\right)_e$, $i, j = 1, \ldots, TR$, of the system responses are considered to be components of the $TR \times TR$-dimensional covariance matrix of system responses, which will be denoted as $\mathbf{C}_{rr}^e \triangleq \left[\text{cov}\left(r_i, r_j\right)_e\right]_{TR \times TR} \triangleq \left[c_{ij}^e\right]_{TR \times TR}$.

3. Known triple correlations, denoted as t_{ijk}^e, for three system responses r_i, r_j, r_k, where $i, j, k = 1, \ldots, TR$:

$$t_{ijk}^e \triangleq \int_{D_e} \left(r_i - r_i^e\right)\left(r_j - r_j^e\right)\left(r_k - r_k^e\right) p_e(\mathbf{r})\, d\mathbf{r}; \quad i, j, k = 1, \ldots, TR; \tag{3.3}$$

4. Known quadruple correlations, denoted as $q_{ijk\ell}^e$, for four system responses r_i, r_j, r_k, r_ℓ, where $i, j, k, \ell = 1, \ldots, TR$:

$$\begin{aligned} q_{ijk\ell}^e &\triangleq \int_{D_e} \left(r_i - r_i^e\right)\left(r_j - r_j^e\right)\left(r_k - r_k^e\right)\left(r_\ell - r_\ell^e\right) p_e(\mathbf{r})\, d\mathbf{r}; \\ &i, j, k, \ell = 1, \ldots, TR. \end{aligned} \tag{3.4}$$

When an unknown distribution, such as $p_e(\mathbf{r})$, needs to be reconstructed from a finite number of its known moments, the principle of maximum entropy (MaxEnt) originally formulated by Jaynes (1957) provides the optimal compatibility with the available information, while simultaneously ensuring minimal spurious information content, yielding an estimate of a probability density with highest uncertainty among all densities satisfying the known moment constraints. According to the MaxEnt principle, such a MaxEnt probability density would satisfy the "available information" provided in Eqs. (3.1)–(3.4), without implying any spurious information or hidden assumptions, if the following conditions are satisfied:

1. $p_e(\mathbf{r})$ maximizes the Shannon (1948) information entropy, S, which is defined:

$$S = -\int_{D_e} p_e(\mathbf{r}) \ln\left[p_e(\mathbf{r})\right] d\mathbf{r}, \tag{3.5}$$

2. $p_e(\mathbf{r})$ satisfies the four "moment-constraints" defined by Eqs. (3.1)–(3.4);
3. $p_e(\mathbf{r})$ satisfies the following normalization condition:

$$\int_{D_e} p_e(\mathbf{r})\, d\mathbf{r} = 1. \tag{3.6}$$

The MaxEnt distribution $p_e(\mathbf{r})$ is obtained as the solution of the variational problem $\partial H(p_e)/\partial p_e = 0$, where the entropy (Lagrangian functional) $H(p_e)$ is defined as follows:

$$\begin{aligned} H(p_e) = & -\int_{D_e} p_e(\mathbf{r})\ln\left[p_e(\mathbf{r})\right]d\mathbf{r} - c_0\left[\int_{D_r} p_e(\mathbf{r})d\mathbf{r} - 1\right] \\ & -\sum_{i=1}^{TR}\beta_i\int_{D_e}\left(r_i - r_i^e\right)p_e(\mathbf{r})d\mathbf{r} - \frac{1}{2}\sum_{i,j=1}^{TR}\lambda_{ij}\left[\int_{D_e}\left(r_i - r_i^e\right)\left(r_j - r_j^e\right)p_e(\mathbf{r})d\mathbf{r} - c_{ij}^e\right] \\ & -\sum_{i,j,k=1}^{TR}\theta_{ijk}\left[\int_{D_e}\left(r_i - r_i^e\right)\left(r_j - r_j^e\right)\left(r_k - r_k^e\right)p_e(\mathbf{r})d\mathbf{r} - t_{ijk}^e\right] \\ & -\sum_{i,j,k,\ell=1}^{TR}\omega_{ijk\ell}\left[\int_{D_e}\left(r_i - r_i^e\right)\left(r_j - r_j^e\right)\left(r_k - r_k^e\right)\left(r_\ell - r_\ell^e\right)p_e(\mathbf{r})d\mathbf{r} - q_{ijk\ell}^e\right]. \end{aligned} \tag{3.7}$$

In Eq. (3.7), the quantities c_0, β_i, λ_{ij}, θ_{ijk}, and $\omega_{ijk\ell}$ denote the respective Lagrange multipliers, and the factor 1/2 has been introduced for subsequent computational convenience. Introducing the variable:

$$\mathbf{z} \triangleq \mathbf{r} - \mathbf{r}^e \triangleq \left(z_1,\ldots,z_{TR}\right)^\dagger;\ \mathbf{r}^e \triangleq \left(r_1^e,\ldots,r_{TR}^e\right)^\dagger;\ z_i \triangleq r_i - r_i^e; \tag{3.8}$$

and solving the equation $\partial H(p_e)/\partial p_e = 0$ yields the following expression for the resulting MaxEnt distribution $p_e(\mathbf{z})$:

$$p_e(\mathbf{z}) = h(\mathbf{z})\left[\int_{D_r} h(\mathbf{z})d\mathbf{z}\right]^{-1}, \tag{3.9}$$

where:

$$h(\mathbf{z}) \triangleq \exp\left\{-\sum_{i=1}^{TR}\beta_i z_i - \frac{1}{2}\sum_{i,j=1}^{TR}\lambda_{ij}z_i z_j - \sum_{i,j,k=1}^{TR}\theta_{ijk}z_i z_j z_k - \sum_{i,j,k,\ell=1}^{TR}\omega_{ijk\ell}\, z_i z_j z_k z_\ell\right\}; \tag{3.10}$$

Re-writing Eqs. (3.1)–(3.4) in terms of the variable $\mathbf{z} \triangleq \mathbf{r} - \mathbf{r}^e$ yields the following formal expressions for the known first four moments of the MaxEnt distribution $p_e(\mathbf{z})$:

$$0 = \int_{D_e} z_i p_e(\mathbf{r})d\mathbf{r},\quad i = 1,\ldots,TR; \tag{3.11}$$

$$c_{ij}^e \triangleq \int_{D_e} z_i z_j p_e(\mathbf{r})d\mathbf{r};\quad i,j = 1,\ldots,TR; \tag{3.12}$$

$$t_{ijk}^e \triangleq \int_{D_e} z_i z_j z_k p_e(\mathbf{r})d\mathbf{r};\quad i,j,k = 1,\ldots,TR; \tag{3.13}$$

$$q_{ijk\ell}^{e} \triangleq \int_{D_e} z_i z_j z_k z_\ell \, p_e(\mathbf{r})\, d\mathbf{r}; \quad i,j,k,\ell = 1,\dots,TR. \tag{3.14}$$

The expression of $p_e(\mathbf{z})$ can be evaluated in closed-form up to the 5th-order in the variables z_i by expanding in Taylor series the 3rd- and 4th-order terms in the expression of $h(\mathbf{z})$, to obtain the following approximate expression:

$$p_e(\mathbf{z}) \cong \frac{f(\mathbf{z})\exp[-g(\mathbf{z})]}{Z(\boldsymbol{\beta})}, \tag{3.15}$$

where the following definitions have been used:

$$g(\mathbf{z}) \triangleq -\sum_{i=1}^{TR} \beta_i z_i - \frac{1}{2}\sum_{i,j=1}^{TR} \lambda_{ij} z_i z_j = -\boldsymbol{\beta}^\dagger \mathbf{z} - \frac{1}{2}\mathbf{z}^\dagger \boldsymbol{\Lambda}\mathbf{z};$$

$$\boldsymbol{\beta} \triangleq (\beta_1,\dots,\beta_{TR})^\dagger; \; \boldsymbol{\Lambda} \triangleq \begin{bmatrix} \lambda_{11} & \bullet & \lambda_{1,TR} \\ \bullet & \bullet & \bullet \\ \lambda_{TR,1} & \bullet & \lambda_{TR,TR} \end{bmatrix}; \tag{3.16}$$

$$f(\mathbf{z}) \triangleq 1 - \sum_{i,j,k=1}^{TR} \theta_{ijk} z_i z_j z_k - \sum_{i,j,k,\ell=1}^{TR} \omega_{ijk\ell}\, z_i z_j z_k z_\ell; \tag{3.17}$$

$$\begin{aligned} Z(\boldsymbol{\beta}) &\triangleq \int_{D_r} f(\mathbf{z})\exp[-g(\mathbf{z})]\, d\mathbf{z} \\ &= F_0(\boldsymbol{\beta}) - \sum_{i,j,k=1}^{TR} \theta_{ijk} F_3^{ijk}(\boldsymbol{\beta}) - \sum_{i,j,k,\ell=1}^{TR} \omega_{ijk\ell}\, F_4^{ijk\ell}(\boldsymbol{\beta}); \end{aligned} \tag{3.18}$$

with the following definitions:

$$F_0(\boldsymbol{\beta}) \triangleq \int_{D_r} \exp\left\{-\boldsymbol{\beta}^\dagger \mathbf{z} - \frac{1}{2}\mathbf{z}^\dagger \boldsymbol{\Lambda}\mathbf{z}\right\} d\mathbf{z}; \tag{3.19}$$

$$F_3^{ijk}(\boldsymbol{\beta}) \triangleq \int_{D_r} z_i z_j z_k \exp\left\{-\sum_{i=1}^{TR} \beta_i z_i - \frac{1}{2}\sum_{i,j=1}^{TR} \lambda_{ij} z_i z_j\right\} d\mathbf{z}; \tag{3.20}$$

$$F_4^{ijk\ell}(\boldsymbol{\beta}) \triangleq \int_{D_r} z_i z_j z_k z_\ell \exp\left\{-\sum_{i=1}^{TR} \beta_i z_i - \frac{1}{2}\sum_{i,j=1}^{TR} \lambda_{ij} z_i z_j\right\} d\mathbf{z}. \tag{3.21}$$

The closed-form expression of the integral shown in Eq. (3.19) is well-known:

$$F_0(\boldsymbol{\beta}) \triangleq \int_{D_r} \exp\left\{-\boldsymbol{\beta}^\dagger \mathbf{z} - \frac{1}{2}\mathbf{z}^\dagger \boldsymbol{\Lambda}\mathbf{z}\right\} d\mathbf{z} = K \exp\left(\frac{1}{2}\boldsymbol{\beta}^\dagger \boldsymbol{\Lambda}^{-1}\boldsymbol{\beta}\right), \tag{3.22}$$

where:

$$K \triangleq \frac{(2\pi)^{TR/2}}{\sqrt{Det(\mathbf{\Lambda})}}; \tag{3.23}$$

It follows from the definitions provided in Eqs. (3.19)–(3.21) that:

$$F_3^{ijk}(\boldsymbol{\beta}) = -\frac{\partial^3 F_0(\boldsymbol{\beta})}{\partial\beta_i\partial\beta_j\partial\beta_k}; \quad F_4^{ijk\ell}(\boldsymbol{\beta}) = \frac{\partial^4 F_0(\boldsymbol{\beta})}{\partial\beta_i\partial\beta_j\partial\beta_k\partial\beta_\ell}. \tag{3.24}$$

Inserting the results obtained in Eqs. (3.19)–(3.24) into Eq. (3.18) yields the following expression:

$$Z(\boldsymbol{\beta}) = F_0(\boldsymbol{\beta}) + \sum_{i,j,k=1}^{TR} \theta_{ijk} \frac{\partial^3 F_0(\boldsymbol{\beta})}{\partial\beta_i\partial\beta_j\partial\beta_k} - \sum_{i,j,k,\ell=1}^{TR} \omega_{ijk\ell} \frac{\partial^4 F_0(\boldsymbol{\beta})}{\partial\beta_i\partial\beta_j\partial\beta_k\partial\beta_\ell}. \tag{3.25}$$

The explicit expressions for the quantities $\partial^3 F_0(\boldsymbol{\beta})/\partial\beta_i\partial\beta_j\partial\beta_k$ and $\partial^4 F_0(\boldsymbol{\beta})/\partial\beta_i\partial\beta_j\partial\beta_k\partial\beta_\ell$ are derived in Appendix 3.A.

Inserting into Eq. (3.25) the expression obtained in Eq. (3.19) for $F_0(\boldsymbol{\beta})$ together with the expressions obtained in Eqs. (3.64) and (3.66) in Appendix 3.A for the quantities $\partial^3 F_0(\boldsymbol{\beta})/\partial\beta_i\partial\beta_j\partial\beta_k$ and $\partial^4 F_0(\boldsymbol{\beta})/\partial\beta_i\partial\beta_j\partial\beta_k\partial\beta_\ell$, respectively, yields the following expression for the normalization integral $Z(\boldsymbol{\beta})$:

$$Z(\boldsymbol{\beta}) = K\left\{1 + \sum_{i,j,k=1}^{TR} \theta_{ijk} G_3^{ijk}(\boldsymbol{\beta}) - \sum_{i,j,k,\ell=1}^{TR} \omega_{ijk\ell}\, G_4^{ijk\ell}(\boldsymbol{\beta})\right\} \exp\left(\frac{1}{2}\boldsymbol{\beta}^\dagger \mathbf{\Lambda}^{-1}\boldsymbol{\beta}\right), \tag{3.26}$$

where the quantities $G_3^{ijk}(\boldsymbol{\beta})$ and $G_4^{ijk\ell}(\boldsymbol{\beta})$ are defined in Appendix 3.A.

In terms of the approximate expression shown in Eq. (3.15) for the distribution $p_e(\mathbf{z})$, the moments defined by Eqs. (3.11)–(3.14) take on the following expressions:

$$0 = \frac{1}{Z(\boldsymbol{\beta})}\int_{D_e} z_i f(\mathbf{z}) \exp[-g(\mathbf{z})]\, d\mathbf{z} = -\frac{1}{Z(\boldsymbol{\beta})}\frac{\partial Z(\boldsymbol{\beta})}{\partial\beta_i}, \quad i = 1,\ldots,TR; \tag{3.27}$$

$$c_{ij}^e = \frac{1}{Z(\boldsymbol{\beta})}\int_{D_e} z_i z_j f(\mathbf{z}) \exp[-g(\mathbf{z})]\, d\mathbf{z} = \frac{1}{Z(\boldsymbol{\beta})}\frac{\partial^2 Z(\boldsymbol{\beta})}{\partial\beta_i\partial\beta_j}; \quad i,j = 1,\ldots,TR; \tag{3.28}$$

$$\begin{aligned} t_{ijk}^e &= \frac{1}{Z(\boldsymbol{\beta})}\int_{D_e} z_i z_j z_k f(\mathbf{z}) \exp[-g(\mathbf{z})]\, d\mathbf{z} \\ &= -\frac{1}{Z(\boldsymbol{\beta})}\frac{\partial^3 Z(\boldsymbol{\beta})}{\partial\beta_i\partial\beta_j\partial\beta_k}; \ i,j,k = 1,\ldots,TR; \end{aligned} \tag{3.29}$$

$$q_{ijk\ell}^{e}=\frac{\int_{D_e} z_i z_j z_k z_\ell f(\mathbf{z})\exp\left[-g(\mathbf{z})\right]d\mathbf{z}}{Z(\boldsymbol{\beta})}$$
$$=\frac{1}{Z(\boldsymbol{\beta})}\frac{\partial^4 Z(\boldsymbol{\beta})}{\partial\beta_i\partial\beta_j\partial\beta_k\partial\beta_\ell};\quad i,j,k,\ell=1,\ldots,TR. \tag{3.30}$$

The relation shown in Eq. (3.27) can be expressed in terms of the normalization integral $Z(\boldsymbol{\beta})$ as follows:

$$0=-\frac{1}{Z(\boldsymbol{\beta})}\frac{\partial Z(\boldsymbol{\beta})}{\partial\beta_i}=-\frac{\partial}{\partial\beta_i}\ln\left[Z(\boldsymbol{\beta})\right]. \tag{3.31}$$

It follows from Eq. (3.26) that:

$$\ln\left[Z(\boldsymbol{\beta})\right]=\frac{1}{2}\boldsymbol{\beta}^{\dagger}\boldsymbol{\Lambda}^{-1}\boldsymbol{\beta}+\ln K$$
$$+\ln\left\{1+\sum_{i,j,k=1}^{TR}\theta_{ijk}G_3^{ijk}(\boldsymbol{\beta})-\sum_{i,j,k,\ell=1}^{TR}\omega_{ijk\ell}G_4^{ijk\ell}(\boldsymbol{\beta})\right\}, \tag{3.32}$$

It follows from Eqs. (3.31) and (3.32) that:

$$0=-\frac{\partial\ln\left[Z(\boldsymbol{\beta})\right]}{\partial\beta_m}=-\sum_{b=1}^{TR}\left(\boldsymbol{\Lambda}^{-1}\right)_{mb}\beta_b$$
$$-\frac{\left[\sum_{i,j,k=1}^{TR}\theta_{ijk}\frac{\partial G_3^{ijk}(\boldsymbol{\beta})}{\partial\beta_m}\right]-\left[\sum_{i,j,k,\ell=1}^{TR}\omega_{ijk\ell}\frac{\partial G_4^{ijk\ell}(\boldsymbol{\beta})}{\partial\beta_m}\right]}{1+\sum_{i,j,k=1}^{TR}\theta_{ijk}G_3^{ijk}(\boldsymbol{\beta})-\sum_{i,j,k,\ell=1}^{TR}\omega_{ijk\ell}G_4^{ijk\ell}(\boldsymbol{\beta})}$$
$$\cong-\sum_{b=1}^{TR}\left(\boldsymbol{\Lambda}^{-1}\right)_{mb}\beta_b-\left[\sum_{i,j,k=1}^{TR}\theta_{ijk}\frac{\partial G_3^{ijk}(\boldsymbol{\beta})}{\partial\beta_m}\right]$$
$$+\left[\sum_{i,j,k,\ell=1}^{TR}\omega_{ijk\ell}\frac{\partial G_4^{ijk\ell}(\boldsymbol{\beta})}{\partial\beta_m}\right]+O\left(\theta_{ijk}^2\right). \tag{3.33}$$

The last approximate equality on the right side of Eq. (3.33) was obtained by expanding the respective denominator in a power series and neglecting the 6th- and higher-order terms.

It follows from the expressions obtained in Eqs. (3.68) and (3.71) of Appendix 3.A that:

$$\frac{\partial G_3^{ijk}(\boldsymbol{\beta})}{\partial\beta_m}=O\left(\boldsymbol{\Lambda}^{-2}\right);\quad \frac{\partial G_4^{ijk\ell}(\boldsymbol{\beta})}{\partial\beta_m}=O\left(\boldsymbol{\Lambda}^{-3}\boldsymbol{\beta}\right). \tag{3.34}$$

It follows from Eqs. (3.33) and (3.34) that if the 3rd-order correlations among the measured system responses are neglected by setting $\theta_{ijk} = 0,\ i,j,k = 1,\ldots,TR$, then the solution of these equations becomes:

$$\boldsymbol{\beta} = \mathbf{0},\ \ if\ \ \theta_{ijk} = 0,\ \ for\ \ i,j,k = 1,\ldots,TR. \tag{3.35}$$

The solution of Eq. (3.33) cannot be obtained analytically in closed-form if the 3rd-order correlations among the measured system responses are not neglected.

Differentiating the relation obtained in Eq. (3.31) with respect to another Lagrange multiplier, β_n, $n = 1,\ldots,TR$, yields the following relation:

$$\begin{aligned}
-\frac{\partial^2 \ln\left[Z(\boldsymbol{\beta})\right]}{\partial\beta_n\partial\beta_m} &= \frac{\partial}{\partial\beta_n}\left[-\frac{1}{Z(\boldsymbol{\beta})}\frac{\partial Z(\boldsymbol{\beta})}{\partial\beta_m}\right] \\
&= \frac{1}{Z^2(\boldsymbol{\beta})}\frac{\partial Z(\boldsymbol{\beta})}{\partial\beta_n}\frac{\partial Z(\boldsymbol{\beta})}{\partial\beta_m} - \frac{1}{Z(\boldsymbol{\beta})}\frac{\partial^2 Z(\boldsymbol{\beta})}{\partial\beta_n\partial\beta_m} \\
&= -\frac{1}{Z(\boldsymbol{\beta})}\int_{D_e} z_n z_m f(\mathbf{z})\exp\left[-g(\mathbf{z})\right]d\mathbf{z} \\
&= -c_{mn}^e;\ \ m,n = 1,\ldots,TR.
\end{aligned} \tag{3.36}$$

The last equality on the right side of Eq. (3.36) has been obtained by using Eqs. (3.31) and (3.35).

On the other hand, differentiating the relation shown in Eq. (3.33) with respect to β_n, $n = 1,\ldots,TR$, yields the following result:

$$\begin{aligned}
-\frac{\partial^2 \ln\left[Z(\boldsymbol{\beta})\right]}{\partial\beta_n\partial\beta_m} &\cong -\left(\boldsymbol{\Lambda}^{-1}\right)_{mn} - \sum_{i,j,k=1}^{TR}\theta_{ijk}\frac{\partial^2 G_3^{ijk}(\boldsymbol{\beta})}{\partial\beta_n\partial\beta_m} \\
&+ \sum_{i,j,k,\ell=1}^{TR}\omega_{ijk\ell}\frac{\partial^2 G_4^{ijk\ell}(\boldsymbol{\beta})}{\partial\beta_n\partial\beta_m} \cong -\left(\boldsymbol{\Lambda}^{-1}\right)_{mn}.
\end{aligned} \tag{3.37}$$

The result obtained on the rightmost side of Eq. (3.37) follows by evaluating all terms at $\boldsymbol{\beta} = \mathbf{0}$ (which implies that 3rd-order correlations were neglected) and by also neglecting the 4th-order correlations by setting $\omega_{ijk\ell} = 0$. The results obtained in Eqs. (3.37) and (3.36) indicate that:

$$\left(\boldsymbol{\Lambda}^{-1}\right)_{mn} = c_{mn}^e;\ \ \boldsymbol{\Lambda}^{-1} = \mathbf{C}^e;\ \ \mathbf{C}^e \triangleq \left(c_{mn}^e\right)_{TR\times TR};\ \ m,n = 1,\ldots,TR, \tag{3.38}$$

where $\mathbf{C}^e \triangleq \left(c_{mn}^e\right)_{TR\times TR}$ denotes the known $TR\times TR$-dimensional covariance matrix of the system responses. It follows from Eqs. (3.38) and (3.23) that:

$$K = (2\pi)^{TR/2}\sqrt{Det\left(\mathbf{C}^e\right)}. \tag{3.39}$$

In view of Eqs. (3.35) and (3.38), neglecting the 3rd- and 4th-order system response correlations while obtaining the results $\boldsymbol{\beta} = \mathbf{0}$ and $\boldsymbol{\Lambda}^{-1} = \mathbf{C}^e$, respectively, is equivalent to neglecting double- and triple-products of measured-system response correlations.

Differentiating the relation shown in Eq. (3.37) with respect to another Lagrange multiplier, β_μ, $\mu = 1,\ldots,TR$, yields the following result:

$$-\frac{\partial^3 \ln[Z(\boldsymbol{\beta})]}{\partial\beta_\mu \partial\beta_n \partial\beta_m} = -\sum_{i,j,k=1}^{TR} \theta_{ijk} \frac{\partial^3 G_3^{ijk}(\boldsymbol{\beta})}{\partial\beta_\mu \partial\beta_n \partial\beta_m} + \sum_{i,j,k,\ell=1}^{TR} \omega_{ijk\ell} \frac{\partial^3 G_4^{ijk\ell}(\boldsymbol{\beta})}{\partial\beta_\mu \partial\beta_n \partial\beta_m}. \tag{3.40}$$

On the other hand, differentiating the relation shown in Eq. (3.36) with respect to β_μ, and evaluating, the resulting expression by using the results obtained in Eqs. (3.35) and (3.38) for the Lagrange multipliers $\boldsymbol{\beta} = \mathbf{0}$ and $\boldsymbol{\Lambda}^{-1} = \mathbf{C}^e$ yields the following result:

$$\begin{aligned}
&-\left\{\frac{\partial^3 \ln[Z(\boldsymbol{\beta})]}{\partial\beta_\mu \partial\beta_n \partial\beta_m}\right\}_{\boldsymbol{\beta}=\mathbf{0}} = \left\{\frac{\partial}{\partial\beta_\mu}\left[\frac{1}{Z^2(\boldsymbol{\beta})}\frac{\partial Z(\boldsymbol{\beta})}{\partial\beta_n}\frac{\partial Z(\boldsymbol{\beta})}{\partial\beta_m} - \frac{1}{Z(\boldsymbol{\beta})}\frac{\partial^2 Z(\boldsymbol{\beta})}{\partial\beta_n \partial\beta_m}\right]\right\}_{\boldsymbol{\beta}=\mathbf{0}} \\
&= \left\{-\frac{2}{Z^3(\boldsymbol{\beta})}\frac{\partial Z(\boldsymbol{\beta})}{\partial\beta_\mu}\frac{\partial Z(\boldsymbol{\beta})}{\partial\beta_n}\frac{\partial Z(\boldsymbol{\beta})}{\partial\beta_m} + \frac{1}{Z^2(\boldsymbol{\beta})}\left[\frac{\partial^2 Z(\boldsymbol{\beta})}{\partial\beta_\mu \partial\beta_n}\frac{\partial Z(\boldsymbol{\beta})}{\partial\beta_m} + \frac{\partial Z(\boldsymbol{\beta})}{\partial\beta_n}\frac{\partial^2 Z(\boldsymbol{\beta})}{\partial\beta_\mu \partial\beta_m}\right]\right\}_{\boldsymbol{\beta}=\mathbf{0}} \\
&+ \left\{\frac{1}{Z^2(\boldsymbol{\beta})}\frac{\partial Z(\boldsymbol{\beta})}{\partial\beta_\mu}\frac{\partial^2 Z(\boldsymbol{\beta})}{\partial\beta_n \partial\beta_m} - \frac{1}{Z(\boldsymbol{\beta})}\frac{\partial^3 Z(\boldsymbol{\beta})}{\partial\beta_\mu \partial\beta_n \partial\beta_m}\right\}_{\boldsymbol{\beta}=\mathbf{0}} \\
&= -\left\{\frac{1}{Z(\boldsymbol{\beta})}\frac{\partial^3 Z(\boldsymbol{\beta})}{\partial\beta_\mu \partial\beta_n \partial\beta_m}\right\}_{\boldsymbol{\beta}=\mathbf{0}} \\
&= \left\{\frac{1}{Z(\boldsymbol{\beta})}\int_{D_e} z_\mu z_n z_m f(\mathbf{z}) \exp[-g(\mathbf{z})]\, d\mathbf{z}\right\}_{\boldsymbol{\beta}=\mathbf{0}} \\
&= t^e_{\mu nm}; \quad \mu, n, m = 1,\ldots,TR.
\end{aligned} \tag{3.41}$$

The last equality shown on the right side of Eq. (3.41) follows from the relation provided in Eq. (3.29).

Evaluating the expression obtained in Eq. (3.40) by using Eqs. (3.70) and (3.79) from Appendix 3.A in conjunction with the results obtained in Eq. (3.35) and (3.38) for the Lagrange multipliers $\boldsymbol{\beta} = \mathbf{0}$ and $\boldsymbol{\Lambda}^{-1} = \mathbf{C}^e$ yields the following relation for determining the Lagrange multipliers θ_{ijk}:

$$\begin{aligned}
&\sum_{i,j,k=1}^{TR} \theta_{ijk} \left\{c^e_{im}\left[c^e_{jn} c^e_{k\mu} + c^e_{j\mu} c^e_{kn}\right] + c^e_{jm}\left[c^e_{in} c^e_{k\mu} + c^e_{i\mu} c^e_{kn}\right] + c^e_{km}\left[c^e_{in} c^e_{j\mu} + c^e_{i\mu} c^e_{jn}\right]\right\} \\
&= -t^e_{\mu nm}; \quad \mu, n, m = 1,\ldots,TR.
\end{aligned} \tag{3.42}$$

Evidently, Eq. (3.42) could be solved numerically, but cannot be solved analytically to obtain a closed-form solution in the general case. Furthermore, the triple cross-correlations $t^e_{\mu nm}, \mu, n, m = 1,...,TR$, are seldom available in practice. Usually, only the 3rd-order self-correlations $t^e_a \triangleq t^e_{aaa}$, for each system response r^e_a, $a = 1,...,TR$, are available in practice for the measured system responses. In this case, Eq. (3.42) will reduce to the following form:

$$\sum_{b=1}^{TR} \tau_{ab}\theta_b = -t^e_a / 6; \quad \tau_{ab} \triangleq \left(c^e_{ab}\right)^3 = \left(c^e_{ba}\right)^3; \quad \theta_b \triangleq \theta_{bbb}; \quad a, b = 1,...,TR. \tag{3.43}$$

Eq. (3.43) can be solved explicitly by inverting the matrix $\mathbf{T} \triangleq (\tau_{ab})$ to obtain:

$$\boldsymbol{\theta} = -\mathbf{T}^{-1}\mathbf{t}^e/6; \, \boldsymbol{\theta} \triangleq (\theta_1,...,\theta_{TR})^\dagger; \; \mathbf{t}^e \triangleq \left(t^e_1,...,t^e_{TR}\right)^\dagger; \mathbf{T} \triangleq (\tau_{ab})_{TR\times TR}. \tag{3.44}$$

In particular, if only the system response variances are available but the 2nd-order correlations among the system responses are negligible or unavailable, then the result obtained in Eq. (3.44) reduces to the following expression for determining the Lagrange multipliers θ_k, $k = 1,...,TR$:

$$\theta_k = -\frac{t^e_k}{6\left(c^e_{kk}\right)^3}; \quad k = 1,...,TR. \tag{3.45}$$

Differentiating the relation provided in Eq. (3.40) with respect to another Lagrange multiplier, β_ν, $\nu = 1,...,TR$, yields the following result:

$$-\frac{\partial^4 \ln[Z(\boldsymbol{\beta})]}{\partial\beta_\nu \partial\beta_\mu \partial\beta_n \partial\beta_m} = \sum_{i,j,k,\ell=1}^{TR} \omega_{ijk\ell} \frac{\partial^4 G_4^{ijk\ell}(\boldsymbol{\beta})}{\partial\beta_\nu \partial\beta_\mu \partial\beta_n \partial\beta_m}; \quad \nu, \mu, n, m = 1,...,TR. \tag{3.46}$$

The expression of $\partial^4 G_4^{ijk\ell}(\boldsymbol{\beta})/\partial\beta_\nu \partial\beta_\mu \partial\beta_n \partial\beta_m$ is provided in Eq. (3.84) in Appendix 3.A.

On the other hand, differentiating the relation shown in Eq. (3.41) with respect to β_ν, and evaluating the resulting expression by using the results obtained in Eqs. (3.35) and (3.38) for the Lagrange multipliers $\boldsymbol{\beta} = \mathbf{0}$ and $\boldsymbol{\Lambda}^{-1} = \mathbf{C}^e$, yields the following expression:

$$\begin{aligned} -\frac{\partial^4 \ln[Z(\boldsymbol{\beta})]}{\partial\beta_\nu \partial\beta_\mu \partial\beta_n \partial\beta_m} &= \frac{\partial}{\partial\beta_\nu}\left\{-\frac{2}{Z^3(\boldsymbol{\beta})}\frac{\partial Z(\boldsymbol{\beta})}{\partial\beta_\mu}\frac{\partial Z(\boldsymbol{\beta})}{\partial\beta_n}\frac{\partial Z(\boldsymbol{\beta})}{\partial\beta_m}\right\} \\ &+ \frac{\partial}{\partial\beta_\nu}\left\{\frac{1}{Z^2(\boldsymbol{\beta})}\left[\frac{\partial^2 Z(\boldsymbol{\beta})}{\partial\beta_\mu \partial\beta_n}\frac{\partial Z(\boldsymbol{\beta})}{\partial\beta_m} + \frac{\partial Z(\boldsymbol{\beta})}{\partial\beta_n}\frac{\partial^2 Z(\boldsymbol{\beta})}{\partial\beta_\mu \partial\beta_m}\right]\right\} \\ &+ \frac{\partial}{\partial\beta_\nu}\left\{\frac{1}{Z^2(\boldsymbol{\beta})}\frac{\partial Z(\boldsymbol{\beta})}{\partial\beta_\mu}\frac{\partial^2 Z(\boldsymbol{\beta})}{\partial\beta_n \partial\beta_m} - \frac{1}{Z(\boldsymbol{\beta})}\frac{\partial^3 Z(\boldsymbol{\beta})}{\partial\beta_\mu \partial\beta_n \partial\beta_m}\right\}. \end{aligned} \tag{3.47}$$

Performing the previous differentiations and evaluating the resulting expression by using the results obtained in Eqs. (3.35) and (3.38) for the Lagrange multipliers $\boldsymbol{\beta} = \mathbf{0}$ and $\boldsymbol{\Lambda}^{-1} = \mathbf{C}^e$ yields the following relation:

$$
\begin{aligned}
-\left\{\frac{\partial^4 \ln\left[Z(\boldsymbol{\beta})\right]}{\partial\beta_\nu \partial\beta_\mu \partial\beta_n \partial\beta_m}\right\}_{\boldsymbol{\beta}=\mathbf{0}} &= -\left\{\frac{1}{Z(\boldsymbol{\beta})}\frac{\partial^4 Z(\boldsymbol{\beta})}{\partial\beta_\nu \partial\beta_\mu \partial\beta_n \partial\beta_m}\right\}_{\boldsymbol{\beta}=\mathbf{0}} \\
&+\left\{\frac{1}{Z^2(\boldsymbol{\beta})}\left[\frac{\partial^2 Z(\boldsymbol{\beta})}{\partial\beta_\mu \partial\beta_n}\frac{\partial^2 Z(\boldsymbol{\beta})}{\partial\beta_\nu \partial\beta_m}+\frac{\partial^2 Z(\boldsymbol{\beta})}{\partial\beta_\nu \partial\beta_n}\frac{\partial^2 Z(\boldsymbol{\beta})}{\partial\beta_\mu \partial\beta_m}+\frac{\partial^2 Z(\boldsymbol{\beta})}{\partial\beta_\nu \partial\beta_\mu}\frac{\partial^2 Z(\boldsymbol{\beta})}{\partial\beta_n \partial\beta_m}\right]\right\}_{\boldsymbol{\beta}=\mathbf{0}}.
\end{aligned}
\tag{3.48}
$$

Using the expressions provided in Eqs. (3.28) and (3.30) to replace the expressions on the right side of Eq. (3.48) yields the following result:

$$
-\left\{\frac{\partial^4 \ln\left[Z(\boldsymbol{\beta})\right]}{\partial\beta_\nu \partial\beta_\mu \partial\beta_n \partial\beta_m}\right\}_{\boldsymbol{\beta}=\mathbf{0}} = -q^e_{\nu\mu nm} + c^e_{\mu n}c^e_{\nu m} + c^e_{\nu n}c^e_{\mu m} + c^e_{\nu\mu}c^e_{nm}. \tag{3.49}
$$

It follows from the results obtained in Eqs. (3.49) and (3.46) that the following relation holds for $\nu,\mu,n,m = 1,\ldots,TR$:

$$
\begin{aligned}
&-q^e_{\nu\mu nm} + c^e_{\mu n}c^e_{\nu m} + c^e_{\nu n}c^e_{\mu m} + c^e_{\nu\mu}c^e_{nm} \\
&= \sum_{i,j,k,\ell=1}^{TR} \omega_{ijk\ell}\left\{\frac{\partial^4 G_4^{ijk\ell}(\boldsymbol{\beta})}{\partial\beta_\nu \partial\beta_\mu \partial\beta_n \partial\beta_m}\right\}_{\boldsymbol{\beta}=\mathbf{0},\,\boldsymbol{\Lambda}^{-1}=\mathbf{C}^e}.
\end{aligned}
\tag{3.50}
$$

The Lagrange multipliers $\omega_{ijk\ell}$ for the 4th-order correlations of the measured system responses are to be obtained as the solution of Eq. (3.50). This equation could be solved numerically but cannot be solved analytically to obtain closed-form solutions for these Lagrange multipliers. In addition, the general 4th-order correlations $q^e_{\nu\mu nm}$ are not expected to be available in practice; at most, the 4th-order self-correlations $q^e_a \triangleq q^e_{aaa}$, for each system response r^e_a, $a = 1,\ldots,TR$, might be available in practice for the measured system responses. In this case, Eq. (3.50) will reduce to the following form for the quantities $\omega_b \triangleq \omega_{bbbb}$ and $G_4^b \triangleq G_4^{bbbb}$:

$$
-q^e_a + 3\left(c^e_{aa}\right)^2 = \sum_{b=1}^{TR} \omega_b \left\{\frac{\partial^4 G_4^b(\boldsymbol{\beta})}{\left(\partial\beta_a\right)^4}\right\}_{\boldsymbol{\beta}=\mathbf{0},\,\boldsymbol{\Lambda}^{-1}=\mathbf{C}^e}. \tag{3.51}
$$

Using the result presented in Eq. (3.84) in Appendix 3.A yields the following expression for the quantity $\partial^4 G_4^b(\boldsymbol{\beta})/\left(\partial\beta_a\right)^4$, for $\boldsymbol{\beta} = \mathbf{0}$ and $\boldsymbol{\Lambda}^{-1} = \mathbf{C}^e$:

$$\left\{\frac{\partial^4 G_4^b(\boldsymbol{\beta})}{(\partial\beta_a)^4}\right\}_{\boldsymbol{\beta}=\mathbf{0},\ \boldsymbol{\Lambda}^{-1}=\mathbf{C}^e} = \left\{c^e_{\ell m}c^e_{in}\left(c^e_{j\mu}c^e_{k\nu}+c^e_{k\mu}c^e_{j\nu}\right)+c^e_{\ell m}c^e_{jn}\left(c^e_{i\mu}c^e_{k\nu}+c^e_{k\mu}c^e_{i\nu}\right)\right.$$
$$+c^e_{\ell m}c^e_{kn}\left(c^e_{i\mu}c^e_{j\nu}+c^e_{j\mu}c^e_{i\nu}\right)+c^e_{im}c^e_{\ell n}\left(c^e_{j\mu}c^e_{k\nu}+c^e_{k\mu}c^e_{j\nu}\right)+c^e_{im}c^e_{jn}\left(c^e_{\ell\mu}c^e_{k\nu}+c^e_{k\mu}c^e_{\ell\nu}\right)$$
$$+c^e_{im}c^e_{kn}\left(c^e_{\ell\mu}c^e_{j\nu}+c^e_{j\mu}c^e_{\ell\nu}\right)+c^e_{jm}c^e_{\ell n}\left(c^e_{i\mu}c^e_{k\nu}+c^e_{k\mu}c^e_{i\nu}\right)+c^e_{jm}c^e_{in}\left(c^e_{\ell\mu}c^e_{k\nu}+c^e_{k\mu}c^e_{\ell\nu}\right) \quad (3.52)$$
$$+c^e_{jm}c^e_{kn}\left(c^e_{\ell\mu}c^e_{i\nu}+c^e_{i\mu}c^e_{\ell\nu}\right)+c^e_{km}c^e_{\ell n}\left(c^e_{i\mu}c^e_{j\nu}+c^e_{j\mu}c^e_{i\nu}\right)$$
$$\left.+c^e_{km}c^e_{in}\left(c^e_{\ell\mu}c^e_{j\nu}+c^e_{j\mu}c^e_{\ell\nu}\right)+c^e_{km}c^e_{jn}\left(c^e_{\ell\mu}c^e_{i\nu}+c^e_{i\mu}c^e_{\ell\nu}\right)\right\}_{m=n=\mu=\nu=a}^{i=j=k=\ell=b} = 24\left(c^e_{ba}\right)^4.$$

Inserting the result obtained in Eq. (3.52) into the right side of Eq. (3.51) yields the following vector-matrix equation for determining the Lagrange multipliers $\omega_b \triangleq \omega_{bbbb}$, $b = 1,...,TR$:

$$\sum_{b=1}^{TR} p_{ab}\omega_b = \left[-q^e_a + 3\left(c^e_{aa}\right)^2\right]\Big/24;\ p_{ab} \triangleq \left(c^e_{ba}\right)^4 = \left(c^e_{ab}\right)^4;\ a = 1,...,TR. \quad (3.53)$$

The solution $\boldsymbol{\omega} \triangleq \left(\omega_1,...,\omega_{TR}\right)^\dagger$ of Eq. (3.53) is obtained in the following form:

$$\boldsymbol{\omega} = \mathbf{P}^{-1}\mathbf{s}/24;\ \mathbf{P} \triangleq \left(p_{ab}\right)_{TR\times TR};\ \mathbf{s} \triangleq \left(s_1,...,s_{TR}\right)^\dagger;$$
$$s_a \triangleq 3\left(c^e_{aa}\right)^2 - q^e_a;\quad a = 1,...,TR. \quad (3.54)$$

In particular, if the (2nd-order) correlations among the measured system responses are negligible or unavailable, then the result obtained in Eq. (3.54) reduces to the following expression for determining the Lagrange multipliers ω_k, $k = 1,...,TR$:

$$\omega_k = \frac{3\left(c^e_{kk}\right)^2 - q^e_k}{24\left(c^e_{kk}\right)^4};\quad k = 1,...,TR. \quad (3.55)$$

The expressions of the Lagrange multipliers obtained in Eqs. (3.35), (3.38), (3.45), and (3.55) are not exact, but have been obtained in closed-forms so they can be used in Eqs. (3.9) and (3.10) to obtain the correspondingly approximate MaxEnt distribution which will be denoted as $p_e^{ME}(\mathbf{r})$ and which has the following expression:

$$p_e^{ME}(\mathbf{z}) = \frac{h_e^{ME}(\mathbf{z})}{\int_{D_r} h_e^{ME}(\mathbf{z})\,d\mathbf{z}};\ h_e^{ME}(\mathbf{z}) \triangleq \exp\left\{-\frac{1}{2}\mathbf{z}^\dagger\boldsymbol{\Lambda}\mathbf{z} - \sum_{i=1}^{TR}\theta_i z_i^3 - \sum_{i=1}^{TR}\omega_i z_i^4\right\}. \quad (3.56)$$

where the Lagrange multipliers $\boldsymbol{\Lambda}$, θ_i and ω_i, $i = 1,...,TR$, have the expressions obtained in Eqs. (3.38), (3.45), and (3.55), respectively. The approximations incurred during the course of obtaining the expression shown in Eq. (3.56) for the 4th-order MaxEnt distribution $p_e^{ME}(\mathbf{r})$ of experimentally measured system responses are summarized:

1. When obtaining the result $\boldsymbol{\beta} = \mathbf{0}$ in Eq. (3.35) for the Lagrange multipliers for the expected/mean values of the system responses, the 3rd-order correlations among the system responses were neglected. This approximation is confirmed, as shown in Appendix 3.B, namely:

$$r_k^{ME} \triangleq \int_{D_e} p_e^{ME}(\mathbf{r})\, d\mathbf{r} = r_k^e, if\; t_{ijk}^e = 0; i,j,k = 1,\ldots,TR. \tag{3.57}$$

2. When obtaining the result $\boldsymbol{\Lambda}^{-1} = \mathbf{C}^e$ in Eq. (3.38) for the Lagrange multipliers for the covariances of the system responses, the 3rd- and 4th-order correlations among the system responses were neglected. This approximation is confirmed by using the expression of the 4th-order MaxEnt distribution $p_e^{ME}(\mathbf{r})$ given in Eq. (3.56), the result shown in Equation (3.57), and the definition of the system responses' covariances, as shown in Appendix 3.B, to obtain the following expression:

$$\begin{aligned} &c_{ij}^{ME} \triangleq \int_{D_e} \left(r_i - r_i^e\right)\left(r_j - r_j^e\right) p_e^e(\mathbf{r})\, d\mathbf{r} = c_{ij}^e; \quad i,j = 1,\ldots,TR; \\ &if \quad t_{ijk}^e = 0; \quad q_{ijk}^e = 0; \quad i,j,k,\ell = 1,\ldots,TR. \end{aligned} \tag{3.58}$$

3. When obtaining the result $\boldsymbol{\theta} = -\mathbf{T}^{-1}\mathbf{t}^e/(6K)$ in Eq. (3.44) for the Lagrange multipliers for the triple correlations of the system responses, only the triple self-correlations among the system responses were considered. This approximation is confirmed, as shown in Appendix 3.B, by using:
   (i) The expression of the approximate 4th-order MaxEnt distribution $p_e^{ME}(\mathbf{r})$ given in Eq. (3.56);
   (ii) The result shown in Equation (3.57); and
   (iii) The definition of the self-correlations of 3rd-order for the system responses.

   Thus, as shown in Appendix 3.B, the following expression is obtained for the triple correlations $t_{ijk}^e;\; i,j,k = 1,\ldots,TR$:

$$t_{ijk}^{ME} \triangleq \int_{D_e} \left(r_i - r_i^e\right)\left(r_j - r_j^e\right)\left(r_k - r_k^e\right) p_e^{ME}(\mathbf{r})\, d\mathbf{r} = t_{ijk}^e; \; i,j,k = 1,\ldots,TR. \tag{3.59}$$

4. When obtaining the result $\boldsymbol{\omega} = \mathbf{P}^{-1}\mathbf{s}/(24K)$ in the equation for the Lagrange multipliers for the quadruple correlations of the measured system responses, only the self-correlations of the 4th-order among the measured system responses were considered. This approximation is confirmed by using the expression of the approximate 4th-order MaxEnt distribution $p_e^{ME}(\mathbf{r})$ given

in Eq. (3.56), the result shown in Eq. (3.57), and the definition of the self-correlations of 3rd-order for the measured system responses to obtain, as shown in Appendix 3.B, the following expression for the quadruple correlations $q^{e}_{ijk\ell}$, $i,j,k,\ell = 1,...,TR$:

$$q^{ME}_{ijk\ell} \triangleq \int_{D_e} \left(r_i - r^e_i\right)\left(r_j - r^e_j\right)\left(r_k - r^e_k\right)\left(r_\ell - r^e_\ell\right) p^{ME}_{e}\left(\mathbf{r}\right) d\mathbf{r} = q^{e}_{ijk\ell}. \quad (3.60)$$

In summary, the results obtained in Eqs. (3.57)–(3.60) indicate that the known/given means, covariances, triple and quadruple correlations for the system responses can all be used as values for the corresponding moments of the approximate 4th-order MaxEnt distribution $p^{ME}_{e}\left(\mathbf{r}\right)$ given in Eq. (3.56), since the approximations incurred when using this correspondence are at least of two orders higher than the approximations incurred for obtaining the moment in question, as follows:

1. The 1st-order moments of $p^{ME}_{e}\left(\mathbf{r}\right)$ have 3rd-order errors by comparison to the measured mean values.
2. The 2nd-order moments of $p^{ME}_{e}\left(\mathbf{r}\right)$ have 4th-order errors by comparison to the measured mean values.
3. The 3rd-order moments of $p^{ME}_{e}\left(\mathbf{r}\right)$ have 5th-order errors by comparison to the measured mean values.
4. The 4th-order moments of $p^{ME}_{e}\left(\mathbf{r}\right)$ have 6th-order errors by comparison to the measured mean values.
5. Notably, if the triple and quadruple correlations are negligible (or unavailable) then the MaxEnt distribution $p^{ME}_{e}\left(\mathbf{r}\right)$ reduces to a multivariate Gaussian with mean $\mathbf{r}^{e}$ and covariance matrix $\mathbf{C}^{e}$.

3.3 CONCLUDING REMARKS

This chapter has presented the closed-form expression for a novel, generic, 4th-order moments-constrained Maximum Entropy (MaxEnt) probability distribution, denoted as $p^{ME}_{e}\left(\mathbf{r}\right)$, which was constructed from the "empirically" or "experimentally" known first four moments (means, covariances, skewness, kurtosis) of an otherwise unknown distribution of a high-dimensional multivariate uncertain quantity of interest, which was called a "system response." This generic 4th-order MaxEnt distribution provides optimal compatibility of the available information while simultaneously ensuring minimal spurious information content, yielding an estimate of a probability density with highest uncertainty among all densities satisfying the known moment constraints. This novel generic 4th-order MaxEnt distribution is of interest in its own right for applications in many areas, including solid-state physics, econometrics, statistical description of gas flows, weather, and climate prediction. In Chapter 4 of this book, this novel generic 4th-order MaxEnt distribution will be used to construct a novel 4th-order predictive modeling methodology aimed at obtaining "best-estimate results with reduced uncertainties" for the first four moments (mean values, covariance, skewness, and kurtosis) of the optimally predicted distribution of model results and calibrated model parameters, by combining 4th-order experimental and computational information, including 4th-order (and higher) sensitivities of computed model system responses to model parameters.

APPENDIX 3.A AUXILIARY COMPUTATIONS FOR CONSTRUCTING THE MOMENT-CONSTRAINED 4TH-ORDER MAXENT DISTRIBUTION

Recall that the following closed-form expression of $F_0(\boldsymbol{\beta})$ was provided in Eq. (3.19) as: $F_0(\boldsymbol{\beta}) = K\exp(\boldsymbol{\beta}^\dagger \boldsymbol{\Lambda}^{-1}\boldsymbol{\beta}/2)$. Differentiating this expression with respect to a Lagrange multiplier β_i yields the following relation:

$$\frac{\partial F_0(\boldsymbol{\beta})}{\partial \beta_i} = K\left[\sum_{a=1}^{TR}\left(\boldsymbol{\Lambda}^{-1}\right)_{ia}\beta_a\right]\exp\left(\frac{1}{2}\boldsymbol{\beta}^\dagger \boldsymbol{\Lambda}^{-1}\boldsymbol{\beta}\right). \tag{3.61}$$

Differentiating the expressions in Eq. (3.61) with respect to a Lagrange multiplier β_j yields the following relation:

$$\begin{aligned}\frac{\partial^2 F_0(\boldsymbol{\beta})}{\partial \beta_i \partial \beta_j} &= K\frac{\partial}{\partial \beta_j}\left\{\left[\sum_{a=1}^{TR}\left(\boldsymbol{\Lambda}^{-1}\right)_{ia}\beta_a\right]\left[\exp\left(\frac{1}{2}\boldsymbol{\beta}^\dagger \boldsymbol{\Lambda}^{-1}\boldsymbol{\beta}\right)\right]\right\} \\ &= KG_2^{ij}(\boldsymbol{\beta})\exp\left(\frac{1}{2}\boldsymbol{\beta}^\dagger \boldsymbol{\Lambda}^{-1}\boldsymbol{\beta}\right);\end{aligned} \tag{3.62}$$

where:

$$G_2^{ij}(\boldsymbol{\beta}) \triangleq \left(\boldsymbol{\Lambda}^{-1}\right)_{ij} + \left[\sum_{a=1}^{TR}\left(\boldsymbol{\Lambda}^{-1}\right)_{ia}\beta_a\right]\left[\sum_{b=1}^{TR}\left(\boldsymbol{\Lambda}^{-1}\right)_{jb}\beta_b\right]. \tag{3.63}$$

Differentiating the expression in Eq. (3.62) with respect to a Lagrange multiplier β_k yields the following relation:

$$\begin{aligned}\frac{\partial^3 F_0(\boldsymbol{\beta})}{\partial \beta_i \partial \beta_j \partial \beta_k} &= K\exp\left(\frac{1}{2}\boldsymbol{\beta}^\dagger \boldsymbol{\Lambda}^{-1}\boldsymbol{\beta}\right)\frac{\partial}{\partial \beta_k}\left\{\left(\boldsymbol{\Lambda}^{-1}\right)_{ij} + \left[\sum_{a=1}^{TR}\left(\boldsymbol{\Lambda}^{-1}\right)_{ia}\beta_a\right]\left[\sum_{b=1}^{TR}\left(\boldsymbol{\Lambda}^{-1}\right)_{jb}\beta_b\right]\right\} \\ &+K\left\{\left(\boldsymbol{\Lambda}^{-1}\right)_{ij} + \left[\sum_{a=1}^{TR}\left(\boldsymbol{\Lambda}^{-1}\right)_{ia}\beta_a\right]\left[\sum_{b=1}^{TR}\left(\boldsymbol{\Lambda}^{-1}\right)_{jb}\beta_b\right]\right\}\frac{\partial}{\partial \beta_k}\left[\exp\left(\frac{1}{2}\boldsymbol{\beta}^\dagger \boldsymbol{\Lambda}^{-1}\boldsymbol{\beta}\right)\right] \\ &= K\left\{\left(\boldsymbol{\Lambda}^{-1}\right)_{ik}\left[\sum_{b=1}^{TR}\left(\boldsymbol{\Lambda}^{-1}\right)_{jb}\beta_b\right] + \left(\boldsymbol{\Lambda}^{-1}\right)_{jk}\left[\sum_{a=1}^{TR}\left(\boldsymbol{\Lambda}^{-1}\right)_{ia}\beta_a\right]\right\}\exp\left(\frac{1}{2}\boldsymbol{\beta}^\dagger \boldsymbol{\Lambda}^{-1}\boldsymbol{\beta}\right) \\ &+K\left\{\left(\boldsymbol{\Lambda}^{-1}\right)_{ij} + \left[\sum_{a=1}^{TR}\left(\boldsymbol{\Lambda}^{-1}\right)_{ia}\beta_a\right]\left[\sum_{b=1}^{TR}\left(\boldsymbol{\Lambda}^{-1}\right)_{jb}\beta_b\right]\right\} \\ &\times\left[\sum_{c=1}^{TR}\left(\boldsymbol{\Lambda}^{-1}\right)_{kc}\beta_c\right]\exp\left(\frac{1}{2}\boldsymbol{\beta}^\dagger \boldsymbol{\Lambda}^{-1}\boldsymbol{\beta}\right) = KG_3^{ijk}(\boldsymbol{\beta})\exp\left(\frac{1}{2}\boldsymbol{\beta}^\dagger \boldsymbol{\Lambda}^{-1}\boldsymbol{\beta}\right);\end{aligned} \tag{3.64}$$

where:

$$\begin{aligned} G_3^{ijk}(\boldsymbol{\beta}) \triangleq\; & \left(\boldsymbol{\Lambda}^{-1}\right)_{ik}\left[\sum_{b=1}^{TR}\left(\boldsymbol{\Lambda}^{-1}\right)_{jb}\beta_b\right]+\left(\boldsymbol{\Lambda}^{-1}\right)_{jk}\left[\sum_{a=1}^{TR}\left(\boldsymbol{\Lambda}^{-1}\right)_{ia}\beta_a\right] \\ & +\left(\boldsymbol{\Lambda}^{-1}\right)_{ij}\left[\sum_{c=1}^{TR}\left(\boldsymbol{\Lambda}^{-1}\right)_{kc}\beta_c\right]+\left[\sum_{a=1}^{TR}\left(\boldsymbol{\Lambda}^{-1}\right)_{ia}\beta_a\right]\left[\sum_{b=1}^{TR}\left(\boldsymbol{\Lambda}^{-1}\right)_{jb}\beta_b\right]\left[\sum_{c=1}^{TR}\left(\boldsymbol{\Lambda}^{-1}\right)_{kc}\beta_c\right]; \end{aligned} \tag{3.65}$$

Differentiating the expression in Eq. (3.64) with respect to a Lagrange multiplier β_ℓ yields the following relation:

$$\begin{aligned} \frac{\partial^4 F_0(\boldsymbol{\beta})}{\partial\beta_i\partial\beta_j\partial\beta_k\partial\beta_c} =\; & K\left\{\left(\boldsymbol{\Lambda}^{-1}\right)_{ij}\left[\sum_{c=1}^{TR}\left(\boldsymbol{\Lambda}^{-1}\right)_{kc}\beta_c\right]+\left(\boldsymbol{\Lambda}^{-1}\right)_{ik}\left[\sum_{b=1}^{TR}\left(\boldsymbol{\Lambda}^{-1}\right)_{jb}\beta_b\right]\right. \\ & +\left(\boldsymbol{\Lambda}^{-1}\right)_{jk}\left[\sum_{a=1}^{TR}\left(\boldsymbol{\Lambda}^{-1}\right)_{ia}\beta_a\right]+\left[\sum_{a=1}^{TR}\left(\boldsymbol{\Lambda}^{-1}\right)_{ia}\beta_a\right]\left[\sum_{b=1}^{TR}\left(\boldsymbol{\Lambda}^{-1}\right)_{jb}\beta_b\right] \\ & \left.\times\left[\sum_{c=1}^{TR}\left(\boldsymbol{\Lambda}^{-1}\right)_{kc}\beta_c\right]\right\}\frac{\partial}{\partial\beta_\ell}\left[\exp\left(\frac{1}{2}\boldsymbol{\beta}^\dagger\boldsymbol{\Lambda}^{-1}\boldsymbol{\beta}\right)\right]+K\exp\left(\frac{1}{2}\boldsymbol{\beta}^\dagger\boldsymbol{\Lambda}^{-1}\boldsymbol{\beta}\right) \\ & \times\frac{\partial}{\partial\beta_\ell}\left\{\left(\boldsymbol{\Lambda}^{-1}\right)_{ij}\left[\sum_{c=1}^{TR}\left(\boldsymbol{\Lambda}^{-1}\right)_{kc}\beta_c\right]+\left(\boldsymbol{\Lambda}^{-1}\right)_{kk}\left[\sum_{b=1}^{TR}\left(\boldsymbol{\Lambda}^{-1}\right)_{jb}\beta_b\right]\right. \\ & \left.+\left(\boldsymbol{\Lambda}^{-1}\right)_{jk}\left[\sum_{a=1}^{TR}\left(\boldsymbol{\Lambda}^{-1}\right)_{ia}\beta_a\right]+\left[\sum_{a=1}^{TR}\left(\boldsymbol{\Lambda}^{-1}\right)_{ia}\beta_a\right]\left[\sum_{b=1}^{TR}\left(\boldsymbol{\Lambda}^{-1}\right)_{jb}\beta_b\right]\left[\sum_{c=1}^{TR}\left(\boldsymbol{\Lambda}^{-1}\right)_{kc}\beta_c\right]\right\} \\ =\; & K\left\{\left(\boldsymbol{\Lambda}^{-1}\right)_{ij}\left[\sum_{c=1}^{TR}\left(\boldsymbol{\Lambda}^{-1}\right)_{kc}\beta_c\right]+\left(\boldsymbol{\Lambda}^{-1}\right)_{ik}\left[\sum_{b=1}^{TR}\left(\boldsymbol{\Lambda}^{-1}\right)_{jb}\beta_b\right]+\left(\boldsymbol{\Lambda}^{-1}\right)_{jk}\left[\sum_{a=1}^{TR}\left(\boldsymbol{\Lambda}^{-1}\right)_{ia}\beta_a\right]\right. \\ & \left.+\left[\sum_{a=1}^{TR}\left(\boldsymbol{\Lambda}^{-1}\right)_{ia}\beta_a\right]\left[\sum_{b=1}^{TR}\left(\boldsymbol{\Lambda}^{-1}\right)_{jb}\beta_b\right]\left[\sum_{c=1}^{TR}\left(\boldsymbol{\Lambda}^{-1}\right)_{kc}\beta_c\right]\right\}\left[\sum_{d=1}^{TR}\left(\boldsymbol{\Lambda}^{-1}\right)_{\ell d}\beta_d\right]\exp\left(\frac{1}{2}\boldsymbol{\beta}^\dagger\boldsymbol{\Lambda}^{-1}\boldsymbol{\beta}\right) \\ & +K\exp\left(\frac{1}{2}\boldsymbol{\beta}^\dagger\boldsymbol{\Lambda}^{-1}\boldsymbol{\beta}\right)\left\{\left(\boldsymbol{\Lambda}^{-1}\right)_{ij}\left(\boldsymbol{\Lambda}^{-1}\right)_{k\ell}+\left(\boldsymbol{\Lambda}^{-1}\right)_{ik}\left(\boldsymbol{\Lambda}^{-1}\right)_{j\ell}+\left(\boldsymbol{\Lambda}^{-1}\right)_{i\ell}\left(\boldsymbol{\Lambda}^{-1}\right)_{jk}\right. \\ & +\left(\boldsymbol{\Lambda}^{-1}\right)_{i\ell}\left[\sum_{b=1}^{TR}\left(\boldsymbol{\Lambda}^{-1}\right)_{jb}\beta_b\right]\left[\sum_{c=1}^{TR}\left(\boldsymbol{\Lambda}^{-1}\right)_{kc}\beta_c\right]+\left(\boldsymbol{\Lambda}^{-1}\right)_{j\ell}\left[\sum_{a=1}^{TR}\left(\boldsymbol{\Lambda}^{-1}\right)_{ia}\beta_a\right]\left[\sum_{c=1}^{TR}\left(\boldsymbol{\Lambda}^{-1}\right)_{kc}\beta_c\right] \\ & \left.+\left(\boldsymbol{\Lambda}^{-1}\right)_{k\ell}\left[\sum_{a=1}^{TR}\left(\boldsymbol{\Lambda}^{-1}\right)_{ia}\beta_a\right]\left[\sum_{b=1}^{TR}\left(\boldsymbol{\Lambda}^{-1}\right)_{jb}\beta_b\right]\right\}=KG_4^{ijk\ell}(\boldsymbol{\beta})\exp\left(\frac{1}{2}\boldsymbol{\beta}^\dagger\boldsymbol{\Lambda}^{-1}\boldsymbol{\beta}\right), \end{aligned} \tag{3.66}$$

where:

$$\begin{aligned} G_4^{ijk\ell}(\boldsymbol{\beta}) \triangleq\; & \left[\sum_{d=1}^{TR}\left(\boldsymbol{\Lambda}^{-1}\right)_{\ell d}\beta_d\right]\left\{\left(\boldsymbol{\Lambda}^{-1}\right)_{ij}\left[\sum_{c=1}^{TR}\left(\boldsymbol{\Lambda}^{-1}\right)_{kc}\beta_c\right]\right. \\ & \left.+\left(\boldsymbol{\Lambda}^{-1}\right)_{ik}\left[\sum_{b=1}^{TR}\left(\boldsymbol{\Lambda}^{-1}\right)_{jb}\beta_b\right]+\left(\boldsymbol{\Lambda}^{-1}\right)_{jk}\left[\sum_{a=1}^{TR}\left(\boldsymbol{\Lambda}^{-1}\right)_{ia}\beta_a\right]\right\} \end{aligned}$$

$$
\begin{aligned}
&+\left[\sum_{d=1}^{TR}\left(\boldsymbol{\Lambda}^{-1}\right)_{\ell d}\beta_d\right]\left[\sum_{a=1}^{TR}\left(\boldsymbol{\Lambda}^{-1}\right)_{ia}\beta_a\right]\left[\sum_{b=1}^{TR}\left(\boldsymbol{\Lambda}^{-1}\right)_{jb}\beta_b\right]\left[\sum_{c=1}^{TR}\left(\boldsymbol{\Lambda}^{-1}\right)_{kc}\beta_c\right] \\
&+\left(\boldsymbol{\Lambda}^{-1}\right)_{ij}\left(\boldsymbol{\Lambda}^{-1}\right)_{k\ell}+\left(\boldsymbol{\Lambda}^{-1}\right)_{ik}\left(\boldsymbol{\Lambda}^{-1}\right)_{j\ell}+\left(\boldsymbol{\Lambda}^{-1}\right)_{i\ell}\left(\boldsymbol{\Lambda}^{-1}\right)_{jk} \\
&+\left(\boldsymbol{\Lambda}^{-1}\right)_{i\ell}\left[\sum_{b=1}^{TR}\left(\boldsymbol{\Lambda}^{-1}\right)_{jb}\beta_b\right]\left[\sum_{c=1}^{TR}\left(\boldsymbol{\Lambda}^{-1}\right)_{kc}\beta_c\right]+\left(\boldsymbol{\Lambda}^{-1}\right)_{j\ell}\left[\sum_{a=1}^{TR}\left(\boldsymbol{\Lambda}^{-1}\right)_{ia}\beta_a\right] \\
&\times\left[\sum_{c=1}^{TR}\left(\boldsymbol{\Lambda}^{-1}\right)_{kc}\beta_c\right]+\left(\boldsymbol{\Lambda}^{-1}\right)_{k\ell}\left[\sum_{a=1}^{TR}\left(\boldsymbol{\Lambda}^{-1}\right)_{ia}\beta_a\right]\left[\sum_{b=1}^{TR}\left(\boldsymbol{\Lambda}^{-1}\right)_{jb}\beta_b\right].
\end{aligned}
\tag{3.67}
$$

The various derivatives of the functions $G_3^{ijk}(\boldsymbol{\beta})$ and $G_4^{ijk\ell}(\boldsymbol{\beta})$ with respect to the Lagrange multipliers $\partial\beta_m$, $m=1,\ldots,TR$, will also be used when evaluating the various derivatives of the normalization integral $Z(\boldsymbol{\beta})$ with respect to these Lagrange multipliers. The derivatives of the function $G_3^{ijk}(\boldsymbol{\beta})$ with respect to the Lagrange multipliers β_m, $m=1,\ldots,TR$, are obtained by using Eq. (3.65), which yields the following expressions:

$$
\begin{aligned}
\frac{\partial G_3^{ijk}(\boldsymbol{\beta})}{\partial\beta_m}&=\left(\boldsymbol{\Lambda}^{-1}\right)_{ik}\left(\boldsymbol{\Lambda}^{-1}\right)_{jm}+\left(\boldsymbol{\Lambda}^{-1}\right)_{jk}\left(\boldsymbol{\Lambda}^{-1}\right)_{im}+\left(\boldsymbol{\Lambda}^{-1}\right)_{ij}\left(\boldsymbol{\Lambda}^{-1}\right)_{km} \\
&+\left(\boldsymbol{\Lambda}^{-1}\right)_{im}\left[\sum_{b=1}^{TR}\left(\boldsymbol{\Lambda}^{-1}\right)_{jb}\beta_b\right]\left[\sum_{c=1}^{TR}\left(\boldsymbol{\Lambda}^{-1}\right)_{kc}\beta_c\right] \\
&+\left[\sum_{a=1}^{TR}\left(\boldsymbol{\Lambda}^{-1}\right)_{ia}\beta_a\right]\left(\boldsymbol{\Lambda}^{-1}\right)_{jm}\left[\sum_{c=1}^{TR}\left(\boldsymbol{\Lambda}^{-1}\right)_{kc}\beta_c\right] \\
&+\left[\sum_{a=1}^{TR}\left(\boldsymbol{\Lambda}^{-1}\right)_{ia}\beta_a\right]\left[\sum_{b=1}^{TR}\left(\boldsymbol{\Lambda}^{-1}\right)_{jb}\beta_b\right]\left(\boldsymbol{\Lambda}^{-1}\right)_{km}.
\end{aligned}
\tag{3.68}
$$

Differentiating the expression in Eq. (3.68) with respect to a Lagrange multiplier β_n yields the following relation:

$$
\begin{aligned}
\frac{\partial^2 G_3^{ijk}(\boldsymbol{\beta})}{\partial\beta_n\partial\beta_m}&=\left(\boldsymbol{\Lambda}^{-1}\right)_{im}\left\{\left(\boldsymbol{\Lambda}^{-1}\right)_{jn}\left[\sum_{c=1}^{TR}\left(\boldsymbol{\Lambda}^{-1}\right)_{kc}\beta_c\right]+\left[\sum_{b=1}^{TR}\left(\boldsymbol{\Lambda}^{-1}\right)_{jb}\beta_b\right]\left(\boldsymbol{\Lambda}^{-1}\right)_{kn}\right\} \\
&+\left(\boldsymbol{\Lambda}^{-1}\right)_{jm}\left\{\left(\boldsymbol{\Lambda}^{-1}\right)_{in}\left[\sum_{c=1}^{TR}\left(\boldsymbol{\Lambda}^{-1}\right)_{kc}\beta_c\right]+\left[\sum_{a=1}^{TR}\left(\boldsymbol{\Lambda}^{-1}\right)_{ia}\beta_a\right]\left(\boldsymbol{\Lambda}^{-1}\right)_{kn}\right\} \\
&+\left(\boldsymbol{\Lambda}^{-1}\right)_{km}\left\{\left(\boldsymbol{\Lambda}^{-1}\right)_{in}\left[\sum_{b=1}^{TR}\left(\boldsymbol{\Lambda}^{-1}\right)_{jb}\beta_b\right]+\left[\sum_{a=1}^{TR}\left(\boldsymbol{\Lambda}^{-1}\right)_{ia}\beta_a\right]\left(\boldsymbol{\Lambda}^{-1}\right)_{jn}\right\}.
\end{aligned}
\tag{3.69}
$$

Differentiating the expression in Eq. (3.69) with respect to a Lagrange multiplier β_μ yields the following relation:

$$\begin{aligned}\frac{\partial^3 G_3^{ijk}(\boldsymbol{\beta})}{\partial\beta_\mu\partial\beta_n\partial\beta_m} &= \left(\boldsymbol{\Lambda}^{-1}\right)_{im}\left\{\left(\boldsymbol{\Lambda}^{-1}\right)_{jn}\left(\boldsymbol{\Lambda}^{-1}\right)_{k\mu}+\left(\boldsymbol{\Lambda}^{-1}\right)_{j\mu}\left(\boldsymbol{\Lambda}^{-1}\right)_{kn}\right\}\\ &+\left(\boldsymbol{\Lambda}^{-1}\right)_{jm}\left\{\left(\boldsymbol{\Lambda}^{-1}\right)_{in}\left(\boldsymbol{\Lambda}^{-1}\right)_{k\mu}+\left(\boldsymbol{\Lambda}^{-1}\right)_{i\mu}\left(\boldsymbol{\Lambda}^{-1}\right)_{kn}\right\}\\ &+\left(\boldsymbol{\Lambda}^{-1}\right)_{km}\left\{\left(\boldsymbol{\Lambda}^{-1}\right)_{in}\left(\boldsymbol{\Lambda}^{-1}\right)_{j\mu}+\left(\boldsymbol{\Lambda}^{-1}\right)_{i\mu}\left(\boldsymbol{\Lambda}^{-1}\right)_{jn}\right\}.\end{aligned} \tag{3.70}$$

The derivatives of the function $G_4^{ijk\ell}(\boldsymbol{\beta})$ with respect to Lagrange multipliers β_m, $m=1,\ldots,TR$, are obtained by using Eq. (3.67). Thus, the first derivative of the function $G_4^{ijk\ell}(\boldsymbol{\beta})$ with respect to Lagrange multipliers β_m, $m=1,\ldots,TR$, has the following expression:

$$\begin{aligned}\frac{\partial G_4^{ijk\ell}(\boldsymbol{\beta})}{\partial\beta_m} &= \left(\boldsymbol{\Lambda}^{-1}\right)_{\ell m}\left\{\left(\boldsymbol{\Lambda}^{-1}\right)_{ij}\left[\sum_{c=1}^{TR}\left(\boldsymbol{\Lambda}^{-1}\right)_{kc}\beta_c\right]+\left(\boldsymbol{\Lambda}^{-1}\right)_{ik}\left[\sum_{b=1}^{TR}\left(\boldsymbol{\Lambda}^{-1}\right)_{jb}\beta_b\right]\right.\\ &\left.+\left(\boldsymbol{\Lambda}^{-1}\right)_{jk}\left[\sum_{a=1}^{TR}\left(\boldsymbol{\Lambda}^{-1}\right)_{ia}\beta_a\right]\right\}+\left[\sum_{d=1}^{TR}\left(\boldsymbol{\Lambda}^{-1}\right)_{\ell d}\beta_d\right]\left\{\left(\boldsymbol{\Lambda}^{-1}\right)_{ij}\left(\boldsymbol{\Lambda}^{-1}\right)_{km}\right.\\ &\left.+\left(\boldsymbol{\Lambda}^{-1}\right)_{ik}\left(\boldsymbol{\Lambda}^{-1}\right)_{jm}+\left(\boldsymbol{\Lambda}^{-1}\right)_{jk}\left(\boldsymbol{\Lambda}^{-1}\right)_{im}\right\}\\ &+\left(\boldsymbol{\Lambda}^{-1}\right)_{\ell m}\left[\sum_{a=1}^{TR}\left(\boldsymbol{\Lambda}^{-1}\right)_{ia}\beta_a\right]\left[\sum_{b=1}^{TR}\left(\boldsymbol{\Lambda}^{-1}\right)_{jb}\beta_b\right]\left[\sum_{c=1}^{TR}\left(\boldsymbol{\Lambda}^{-1}\right)_{kc}\beta_c\right]\\ &+\left[\sum_{d=1}^{TR}\left(\boldsymbol{\Lambda}^{-1}\right)_{\ell d}\beta_d\right]\left(\boldsymbol{\Lambda}^{-1}\right)_{im}\left[\sum_{b=1}^{TR}\left(\boldsymbol{\Lambda}^{-1}\right)_{jb}\beta_b\right]\left[\sum_{c=1}^{TR}\left(\boldsymbol{\Lambda}^{-1}\right)_{kc}\beta_c\right]\\ &+\left[\sum_{d=1}^{TR}\left(\boldsymbol{\Lambda}^{-1}\right)_{\ell d}\beta_d\right]\left[\sum_{a=1}^{TR}\left(\boldsymbol{\Lambda}^{-1}\right)_{ia}\beta_a\right]\left(\boldsymbol{\Lambda}^{-1}\right)_{jm}\left[\sum_{c=1}^{TR}\left(\boldsymbol{\Lambda}^{-1}\right)_{kc}\beta_c\right]\\ &+\left[\sum_{d=1}^{TR}\left(\boldsymbol{\Lambda}^{-1}\right)_{\ell d}\beta_d\right]\left[\sum_{a=1}^{TR}\left(\boldsymbol{\Lambda}^{-1}\right)_{ia}\beta_a\right]\left[\sum_{b=1}^{TR}\left(\boldsymbol{\Lambda}^{-1}\right)_{jb}\beta_b\right]\left(\boldsymbol{\Lambda}^{-1}\right)_{km}\\ &+\left(\boldsymbol{\Lambda}^{-1}\right)_{i\ell}\left\{\left(\boldsymbol{\Lambda}^{-1}\right)_{jm}\left[\sum_{c=1}^{TR}\left(\boldsymbol{\Lambda}^{-1}\right)_{kc}\beta_c\right]+\left[\sum_{b=1}^{TR}\left(\boldsymbol{\Lambda}^{-1}\right)_{jb}\beta_b\right]\left(\boldsymbol{\Lambda}^{-1}\right)_{km}\right\}\\ &+\left(\boldsymbol{\Lambda}^{-1}\right)_{j\ell}\left\{\left(\boldsymbol{\Lambda}^{-1}\right)_{im}\left[\sum_{c=1}^{TR}\left(\boldsymbol{\Lambda}^{-1}\right)_{kc}\beta_c\right]+\left[\sum_{a=1}^{TR}\left(\boldsymbol{\Lambda}^{-1}\right)_{ia}\beta_a\right]\left(\boldsymbol{\Lambda}^{-1}\right)_{km}\right\}\\ &+\left(\boldsymbol{\Lambda}^{-1}\right)_{k\ell}\left\{\left(\boldsymbol{\Lambda}^{-1}\right)_{im}\left[\sum_{b=1}^{TR}\left(\boldsymbol{\Lambda}^{-1}\right)_{jb}\beta_b\right]+\left[\sum_{a=1}^{TR}\left(\boldsymbol{\Lambda}^{-1}\right)_{ia}\beta_a\right]\left(\boldsymbol{\Lambda}^{-1}\right)_{jm}\right\}.\end{aligned} \tag{3.71}$$

Differentiating Equation (3.71) with respect to Lagrange multiplier β_n $n=1,\ldots,TR$, yields the following relation:

$$\frac{\partial^2 G_4^{ijk\ell}(\boldsymbol{\beta})}{\partial\beta_n\partial\beta_m} = X_1+X_2+X_3+X_4+X_5+X_6, \tag{3.72}$$

where:

$$X_1 \triangleq \left(\mathbf{\Lambda}^{-1}\right)_{\ell m}\left\{\left(\mathbf{\Lambda}^{-1}\right)_{ij}\left(\mathbf{\Lambda}^{-1}\right)_{kn}+\left(\mathbf{\Lambda}^{-1}\right)_{ik}\left(\mathbf{\Lambda}^{-1}\right)_{jn}+\left(\mathbf{\Lambda}^{-1}\right)_{jk}\left(\mathbf{\Lambda}^{-1}\right)_{in}\right\} \\ +\left(\mathbf{\Lambda}^{-1}\right)_{\ell n}\left\{\left(\mathbf{\Lambda}^{-1}\right)_{ij}\left(\mathbf{\Lambda}^{-1}\right)_{km}+\left(\mathbf{\Lambda}^{-1}\right)_{ik}\left(\mathbf{\Lambda}^{-1}\right)_{jm}+\left(\mathbf{\Lambda}^{-1}\right)_{jk}\left(\mathbf{\Lambda}^{-1}\right)_{im}\right\}, \tag{3.73}$$

$$X_2 \triangleq \left(\mathbf{\Lambda}^{-1}\right)_{\ell m}\left\{\left(\mathbf{\Lambda}^{-1}\right)_{in}\left[\sum_{b=1}^{TR}\left(\mathbf{\Lambda}^{-1}\right)_{jb}\beta_b\right]\left[\sum_{c=1}^{TR}\left(\mathbf{\Lambda}^{-1}\right)_{kc}\beta_c\right] \right. \\ +\left[\sum_{a=1}^{TR}\left(\mathbf{\Lambda}^{-1}\right)_{ia}\beta_a\right]\left(\mathbf{\Lambda}^{-1}\right)_{jn}\left[\sum_{c=1}^{TR}\left(\mathbf{\Lambda}^{-1}\right)_{kc}\beta_c\right] \\ \left. +\left[\sum_{a=1}^{TR}\left(\mathbf{\Lambda}^{-1}\right)_{ia}\beta_a\right]\left[\sum_{b=1}^{TR}\left(\mathbf{\Lambda}^{-1}\right)_{jb}\beta_b\right]\left(\mathbf{\Lambda}^{-1}\right)_{kn}\right\}, \tag{3.74}$$

$$X_3 \triangleq \left(\mathbf{\Lambda}^{-1}\right)_{im}\left\{\left(\mathbf{\Lambda}^{-1}\right)_{\ell n}\left[\sum_{b=1}^{TR}\left(\mathbf{\Lambda}^{-1}\right)_{jb}\beta_b\right]\left[\sum_{c=1}^{TR}\left(\mathbf{\Lambda}^{-1}\right)_{kc}\beta_c\right] \right. \\ +\left[\sum_{d=1}^{TR}\left(\mathbf{\Lambda}^{-1}\right)_{\ell d}\beta_d\right]\left(\mathbf{\Lambda}^{-1}\right)_{jn}\left[\sum_{c=1}^{TR}\left(\mathbf{\Lambda}^{-1}\right)_{kc}\beta_c\right] \\ \left. +\left[\sum_{d=1}^{TR}\left(\mathbf{\Lambda}^{-1}\right)_{\ell d}\beta_d\right]\left[\sum_{b=1}^{TR}\left(\mathbf{\Lambda}^{-1}\right)_{jb}\beta_b\right]\left(\mathbf{\Lambda}^{-1}\right)_{kn}\right\}, \tag{3.75}$$

$$X_4 \triangleq \left(\mathbf{\Lambda}^{-1}\right)_{jm}\left\{\left(\mathbf{\Lambda}^{-1}\right)_{\ell n}\left[\sum_{a=1}^{TR}\left(\mathbf{\Lambda}^{-1}\right)_{ia}\beta_a\right]\left[\sum_{c=1}^{TR}\left(\mathbf{\Lambda}^{-1}\right)_{kc}\beta_c\right] \right. \\ +\left[\sum_{d=1}^{TR}\left(\mathbf{\Lambda}^{-1}\right)_{\ell d}\beta_d\right]\left(\mathbf{\Lambda}^{-1}\right)_{in}\left[\sum_{c=1}^{TR}\left(\mathbf{\Lambda}^{-1}\right)_{kc}\beta_c\right] \\ \left. +\left[\sum_{d=1}^{TR}\left(\mathbf{\Lambda}^{-1}\right)_{\ell d}\beta_d\right]\left[\sum_{a=1}^{TR}\left(\mathbf{\Lambda}^{-1}\right)_{ia}\beta_a\right]\left(\mathbf{\Lambda}^{-1}\right)_{kn}\right\}, \tag{3.76}$$

$$X_5 \triangleq \left(\mathbf{\Lambda}^{-1}\right)_{km}\left\{\left(\mathbf{\Lambda}^{-1}\right)_{\ell n}\left[\sum_{a=1}^{TR}\left(\mathbf{\Lambda}^{-1}\right)_{ia}\beta_a\right]\left[\sum_{b=1}^{TR}\left(\mathbf{\Lambda}^{-1}\right)_{jb}\beta_b\right] \right. \\ +\left[\sum_{d=1}^{TR}\left(\mathbf{\Lambda}^{-1}\right)_{\ell d}\beta_d\right]\left(\mathbf{\Lambda}^{-1}\right)_{in}\left[\sum_{b=1}^{TR}\left(\mathbf{\Lambda}^{-1}\right)_{jb}\beta_b\right] \\ \left. +\left[\sum_{d=1}^{TR}\left(\mathbf{\Lambda}^{-1}\right)_{\ell d}\beta_d\right]\left[\sum_{a=1}^{TR}\left(\mathbf{\Lambda}^{-1}\right)_{ia}\beta_a\right]\left(\mathbf{\Lambda}^{-1}\right)_{jn}\right\}, \tag{3.77}$$

$$X_6 \triangleq \left(\mathbf{\Lambda}^{-1}\right)_{i\ell}\left[\left(\mathbf{\Lambda}^{-1}\right)_{jm}\left(\mathbf{\Lambda}^{-1}\right)_{kn}+\left(\mathbf{\Lambda}^{-1}\right)_{jn}\left(\mathbf{\Lambda}^{-1}\right)_{km}\right] \\ +\left(\mathbf{\Lambda}^{-1}\right)_{j\ell}\left[\left(\mathbf{\Lambda}^{-1}\right)_{im}\left(\mathbf{\Lambda}^{-1}\right)_{kn}+\left(\mathbf{\Lambda}^{-1}\right)_{im}\left(\mathbf{\Lambda}^{-1}\right)_{km}\right] \\ +\left(\mathbf{\Lambda}^{-1}\right)_{k\ell}\left[\left(\mathbf{\Lambda}^{-1}\right)_{im}\left(\mathbf{\Lambda}^{-1}\right)_{jn}+\left(\mathbf{\Lambda}^{-1}\right)_{in}\left(\mathbf{\Lambda}^{-1}\right)_{jm}\right]. \tag{3.78}$$

Differentiating Eq. (3.72) with respect to Lagrange multiplier β_μ $\mu = 1,\ldots,TR$, yields the following relation:

$$\frac{\partial^3 G_4^{ijk\ell}(\boldsymbol{\beta})}{\partial\beta_\mu\partial\beta_n\partial\beta_m} = Y_1 + Y_2 + Y_3 + Y_4, \tag{3.79}$$

where:

$$\begin{aligned}
Y_1 \triangleq & \left(\boldsymbol{\Lambda}^{-1}\right)_{\ell m}\left(\boldsymbol{\Lambda}^{-1}\right)_{in}\left[\left(\boldsymbol{\Lambda}^{-1}\right)_{j\mu}\sum_{c=1}^{TR}\left(\boldsymbol{\Lambda}^{-1}\right)_{kc}\beta_c + \left(\boldsymbol{\Lambda}^{-1}\right)_{k\mu}\sum_{b=1}^{TR}\left(\boldsymbol{\Lambda}^{-1}\right)_{jb}\beta_b\right] \\
& + \left(\boldsymbol{\Lambda}^{-1}\right)_{\ell m}\left(\boldsymbol{\Lambda}^{-1}\right)_{jn}\left[\left(\boldsymbol{\Lambda}^{-1}\right)_{i\mu}\sum_{c=1}^{TR}\left(\boldsymbol{\Lambda}^{-1}\right)_{kc}\beta_c + \left(\boldsymbol{\Lambda}^{-1}\right)_{k\mu}\sum_{a=1}^{TR}\left(\boldsymbol{\Lambda}^{-1}\right)_{ia}\beta_a\right] \\
& + \left(\boldsymbol{\Lambda}^{-1}\right)_{\ell m}\left(\boldsymbol{\Lambda}^{-1}\right)_{kn}\left[\left(\boldsymbol{\Lambda}^{-1}\right)_{i\mu}\sum_{b=1}^{TR}\left(\boldsymbol{\Lambda}^{-1}\right)_{jb}\beta_b + \left(\boldsymbol{\Lambda}^{-1}\right)_{j\mu}\sum_{a=1}^{TR}\left(\boldsymbol{\Lambda}^{-1}\right)_{ia}\beta_a\right],
\end{aligned} \tag{3.80}$$

$$\begin{aligned}
Y_2 \triangleq & \left(\boldsymbol{\Lambda}^{-1}\right)_{im}\left(\boldsymbol{\Lambda}^{-1}\right)_{\ell n}\left[\left(\boldsymbol{\Lambda}^{-1}\right)_{j\mu}\sum_{c=1}^{TR}\left(\boldsymbol{\Lambda}^{-1}\right)_{kc}\beta_c + \left(\boldsymbol{\Lambda}^{-1}\right)_{k\mu}\sum_{b=1}^{TR}\left(\boldsymbol{\Lambda}^{-1}\right)_{jb}\beta_b\right] \\
& + \left(\boldsymbol{\Lambda}^{-1}\right)_{im}\left(\boldsymbol{\Lambda}^{-1}\right)_{jn}\left[\left(\boldsymbol{\Lambda}^{-1}\right)_{\ell\mu}\sum_{c=1}^{TR}\left(\boldsymbol{\Lambda}^{-1}\right)_{kc}\beta_c + \left(\boldsymbol{\Lambda}^{-1}\right)_{k\mu}\sum_{d=1}^{TR}\left(\boldsymbol{\Lambda}^{-1}\right)_{\ell d}\beta_d\right] \\
& + \left(\boldsymbol{\Lambda}^{-1}\right)_{im}\left(\boldsymbol{\Lambda}^{-1}\right)_{kn}\left[\left(\boldsymbol{\Lambda}^{-1}\right)_{\ell\mu}\sum_{b=1}^{TR}\left(\boldsymbol{\Lambda}^{-1}\right)_{jb}\beta_b + \left(\boldsymbol{\Lambda}^{-1}\right)_{j\mu}\sum_{d=1}^{TR}\left(\boldsymbol{\Lambda}^{-1}\right)_{\ell d}\beta_d\right],
\end{aligned} \tag{3.81}$$

$$\begin{aligned}
Y_3 \triangleq & \left(\boldsymbol{\Lambda}^{-1}\right)_{jm}\left(\boldsymbol{\Lambda}^{-1}\right)_{\ell n}\left[\left(\boldsymbol{\Lambda}^{-1}\right)_{i\mu}\sum_{c=1}^{TR}\left(\boldsymbol{\Lambda}^{-1}\right)_{kc}\beta_c + \left(\boldsymbol{\Lambda}^{-1}\right)_{k\mu}\sum_{a=1}^{TR}\left(\boldsymbol{\Lambda}^{-1}\right)_{ia}\beta_a\right] \\
& + \left(\boldsymbol{\Lambda}^{-1}\right)_{jm}\left(\boldsymbol{\Lambda}^{-1}\right)_{in}\left[\left(\boldsymbol{\Lambda}^{-1}\right)_{\ell\mu}\sum_{c=1}^{TR}\left(\boldsymbol{\Lambda}^{-1}\right)_{kc}\beta_c + \left(\boldsymbol{\Lambda}^{-1}\right)_{k\mu}\sum_{d=1}^{TR}\left(\boldsymbol{\Lambda}^{-1}\right)_{\ell d}\beta_d\right] \\
& + \left(\boldsymbol{\Lambda}^{-1}\right)_{jm}\left(\boldsymbol{\Lambda}^{-1}\right)_{kn}\left[\left(\boldsymbol{\Lambda}^{-1}\right)_{\ell\mu}\sum_{a=1}^{TR}\left(\boldsymbol{\Lambda}^{-1}\right)_{ia}\beta_a + \left(\boldsymbol{\Lambda}^{-1}\right)_{i\mu}\sum_{d=1}^{TR}\left(\boldsymbol{\Lambda}^{-1}\right)_{\ell d}\beta_d\right],
\end{aligned} \tag{3.82}$$

$$\begin{aligned}
Y_4 \triangleq & \left(\boldsymbol{\Lambda}^{-1}\right)_{km}\left(\boldsymbol{\Lambda}^{-1}\right)_{\ell n}\left[\left(\boldsymbol{\Lambda}^{-1}\right)_{i\mu}\sum_{b=1}^{TR}\left(\boldsymbol{\Lambda}^{-1}\right)_{jb}\beta_b + \left(\boldsymbol{\Lambda}^{-1}\right)_{j\mu}\sum_{a=1}^{TR}\left(\boldsymbol{\Lambda}^{-1}\right)_{ia}\beta_a\right] \\
& + \left(\boldsymbol{\Lambda}^{-1}\right)_{km}\left(\boldsymbol{\Lambda}^{-1}\right)_{in}\left[\left(\boldsymbol{\Lambda}^{-1}\right)_{\ell\mu}\sum_{b=1}^{TR}\left(\boldsymbol{\Lambda}^{-1}\right)_{jb}\beta_b + \left(\boldsymbol{\Lambda}^{-1}\right)_{j\mu}\sum_{d=1}^{TR}\left(\boldsymbol{\Lambda}^{-1}\right)_{\ell d}\beta_d\right] \\
& + \left(\boldsymbol{\Lambda}^{-1}\right)_{km}\left(\boldsymbol{\Lambda}^{-1}\right)_{jn}\left[\left(\boldsymbol{\Lambda}^{-1}\right)_{\ell\mu}\sum_{a=1}^{TR}\left(\boldsymbol{\Lambda}^{-1}\right)_{ia}\beta_a + \left(\boldsymbol{\Lambda}^{-1}\right)_{i\mu}\sum_{d=1}^{TR}\left(\boldsymbol{\Lambda}^{-1}\right)_{\ell d}\beta_d\right].
\end{aligned} \tag{3.83}$$

Differentiating Eq. (3.79) with respect to Lagrange multiplier β_ν, $\nu = 1,\ldots,TR$, yields the following relation:

$$\frac{\partial^4 G_4^{ijk\ell}(\boldsymbol{\beta})}{\partial\beta_\nu\partial\beta_\mu\partial\beta_n\partial\beta_m} = W_1 + W_2 + W_3 + W_4, \tag{3.84}$$

where:

$$W_1 \triangleq \left(\mathbf{\Lambda}^{-1}\right)_{\ell m}\left(\mathbf{\Lambda}^{-1}\right)_{in}\left[\left(\mathbf{\Lambda}^{-1}\right)_{j\mu}\left(\mathbf{\Lambda}^{-1}\right)_{k\nu}+\left(\mathbf{\Lambda}^{-1}\right)_{k\mu}\left(\mathbf{\Lambda}^{-1}\right)_{j\nu}\right]$$
$$+\left(\mathbf{\Lambda}^{-1}\right)_{\ell m}\left(\mathbf{\Lambda}^{-1}\right)_{jn}\left[\left(\mathbf{\Lambda}^{-1}\right)_{i\mu}\left(\mathbf{\Lambda}^{-1}\right)_{k\nu}+\left(\mathbf{\Lambda}^{-1}\right)_{k\mu}\left(\mathbf{\Lambda}^{-1}\right)_{i\nu}\right] \tag{3.85}$$
$$+\left(\mathbf{\Lambda}^{-1}\right)_{\ell m}\left(\mathbf{\Lambda}^{-1}\right)_{kn}\left[\left(\mathbf{\Lambda}^{-1}\right)_{i\mu}\left(\mathbf{\Lambda}^{-1}\right)_{j\nu}+\left(\mathbf{\Lambda}^{-1}\right)_{j\mu}\left(\mathbf{\Lambda}^{-1}\right)_{i\nu}\right],$$

$$W_2 \triangleq \left(\mathbf{\Lambda}^{-1}\right)_{im}\left(\mathbf{\Lambda}^{-1}\right)_{\ell n}\left[\left(\mathbf{\Lambda}^{-1}\right)_{j\mu}\left(\mathbf{\Lambda}^{-1}\right)_{k\nu}+\left(\mathbf{\Lambda}^{-1}\right)_{k\mu}\left(\mathbf{\Lambda}^{-1}\right)_{j\nu}\right]$$
$$+\left(\mathbf{\Lambda}^{-1}\right)_{im}\left(\mathbf{\Lambda}^{-1}\right)_{jn}\left[\left(\mathbf{\Lambda}^{-1}\right)_{\ell\mu}\left(\mathbf{\Lambda}^{-1}\right)_{k\nu}+\left(\mathbf{\Lambda}^{-1}\right)_{k\mu}\left(\mathbf{\Lambda}^{-1}\right)_{\ell\nu}\right] \tag{3.86}$$
$$+\left(\mathbf{\Lambda}^{-1}\right)_{im}\left(\mathbf{\Lambda}^{-1}\right)_{kn}\left[\left(\mathbf{\Lambda}^{-1}\right)_{\ell\mu}\left(\mathbf{\Lambda}^{-1}\right)_{j\nu}+\left(\mathbf{\Lambda}^{-1}\right)_{j\mu}\left(\mathbf{\Lambda}^{-1}\right)_{\ell\nu}\right],$$

$$W_3 \triangleq \left(\mathbf{\Lambda}^{-1}\right)_{jm}\left(\mathbf{\Lambda}^{-1}\right)_{\ell n}\left[\left(\mathbf{\Lambda}^{-1}\right)_{i\mu}\left(\mathbf{\Lambda}^{-1}\right)_{k\nu}+\left(\mathbf{\Lambda}^{-1}\right)_{k\mu}\left(\mathbf{\Lambda}^{-1}\right)_{i\nu}\right]$$
$$+\left(\mathbf{\Lambda}^{-1}\right)_{jm}\left(\mathbf{\Lambda}^{-1}\right)_{in}\left[\left(\mathbf{\Lambda}^{-1}\right)_{\ell\mu}\left(\mathbf{\Lambda}^{-1}\right)_{k\nu}+\left(\mathbf{\Lambda}^{-1}\right)_{k\mu}\left(\mathbf{\Lambda}^{-1}\right)_{\ell\nu}\right] \tag{3.87}$$
$$+\left(\mathbf{\Lambda}^{-1}\right)_{jm}\left(\mathbf{\Lambda}^{-1}\right)_{kn}\left[\left(\mathbf{\Lambda}^{-1}\right)_{\ell\mu}\left(\mathbf{\Lambda}^{-1}\right)_{i\nu}+\left(\mathbf{\Lambda}^{-1}\right)_{i\mu}\left(\mathbf{\Lambda}^{-1}\right)_{\ell\nu}\right],$$

$$W_4 \triangleq \left(\mathbf{\Lambda}^{-1}\right)_{km}\left(\mathbf{\Lambda}^{-1}\right)_{\ell n}\left[\left(\mathbf{\Lambda}^{-1}\right)_{i\mu}\left(\mathbf{\Lambda}^{-1}\right)_{j\nu}+\left(\mathbf{\Lambda}^{-1}\right)_{j\mu}\left(\mathbf{\Lambda}^{-1}\right)_{i\nu}\right]$$
$$+\left(\mathbf{\Lambda}^{-1}\right)_{km}\left(\mathbf{\Lambda}^{-1}\right)_{in}\left[\left(\mathbf{\Lambda}^{-1}\right)_{\ell\mu}\left(\mathbf{\Lambda}^{-1}\right)_{j\nu}+\left(\mathbf{\Lambda}^{-1}\right)_{j\mu}\left(\mathbf{\Lambda}^{-1}\right)_{\ell\nu}\right] \tag{3.88}$$
$$+\left(\mathbf{\Lambda}^{-1}\right)_{km}\left(\mathbf{\Lambda}^{-1}\right)_{jn}\left[\left(\mathbf{\Lambda}^{-1}\right)_{\ell\mu}\left(\mathbf{\Lambda}^{-1}\right)_{i\nu}+\left(\mathbf{\Lambda}^{-1}\right)_{i\mu}\left(\mathbf{\Lambda}^{-1}\right)_{\ell\nu}\right].$$

APPENDIX 3.B APPROXIMATIONS INHERENT TO THE 4TH-ORDER MAXIMUM ENTROPY DISTRIBUTION $p_e^{ME}(\mathbf{r})$

As shown in Eq. (3.56), the expression of the approximate 4th-order MaxEnt probability distribution function $p_e^{ME}(\mathbf{r})$ for the responses has the following form:

$$p_e^{ME}(\mathbf{z}) = \frac{h_e^{ME}(\mathbf{z})}{\int_{D_r} h_e^{ME}(\mathbf{z})\,d\mathbf{z}} \cong \frac{1}{Z_e^{ME}} f_e^{ME}(\mathbf{r})\exp\left[-g_e^{ME}(\mathbf{r})\right], \tag{3.89}$$

where:

$$g_e^{ME}(\mathbf{r}) \triangleq \exp\left[-\frac{1}{2}\left(\mathbf{r}-\mathbf{r}^e\right)^\dagger\left(\mathbf{C}^e\right)^{-1}\left(\mathbf{r}-\mathbf{r}^e\right)\right]; \tag{3.90}$$

$$f_e^{ME}(\mathbf{r}) \triangleq 1-\sum_{i=1}^{TR}\theta_i\left(r_i-r_i^e\right)^3-\sum_{i=1}^{TR}\omega_i\left(r_i-r_i^e\right)^4; \tag{3.91}$$

$$Z_e^{ME} = \int_{D_r} f_e^{ME}(\mathbf{r})\exp\left[-g_e^{ME}(\mathbf{r})\right]d\mathbf{r} = K\left(1-\boldsymbol{\omega}^{\dagger}\mathbf{g}_4\right) = K\left[1-3\sum_{i=1}^{TR}\omega_i\left(c_{ii}^e\right)^2\right]. \quad (3.92)$$

The MaxEnt probability distribution function $p_e^{ME}(\mathbf{r})$ is properly normalized, as shown in the following:

$$\begin{aligned}\int_{D_r} p_e^{ME}(\mathbf{r})d\mathbf{r} &= K^{-1}\left[1-3\sum_{i=1}^{TR}\omega_i\left(c_{ii}^e\right)^2\right]^{-1}\left\{\int_{D_r}\exp\left[-\frac{1}{2}\left(\mathbf{r}-\mathbf{r}^e\right)^{\dagger}\left(\mathbf{C}^e\right)^{-1}\left(\mathbf{r}-\mathbf{r}^e\right)\right]d\mathbf{r}\right.\\ &-\sum_{i=1}^{TR}\theta_i\int_{D_r}\left(r_i-r_i^e\right)^3\exp\left[-\frac{1}{2}\left(\mathbf{r}-\mathbf{r}^e\right)^{\dagger}\left(\mathbf{C}^e\right)^{-1}\left(\mathbf{r}-\mathbf{r}^e\right)\right]d\mathbf{r}\\ &\left.-\sum_{i=1}^{TR}\omega_i\int_{D_r}\left(r_i-r_i^e\right)^4\exp\left[-\frac{1}{2}\left(\mathbf{r}-\mathbf{r}^e\right)^{\dagger}\left(\mathbf{C}^e\right)^{-1}\left(\mathbf{r}-\mathbf{r}^e\right)\right]d\mathbf{r}\right\}\\ &= K^{-1}\left[1-3\sum_{i=1}^{TR}\omega_i\left(c_{ii}^e\right)^2\right]^{-1}\left[K-0-3K\sum_{i=1}^{TR}\omega_i\left(c_{ii}^e\right)^2\right] = 1.\end{aligned} \quad (3.93)$$

The 1st-order moments r_i^{ME}, $i=1,\ldots,TR$, of $p_e^{ME}(\mathbf{r})$ are given by the following expression:

$$\begin{aligned}r_i^{ME} &\triangleq \int_{D_r} r_i p_e^{ME}(\mathbf{r})d\mathbf{r} = K^{-1}\left[1-3\sum_{a=1}^{TR}\omega_a\left(c_{aa}^e\right)^2\right]^{-1}\\ &\times\left\{\int_{D_r} r_k\exp\left[-\frac{1}{2}\left(\mathbf{r}-\mathbf{r}^e\right)^{\dagger}\left(\mathbf{C}^e\right)^{-1}\left(\mathbf{r}-\mathbf{r}^e\right)\right]d\mathbf{r}\right.\\ &-\sum_{a=1}^{TR}\theta_a\int_{D_r} r_i\left(r_a-r_a^e\right)^3\exp\left[-\frac{1}{2}\left(\mathbf{r}-\mathbf{r}^e\right)^{\dagger}\left(\mathbf{C}^e\right)^{-1}\left(\mathbf{r}-\mathbf{r}^e\right)\right]d\mathbf{r}\\ &\left.-\sum_{a=1}^{TR}\omega_a\int_{D_r} r_i\left(r_a-r_a^e\right)^4\exp\left[-\frac{1}{2}\left(\mathbf{r}-\mathbf{r}^e\right)^{\dagger}\left(\mathbf{C}^e\right)^{-1}\left(\mathbf{r}-\mathbf{r}^e\right)\right]d\mathbf{r}\right\}\\ &= K^{-1}\left[1-3\sum_{a=1}^{TR}\omega_a\left(c_{aa}^e\right)^2\right]^{-1}\left\{r_i^e\int_{D_r}\exp\left[-\frac{1}{2}\mathbf{z}^{\dagger}\left(\mathbf{C}^e\right)^{-1}\mathbf{z}\right]d\mathbf{z}\right.\\ &-\int_{D_r} z_i\exp\left[-\frac{1}{2}\mathbf{z}^{\dagger}\left(\mathbf{C}^e\right)^{-1}\mathbf{z}\right]d\mathbf{z}-\sum_{a=1}^{TR}\theta_a r_i^e\int_{D_r} z_a^3\exp\left[-\frac{1}{2}\mathbf{z}^{\dagger}\left(\mathbf{C}^e\right)^{-1}\mathbf{z}\right]d\mathbf{z}\\ &-\sum_{a=1}^{TR}\theta_a\int_{D_r} z_i z_a^3\exp\left[-\frac{1}{2}\mathbf{z}^{\dagger}\left(\mathbf{C}^e\right)^{-1}\mathbf{z}\right]d\mathbf{z}-\sum_{a=1}^{TR}\omega_a r_i^e\int_{D_r} z_a^4\exp\left[-\frac{1}{2}\mathbf{z}^{\dagger}\left(\mathbf{C}^e\right)^{-1}\mathbf{z}\right]d\mathbf{z}\\ &\left.-\sum_{a=1}^{TR}\omega_a\int_{D_r} z_i z_a^4\exp\left[-\frac{1}{2}\mathbf{z}^{\dagger}\left(\mathbf{C}^e\right)^{-1}\mathbf{z}\right]d\mathbf{z}\right\} = K^{-1}\left[1-3\sum_{a=1}^{TR}\omega_a\left(c_{aa}^e\right)^2\right]^{-1}\end{aligned} \quad (3.94)$$

$$\times\left\{Kr_i^e-0-0-O\left[K\sum_{a=1}^{TR}\theta_a\left(c_{aa}^e\right)^2\right]-3K\sum_{a=1}^{TR}\omega_a r_i^e\left(c_{aa}^e\right)^2-0\right\}$$

$$=K^{-1}\left[1-3\sum_{a=1}^{TR}\omega_a\left(c_{aa}^e\right)^2\right]^{-1}K\left[1-3\sum_{a=1}^{TR}\omega_a\left(c_{aa}^e\right)^2\right]r_i^e$$

$$-K^{-1}\left[1-3\sum_{a=1}^{TR}\omega_a\left(c_{aa}^e\right)^2\right]^{-1}\times K\times O\left[\sum_{a=1}^{TR}t_a^e\left(c_{aa}^e\right)^{-3}\left(c_{aa}^e\right)^2\right]=r_i^e+O\left(t_i^e/c_{ii}^e\right).$$

As indicated by the result obtained in Eq. (3.94), if the 3rd-order correlations t_{ijk}^e among the measured responses are all neglected, then the MaxEnt moments r_i^{ME} coincide with the measured moment of the i-th response, r_i^e (i.e., $r_i^{ME}=r_i^e$ if $t_{ijk}^e=0$).

Within the approximation $r_i^{ME}=r_i^e$, the 2nd-order moments c_{ij}^{ME}, $i,j=1,\ldots,TR$, of $p_e^{ME}(\mathbf{r})$ are given by the following expression:

$$c_{ij}^{ME}\triangleq\int_{D_e}\left(r_i-r_i^e\right)\left(r_j-r_j^e\right)p_e^{ME}(\mathbf{r})\,d\mathbf{r}=K^{-1}\left[1-3\sum_{a=1}^{TR}\omega_a\left(c_{aa}^e\right)^2\right]^{-1}$$

$$\times\left\{\int_{D_r}z_iz_j\exp\left[-\frac{1}{2}\mathbf{z}^\dagger\left(\mathbf{C}^e\right)^{-1}\mathbf{z}\right]d\mathbf{z}-\sum_{a=1}^{TR}\theta_a\int_{D_r}z_iz_jz_a^3\exp\left[-\frac{1}{2}\mathbf{z}^\dagger\left(\mathbf{C}^e\right)^{-1}\mathbf{z}\right]d\mathbf{z}\right.\qquad(3.95)$$

$$\left.-\sum_{a=1}^{TR}\omega_a\int_{D_r}z_iz_jz_a^4\exp\left[-\frac{1}{2}\mathbf{z}^\dagger\left(\mathbf{C}^e\right)^{-1}\mathbf{z}\right]d\mathbf{z}\right\}=K^{-1}\left[1-3\sum_{a=1}^{TR}\omega_a\left(c_{aa}^e\right)^2\right]^{-1}$$

$$\times K\left[1-3\sum_{a=1}^{TR}\omega_a\left(c_{aa}^e\right)^2\right]\left\{c_{ij}^e-0-O\left(\omega_a\right)\right\}=c_{ij}^e;\quad i,j=1,\ldots,TR.$$

Within the approximation $r_i^{ME}=r_i^e$, the 3rd-order moments t_{ijk}^{ME}, $i,j,k=1,\ldots,TR$, of $p_e^{ME}(\mathbf{r})$ are given by the following expression:

$$t_{ijk}^{ME}\triangleq\int_{D_e}\left(r_i-r_i^e\right)\left(r_j-r_j^e\right)\left(r_k-r_k^e\right)p_e^{ME}(\mathbf{r})\,d\mathbf{r}=K^{-1}\left[1-3\sum_{a=1}^{TR}\omega_a\left(c_{aa}^e\right)^2\right]^{-1}$$

$$\times\left\{\int_{D_r}z_iz_jz_k\exp\left[-\frac{1}{2}\mathbf{z}^\dagger\left(\mathbf{C}^e\right)^{-1}\mathbf{z}\right]d\mathbf{z}-\sum_{a=1}^{TR}\theta_a\int_{D_r}z_iz_jz_kz_a^3\exp\left[-\frac{1}{2}\mathbf{z}^\dagger\left(\mathbf{C}^e\right)^{-1}\mathbf{z}\right]d\mathbf{z}\right.\qquad(3.96)$$

$$\left.-\sum_{a=1}^{TR}\omega_a\int_{D_r}z_iz_jz_kz_a^4\exp\left[-\frac{1}{2}\mathbf{z}^\dagger\left(\mathbf{C}^e\right)^{-1}\mathbf{z}\right]d\mathbf{z}\right\}=t_{ijk}^e+HOT,$$

where the acronym "*HOT*" denotes "higher-order terms."

Within the approximation $r_i^{ME}=r_i^e$, the 4th-order moments q_{ijk}^e, $i,j,k,\ell=1,\ldots,TR$, of $p_e^{ME}(\mathbf{r})$ are given by the following expression:

$$
\begin{aligned}
q_{ijk\ell}^{ME} &\triangleq \int_{D_e} \left(r_i - r_i^e\right)\left(r_j - r_j^e\right)\left(r_k - r_k^e\right)\left(r_\ell - r_\ell^e\right) p_e^{ME}(\mathbf{r})\, d\mathbf{r} \\
&= K^{-1}\left[1 - 3\sum_{a=1}^{TR} \omega_a \left(c_{aa}^e\right)^2\right]^{-1} \left\{ \int_{D_r} z_i z_j z_k z_\ell \exp\left[-\frac{1}{2}\mathbf{z}^\dagger \left(\mathbf{C}^e\right)^{-1} \mathbf{z}\right] d\mathbf{z} \right. \\
&- \sum_{a=1}^{TR} \theta_a \int_{D_r} z_i z_j z_k z_\ell z_a^3 \exp\left[-\frac{1}{2}\mathbf{z}^\dagger \left(\mathbf{C}^e\right)^{-1} \mathbf{z}\right] d\mathbf{z} \\
&\left. - \sum_{a=1}^{TR} \omega_a \int_{D_r} z_i z_j z_k z_\ell z_a^4 \exp\left[-\frac{1}{2}\mathbf{z}^\dagger \left(\mathbf{C}^e\right)^{-1} \mathbf{z}\right] d\mathbf{z} \right\} = q_{ijk\ell}^e + HOT,
\end{aligned} \tag{3.97}
$$

where the acronym "*HOT*" denotes "higher-order terms."

4 4th-BERRU-PM
4th-Order Maxent Predictive Modeling Methodology for Combining Measurements with Computations to Obtain Best-Estimate Results with Reduced Predicted Uncertainties

4.1 INTRODUCTION

Scientific progress stems from the judicious combination of experimental information with results produced by computational models, neither of which is perfectly accurate. On the one hand, computations are afflicted by errors stemming from numerical procedures, uncertain model parameters, boundary/initial conditions and/or imperfectly known physical processes. On the other hand, results of measurements are afflicted by experimental errors which means that around any reported experimental value there always exists a range of values that may also be plausibly representative of the true but unknown value of the measured quantity. Extracting "best-estimate" values for model parameters and predicted results, together with "best-estimate" uncertainties for these parameters and results requires the combination of experimental and computational data and their uncertainties.

The newly developed "2nd-BERRU-PM" methodology presented in Chapter 1 can incorporate arbitrarily high sensitivities, thereby vastly generalizing extant data adjustment and data assimilation methodologies. The essential contributions of the 2nd- and higher-order sensitivities for reducing predicted uncertainties in the model responses have been illustrated in Chapter 2 by applying the 2nd-BERRU-PM methodology to the neutron leakage response of the polyethylene-reflected plutonium (PERP) OECD/NEA reactor physics benchmark which was modeled using the neutron transport equation involving 21976 imprecisely known parameters, the solution of which is representative of "large-scale computations."

Although the 2nd-BERRU-PM methodology can incorporate arbitrarily high-order sensitivities of the model response with respect to the imprecisely known parameters underlying the computational model, this methodology is limited to

DOI: 10.1201/9781003478119-4

considering just 2nd-order moments (hence the designation "2nd-") of the experimentally measured responses. Also, the "output" produced by the 2nd-BERRU-PM methodology is limited to yielding optimal best-estimate values for the means and covariances (i.e., the 1st- and 2nd-order moments) of the best-estimate predicted distribution of responses and parameters. Since skewness and kurtosis play an essential role in determining the asymmetries of distributions, it is important to generalize the 2nd-BERRU-PM methodology to enable the incorporation of 3rd- and 4th-order moments of measured and computed responses, as well as to enable the computation of skewness and kurtosis of the best-estimate predicted posterior distribution of calibrated model parameters and responses. Such a generalization has now been enabled by the development of the closed-form expression of the 4th-order moment-constrained MaxEnt presented in Chapter 3, which underlies the development of the 4th-BERRU-PM ("*4th*-Order *B*est-*E*stimate *R*esults with *R*educed *U*ncertainties *P*redictive *M*odeling") methodology to be presented in this chapter. The 4th-BERRU-PM methodology has been conceived by Cacuci (2023e) and encompasses the following components:

1. A mathematical model of a physical system, comprising linear and/or nonlinear equations that relate the system's independent variables and parameters to the system's state (i.e., dependent) variables;
2. Arbitrarily high-order sensitivities and moments of the distribution of model parameters, which will however be explicitly used only up to 4th-order in the derivation of the practically applicable end results produced by the 4th-BERRU-PM methodology;
3. One or several computational results, customarily referred to as system responses (or objective functions, or indices of performance), which are computed using the mathematical model; and
4. The first four moments (mean values, variances/covariances, skewness, and kurtosis) of the distribution of experimentally measured responses.

This chapter is structured as follows: Section 4.2 presents the 4th-order moment-constrained MaxEnt probabilistic representation of the joint distribution of computational model parameters and computed responses. This joint distribution is constructed by following the same procedure as in Chapter 3 but incorporating 4th-order moments (means, variances/covariances, skewness, and kurtosis) of model parameters, measured, and computed model responses. Section 4.3 presents the mathematical framework and the output (best-estimate results with reduced uncertainties) of the 4th-BERRU-PM methodology. Concluding remarks are presented in Section 4.4.

4.2 CONSTRUCTION OF THE MOMENTS-CONSTRAINED 4TH-ORDER MAXENT DISTRIBUTION OF THE COMPUTATIONAL MODEL PARAMETERS AND RESPONSES

The 4th-order moment-constrained MaxEnt joint distribution of model parameters and computed responses will encompass the following moments:

1. The vector $\boldsymbol{\alpha}^0 \triangleq \left(\alpha_1^0,...,\alpha_{TP}^0\right)^\dagger$ of expected/nominal values of the model parameters, where α_i^0 denotes the nominal/expected value of the parameter $i = 1,\ldots,TP$;
2. The parameter covariance matrix, $\mathbf{C}_{\alpha\alpha} \triangleq \left[c_{ij}^{\alpha}\right]_{TP\times TP}$, where the quantity c_{ij}^{α} denotes the covariance of two model parameters, α_i and α_j, where $i,j = 1,\ldots,TP$;
3. The triple correlations of three model parameters α_i, α_j, and α_k, denoted as t_{ijk}^{α}, where $i,j,k = 1,\ldots,TP$;
4. The quadruple correlations of four model parameters α_i, α_j, α_k, and α_ℓ, denoted as $q_{ijk\ell}^{\alpha}$, where $i,j,k,\ell = 1,\ldots,TR$;
5. The vector of expected values of the computed responses, $\mathbf{E}_c(\mathbf{r}) \triangleq \left[E_c(r_1),...,E_c(r_k),...,E_c(r_{TR})\right]^\dagger$, where $E_c(r_k)$ denotes the expectation value of a computed response $r_k(\boldsymbol{\alpha})$, for $k = 1,\ldots,TP$, the expression of which is presented in Chapter 1;
6. The parameter-response computed correlation matrix, $\mathbf{C}_{\alpha r}^{c} \triangleq \left[\text{cor}(\alpha_i,r_k)\right]_{TP\times TR}$, where $\text{cor}(\alpha_i,r_k)$ denotes the correlation between a parameter α_i and a computed response r_k, for $i = 1,\ldots,TP$ and $k = 1,\ldots,TR$, the expression of which is presented in Chapter 1;
7. The covariance matrix of computed responses, $\mathbf{C}_{rr}^{c} \triangleq \left[\text{cov}(r_k,r_k)\right]_{TR\times TR}$, where $\text{cov}(r_k,r_k)$ denotes the covariance between two computed responses r_k and r_ℓ, for $k,\ell = 1,...,TR$, the expression of which is presented in Chapter 1;
8. The triple correlations, $\mu_3(r_k,r_\ell,r_m)$, among three computed responses, r_k, r_ℓ, and r_m, for $k,\ell,m = 1,...,TR$, the expression of which is presented in Chapter 1;
9. The quadruple correlations, $\mu_4(r_k,r_\ell,r_m,r_n)$, among four computed responses, r_k, r_ℓ, r_m, and r_n, for $k,\ell,m,n = 1,...,TR$, the expression of which is presented in Chapter 1.

The *4th-order* "MaxEnt probability density of the joint model parameters and computed responses" to be constructed using the aforementioned information will be denoted as $p_c^{ME}(\boldsymbol{\alpha},\mathbf{r})$, where the subscript "*c*" indicates "*computational model*," while the superscript "*ME*" indicates "*MaxEnt*" approximation. To ensure optimal compatibility with the previous information while simultaneously ensuring minimal spurious information content, the probability density $p_c^{ME}(\boldsymbol{\alpha},\mathbf{r})$ would need to satisfy the following conditions:

1. $p_c^{ME}(\boldsymbol{\alpha},\mathbf{r})$ maximizes the Shannon (1948) information entropy, S, as defined:

$$S = -\int_D p_c^{ME}(\boldsymbol{\alpha},\mathbf{r}) \ln\left[p_c^{ME}(\boldsymbol{\alpha},\mathbf{r})\right] d\boldsymbol{\alpha}\, d\mathbf{r}; \tag{4.1}$$

2. $p_c^{ME}(\boldsymbol{\alpha},\mathbf{r})$ satisfies the "moments constraints" enumerated in items one through nine in the previous model;

3. $p_c^{ME}(\boldsymbol{\alpha},\mathbf{r})$ satisfies the normalization condition:

$$\int_D p_c^{ME}(\boldsymbol{\alpha},\mathbf{r})\,d\boldsymbol{\alpha}d\mathbf{r}=1. \tag{4.2}$$

It was shown in Chapter 3 that attempting to include the triple and quadruple cross-correlations among experimentally/empirically obtained responses leads to insurmountable computational difficulties for little, if any, gain in accuracy. For this reason, only the self-correlations of orders three and four were included, in Chapter 3, in the practically computable expression of the 4th-order MaxEnt distribution of the empirically obtained responses. The same approximations will be used in this chapter in constructing the distribution $p_c^{ME}(\boldsymbol{\alpha},\mathbf{r})$, namely only the self-correlations of 3rd- and 4th-order will be included. Thus, the MaxEnt distribution $p_c^{ME}(\boldsymbol{\alpha},\mathbf{r})$ will be obtained as the solution of the variational problem $\partial H\left(p_c^{ME}\right)/\partial p_c^{ME}=0$ where the entropy (Lagrangian functional) $H\left(p_c^{ME}\right)$ is defined as follows:

$$\begin{aligned}
H\left(p_c^{ME}\right)&=-\int_D p_c^{ME}(\boldsymbol{\alpha},\mathbf{r})\ln\left[p_c^{ME}(\boldsymbol{\alpha},\mathbf{r})\right]d\boldsymbol{\alpha}d\mathbf{r}-\lambda_0\left[\int_D p_c^{ME}(\boldsymbol{\alpha},\mathbf{r})\,d\boldsymbol{\alpha}d\mathbf{r}-1\right]\\
&-\sum_{k=1}^{TR}a_k^{(1)}\left[\int_D(\delta r_k)\,p_c^{ME}(\boldsymbol{\alpha},\mathbf{r})\,d\boldsymbol{\alpha}d\mathbf{r}\right]-\sum_{i=1}^{TP}a_i^{(2)}\left[\int_D(\delta\alpha_i)\,p_c^{ME}(\boldsymbol{\alpha},\mathbf{r})\,d\boldsymbol{\alpha}d\mathbf{r}\right]\\
&-\frac{1}{2}\sum_{k=1}^{TR}\sum_{\ell=1}^{TR}b_{k\ell}^{(11)}\left[\int_D(\delta r_k)(\delta r_\ell)\,p_c^{ME}(\boldsymbol{\alpha},\mathbf{r})\,d\boldsymbol{\alpha}d\mathbf{r}-\operatorname{cov}(r_k,r_\ell)\right]\\
&-\sum_{k=1}^{TR}\sum_{i=1}^{TP}b_{ki}^{(12)}\left[\int_D(\delta\alpha_i)(\delta r_k)\,p_c^{ME}(\boldsymbol{\alpha},\mathbf{r})\,d\boldsymbol{\alpha}d\mathbf{r}-\operatorname{cor}(\alpha_i,r_k)\right]\\
&-\frac{1}{2}\sum_{i=1}^{TP}\sum_{j=1}^{TP}b_{ij}^{(22)}\left[\int_D(\delta\alpha_i)(\delta\alpha_j)\,p_c^{ME}(\boldsymbol{\alpha},\mathbf{r})\,d\boldsymbol{\alpha}d\mathbf{r}-c_{ij}^{\alpha}\right]\\
&-\sum_{k=1}^{TR}\psi_k^{(1)}\left[\int_D(\delta r_k)^3\,p_c^{ME}(\boldsymbol{\alpha},\mathbf{r})\,d\boldsymbol{\alpha}d\mathbf{r}-\mu_3(r_k)\right]-\sum_{i=1}^{TP}\psi_i^{(2)}\left[\int_D(\delta\alpha_i)^3\,p_c^{ME}(\boldsymbol{\alpha},\mathbf{r})\,d\boldsymbol{\alpha}d\mathbf{r}-t_i^{\alpha}\right]\\
&-\sum_{k=1}^{TR}\chi_k^{(1)}\left[\int_D(\delta r_k)^4\,p_c^{ME}(\boldsymbol{\alpha},\mathbf{r})\,d\boldsymbol{\alpha}d\mathbf{r}-\mu_4(r_k)\right]-\sum_{i=1}^{TP}\chi_i^{(2)}\left[\int_D(\delta\alpha_i)^4\,p_c^{ME}(\boldsymbol{\alpha},\mathbf{r})\,d\boldsymbol{\alpha}d\mathbf{r}-q_i^{\alpha}\right].
\end{aligned} \tag{4.3}$$

In Eq. (4.3), the quantities $a_k^{(1)}$, $a_i^{(2)}$, $b_{k\ell}^{(11)}$,$b_{ki}^{(12)}$, $b_{ij}^{(22)}$, $\psi_k^{(1)}$, $\psi_i^{(2)}$, $\chi_k^{(1)}$, and $\chi_i^{(2)}$ denote the respective Lagrange multipliers, and the factors 1/2 have been introduced for subsequent computational convenience.

Solving the equation $\partial H\left(p_c^{ME}\right)/\partial p_c^{ME}=0$ yields the following expression for the resulting MaxEnt distribution $p_c^{ME}(\boldsymbol{\alpha},\mathbf{r})$:

$$p_c^{ME}(\boldsymbol{\alpha},\mathbf{r})=Z_c^{-1}\varphi(\boldsymbol{\alpha},\mathbf{r});\quad Z_c\triangleq\int_D\varphi(\boldsymbol{\alpha},\mathbf{r})\,d\boldsymbol{\alpha}\,d\mathbf{r}; \tag{4.4}$$

where:

$$\varphi(\boldsymbol{\alpha},\mathbf{r}) \triangleq \exp\left\{-\sum_{k=1}^{TR} a_k^{(1)}(\delta r_k) - \sum_{i=1}^{TP} a_i^{(2)}(\delta\alpha_i) - \frac{1}{2}\sum_{k=1}^{TR}\sum_{\ell=1}^{TR} b_{k\ell}^{(11)}(\delta r_k)(\delta r_\ell)\right.$$
$$-\sum_{k=1}^{TR}\sum_{i=1}^{TP} b_{ki}^{(12)}(\delta\alpha_i)(\delta r_k) - \frac{1}{2}\sum_{i=1}^{TP}\sum_{j=1}^{TP} b_{ij}^{(22)}(\delta\alpha_i)(\delta\alpha_j) - \sum_{k=1}^{TR}\psi_k^{(1)}(\delta r_k)^3 \tag{4.5}$$
$$\left.-\sum_{i=1}^{TP}\psi_i^{(2)}(\delta\alpha_i)^3 - \sum_{k=1}^{TR}\chi_k^{(1)}(\delta r_k)^4 - \sum_{i=1}^{TP}\chi_i^{(2)}(\delta\alpha_i)^4\right\}.$$

The Lagrange multipliers which appear in Eq. (4.5) are determined by following the same conceptual procedure as was used to determine the Lagrange multipliers in Chapter 3, by expanding the 3rd- and 4th-order terms in parameter and response variations in a Taylor series, to obtain the following relation:

$$\varphi(\boldsymbol{\alpha},\mathbf{r}) \cong \varphi_2(\mathbf{u})\exp\left[-\varphi_1(\mathbf{u})\right], \tag{4.6}$$

where the following definitions have been used:

$$\varphi_1(\mathbf{u}) \triangleq \mathbf{a}^\dagger\mathbf{u} + \frac{1}{2}\mathbf{u}^\dagger\mathbf{B}\mathbf{u}, \tag{4.7}$$

$$\varphi_2(\mathbf{u}) \triangleq 1 - \sum_{i=1}^{TR}\psi_k^{(1)}(\delta r_k)^3 - \sum_{i=1}^{TP}\psi_i^{(2)}(\delta\alpha_i)^3 - \sum_{k=1}^{TR}\chi_k^{(1)}(\delta r_k)^4 - \sum_{i=1}^{TP}\chi_i^{(2)}(\delta\alpha_i)^4, \tag{4.8}$$

and where the various vectors and matrices are defined as follows:

$$\mathbf{u} = \begin{pmatrix}\mathbf{r} - \mathbf{E}_c(\mathbf{r}) \\ \boldsymbol{\alpha} - \boldsymbol{\alpha}^0\end{pmatrix};\ \mathbf{a} \triangleq \begin{pmatrix}\mathbf{a}^{(1)} \\ \mathbf{a}^{(2)}\end{pmatrix};\ \mathbf{a}^{(1)} \triangleq \begin{pmatrix} a_1^{(1)} \\ \bullet \\ a_{TR}^{(1)}\end{pmatrix};\ \mathbf{a}^{(2)} \triangleq \begin{pmatrix} a_1^{(2)} \\ \bullet \\ a_{TP}^{(1)}\end{pmatrix}; \tag{4.9}$$

$$\mathbf{B} \triangleq \begin{pmatrix}\mathbf{B}^{(11)} & \mathbf{B}^{(12)} \\ \left[\mathbf{B}^{(12)}\right]^\dagger & \mathbf{B}^{(22)}\end{pmatrix};\ \mathbf{B}^{(11)} \triangleq \begin{pmatrix} b_{11}^{(11)} & \bullet & b_{1,TR}^{(11)} \\ \bullet & b_{k\ell}^{(11)} & \bullet \\ b_{TR,1}^{(11)} & \bullet & b_{TR,TR}^{(11)}\end{pmatrix};$$
$$\mathbf{B}^{(12)} \triangleq \begin{pmatrix} b_{11}^{(12)} & \bullet & b_{1,TP}^{(12)} \\ \bullet & b_{k\ell}^{(12)} & \bullet \\ b_{TR,1}^{(12)} & \bullet & b_{TR,TP}^{(12)}\end{pmatrix};\ \mathbf{B}^{(22)} \triangleq \begin{pmatrix} b_{11}^{(22)} & \bullet & b_{1,TP}^{(22)} \\ \bullet & b_{k\ell}^{(22)} & \bullet \\ b_{TP,1}^{(22)} & \bullet & b_{TP,TP}^{(22)}\end{pmatrix}. \tag{4.10}$$

Using Eqs. (4.6)–(4.10) in Eq. (4.4) yields the following approximate expression for the normalization integral $Z_c(\mathbf{a})$:

$$Z_c(\mathbf{a}) \cong \int_D \varphi_2(\mathbf{u}) \exp[-\varphi_1(\mathbf{u})]\, d\mathbf{u} = \Phi(\mathbf{a}) - \sum_{k=1}^{TR} \psi_k^{(1)} \Psi_k^{(1)}(\mathbf{a}) - \sum_{i=1}^{TP} \psi_i^{(2)} \Psi_i^{(2)}(\mathbf{a}) - \sum_{k=1}^{TR} \chi_k^{(1)} \mathrm{X}_k^{(1)}(\mathbf{a}) - \sum_{i=1}^{TP} \chi_i^{(2)} \mathrm{X}_i^{(2)}(\mathbf{a}), \tag{4.11}$$

where:

$$\Phi(\mathbf{a}) \triangleq \int_D \exp\left(-\mathbf{a}^\dagger \mathbf{u} - \frac{1}{2}\mathbf{u}^\dagger \mathbf{B}\mathbf{u}\right) d\mathbf{u} = K_c \exp\left(\frac{1}{2}\mathbf{a}^\dagger \mathbf{B}^{-1}\mathbf{a}\right); \tag{4.12}$$

$$K_c \triangleq \frac{(2\pi)^{(TR+TP)/2}}{\sqrt{Det(\mathbf{B})}}; \tag{4.13}$$

$$\Psi_k^{(1)}(\mathbf{a}) \triangleq \int_D (\delta r_k)^3 \exp\left(-\mathbf{a}^\dagger \mathbf{u} - \frac{1}{2}\mathbf{u}^\dagger \mathbf{B}\mathbf{u}\right) d\mathbf{u};\ k = 1,\ldots,TR; \tag{4.14}$$

$$\Psi_i^{(2)}(\mathbf{a}) \triangleq \int_D (\delta \alpha_i)^3 \exp\left(-\mathbf{a}^\dagger \mathbf{u} - \frac{1}{2}\mathbf{u}^\dagger \mathbf{B}\mathbf{u}\right) d\mathbf{u};\ i = 1,\ldots,TP; \tag{4.15}$$

$$\mathrm{X}_k^{(1)}(\mathbf{a}) \triangleq \int_D (\delta r_k)^4 \exp\left(-\mathbf{a}^\dagger \mathbf{u} - \frac{1}{2}\mathbf{u}^\dagger \mathbf{B}\mathbf{u}\right) d\mathbf{u};\ k = 1,\ldots,TR; \tag{4.16}$$

$$\mathrm{X}_i^{(2)}(\mathbf{a}) \triangleq \int_D (\delta \alpha_i)^4 \exp\left(-\mathbf{a}^\dagger \mathbf{u} - \frac{1}{2}\mathbf{u}^\dagger \mathbf{B}\mathbf{u}\right) d\mathbf{u};\ i = 1,\ldots,TP. \tag{4.17}$$

The Lagrange multipliers are determined by differentiating the "free energy" $[-\ln Z_c(\mathbf{a})]$ with respect to the components of the vector of Lagrange multipliers $\mathbf{a} \triangleq \left(\mathbf{a}^{(1)}, \mathbf{a}^{(2)}\right)^\dagger$. For the sake of consistency, the same approximations which have been made in Chapter 3 will also be employed for determining the Lagrange multipliers which appear in Eq. (4.5). Thus, using Eq. (4.12) and neglecting, for the present case, the corresponding 3rd-and 4th-order correlations in Eq. (4.11) yields the following relations:

$$\frac{\partial[-\ln Z_c(\mathbf{a})]}{\partial a_k^{(1)}} = \frac{1}{Z_c}\int_D (\delta r_k) \exp\left(-\mathbf{a}^\dagger \mathbf{u} - \frac{1}{2}\mathbf{u}^\dagger \mathbf{B}\mathbf{u}\right) d\mathbf{u} = 0 = \left(\mathbf{B}^{-1}\mathbf{a}\right)_k;\ k = 1,\ldots,TR; \tag{4.18}$$

$$\frac{\partial[-\ln Z_c(\mathbf{a})]}{\partial a_i^{(2)}} = \frac{1}{Z_c}\int_D (\delta \alpha_i) \exp\left(-\mathbf{a}^\dagger \mathbf{u} - \frac{1}{2}\mathbf{u}^\dagger \mathbf{B}\mathbf{u}\right) d\mathbf{u} = 0 = \left(\mathbf{B}^{-1}\mathbf{a}\right)_i;\ i = 1,\ldots,TP. \tag{4.19}$$

It follows from Eqs. (4.18) and (4.19) that the following result holds when the triple correlations are neglected:

$$\mathbf{a} = \mathbf{0}. \tag{4.20}$$

Differentiating the relations provided in Eqs. (4.18) and (4.19) yields the following relations:

$$\frac{\partial^2\left[-\ln Z_c(\mathbf{a})\right]}{\partial a_j^{(1)}\partial a_k^{(1)}} = -\operatorname{cov}\left(r_j, r_k\right); j,k = 1,\ldots,TR; \tag{4.21}$$

$$\frac{\partial^2\left[-\ln Z_c(\mathbf{a})\right]}{\partial a_i^{(2)}\partial a_k^{(1)}} = -cor\left(\alpha_i, r_k\right); \quad i = 1,\ldots,TP; \quad k = 1,\ldots,TR; \tag{4.22}$$

$$\frac{\partial^2\left[-\ln Z_c(\mathbf{a})\right]}{\partial a_i^{(2)}\partial a_j^{(2)}} = -\operatorname{cov}\left(\alpha_i, \alpha_j\right); \quad i,j = 1,\ldots,TP. \tag{4.23}$$

The results obtained in Eqs. (4.21)–(4.23) can be collectively written in vector-matrix form as follows:

$$\begin{gathered}\frac{\partial^2\left[-\ln Z_c(\mathbf{a})\right]}{\partial \mathbf{a}\,\partial \mathbf{a}} = -\mathbf{C}_c; \ \mathbf{C}_c \triangleq \begin{pmatrix} \mathbf{C}_{rr}^c & \mathbf{C}_{r\alpha}^c \\ \mathbf{C}_{\alpha r}^c & \mathbf{C}_{\alpha\alpha} \end{pmatrix}; \ \mathbf{C}_{rr}^c \triangleq \left[\operatorname{cov}\left(r_j, r_k\right)\right]_{TR\times TR}; \\ \mathbf{C}_{r\alpha}^c \triangleq \left[cor\left(r_k, \alpha_i\right)\right]_{TR\times TP} = \left(\mathbf{C}_{\alpha r}^c\right)^\dagger; \quad \mathbf{C}_{\alpha\alpha} \triangleq \left[\operatorname{cov}\left(\alpha_i, \alpha_j\right)\right]_{TP\times TP}.\end{gathered} \tag{4.24}$$

On the other hand, it follows from Eq. (4.11) that the following result holds when the quadruple correlations are neglected:

$$\frac{\partial^2\left[-\ln Z_c(\mathbf{a})\right]}{\partial \mathbf{a}\,\partial \mathbf{a}} = -\mathbf{B}^{-1}. \tag{4.25}$$

The relations obtained in Eqs. (4.24) and (4.25) imply the following relation:

$$\mathbf{B}^{-1} = \mathbf{C}_c. \tag{4.26}$$

Using the result obtained in Eq. (4.26) simplifies the derivations of the expressions of the functions $\Psi_k^{(1)}(\mathbf{a})$, $X_k^{(1)}(\mathbf{a})$, $\Psi_i^{(2)}(\mathbf{a})$, and $X_i^{(2)}(\mathbf{a})$, which are performed in Appendix 4.A. The final expressions obtained in Appendix 4.A are as follows:

$$\Psi_k^{(1)}(\mathbf{a}) = -K_c\Phi_{k,1}^{(3)}\exp\left[\eta(\mathbf{a})\right]; \quad k = 1,\ldots,TR; \tag{4.27}$$

$$X_k^{(1)}(\mathbf{a}) = K_c\Phi_{k,1}^{(4)}\exp\left[\eta(\mathbf{a})\right]; \quad k = 1,\ldots,TR; \tag{4.28}$$

$$\Psi_i^{(2)}(\mathbf{a}) = -K_c \Phi_{i,2}^{(3)} \exp[\eta(\mathbf{a})]; \; i = 1,...,TP; \tag{4.29}$$

$$X_i^{(2)}(\mathbf{a}) = K_c \Phi_{i,2}^{(4)} \exp[\eta(\mathbf{a})]; \; i = 1,...,TP; \tag{4.30}$$

where the following definitions were used:

$$\eta(\mathbf{a}) \triangleq \frac{1}{2}\mathbf{a}^\dagger \mathbf{C}_c \mathbf{a} = \frac{1}{2}\mathbf{a}^\dagger \mathbf{B}^{-1}\mathbf{a}; \tag{4.31}$$

$$\Phi_{k,1}^{(3)} \triangleq 3\Phi_{k,1}^{(1)}(\mathbf{a})\operatorname{var}(r_k, r_k) + \left[\Phi_{k,1}^{(1)}(\mathbf{a})\right]^3; \tag{4.32}$$

$$\Phi_{k,1}^{(4)} \triangleq 3\left[\operatorname{var}(r_k, r_k)\right]^2 + 6\left[\Phi_{k,1}^{(1)}(\mathbf{a})\right]^2 \operatorname{var}(r_k, r_k) + \left[\Phi_{k,1}^{(1)}(\mathbf{a})\right]^4; \tag{4.33}$$

$$\Phi_{i,2}^{(3)} \triangleq 3\Phi_{i,2}^{(1)}(\mathbf{a})\operatorname{var}(\alpha_i, \alpha_i) + \left[\Phi_{i,2}^{(1)}(\mathbf{a})\right]^3; \tag{4.34}$$

$$\Phi_{i,2}^{(4)} \triangleq 3\left[\operatorname{var}(\alpha_i, \alpha_i)\right]^2 + 6\left[\Phi_{i,2}^{(1)}(\mathbf{a})\right]^2 \operatorname{var}(\alpha_i, \alpha_i) + \left[\Phi_{i,2}^{(1)}(\mathbf{a})\right]^4; \tag{4.35}$$

$$\Phi_{k,1}^{(1)}(\mathbf{a}) \triangleq \sum_{j=1}^{TR} \operatorname{cov}(r_j, r_k) a_j^{(1)} + \sum_{j=1}^{TP} \operatorname{cor}(\alpha_j, r_k) a_j^{(2)}; \tag{4.36}$$

$$\Phi_{i,2}^{(1)} \triangleq \sum_{j=1}^{TP} a_j^{(1)} \operatorname{cor}(\alpha_i, r_j) + \sum_{j=1}^{TP} a_j^{(2)} \operatorname{cov}(\alpha_i, \alpha_j). \tag{4.37}$$

Using the results obtained in Eqs. (4.27)–(4.30) into Eq. (4.11) yields the following expression for the normalization integral $Z_c(\mathbf{a})$:

$$\begin{aligned} Z_c(\mathbf{a}) = K_c \Bigg\{ & 1 + \sum_{k=1}^{TR} \psi_k^{(1)} \Phi_{k,1}^{(3)}(\mathbf{a}) + \sum_{i=1}^{TP} \psi_i^{(2)} \Phi_{i,2}^{(3)}(\mathbf{a}) \\ & - \sum_{k=1}^{TR} \chi_k^{(1)} \Phi_{k,1}^{(4)}(\mathbf{a}) - \sum_{i=1}^{TP} \chi_i^{(2)} \Phi_{i,2}^{(4)}(\mathbf{a}) \Bigg\} \exp[\eta(\mathbf{a})]. \end{aligned} \tag{4.38}$$

It follows from the expression provided in Eq. (4.38) that:

$$\begin{aligned} \ln Z_c(\mathbf{a}) = \ln K_c + \eta(\mathbf{a}) + \ln \Bigg\{ & 1 + \sum_{k=1}^{TR} \psi_k^{(1)} \Phi_{k,1}^{(3)}(\mathbf{a}) + \sum_{i=1}^{TP} \psi_i^{(2)} \Phi_{i,2}^{(3)}(\mathbf{a}) \\ & - \sum_{k=1}^{TR} \chi_k^{(1)} \Phi_{k,1}^{(4)}(\mathbf{a}) - \sum_{i=1}^{TP} \chi_i^{(2)} \Phi_{i,2}^{(4)}(\mathbf{a}) \Bigg\}. \end{aligned} \tag{4.39}$$

The Lagrange multipliers $\psi_k^{(1)}$, $\chi_k^{(1)}$, for $(k = 1,...,TR)$, are obtained by using Eq. (4.39) to determine the 3rd-order derivatives of $[-\ln Z_c(\mathbf{a})]$ with respect to components of the vector of Lagrange multipliers $\mathbf{a} \triangleq \left(\mathbf{a}^{(1)}, \mathbf{a}^{(2)}\right)^\dagger$. The procedure for determining these 3rd-order derivatives is the same as was used to obtain the corresponding expression presented in Chapter 3. In particular, the 3rd-order derivatives of $\left[-\ln Z_c(\mathbf{a})\right]$ with respect to components of the vector of Lagrange multipliers $\mathbf{a}^{(1)}$ are obtained (see also Appendix 4.A) in the following form:

$$-\left\{\frac{\partial^3 \ln Z_c(\mathbf{a})}{\partial a_k^{(1)}\partial a_k^{(1)}\partial a_k^{(1)}}\right\}_{\mathbf{a}=\mathbf{0}} = -\left\{\frac{1}{Z_c(\mathbf{a})}\frac{\partial^3 Z_c(\mathbf{a})}{\partial a_k^{(1)}\partial a_k^{(1)}\partial a_k^{(1)}}\right\}_{\mathbf{a}=\mathbf{0}} = \mu_3(r_k)$$
$$\cong -\sum_{j=1}^{TR}\psi_j^{(1)}\frac{\partial^3\Phi_{j,1}^{(3)}(\mathbf{a})}{\partial a_k^{(1)}\partial a_k^{(1)}\partial a_k^{(1)}} = -6\psi_k^{(1)}\left[\operatorname{var}(r_k,r_k)\right]^3;\quad k=1,\ldots,TR. \tag{4.40}$$

It follows from Eq. (4.40) that the Lagrange multipliers $\psi_k^{(1)}$, $k=1,\ldots,TR$, have the following expressions:

$$\psi_k^{(1)} = -\frac{\mu_3(r_k)}{6\left[\operatorname{var}(r_k,r_k)\right]^3};\quad k=1,\ldots,TR. \tag{4.41}$$

The 3rd-order derivatives of $\left[-\ln Z_c(\mathbf{a})\right]$ with respect to components of the vector of Lagrange multipliers $\mathbf{a}^{(2)}$ are obtained by following the same procedure as was used for obtaining the expression shown in Eq. (4.41), which yields the following results:

$$-\left\{\frac{\partial^3 \ln Z_c(\mathbf{a})}{\partial a_i^{(2)}\partial a_i^{(2)}\partial a_i^{(2)}}\right\}_{\mathbf{a}=\mathbf{0}} = t_i^{\alpha} \cong -6\psi_i^{(2)}\left(c_{ii}^{\alpha}\right)^3;\quad i=1,\ldots,TP; \tag{4.42}$$

which implies, in turn, that:

$$\psi_i^{(2)} = -\frac{t_i^{\alpha}}{6\left(c_{ii}^{\alpha}\right)^3};\quad i=1,\ldots,TP. \tag{4.43}$$

As expected, the results obtained in Eqs. (4.41) and (4.43) correspond to those previously obtained in Chapter 3, thus confirming that the approximations incurred in the course of determining the Lagrange multipliers for the 3rd-order self-correlations among responses and/or parameters are consistent with each other.

The Lagrange multipliers $\psi_i^{(2)}$ and $\chi_i^{(2)}$ $(i=1,\ldots,TP)$ are obtained by using Eq. (4.38) to determine the 4th-order derivatives of $\left[-\ln Z_c(\mathbf{a})\right]$ with respect to components of the vector of Lagrange multipliers $\mathbf{a} \triangleq \left(\mathbf{a}^{(1)},\mathbf{a}^{(2)}\right)^{\dagger}$. The procedure for determining these 4th-order derivatives is the same as was used in Chapter 3. In particular, the 4th-order derivatives of $\left[-\ln Z_c(\mathbf{a})\right]$ with respect to components of the vector of Lagrange multipliers $\mathbf{a}^{(1)}$ are obtained (see also Appendix 4.A) in the following form:

$$-\left\{\frac{\partial^4 \ln Z_c(\mathbf{a})}{\partial a_k^{(1)}\partial a_k^{(1)}\partial a_k^{(1)}\partial a_k^{(1)}}\right\}_{\mathbf{a}=\mathbf{0}} = -\left\{\frac{1}{Z_c(\mathbf{a})}\frac{\partial^4 Z_c(\mathbf{a})}{\partial a_k^{(1)}\partial a_k^{(1)}\partial a_k^{(1)}\partial a_k^{(1)}}\right\}_{\mathbf{a}=\mathbf{0}}$$

$$+\left\{\frac{3}{\left[Z_c(\mathbf{a})\right]^2}\left[\frac{\partial^2 Z_c(\mathbf{a})}{\partial a_k^{(1)}\partial a_k^{(1)}}\right]^2\right\}_{\boldsymbol{\beta}=\mathbf{0}} = -\mu_4(r_k)+3\left[\mathrm{var}(r_k,r_k)\right]^2 \tag{4.44}$$

$$\cong \sum_{j=1}^{TR}\chi_j^{(1)}\frac{\partial^4 \Phi_{j,1}^{(4)}(\mathbf{a})}{\partial a_k^{(1)}\partial a_k^{(1)}\partial a_k^{(1)}\partial a_k^{(1)}} = 24\chi_k^{(1)}\left[\mathrm{var}(r_k,r_k)\right]^4; \quad k=1,\ldots,TR.$$

It follows from Eq. (4.44) that the Lagrange multipliers $\chi_k^{(1)}$, $k=1,\ldots,TR$, have the following expressions:

$$\chi_k^{(1)} = \frac{3\left[\mathrm{var}(r_k,r_k)\right]^2 - \mu_4(r_k)}{24\left[\mathrm{var}(r_k,r_k)\right]^4}; \quad k=1,\ldots,TR. \tag{4.45}$$

The 4th-order derivatives of $\left[-\ln Z_c(\mathbf{a})\right]$ with respect to components of the vector of Lagrange multipliers $\mathbf{a}^{(2)}$ are obtained by following the same procedure as was used for obtaining the expression shown in Eq. (4.45), which ultimately yields the following result:

$$\chi_i^{(2)} = \frac{3\left(c_{ii}^{\alpha}\right)^2 - q_i^{\alpha}}{24\left(c_{ii}^{\alpha}\right)^4}; \quad i=1,\ldots,TP. \tag{4.46}$$

As expected, the results obtained in Eqs. (4.45) and (4.46) correspond to those previously obtained in Chapter 3, thus confirming that the approximations incurred in the course of determining the Lagrange multipliers for the 4th-order self-correlations among responses and/or parameters are consistent with each other.

Collecting the results obtained in Eqs. (4.4), (4.5), (4.9), (4.20), (4.26), (4.41),(4.43), (4.45), and (4.46) yields the following expression for the 4th-order MaxEnt joint distribution, $p_c^{ME}(\boldsymbol{\alpha},\mathbf{r})$, of the computed model responses and parameters:

$$p_c^{ME}(\boldsymbol{\alpha},\mathbf{r}) = Z_c^{-1}\varphi(\boldsymbol{\alpha},\mathbf{r}); \quad Z_c \triangleq \int_D \varphi(\boldsymbol{\alpha},\mathbf{r})\,d\boldsymbol{\alpha}\,d\mathbf{r}; \tag{4.47}$$

where:

$$\varphi(\boldsymbol{\alpha},\mathbf{r}) = \exp\left\{-\frac{1}{2}\begin{pmatrix}\mathbf{r}-\mathbf{E}_c(\mathbf{r})\\ \boldsymbol{\alpha}-\boldsymbol{\alpha}^0\end{pmatrix}^{\dagger}\begin{pmatrix}\mathbf{C}_{rr}^c & \mathbf{C}_{r\alpha}^c\\ \mathbf{C}_{\alpha r}^c & \mathbf{C}_{\alpha\alpha}\end{pmatrix}^{-1}\begin{pmatrix}\mathbf{r}-\mathbf{E}_c(\mathbf{r})\\ \boldsymbol{\alpha}-\boldsymbol{\alpha}^0\end{pmatrix}\right.$$

$$-\sum_{i=1}^{TP}\psi_i^{(2)}\left(\alpha_i-\alpha_i^0\right)^3 - \sum_{k=1}^{TR}\psi_k^{(1)}\left[r_k(\boldsymbol{\alpha})-E_c(r_k)\right]^3 \tag{4.48}$$

$$\left.-\sum_{i=1}^{TP}\chi_i^{(2)}\left(\alpha_i-\alpha_i^0\right)^4 - \sum_{k=1}^{TR}\chi_k^{(1)}\left[r_k(\boldsymbol{\alpha})-E_c(r_k)\right]^4\right\}.$$

Notably, if the triple and quadruple correlations are negligible (or unavailable) then the MaxEnt distribution $p_c^{ME}(\mathbf{r})$, presented in Eq. (4.48), of the computed model responses and parameters reduces to a multivariate Gaussian with mean $\left[\mathbf{E}_c(\mathbf{r}), \boldsymbol{\alpha}^0\right]$ and covariance matrix $\mathbf{C}_c$.

4.3 MATHEMATICAL FRAMEWORK OF THE 4TH-BERRU-PM METHODOLOGY FOR OBTAINING BEST-ESTIMATE RESULTS WITH REDUCED UNCERTAINTIES

This section presents the mathematical and physical considerations leading to the development of the "4th-BERRU-PM" methodology. This acronym stands for the "4th-Order Best-Estimate Results with Reduced Uncertainties –Predictive Modeling." This methodology *probabilistically* incorporates, using Bayes' Theorem, the 4th-order moments-constrained MaxEnt distribution representing the computational model, which was obtained in Section 4.2, above, with the 4th-order moments-constrained MaxEnt distribution of measured responses which was obtained in Chapter 3.

The known moments of the experimentally/empirically obtained responses were defined in Chapter 3 and are provided next for convenient reference, as follows:

1. Mean/expectation values, r_i^e, for the empirical responses r_i; $i = 1,\ldots,TR$:

$$r_i^e \triangleq \int_{D_e} r_i p_e^{ME}(\mathbf{z})\, d\mathbf{r}; \ \mathbf{r}^e \triangleq \left(r_1^e,\ldots,r_i^e,\ldots,r_{TR}^e\right)^\dagger; \ i = 1,\ldots,TR. \tag{4.49}$$

2. Covariances, c_{ij}^e, for two empirical r_i and r_j; $i,j = 1,\ldots,TR$:

$$c_{ij}^e \triangleq \text{cov}\left(r_i, r_j\right)_e \triangleq \int_{D_e} \left(r_i - r_i^e\right)\left(r_j - r_j^e\right) p_e^{ME}(\mathbf{z})\, d\mathbf{r}; \ \ i,j = 1,\ldots,TR; \tag{4.50}$$

   The covariances $\text{cov}\left(r_i, r_j\right)_e$, $i,j = 1,\ldots,TR$, of the empirical responses are considered to be components of the $TR \times TR$-dimensional covariance matrix of empirical responses, which will be denoted as $\mathbf{C}_{rr}^e \triangleq \left[\text{cov}\left(r_i, r_j\right)_e\right]_{TR \times TR} \triangleq \left[c_{ij}^e\right]_{TR \times TR}$.

3. Triple correlations, t_{ijk}^e, for three empirical responses r_i, r_j and r_k; $i,j,k = 1,\ldots,TR$:

$$t_{ijk}^e \triangleq \int_{D_e} \left(r_i - r_i^e\right)\left(r_j - r_j^e\right)\left(r_k - r_k^e\right) p_e^{ME}(\mathbf{z})\, d\mathbf{r}; \ \ t_k^e \triangleq t_{kkk}^e; \\ i,j,k = 1,\ldots,TR; \tag{4.51}$$

4. Quadruple correlations, $q_{ijk\ell}^e$, for four empirical responses r_i, r_j, r_k, r_ℓ; $i,j,k,\ell = 1,\ldots,TR$:

$$q_{ijk\ell}^e \triangleq \int_{D_e} \left(r_i - r_i^e\right)\left(r_j - r_j^e\right)\left(r_k - r_k^e\right)\left(r_\ell - r_\ell^e\right) p_e^{ME}(\mathbf{z})\, d\mathbf{r}; \\ q_k^e \triangleq q_{kkkk}^e; \ i,j,k,\ell = 1,\ldots,TR. \tag{4.52}$$

Recall that the following expression was obtained in Chapter 3 for the 4th-order moments-constrained MaxEnt distribution, denoted as $p_e^{ME}(\mathbf{z})$, of experimentally/empirically obtained responses $\mathbf{r}^e \triangleq (r_1^e,...,r_{TR}^e)^\dagger$:

$$p_e^{ME}(\mathbf{z}) = \frac{\exp\left\{\begin{bmatrix} -\frac{1}{2}(\mathbf{r}-\mathbf{r}^e)^\dagger \mathbf{\Lambda}(\mathbf{r}-\mathbf{r}^e) - \sum_{i=1}^{TR}\theta_i (r_i - r_i^e)^3 \\ -\sum_{i=1}^{TR}\omega_i (r_i - r_i^e)^4 \end{bmatrix}\right\}}{\int_{D_r}\exp\left\{\begin{bmatrix} -\frac{1}{2}(\mathbf{r}-\mathbf{r}^e)^\dagger \mathbf{\Lambda}(\mathbf{r}-\mathbf{r}^e) - \sum_{i=1}^{TR}\theta_i (r_i - r_i^e)^3 \\ -\sum_{i=1}^{TR}\omega_i (r_i - r_i^e)^4 \end{bmatrix}\right\} d\mathbf{r}}. \tag{4.53}$$

The Lagrange multipliers $\mathbf{\Lambda}$, θ_i and ω_i, $i = 1,...,TR$, which appear in Eq. (4.53) have the following expressions obtained in Chapter 3:

1. The matrix of Lagrange multipliers $\mathbf{\Lambda} \triangleq (\lambda_{ij})$; $i, j = 1,...,TR$, was expressed in terms of the covariances of the measured responses as follows:

$$\left(\mathbf{\Lambda}^{-1}\right)_{mn} = c_{mn}^e;\ \mathbf{\Lambda}^{-1} = \mathbf{C}^e;\ \mathbf{C}^e \triangleq \left(c_{mn}^e\right)_{TR\times TR};\ m,n = 1,...,TR; \tag{4.54}$$

   where $\mathbf{C}^e \triangleq \left(c_{mn}^e\right)_{TR\times TR}$ denotes the known $TR \times TR$-dimensional covariance matrix of the system responses.
2. The vector $\boldsymbol{\theta} \triangleq (\theta_1,...,\theta_{TR})^\dagger$ was obtained in terms of the triple self-correlations $\mathbf{t}^e \triangleq (t_1^e,...,t_{TR}^e)^\dagger$ and the covariances (c_{mn}^e) of the measured responses having the following expression:

$$\boldsymbol{\theta} = -\mathbf{T}^{-1}\mathbf{t}^e/6;\ \mathbf{T} \triangleq (\tau_{mn})_{TR\times TR};\ \tau_{mn} \triangleq \left(c_{mn}^e\right)^3 = \left(c_{nm}^e\right)^3; \quad m,n = 1,...,TR. \tag{4.55}$$

3. The vector $\boldsymbol{\omega} \triangleq (\omega_1,...,\omega_{TR})^\dagger$ is obtained in terms of the quadruple self-correlations q_m^e, $m = 1,...,TR$, and the covariances (c_{mn}^e) of the measured responses as follows:

$$\boldsymbol{\omega} = \mathbf{P}^{-1}\mathbf{s}/24;\ \mathbf{P} \triangleq (p_{mn})_{TR\times TR};\ p_{mn} \triangleq \left(c_{mn}^e\right)^4 = \left(c_{mn}^e\right)^4; \quad m,n = 1,...,TR\ \ \mathbf{s} \triangleq (s_1,...,s_{TR})^\dagger;\ s_m \triangleq 3\left(c_{mm}^e\right)^2 - q_m^e;\ m = 1,...,TR. \tag{4.56}$$

In particular, if only the response variances are available but the (other) 2nd-order correlations among the responses are negligible or unavailable, then the result obtained in Eq. (4.55) reduces to the following simple expression for determining the Lagrange multipliers θ_k, $k = 1,...,TR$:

$$\theta_k = -\frac{t_k^e}{6\left(c_{kk}^e\right)^3}; \quad k = 1,...,TR. \tag{4.57}$$

Similarly, if the (2nd-order) correlations among the measured system responses are negligible or unavailable, then the result obtained in Eq. (4.56) reduces to the following simple expression for determining the Lagrange multipliers ω_k, $k = 1,...,TR$:

$$\omega_k = \frac{3\left(c_{kk}^e\right)^2 - q_k^e}{24\left(c_{kk}^e\right)^4}; \quad k = 1,...,TR. \tag{4.58}$$

Using Bayes' Theorem to combine Eqs. (4.53) and (4.47) yield the following expression for the "best-estimate" posterior distribution of all available computational and experimental information, which will be denoted as $p_4^{be}(\mathbf{r},\boldsymbol{\alpha})$, where the superscript "*be*" denotes "*best-estimate*" while the subscript "*4*" denotes "*4th-order*":

$$\begin{gathered} p_4^{be}(\mathbf{r},\boldsymbol{\alpha}) \triangleq p_e^{ME}(\mathbf{r})\, p_c^{ME}(\mathbf{r},\boldsymbol{\alpha}) = \frac{\exp\left[-Q(\mathbf{r},\boldsymbol{\alpha})\right]}{Z_p}, \\ Z_p \triangleq \int_D \exp\left[-Q(\mathbf{r},\boldsymbol{\alpha})\right] d\boldsymbol{\alpha}\, d\mathbf{r}, \end{gathered} \tag{4.59}$$

where:

$$\begin{aligned} Q(\mathbf{r},\boldsymbol{\alpha}) &\triangleq \frac{1}{2}\left(\mathbf{r}-\mathbf{r}^e\right)^\dagger \left(\mathbf{C}^e\right)^{-1}\left(\mathbf{r}-\mathbf{r}^e\right) + \sum_{i=1}^{TR}\theta_i\left(r_i - r_i^e\right)^3 + \sum_{i=1}^{TR}\omega_i\left(r_i - r_i^e\right)^4 \\ &+ \frac{1}{2}\begin{pmatrix} \mathbf{r}-\mathbf{E}_c(\mathbf{r}) \\ \boldsymbol{\alpha}-\boldsymbol{\alpha}^0 \end{pmatrix}^\dagger \begin{pmatrix} \mathbf{C}_{rr}^c & \mathbf{C}_{r\alpha}^c \\ \mathbf{C}_{\alpha r}^c & \mathbf{C}_{\alpha\alpha} \end{pmatrix}^{-1} \begin{pmatrix} \mathbf{r}-\mathbf{E}_c(\mathbf{r}) \\ \boldsymbol{\alpha}-\boldsymbol{\alpha}^0 \end{pmatrix} + \sum_{i=1}^{TP}\psi_i^{(2)}\left(\alpha_i - \alpha_i^0\right)^3 \\ &+ \sum_{k=1}^{TR}\psi_k^{(1)}\left[r_k(\boldsymbol{\alpha}) - E_c(r_k)\right]^3 + \sum_{i=1}^{TP}\chi_i^{(2)}\left(\alpha_i - \alpha_i^0\right)^4 + \sum_{k=1}^{TR}\chi_k^{(1)}\left[r_k(\boldsymbol{\alpha}) - E_c(r_k)\right]^4. \end{aligned} \tag{4.60}$$

The moments of the best-estimate distribution $p_4^{be}(\mathbf{r},\boldsymbol{\alpha})$ will be expressed in terms of ratios of integrals of the following form:

$$I_1 \triangleq \int_D \exp\left[-Q(\mathbf{r},\boldsymbol{\alpha})\right] d\boldsymbol{\alpha}\, d\mathbf{r};\; I_2 \triangleq \int_D A(\mathbf{r},\boldsymbol{\alpha}) \exp\left[-Q(\mathbf{r},\boldsymbol{\alpha})\right] d\boldsymbol{\alpha}\, d\mathbf{r}. \tag{4.61}$$

Integrals such as those appearing in Eq. (4.61), where $A(\mathbf{r},\boldsymbol{\alpha})$ is a specified function, can be evaluated accurately, within a user defined degree of accuracy, by using the saddle-point (Laplace) method. The integrals I_1 and I_2 have identical saddle points, which will be denoted as $(\mathbf{r}_s,\boldsymbol{\alpha}_s)$, and which are defined as the points where the gradient of the functional $Q(\mathbf{r},\boldsymbol{\alpha})$ vanishes in the phase-space $(\mathbf{r},\boldsymbol{\alpha})$, in other words:

$$\frac{\partial Q(\mathbf{r},\boldsymbol{\alpha})}{\partial \mathbf{r}} = \mathbf{0}, \quad \frac{\partial Q(\mathbf{r},\boldsymbol{\alpha})}{\partial \boldsymbol{\alpha}} = \mathbf{0}, \quad at\ (\mathbf{r},\boldsymbol{\alpha}) = (\mathbf{r}_s,\boldsymbol{\alpha}_s). \tag{4.62}$$

In order to solve the systems of equations represented by Eq. (4.62), it is convenient to commence with the component-wise representation of the quadratic form $Q(\mathbf{r},\boldsymbol{\alpha})$ defined in Eq. (4.60), which is as follows:

$$\begin{aligned} Q(\mathbf{r},\boldsymbol{\alpha}) &\triangleq \frac{1}{2}\sum_{i,j=1}^{TR}\lambda_{ij}\left(r_i-r_i^e\right)\left(r_i-r_j^e\right)+\sum_{i=1}^{TR}\theta_i\left(r_i-r_i^e\right)^3+\sum_{i=1}^{TR}\omega_i\left(r_i-r_i^e\right)^4 \\ &+\frac{1}{2}\sum_{k=1}^{TR}\sum_{\ell=1}^{TR}b_{k\ell}^{(11)}\left[r_k-E_c\left(r_k\right)\right]\left[r_\ell-E_c\left(r_\ell\right)\right]+\sum_{k=1}^{TR}\sum_{i=1}^{TP}b_{ki}^{(12)}\left(\alpha_i-\alpha_i^0\right)\left[r_k-E_c\left(r_k\right)\right] \\ &+\frac{1}{2}\sum_{i=1}^{TP}\sum_{j=1}^{TP}b_{ij}^{(22)}\left(\alpha_i-\alpha_i^0\right)\left(\alpha_j-\alpha_j^0\right)+\sum_{i=1}^{TP}\psi_i^{(2)}\left(\alpha_i-\alpha_i^0\right)^3 \\ &+\sum_{k=1}^{TR}\psi_k^{(1)}\left[r_k-E_c\left(r_k\right)\right]^3+\sum_{i=1}^{TP}\chi_i^{(2)}\left(\alpha_i-\alpha_i^0\right)^4+\sum_{k=1}^{TR}\chi_k^{(1)}\left[r_k-E_c\left(r_k\right)\right]^4. \end{aligned} \tag{4.63}$$

Using the previous component form expression of $Q(\mathbf{r},\boldsymbol{\alpha})$ in Eq. (4.62) yields the following relations which hold at the saddle-point $(\mathbf{r},\boldsymbol{\alpha})=(\mathbf{r}_s,\boldsymbol{\alpha}_s)$:

$$\begin{aligned} \frac{\partial Q(\mathbf{r},\boldsymbol{\alpha})}{\partial r_\mu} &= 0=\sum_{j=1}^{TR}\lambda_{\mu j}\left(r_j-r_j^e\right)+3\theta_\mu\left(r_\mu-r_\mu^e\right)^2+4\omega_\mu\left(r_\mu-r_\mu^e\right)^3 \\ &+\sum_{\ell=1}^{TR}b_{\mu\ell}^{(11)}\left[r_\ell-E_c\left(r_\ell\right)\right]+\sum_{i=1}^{TP}b_{\mu i}^{(12)}\left(\alpha_i-\alpha_i^0\right)+3\psi_\mu^{(1)}\left[r_\mu-E_c\left(r_\mu\right)\right]^2 \\ &+4\chi_\mu^{(1)}\left[r_\mu-E_c\left(r_\mu\right)\right]^3;\quad \mu=1,\dots,TR. \end{aligned} \tag{4.64}$$

$$\begin{aligned} \frac{\partial Q(\mathbf{r},\boldsymbol{\alpha})}{\partial \alpha_\nu} &= 0=\sum_{j=1}^{TP}b_{\nu j}^{(22)}\left(\alpha_j-\alpha_j^0\right)+\sum_{k=1}^{TR}b_{\nu k}^{(12)}\left[r_k-E_c\left(r_k\right)\right] \\ &+3\psi_\nu^{(2)}\left(\alpha_\nu-\alpha_\nu^0\right)^2+4\chi_\nu^{(2)}\left(\alpha_\nu-\alpha_\nu^0\right)^3;\quad \nu=1,\dots,TP. \end{aligned} \tag{4.65}$$

The relations expressed by Eqs. (4.64) and (4.65) can be arranged in the following matrix vector form which holds at the saddle-point $(\mathbf{r},\boldsymbol{\alpha})=(\mathbf{r}_s,\boldsymbol{\alpha}_s)$:

$$\begin{pmatrix} \mathbf{B}^{(11)} & \mathbf{B}^{(12)} \\ \left[\mathbf{B}^{(12)}\right]^\dagger & \mathbf{B}^{(22)} \end{pmatrix}\begin{pmatrix} \mathbf{r}-\mathbf{E}_c(\mathbf{r}) \\ \boldsymbol{\alpha}-\boldsymbol{\alpha}^0 \end{pmatrix}+\begin{pmatrix} \boldsymbol{\Lambda}\left(\mathbf{r}-\mathbf{r}^e\right) \\ \mathbf{0} \end{pmatrix}+\begin{pmatrix} \mathbf{v}(\mathbf{r}) \\ \mathbf{w}(\boldsymbol{\alpha}) \end{pmatrix}=\begin{pmatrix} \mathbf{0} \\ \mathbf{0} \end{pmatrix}, \tag{4.66}$$

where:

$$\mathbf{v}(\mathbf{r})\triangleq\left[v_1(\mathbf{r}),\dots,v_{TR}(\mathbf{r})\right]^\dagger;\quad \mathbf{w}(\boldsymbol{\alpha})\triangleq\left[w_1(\boldsymbol{\alpha}),\dots,w_{TP}(\boldsymbol{\alpha})\right]^\dagger, \tag{4.67}$$

with

$$\begin{aligned} v_\mu(\mathbf{r}) &\triangleq 3\psi_\mu^{(1)}\left[r_\mu-E_c\left(r_\mu\right)\right]^2+4\chi_\mu^{(1)}\left[r_\mu-E_c\left(r_\mu\right)\right]^3 \\ &+3\theta_\mu\left(r_\mu-r_\mu^e\right)^2+4\omega_\mu\left(r_\mu-r_\mu^e\right)^3;\quad \mu=1,\dots,TR; \end{aligned} \tag{4.68}$$

$$w_\nu(\boldsymbol{\alpha}) \triangleq 3\psi_\nu^{(2)}\left(\alpha_\nu - \alpha_\nu^0\right)^2 + 4\chi_\nu^{(2)}\left(\alpha_\nu - \alpha_\nu^0\right)^3; \quad \nu = 1,\dots,TP. \tag{4.69}$$

Multiplying Eq. (4.66) on the left by the matrix $\mathbf{B}^{-1} = \mathbf{C}_c$ .yields the following equation satisfied at the saddle-point $(\mathbf{r},\boldsymbol{\alpha}) = (\mathbf{r}_s,\boldsymbol{\alpha}_s)$:

$$\begin{pmatrix} \mathbf{r} - \mathbf{E}_c(\mathbf{r}) \\ \boldsymbol{\alpha} - \boldsymbol{\alpha}^0 \end{pmatrix} + \begin{pmatrix} \mathbf{C}_{rr}^c & \mathbf{C}_{r\alpha}^c \\ \mathbf{C}_{\alpha r}^c & \mathbf{C}_{\alpha\alpha} \end{pmatrix} \begin{pmatrix} \boldsymbol{\Lambda}(\mathbf{r} - \mathbf{r}^e) \\ \mathbf{0} \end{pmatrix} + \begin{pmatrix} \mathbf{C}_{rr}^c & \mathbf{C}_{r\alpha}^c \\ \mathbf{C}_{\alpha r}^c & \mathbf{C}_{\alpha\alpha} \end{pmatrix} \begin{pmatrix} \mathbf{v}(\mathbf{r}) \\ \mathbf{w}(\boldsymbol{\alpha}) \end{pmatrix} = \begin{pmatrix} \mathbf{0} \\ \mathbf{0} \end{pmatrix}. \tag{4.70}$$

Carrying out the matrix vector multiplications in Eq. (4.70) yields the following (vector-valued) equations:

$$\mathbf{r} - \mathbf{E}_c(\mathbf{r}) + \mathbf{C}_{rr}^c\boldsymbol{\Lambda}(\mathbf{r} - \mathbf{r}^e) + \mathbf{C}_{rr}^c\mathbf{v}(\mathbf{r}) + \mathbf{C}_{r\alpha}^c\mathbf{w}(\boldsymbol{\alpha}) = \mathbf{0}, \tag{4.71}$$

$$\boldsymbol{\alpha} - \boldsymbol{\alpha}^0 + \mathbf{C}_{\alpha r}^c\boldsymbol{\Lambda}(\mathbf{r} - \mathbf{r}^e) + \mathbf{C}_{\alpha r}^c\mathbf{v}(\mathbf{r}) + \mathbf{C}_{\alpha\alpha}\mathbf{w}(\boldsymbol{\alpha}) = \mathbf{0}. \tag{4.72}$$

It is convenient to rearrange Eq. (4.71) into the following form:

$$\left(\mathbf{I}_r + \mathbf{C}_{rr}^c\boldsymbol{\Lambda}\right)(\mathbf{r} - \mathbf{r}^e) + \mathbf{r}^e - \mathbf{E}_c(\mathbf{r}) + \mathbf{C}_{rr}^c\mathbf{v}(\mathbf{r}) + \mathbf{C}_{r\alpha}^c\mathbf{w}(\boldsymbol{\alpha}) = \mathbf{0}. \tag{4.73}$$

where $\mathbf{I}_r$ denotes the TR-dimensional identity matrix. Multiplying Eq. (4.73) on the left by the matrix $\left(\mathbf{I}_r + \mathbf{C}_{rr}^c\boldsymbol{\Lambda}\right)^{-1}$ and recalling that $\boldsymbol{\Lambda}^{-1} = \mathbf{C}^e$ transforms this equation into the following form at the saddle-point $(\mathbf{r},\boldsymbol{\alpha}) = (\mathbf{r}_s,\boldsymbol{\alpha}_s)$:

$$\begin{aligned} &\mathbf{r} - \mathbf{r}^e - \mathbf{C}^e\left(\mathbf{C}^e + \mathbf{C}_{rr}^c\right)^{-1}\left[\mathbf{E}_c(\mathbf{r}) - \mathbf{r}^e\right] \\ &+ \mathbf{C}^e\left(\mathbf{C}^e + \mathbf{C}_{rr}^c\right)^{-1}\left[\mathbf{C}_{rr}^c\mathbf{v}(\mathbf{r}) + \mathbf{C}_{r\alpha}^c\mathbf{w}(\boldsymbol{\alpha})\right] = \mathbf{0}. \end{aligned} \tag{4.74}$$

Using Eq. (4.74) to replace the quantity $(\mathbf{r} - \mathbf{r}^e)$ in Eq. (4.72) transforms the latter equation into the following form, which holds at the saddle-point $(\mathbf{r},\boldsymbol{\alpha}) = (\mathbf{r}_s,\boldsymbol{\alpha}_s)$:

$$\begin{aligned} &\boldsymbol{\alpha} - \boldsymbol{\alpha}^0 + \mathbf{C}_{\alpha r}^c\left(\mathbf{C}^e + \mathbf{C}_{rr}^c\right)^{-1}\left[\mathbf{E}_c(\mathbf{r}) - \mathbf{r}^e\right] + \mathbf{C}_{\alpha r}^c\left[\mathbf{I}_r - \left(\mathbf{C}^e + \mathbf{C}_{rr}^c\right)^{-1}\mathbf{C}_{rr}^c\right]\mathbf{v}(\mathbf{r}) \\ &+ \left[\mathbf{C}_{\alpha\alpha} - \mathbf{C}_{\alpha r}^c\left(\mathbf{C}^e + \mathbf{C}_{rr}^c\right)^{-1}\mathbf{C}_{r\alpha}^c\right]\mathbf{w}(\boldsymbol{\alpha}) = \mathbf{0}. \end{aligned} \tag{4.75}$$

It is important to note that if the triple and quadruple computed model response and parameter correlations are neglected in Eqs. (4.74) and (4.75), then $\mathbf{v}(\mathbf{r}) = \mathbf{0}$ and $\mathbf{w}(\boldsymbol{\alpha}) = \mathbf{0}$, so these equations can be solved exactly to obtain the following results for the components of the saddle-point $(\mathbf{r}_s,\boldsymbol{\alpha}_s)$:

$$\mathbf{r}_s^{(0)} = \mathbf{r}^e + \mathbf{C}^e\left(\mathbf{C}^e + \mathbf{C}_{rr}^c\right)^{-1}\left[\mathbf{E}_c(\mathbf{r}) - \mathbf{r}^e\right], \tag{4.76}$$

$$\boldsymbol{\alpha}_s^{(0)} = \boldsymbol{\alpha}^0 - \mathbf{C}_{\alpha r}^c \left(\mathbf{C}^e + \mathbf{C}_{rr}^c\right)^{-1} \left[\mathbf{E}_c(\mathbf{r}) - \mathbf{r}^e\right]. \tag{4.77}$$

The superscript "(0)" has been used to indicate that the quantities $\mathbf{r}_s^{(0)}$ and $\boldsymbol{\alpha}_s^{(0)}$ do *not* account for the triple and quadruple correlations among computed model responses, but do take into account the triple and quadruple correlations among model parameters which may have been used to compute the vector of mean values of the computed responses, $\mathbf{E}_c(\mathbf{r})$, and the matrices $\mathbf{C}_{rr}^c$ and $\mathbf{C}_{\alpha r}^c$. The quantities $\mathbf{r}_s^{(0)}$ and $\boldsymbol{\alpha}_s^{(0)}$ fully account for all 1st- and 2nd-order correlations among computed and measured responses and model parameters.

If the triple and quadruple correlations among computed model responses, and among model parameters are not neglected, then $\mathbf{v}(\mathbf{r}) \neq \mathbf{0}$ and $\mathbf{w}(\boldsymbol{\alpha}) \neq \mathbf{0}$, in which case the coupled system consisting of Eqs. (4.74) and (4.75), can be solved by using Newton's iteration method to determine the saddle-point $(\mathbf{r}, \boldsymbol{\alpha}) = (\mathbf{r}_s, \boldsymbol{\alpha}_s)$, as follows:

$$\begin{pmatrix} \mathbf{r}_s^{(n+1)} \\ \boldsymbol{\alpha}_s^{(n+1)} \end{pmatrix} = \begin{pmatrix} \mathbf{r}_s^{(n)} \\ \boldsymbol{\alpha}_s^{(n)} \end{pmatrix} - \left[D\xi\left(\mathbf{r}_s^{(n)}, \boldsymbol{\alpha}_s^{(n)}\right)\right]^{-1} \xi\left(\mathbf{r}_s^{(n)}, \boldsymbol{\alpha}_s^{(n)}\right), \quad n = 0, 1, ... \tag{4.78}$$

where:

$$\xi(\mathbf{r}, \boldsymbol{\alpha}) \triangleq \begin{pmatrix} \xi_1(\mathbf{r}, \boldsymbol{\alpha}) \\ \xi_2(\mathbf{r}, \boldsymbol{\alpha}) \end{pmatrix}; \quad D\xi(\mathbf{r}, \boldsymbol{\alpha}) \triangleq \begin{pmatrix} \partial\xi_1(\mathbf{r}, \boldsymbol{\alpha})/\partial\mathbf{r} & \partial\xi_1(\mathbf{r}, \boldsymbol{\alpha})/\partial\boldsymbol{\alpha} \\ \partial\xi_2(\mathbf{r}, \boldsymbol{\alpha})/\partial\mathbf{r} & \partial\xi_2(\mathbf{r}, \boldsymbol{\alpha})/\partial\boldsymbol{\alpha} \end{pmatrix}; \tag{4.79}$$

with:

$$\xi_1(\mathbf{r}, \boldsymbol{\alpha}) \triangleq \mathbf{r} - \mathbf{r}_s^{(0)} + \mathbf{C}^e \left(\mathbf{C}^e + \mathbf{C}_{rr}^c\right)^{-1} \left[\mathbf{C}_{rr}^c \mathbf{v}(\mathbf{r}) + \mathbf{C}_{r\alpha}^c \mathbf{w}(\boldsymbol{\alpha})\right]; \tag{4.80}$$

$$\begin{aligned} \xi_2(\mathbf{r}, \boldsymbol{\alpha}) &\triangleq \boldsymbol{\alpha} - \boldsymbol{\alpha}_s^{(0)} + \mathbf{C}_{\alpha r}^c \left[\mathbf{I}_r - \left(\mathbf{C}^e + \mathbf{C}_{rr}^c\right)^{-1} \mathbf{C}_{rr}^c\right] \mathbf{v}(\mathbf{r}) \\ &+ \left[\mathbf{C}_{\alpha\alpha} - \mathbf{C}_{\alpha r}^c \left(\mathbf{C}^e + \mathbf{C}_{rr}^c\right)^{-1} \mathbf{C}_{r\alpha}^c\right] \mathbf{w}(\boldsymbol{\alpha}); \end{aligned} \tag{4.81}$$

$$\frac{\partial\xi_1(\mathbf{r}, \boldsymbol{\alpha})}{\partial\mathbf{r}} = \mathbf{I}_r + \mathbf{C}^e \left(\mathbf{C}^e + \mathbf{C}_{rr}^c\right)^{-1} \mathbf{C}_{rr}^c \frac{\partial\mathbf{v}(\mathbf{r})}{\partial\mathbf{r}}; \tag{4.82}$$

$$\frac{\partial\xi_1(\mathbf{r}, \boldsymbol{\alpha})}{\partial\boldsymbol{\alpha}} = \mathbf{C}^e \left(\mathbf{C}^e + \mathbf{C}_{rr}^c\right)^{-1} \mathbf{C}_{r\alpha}^c \frac{\partial\mathbf{w}(\boldsymbol{\alpha})}{\partial\boldsymbol{\alpha}}; \tag{4.83}$$

$$\frac{\partial\xi_2(\mathbf{r}, \boldsymbol{\alpha})}{\partial\mathbf{r}} = \mathbf{C}_{\alpha r}^c \left[\mathbf{I}_r - \left(\mathbf{C}^e + \mathbf{C}_{rr}^c\right)^{-1} \mathbf{C}_{rr}^c\right] \frac{\partial\mathbf{v}(\mathbf{r})}{\partial\mathbf{r}} \tag{4.84}$$

$$\frac{\partial\xi_2(\mathbf{r}, \boldsymbol{\alpha})}{\partial\boldsymbol{\alpha}} = \mathbf{I}_\alpha + \left[\mathbf{C}_{\alpha\alpha} - \mathbf{C}_{\alpha r}^c \left(\mathbf{C}^e + \mathbf{C}_{rr}^c\right)^{-1} \mathbf{C}_{r\alpha}^c\right] \frac{\partial\mathbf{w}(\boldsymbol{\alpha})}{\partial\boldsymbol{\alpha}}. \tag{4.85}$$

The matrix $\mathbf{I}_\alpha$ in Eq. (4.85), denotes the TP-dimensional identity matrix.

Using the definitions provided in Eqs. (4.67) and (4.68) enable the computation of the $TR \times TR$-dimensional diagonal matrix $\mathbf{V}(\mathbf{r}) \triangleq \partial \mathbf{v}(\mathbf{r})/\partial \mathbf{r}$, having zero-valued non-diagonal elements and non-zero diagonal elements as defined here:

$$\mathbf{V}(\mathbf{r}) \triangleq \frac{\partial \mathbf{v}(\mathbf{r})}{\partial \mathbf{r}} = Diag\left\{6\psi_\mu^{(1)}\left[r_\mu - E_c(r_\mu)\right] + 12\chi_\mu^{(1)}\left[r_\mu - E_c(r_\mu)\right]^2 + 6\theta_\mu\left(r_\mu - r_\mu^e\right) + 12\omega_\mu\left(r_\mu - r_\mu^e\right)^2\right\}; \quad \mu = 1,\dots,TR. \tag{4.86}$$

Notably, the components of the matrix $\mathbf{V}(\mathbf{r})$ contain 1st-order terms in the triple and quadruple self-correlations of the computed model responses.

Similarly, using the definitions provided in Equations (4.67) and (4.69) enables the computation of the $TP \times TP$-dimensional diagonal matrix $\mathbf{W}(\boldsymbol{\alpha}) \triangleq \partial \mathbf{w}(\boldsymbol{\alpha})/\partial \boldsymbol{\alpha}$, having zero-valued non-diagonal elements and non-zero diagonal elements as defined here:

$$\mathbf{W}(\boldsymbol{\alpha}) \triangleq \frac{\partial \mathbf{w}(\boldsymbol{\alpha})}{\partial \boldsymbol{\alpha}} = Diag\left[6\psi_\nu^{(2)}\left(\alpha_\nu - \alpha_\nu^0\right) + 12\chi_\nu^{(2)}\left(\alpha_\nu - \alpha_\nu^0\right)^2\right]; \quad \nu = 1,\dots,TP. \tag{4.87}$$

Notably, the components of the matrix $\mathbf{W}(\boldsymbol{\alpha})$ contain 1st-order terms in the triple and quadruple self-correlations of the model parameters.

It follows from the expressions obtained in Eqs. (4.82)–(4.87) that:

$$\left[D\xi(\mathbf{r},\boldsymbol{\alpha})\right]^{-1} = \left[\begin{pmatrix} \mathbf{I}_r & \mathbf{0} \\ \mathbf{0} & \mathbf{I}_\alpha \end{pmatrix} + \mathbf{M}(\mathbf{r},\boldsymbol{\alpha})\right]^{-1} \cong \begin{pmatrix} \mathbf{I}_r & \mathbf{0} \\ \mathbf{0} & \mathbf{I}_\alpha \end{pmatrix} - \mathbf{M}(\mathbf{r},\boldsymbol{\alpha}) + H.O.T., \tag{4.88}$$

where the acronym "*HOT*" denotes "higher-order terms" that comprise 2nd-order (and higher-) powers of the triple and quadruple computed model responses and parameters, and where the matrix $\mathbf{M}(\mathbf{r},\boldsymbol{\alpha})$ is defined here:

$$\begin{aligned}
&\mathbf{M}(\mathbf{r},\boldsymbol{\alpha}) \triangleq \begin{pmatrix} \mathbf{M}_{11}(\mathbf{r},\boldsymbol{\alpha}) & \mathbf{M}_{12}(\mathbf{r},\boldsymbol{\alpha}) \\ \mathbf{M}_{21}(\mathbf{r},\boldsymbol{\alpha}) & \mathbf{M}_{22}(\mathbf{r},\boldsymbol{\alpha}) \end{pmatrix}; \\
&\mathbf{M}_{11}(\mathbf{r},\boldsymbol{\alpha}) \triangleq \mathbf{C}^e\left(\mathbf{C}^e + \mathbf{C}_{rr}^c\right)^{-1}\mathbf{C}_{rr}^c\mathbf{V}(\mathbf{r}); \\
&\mathbf{M}_{12}(\mathbf{r},\boldsymbol{\alpha}) \triangleq \mathbf{C}^e\left(\mathbf{C}^e + \mathbf{C}_{rr}^c\right)^{-1}\mathbf{C}_{r\alpha}^c\mathbf{W}(\boldsymbol{\alpha}); \\
&\mathbf{M}_{21}(\mathbf{r},\boldsymbol{\alpha}) \triangleq \mathbf{C}_{\alpha r}^c\left[\mathbf{I}_r - \left(\mathbf{C}^e + \mathbf{C}_{rr}^c\right)^{-1}\mathbf{C}_{rr}^c\right]\mathbf{V}(\mathbf{r}); \\
&\mathbf{M}_{22}(\mathbf{r},\boldsymbol{\alpha}) \triangleq \left[\mathbf{C}_{\alpha\alpha} - \mathbf{C}_{\alpha r}^c\left(\mathbf{C}^e + \mathbf{C}_{rr}^c\right)^{-1}\mathbf{C}_{r\alpha}^c\right]\mathbf{W}(\boldsymbol{\alpha}).
\end{aligned} \tag{4.89}$$

Inserting the expression obtained in Eq. (4.88) into Eq. (4.78), and using the expressions provided in Eqs. (4.80), (4.81), and (4.89), yield the following form for the Newton iteration for determining the saddle-point $(\mathbf{r},\boldsymbol{\alpha}) = (\mathbf{r}_s,\boldsymbol{\alpha}_s)$:

$$\begin{pmatrix} \mathbf{r}_s^{(n+1)} \\ \boldsymbol{\alpha}_s^{(n+1)} \end{pmatrix} \cong \begin{pmatrix} \mathbf{r}_s^{(n)} \\ \boldsymbol{\alpha}_s^{(n)} \end{pmatrix} - \begin{pmatrix} \xi_1\left(\mathbf{r}_s^{(n)},\boldsymbol{\alpha}_s^{(n)}\right) \\ \xi_2\left(\mathbf{r}_s^{(n)},\boldsymbol{\alpha}_s^{(n)}\right) \end{pmatrix} + \mathbf{M}\left(\mathbf{r}_s^{(n)},\boldsymbol{\alpha}_s^{(n)}\right) \begin{pmatrix} \xi_1\left(\mathbf{r}_s^{(n)},\boldsymbol{\alpha}_s^{(n)}\right) \\ \xi_2\left(\mathbf{r}_s^{(n)},\boldsymbol{\alpha}_s^{(n)}\right) \end{pmatrix}, \quad n=0,1,.... \tag{4.90}$$

Inserting the expressions obtained in Eqs. (4.80), (4.81), and (4.89) into Eq. (4.90) yields the following expression for the Newton iteration aimed at determining the saddle-point $(\mathbf{r}_s,\boldsymbol{\alpha}_s)$, for $n=0,1,....$

$$\begin{aligned} \begin{pmatrix} \mathbf{r}_s^{(n+1)} \\ \boldsymbol{\alpha}_s^{(n+1)} \end{pmatrix} &\cong \begin{pmatrix} \mathbf{r}_s^{(0)} + \mathbf{p}_1\left(\mathbf{r}_s^{(n)},\boldsymbol{\alpha}_s^{(n)}\right) \\ \boldsymbol{\alpha}_s^{(0)} + \mathbf{p}_2\left(\mathbf{r}_s^{(n)},\boldsymbol{\alpha}_s^{(n)}\right) \end{pmatrix} \\ &+ \begin{pmatrix} \mathbf{M}_{11}\left(\mathbf{r}_s^{(n)},\boldsymbol{\alpha}_s^{(n)}\right) & \mathbf{M}_{12}\left(\mathbf{r}_s^{(n)},\boldsymbol{\alpha}_s^{(n)}\right) \\ \mathbf{M}_{21}\left(\mathbf{r}_s^{(n)},\boldsymbol{\alpha}_s^{(n)}\right) & \mathbf{M}_{22}\left(\mathbf{r}_s^{(n)},\boldsymbol{\alpha}_s^{(n)}\right) \end{pmatrix} \begin{pmatrix} \mathbf{r}_s^{(n)} - \mathbf{r}_s^{(0)} \\ \boldsymbol{\alpha}_s^{(n)} - \boldsymbol{\alpha}_s^{(0)} \end{pmatrix}, \end{aligned} \tag{4.91}$$

where:

$$\mathbf{p}_1\left(\mathbf{r}_s^{(n)},\boldsymbol{\alpha}_s^{(n)}\right) \triangleq \mathbf{C}^e\left(\mathbf{C}^e + \mathbf{C}_{rr}^c\right)^{-1}\left[\mathbf{C}_{rr}^c\mathbf{v}\left(\mathbf{r}_s^{(n)}\right) + \mathbf{C}_{r\alpha}^c\mathbf{w}\left(\boldsymbol{\alpha}_s^{(n)}\right)\right]; \tag{4.92}$$

$$\begin{aligned} \mathbf{p}_2\left(\mathbf{r}_s^{(n)},\boldsymbol{\alpha}_s^{(n)}\right) &\triangleq \mathbf{C}_{\alpha r}^c\left[\mathbf{I}_r - \left(\mathbf{C}^e + \mathbf{C}_{rr}^c\right)^{-1}\mathbf{C}_{rr}^c\right]\mathbf{v}\left(\mathbf{r}_s^{(n)}\right) \\ &+ \left[\mathbf{C}_{\alpha\alpha} - \mathbf{C}_{\alpha r}^c\left(\mathbf{C}^e + \mathbf{C}_{rr}^c\right)^{-1}\mathbf{C}_{r\alpha}^c\right]\mathbf{w}\left(\boldsymbol{\alpha}_s^{(n)}\right). \end{aligned} \tag{4.93}$$

The starting point, $\left(\mathbf{r}_s^{(0)},\boldsymbol{\alpha}_s^{(0)}\right)$, for the Newton iteration is provided by Eqs. (4.76) and (4.77), which are exact if the triple and quadruple response correlations are neglected. Setting $n=0$ in Eq. (4.91) provides the following form for the first Newton iteration:

$$\begin{pmatrix} \mathbf{r}_s^{(1)} \\ \boldsymbol{\alpha}_s^{(1)} \end{pmatrix} = \begin{pmatrix} \mathbf{r}_s^{(0)} + \mathbf{p}_1\left(\mathbf{r}_s^{(0)},\boldsymbol{\alpha}_s^{(0)}\right) \\ \boldsymbol{\alpha}_s^{(0)} + \mathbf{p}_2\left(\mathbf{r}_s^{(0)},\boldsymbol{\alpha}_s^{(0)}\right) \end{pmatrix}. \tag{4.94}$$

Notably, the contributions stemming from the matrix $\mathbf{M}\left(\mathbf{r}_s^{(n)},\boldsymbol{\alpha}_s^{(n)}\right)$ are nullified for the first-iterate $\left(\mathbf{r}_s^{(1)},\boldsymbol{\alpha}_s^{(1)}\right)$ of the saddle-point $(\mathbf{r}_s,\boldsymbol{\alpha}_s)$ but the 3rd- and 4th-order response correlations do contribute through the terms $\mathbf{p}_1\left(\mathbf{r}_s^{(0)},\boldsymbol{\alpha}_s^{(0)}\right)$ and $\mathbf{p}_2\left(\mathbf{r}_s^{(0)},\boldsymbol{\alpha}_s^{(0)}\right)$, respectively, which have the following specific expressions:

$$\mathbf{p}_1\left(\mathbf{r}_s^{(0)},\boldsymbol{\alpha}_s^{(0)}\right) = \mathbf{C}^e\left(\mathbf{C}^e + \mathbf{C}_{rr}^c\right)^{-1}\left[\mathbf{C}_{rr}^c\mathbf{v}\left(\mathbf{r}_s^{(0)}\right) + \mathbf{C}_{r\alpha}^c\mathbf{w}\left(\boldsymbol{\alpha}_s^{(0)}\right)\right]; \tag{4.95}$$

$$\begin{aligned} \mathbf{p}_2\left(\mathbf{r}_s^{(0)},\boldsymbol{\alpha}_s^{(0)}\right) &= \mathbf{C}_{\alpha r}^c\left[\mathbf{I}_r - \left(\mathbf{C}^e + \mathbf{C}_{rr}^c\right)^{-1}\mathbf{C}_{rr}^c\right]\mathbf{v}\left(\mathbf{r}_s^{(0)}\right) \\ &+ \left[\mathbf{C}_{\alpha\alpha} - \mathbf{C}_{\alpha r}^c\left(\mathbf{C}^e + \mathbf{C}_{rr}^c\right)^{-1}\mathbf{C}_{r\alpha}^c\right]\mathbf{w}\left(\pm_s^{(0)}\right). \end{aligned} \tag{4.96}$$

$$\mathbf{v}\left(\mathbf{r}_s^{(0)}\right) \triangleq \left[v_1\left(\mathbf{r}_s^{(0)}\right),\ldots,v_{TR}\left(\mathbf{r}_s^{(0)}\right)\right]^\dagger; \quad \mathbf{w}\left(\boldsymbol{\alpha}_s^{(0)}\right) \triangleq \left[w_1\left(\boldsymbol{\alpha}_s^{(0)}\right),\ldots,w_{TP}\left(\boldsymbol{\alpha}_s^{(0)}\right)\right]^\dagger; \tag{4.97}$$

$$\begin{aligned} v_k\left(\mathbf{r}_s^{(0)}\right) \triangleq & -\frac{\mu_3(r_k)}{2\left[\operatorname{var}(r_k,r_k)\right]^3}\left[r_k^{(0)} - E_c(r_k)\right]^2 - \frac{t_k^e}{2\left(c_{kk}^e\right)^3}\left(r_k^{(0)} - r_k^e\right)^2 \\ & + \frac{3\left[\operatorname{var}(r_k,r_k)\right]^2 - \mu_4(r_k)}{6\left[\operatorname{var}(r_k,r_k)\right]^4}\left[r_k^{(0)} - E_c(r_k)\right]^3 + \frac{3\left(c_{kk}^e\right)^2 - q_k^e}{6\left(c_{kk}^e\right)^4}\left(r_k^{(0)} - r_k^e\right)^3; \end{aligned} \tag{4.98}$$

$$\begin{aligned} w_i\left(\boldsymbol{\alpha}_s^{(0)}\right) \triangleq & -\frac{t_i^\alpha}{2\left(c_{ii}^\alpha\right)^3}\left(\alpha_i^{(0)} - \alpha_i^0\right)^2 \\ & + \frac{3\left(c_{ii}^\alpha\right)^2 - q_i^\alpha}{6\left(c_{ii}^\alpha\right)^4}\left(\alpha_i^{(0)} - \alpha_i^0\right)^3; \quad i = 1,\ldots,TP. \end{aligned} \tag{4.99}$$

The moments of the best-estimate 4th-order distribution $p_4^{be}(\mathbf{r},\boldsymbol{\alpha})$ can now be determined by using the saddle-point expression obtained in Eq. (4.94), if the 1st-order contributions from the triple and quadruple correlations of the measured and computed responses suffice. On the other hand, if higher-order contributions from these triple and quadruple response correlations are deemed to be important, they can be taken into account in subsequent Newton-iterates of Eq. (4.91).

The moments of the best-estimate 4th-order distribution $p_4^{be}(\mathbf{r},\boldsymbol{\alpha})$ can now be obtained in closed-form by using the generic representation of the integrals defined in Eq. (4.61) in conjunction with Eq. (4.91). The expressions of these moments are provided in the remainder of this chapter.

4.3.1 Best-Estimate 4th-Order Expression of the Vector of Mean Values of the Predicted Responses

The vector of best-estimate mean values of responses is denoted as $\mathbf{r}^{be} \triangleq \left(r_1^{be},\ldots,r_{TR}^{be}\right)^\dagger$ and are obtained as follows:

$$\mathbf{r}^{be} \triangleq \int_D \mathbf{r}\, p_4^{be}(\mathbf{r},\boldsymbol{\alpha})\, d\boldsymbol{\alpha}\, d\mathbf{r} \cong \mathbf{r}_s^{(n+1)}. \tag{4.100}$$

The lowest approximation of the 4th-order expression for the best-estimate predicted mean values of the responses which contains the contributions from the triple and quadruple response and parameter correlations, will be denoted as $\mathbf{r}_1^{be}$ and is provided by Eq. (4.94), namely:

$$\mathbf{r}_1^{be} = \mathbf{r}_s^{(0)} + \mathbf{p}_1\left(\mathbf{r}_s^{(0)},\boldsymbol{\alpha}_s^{(0)}\right) = \mathbf{r}^e + \mathbf{C}^e\left(\mathbf{C}^e + \mathbf{C}_{rr}^c\right)^{-1}\left[\mathbf{E}_c(\mathbf{r}) - \mathbf{r}^e\right] + \mathbf{p}_1\left(\mathbf{r}_s^{(0)},\boldsymbol{\alpha}_s^{(0)}\right). \tag{4.101}$$

4.3.2 Best-Estimate 4th-Order Expression of the Vector of Mean Values of the Predicted Calibrated Model Parameters

The vector of best-estimate calibrated model parameters is denoted as $\boldsymbol{\alpha}^{be} \triangleq \left(\alpha_1^{be},...,\alpha_{TP}^{be}\right)^{\dagger}$ and are obtained as follows:

$$\boldsymbol{\alpha}^{be} \triangleq \int_D \boldsymbol{\alpha} p_4^{be}(\mathbf{r},\boldsymbol{\alpha})\, d\boldsymbol{\alpha}\, d\mathbf{r} \cong \boldsymbol{\alpha}_s^{(n+1)}. \tag{4.102}$$

The lowest approximation of the 4th-order predicted calibrated model parameters, which contains the contributions from the triple and quadruple response and parameter correlations, will be denoted as $\boldsymbol{\alpha}_1^{be}$ and is provided by Eq. (4.94), namely:

$$\begin{aligned} \boldsymbol{\alpha}_1^{be} &= \boldsymbol{\alpha}_s^{(0)} + \mathbf{p}_2\left(\mathbf{r}_s^{(0)},\boldsymbol{\alpha}_s^{(0)}\right) = \boldsymbol{\alpha}^0 \\ &-\mathbf{C}_{\alpha r}^{c}\left(\mathbf{C}^{e}+\mathbf{C}_{rr}^{c}\right)^{-1}\left[\mathbf{E}_c(\mathbf{r})-\mathbf{r}^{e}\right]+\mathbf{p}_2\left(\mathbf{r}_s^{(0)},\boldsymbol{\alpha}_s^{(0)}\right). \end{aligned} \tag{4.103}$$

4.3.3 Best-Estimate 4th-Order Expression of the Covariance Matrix of the Predicted Responses

The covariance matrix of the predicted responses is denoted as $\mathbf{C}_{rr}^{be}$ and is defined as follows:

$$\mathbf{C}_{rr}^{be} \triangleq \int_D \left(\mathbf{r}-\mathbf{r}^{be}\right)\left(\mathbf{r}-\mathbf{r}^{be}\right)^{\dagger} p_4^{be}(\mathbf{r},\boldsymbol{\alpha})\, d\boldsymbol{\alpha}\, d\mathbf{r}. \tag{4.104}$$

The lowest approximation for $\mathbf{C}_{rr}^{be}$ is denoted as $\left(\mathbf{C}_{rr}^{be}\right)_1$ and is determined as follows:

$$\begin{aligned} \left(\mathbf{C}_{rr}^{be}\right)_1 &\triangleq \int_D \left(\mathbf{r}-\mathbf{r}_1^{be}\right)\left(\mathbf{r}-\mathbf{r}_1^{be}\right)^{\dagger} p_4^{be}(\mathbf{r},\boldsymbol{\alpha})\, d\boldsymbol{\alpha}\, d\mathbf{r} \\ &= \int_D \left[\mathbf{r}-\mathbf{r}_s^{(0)}-\mathbf{p}_1\left(\mathbf{r}_s^{(0)},\boldsymbol{\alpha}_s^{(0)}\right)\right]\left[\mathbf{r}-\mathbf{r}_s^{(0)}-\mathbf{p}_1\left(\mathbf{r}_s^{(0)},\boldsymbol{\alpha}_s^{(0)}\right)\right]^{\dagger} p_4^{be}(\mathbf{r},\boldsymbol{\alpha})\, d\boldsymbol{\alpha}\, d\mathbf{r} \\ &= \left(\mathbf{C}_{rr}^{be}\right)_0 + O\left[\mu_3^2(r_k)\right] + O\left[\left(t_k^e\right)^2\right] + O\left[\mu_4^2(r_k)\right] + O\left[\left(q_k^e\right)^2\right], \end{aligned} \tag{4.105}$$

where:

$$\begin{aligned} \left(\mathbf{C}_{rr}^{be}\right)_0 &\triangleq \int_D \left(\mathbf{r}-\mathbf{r}_s^{(0)}\right)\left(\mathbf{r}-\mathbf{r}_s^{(0)}\right)^{\dagger} p_4^{be}(\mathbf{r},\boldsymbol{\alpha})\, d\boldsymbol{\alpha}\, d\mathbf{r} \\ &= \mathbf{C}_{rr}^{e} - \mathbf{C}_{rr}^{e}\left(\mathbf{C}_{rr}^{e}+\mathbf{C}_{rr}^{c}\right)^{-1}\mathbf{C}_{rr}^{e}. \end{aligned} \tag{4.106}$$

The 2nd-order terms involving the "squared triple correlations" and "squared quadruple correlations" shown in Eq. (4.105) arise from terms involving the quantity

$\mathbf{p}_1\left(\mathbf{r}_s^{(0)},\boldsymbol{\alpha}_s^{(0)}\right)$ and are expected to be negligible by comparison to the leading term $\left(\mathbf{C}_{rr}^{be}\right)_0$. Not explicitly shown in Eq. (4.105) are terms involving products of triple correlations among parameters and responses, which also stem from the quantity $\mathbf{p}_1\left(\mathbf{r}_s^{(0)},\boldsymbol{\alpha}_s^{(0)}\right)$ and which are also negligible by comparison to the leading term $\left(\mathbf{C}_{rr}^{be}\right)_0$.

The expression provided in Eq. (4.106) has been obtained by using Eq. (4.76) and the result:

$$\begin{aligned}&\int_D \left[\mathbf{E}_c(\mathbf{r})-\mathbf{r}^e\right]\left[\mathbf{E}_c(\mathbf{r})-\mathbf{r}^e\right]^\dagger p_4^{be}(\mathbf{r},\boldsymbol{\alpha})\,d\boldsymbol{\alpha}\,d\mathbf{r}\\&=\int_D \left[\mathbf{r}-\mathbf{r}^e-\mathbf{r}+\mathbf{E}_c(\mathbf{r})\right]\left[\mathbf{r}-\mathbf{r}^e-\mathbf{r}+\mathbf{E}_c(\mathbf{r})\right]^\dagger p_4^{be}(\mathbf{r},\boldsymbol{\alpha})\,d\boldsymbol{\alpha}\,d\mathbf{r}=\mathbf{C}_{rr}^e+\mathbf{C}_{rr}^c.\end{aligned} \tag{4.107}$$

4.3.4 Best-Estimate 4th-Order Expression of the Covariance Matrix of the Calibrated Model Parameters

The covariance matrix of the predicted calibrated parameters is denoted as $\mathbf{C}_{\alpha\alpha}^{be}$ and is defined as follows:

$$\mathbf{C}_{\alpha\alpha}^{be} \triangleq \int_D \left(\boldsymbol{\alpha}-\boldsymbol{\alpha}^{be}\right)\left(\boldsymbol{\alpha}-\boldsymbol{\alpha}^{be}\right)^\dagger p_4^{be}(\mathbf{r},\boldsymbol{\alpha})\,d\boldsymbol{\alpha}\,d\mathbf{r}. \tag{4.108}$$

The lowest order expression for $\mathbf{C}_{\alpha\alpha}^{be}$ will be denoted as $\left(\mathbf{C}_{\alpha\alpha}^{be}\right)_1$ and is determined as follows:

$$\begin{aligned}\left(\mathbf{C}_{\alpha\alpha}^{be}\right)_1 &\triangleq \int_D \left(\boldsymbol{\alpha}-\boldsymbol{\alpha}_1^{be}\right)\left(\boldsymbol{\alpha}-\boldsymbol{\alpha}_1^{be}\right)^\dagger p_4^{be}(\mathbf{r},\boldsymbol{\alpha})\,d\boldsymbol{\alpha}\,d\mathbf{r}\\&=\int_D \left[\boldsymbol{\alpha}-\boldsymbol{\alpha}_s^{(0)}-\mathbf{p}_2\left(\mathbf{r}_s^{(0)},\boldsymbol{\alpha}_s^{(0)}\right)\right]\left[\boldsymbol{\alpha}-\boldsymbol{\alpha}_s^{(0)}-\mathbf{p}_2\left(\mathbf{r}_s^{(0)},\boldsymbol{\alpha}_s^{(0)}\right)\right]^\dagger p_4^{be}(\mathbf{r},\boldsymbol{\alpha})\,d\boldsymbol{\alpha}\,d\mathbf{r}\\&=\left(\mathbf{C}_{\alpha\alpha}^{be}\right)_0+O\left[\left(t_i^{\alpha}\right)^2\right]+O\left[\left(q_i^{\alpha}\right)^2\right],\end{aligned} \tag{4.109}$$

where:

$$\begin{aligned}\left(\mathbf{C}_{\alpha\alpha}^{be}\right)_0 &\triangleq \int_D \left(\boldsymbol{\alpha}-\boldsymbol{\alpha}_s^{(0)}\right)\left(\boldsymbol{\alpha}-\boldsymbol{\alpha}_s^{(0)}\right)^\dagger p_4^{be}(\mathbf{r},\boldsymbol{\alpha})\,d\boldsymbol{\alpha}\,d\mathbf{r}\\&=\mathbf{C}_{\alpha\alpha}-\mathbf{C}_{\alpha r}^c\left(\mathbf{C}_{rr}^e+\mathbf{C}_{rr}^c\right)^{-1}\mathbf{C}_{r\alpha}^c.\end{aligned} \tag{4.110}$$

The 2nd-order terms involving the "squared triple correlations" and "squared quadruple correlations" shown in Eq. (4.109) arise from terms involving the quantity $\mathbf{p}_2\left(\mathbf{r}_s^{(0)},\boldsymbol{\alpha}_s^{(0)}\right)$ and are expected to be negligible by comparison to the leading term $\left(\mathbf{C}_{\alpha\alpha}^{be}\right)_0$. Not explicitly shown in Eq. (4.109) are terms involving products of triple correlations among parameters and responses, which also stem from the quantity

$\mathbf{p}_2\left(\mathbf{r}_s^{(0)},\boldsymbol{\alpha}_s^{(0)}\right)$ and which are also negligible by comparison to the leading term $\left(\mathbf{C}_{\alpha\alpha}^{be}\right)_0$. The expression provided in Eq. (4.110) has been obtained by using Eqs. (4.77) and (4.107).

4.3.5 Best-Estimate 4th-Order Expression of the Correlation Matrix of the Predicted Responses and Calibrated Model Parameters

The correlation matrix of the predicted best-estimate responses and calibrated model parameters is denoted as $\mathbf{C}_{\alpha r}^{be}$ and is defined as follows:

$$\mathbf{C}_{\alpha r}^{be} \triangleq \int_D \left(\boldsymbol{\alpha}-\boldsymbol{\alpha}^{be}\right)\left(\mathbf{r}-\mathbf{r}^{be}\right)^{\dagger} p_4^{be}\left(\mathbf{r},\boldsymbol{\alpha}\right)\, d\boldsymbol{\alpha}\, d\mathbf{r}. \tag{4.111}$$

The lowest approximation for $\mathbf{C}_{\alpha r}^{be}$ will be denoted as $\left(\mathbf{C}_{\alpha r}^{be}\right)_1$ and is determined as follows:

$$\begin{aligned}\left(\mathbf{C}_{\alpha r}^{be}\right)_1 &\triangleq \int_D \left(\boldsymbol{\alpha}-\boldsymbol{\alpha}_1^{be}\right)\left(\mathbf{r}-\mathbf{r}_1^{be}\right)^{\dagger} p_4^{be}\left(\mathbf{r},\boldsymbol{\alpha}\right) d\boldsymbol{\alpha}\, d\mathbf{r}\\ &= \int_D \left[\boldsymbol{\alpha}-\boldsymbol{\alpha}_s^{(0)}-\mathbf{p}_2\left(\mathbf{r}_s^{(0)},\boldsymbol{\alpha}_s^{(0)}\right)\right]\left[\mathbf{r}-\mathbf{r}_s^{(0)}-\mathbf{p}_1\left(\mathbf{r}_s^{(0)},\boldsymbol{\alpha}_s^{(0)}\right)\right]^{\dagger} p_4^{be}\left(\mathbf{r},\boldsymbol{\alpha}\right) d\boldsymbol{\alpha}\, d\mathbf{r}\\ &= \left(\mathbf{C}_{\alpha r}^{be}\right)_0 + HOT,\end{aligned} \tag{4.112}$$

where:

$$\begin{aligned}\left(\mathbf{C}_{\alpha r}^{be}\right)_0 &\triangleq \int_D \left(\boldsymbol{\alpha}-\boldsymbol{\alpha}_s^{(0)}\right)\left(\mathbf{r}-\mathbf{r}_s^{(0)}\right)^{\dagger} p_4^{be}\left(\mathbf{r},\boldsymbol{\alpha}\right) d\boldsymbol{\alpha}\, d\mathbf{r}\\ &= -\mathbf{C}_{\alpha r}^{c}\left(\mathbf{C}_{rr}^{e}+\mathbf{C}_{rr}^{c}\right)^{-1}\mathbf{C}_{rr}^{e}.\end{aligned} \tag{4.113}$$

In Eq. (4.112), "*HOT*" denotes "higher-order terms" involving products of triple and quadruple correlations which arise from terms involving the quantities $\mathbf{p}_1\left(\mathbf{r}_s^{(0)},\boldsymbol{\alpha}_s^{(0)}\right)$ and $\mathbf{p}_2\left(\mathbf{r}_s^{(0)},\boldsymbol{\alpha}_s^{(0)}\right)$, and which are expected to be negligible by comparison to the leading term $\left(\mathbf{C}_{rr}^{be}\right)_0$. The expression provided in Eq. (4.110) has been obtained by using Eqs. (4.77) and (4.107), and by recalling that the parameters are uncorrelated to the measured responses, which implies that $\int_D \left(\mathbf{r}-\mathbf{r}^{e}\right)\left(\boldsymbol{\alpha}-\boldsymbol{\alpha}^{0}\right)^{\dagger} p_4^{be}\left(\mathbf{r},\boldsymbol{\alpha}\right)\, d\boldsymbol{\alpha}\, d\mathbf{r} = 0$.

The transposed matrix $\mathbf{C}_{r\alpha}^{be} = \left(\mathbf{C}_{\alpha r}^{be}\right)^{\dagger}$ is defined:

$$\mathbf{C}_{r\alpha}^{be} \triangleq \int_D \left(\mathbf{r}-\mathbf{r}^{be}\right)\left(\boldsymbol{\alpha}-\boldsymbol{\alpha}^{be}\right)^{\dagger} p_4^{be}\left(\mathbf{r},\boldsymbol{\alpha}\right)\, d\boldsymbol{\alpha}\, d\mathbf{r}. \tag{4.114}$$

Following the same steps as those leading to the result obtained in Eq. (4.112) yields the following expression for $\left(\mathbf{C}_{r\alpha}^{be}\right)_1$:

$$
\begin{aligned}
\left(\mathbf{C}_{r\alpha}^{be}\right)_1 &\triangleq \int_D \left(\mathbf{r}-\mathbf{r}_1^{be}\right)\left(\boldsymbol{\alpha}-\boldsymbol{\alpha}_1^{be}\right)^\dagger p_4^{be}(\mathbf{r},\boldsymbol{\alpha})\, d\boldsymbol{\alpha}\, d\mathbf{r} \\
&= \int_D \left[\mathbf{r}-\mathbf{r}_s^{(0)}-\mathbf{p}_1\left(\mathbf{r}_s^{(0)},\boldsymbol{\alpha}_s^{(0)}\right)\right]\left[\boldsymbol{\alpha}-\boldsymbol{\alpha}_s^{(0)}-\mathbf{p}_2\left(\mathbf{r}_s^{(0)},\boldsymbol{\alpha}_s^{(0)}\right)\right]^\dagger p_4^{be}(\mathbf{r},\boldsymbol{\alpha})\, d\boldsymbol{\alpha}\, d\mathbf{r} \qquad (4.115)\\
&= \left(\mathbf{C}_{r\alpha}^{be}\right)_0 + HOT,
\end{aligned}
$$

where:

$$
\left(\mathbf{C}_{r\alpha}^{be}\right)_0 \triangleq \int_D \left(\mathbf{r}-\mathbf{r}_s^{(0)}\right)\left(\boldsymbol{\alpha}-\boldsymbol{\alpha}_s^{(0)}\right)^\dagger p_4^{be}(\mathbf{r},\boldsymbol{\alpha})\, d\boldsymbol{\alpha}\, d\mathbf{r} = -\mathbf{C}_{rr}^{e}\left(\mathbf{C}_{rr}^{e}+\mathbf{C}_{rr}^{c}\right)^{-1}\mathbf{C}_{r\alpha}^{c}. \qquad (4.116)
$$

4.3.6 Best-Estimate 4th-Order Expression of the Triple Correlations among Predicted Responses and Calibrated Model Parameters

Four types of triple correlations can occur among the predicted responses and model parameters. The 4th-order expressions corresponding to each of these four types of triple correlations are provided.

1. The best-estimate predicted triple correlation among three best-estimate predicted responses, $r_k^{be}, r_\ell^{be}, r_m^{be}$, are denoted as $\mu_3^{be}\left(r_k^{be}, r_\ell^{be}, r_m^{be}\right)$ for $k,\ell,m = 1,\ldots,TR$, and are defined as follows:

$$
\mu_3^{be}\left(r_k^{be}, r_\ell^{be}, r_m^{be}\right) \triangleq \int_D \left(r_k - r_k^{be}\right)\left(r_\ell - r_\ell^{be}\right)\left(r_m - r_m^{be}\right) p_4^{be}(\mathbf{r},\boldsymbol{\alpha})\, d\boldsymbol{\alpha}\, d\mathbf{r}. \qquad (4.117)
$$

The lowest approximation expression for $\mu_3^{be}\left(r_k^{be}, r_\ell^{be}, r_m^{be}\right)$ will be denoted as $\left[\mu_3^{be}\left(r_k^{be}, r_\ell^{be}, r_m^{be}\right)\right]_1$ and is determined by using the expression of the best-estimate response obtained in Eq. (4.101) in Eq. (4.117), to obtain the following result:

$$
\begin{aligned}
&\left[\mu_3^{be}\left(r_k^{be}, r_\ell^{be}, r_m^{be}\right)\right]_1 \\
&= \int_D \left(\delta r_k - \rho_k\right)\left(\delta r_\ell - \rho_\ell\right)\left(\delta r_m - \rho_m\right) p_4^{be}(\mathbf{r},\boldsymbol{\alpha})\, d\boldsymbol{\alpha}\, d\mathbf{r}, \qquad (4.118)
\end{aligned}
$$

where:

$$
\delta r_k \triangleq r_k - r_k^{e};\quad \rho_k \triangleq \left\{\mathbf{C}^{e}\left(\mathbf{C}^{e}+\mathbf{C}_{rr}^{c}\right)^{-1}\left[\mathbf{E}_c(\mathbf{r})-\mathbf{r}^{e}\right]+\mathbf{p}_1\left(\mathbf{r}_s^{(0)},\boldsymbol{\alpha}_s^{(0)}\right)\right\}_k. \qquad (4.119)
$$

Recalling the definition provided in Eq. (4.59), recalling the definitions for the moments of the experimentally measured responses provided in Eqs. (4.49)–(4.52), and performing the integration in Eq. (4.118) yields the following result for the lowest approximation, denoted as $\left[\mu_3^{be}\left(r_k^{be}, r_\ell^{be}, r_m^{be}\right)\right]_1$, of the best-estimate triple correlations for the best-estimate responses:

$$\left[\mu_3^{be}\left(r_k^{be},r_\ell^{be},r_m^{be}\right)\right]_1 = t_{ijk}^e - \rho_\ell c_{km}^e - \rho_m c_{k\ell}^e - \rho_k c_{\ell m}^e - \rho_k \rho_\ell \rho_m;\quad k,\ell,m = 1,\ldots,TR. \tag{4.120}$$

2. The best-estimate predicted triple correlations among a best-estimate calibrated model parameter, α_k^{be}, and two best-estimate predicted responses, r_ℓ^{be}, r_m^{be}, are denoted as $\mu_3^{be}\left(\alpha_k^{be},r_\ell^{be},r_m^{be}\right)$ for $k=1,\ldots,TP$ and $\ell,m=1,\ldots,TR$, and are defined as follows:

$$\mu_3^{be}\left(\alpha_k^{be},r_\ell^{be},r_m^{be}\right) \triangleq \int_D \left(\alpha_k - \alpha_k^{be}\right)\left(r_\ell - r_\ell^{be}\right)\left(r_m - r_m^{be}\right) p_4^{be}(\mathbf{r},\boldsymbol{\alpha})\,d\boldsymbol{\alpha}\,d\mathbf{r}. \tag{4.121}$$

The lowest approximation for $\mu_3^{be}\left(\alpha_k^{be},r_\ell^{be},r_m^{be}\right)$ will be denoted as $\left[\mu_3^{be}\left(\alpha_k^{be},r_\ell^{be},r_m^{be}\right)\right]_1$ and is determined by using the corresponding expressions of the best-estimate responses and parameters, to obtain the following result:

$$\mu_3^{be}\left(\alpha_k^{be},r_\ell^{be},r_m^{be}\right) = \int_D \left(\delta\alpha_k + \beta_k\right)\left(\delta r_\ell - \rho_\ell\right)\left(\delta r_m - \rho_m\right) p_4^{be}(\mathbf{r},\boldsymbol{\alpha})\,d\boldsymbol{\alpha}\,d\mathbf{r}, \tag{4.122}$$

where:

$$\delta\alpha_k \triangleq \alpha_k - \alpha_k^0;\quad \beta_k \triangleq \left\{\mathbf{C}_{\alpha r}^c\left(\mathbf{C}^e + \mathbf{C}_{rr}^c\right)^{-1}\left[\mathbf{E}_c(\mathbf{r}) - \mathbf{r}^e\right] + \mathbf{p}_2\left(\mathbf{r}_s^{(0)},\boldsymbol{\alpha}_s^{(0)}\right)\right\}_k. \tag{4.123}$$

Performing the integration in Eq. (4.122) and noting that the model parameters are uncorrelated with the experimentally measured responses yield the following result:

$$\mu_3^{be}\left(\alpha_k^{be},r_\ell^{be},r_m^{be}\right) = \beta_k\left(c_{\ell m}^e + \rho_\ell\rho_m\right). \tag{4.124}$$

3. The best-estimate predicted triple correlations among two best-estimate calibrated model parameters, $\alpha_k^{be},\alpha_\ell^{be}$, and a best-estimate predicted response, r_m^{be}, are denoted as $\mu_3^{be}\left(\alpha_k^{be},\alpha_\ell^{be},r_m^{be}\right)$ for $k,\ell=1,\ldots,TP$ and $m=1,\ldots,TR$, and are defined as follows:

$$\mu_3^{be}\left(\alpha_k^{be},\alpha_\ell^{be},r_m^{be}\right) \triangleq \int_D \left(\alpha_k - \alpha_k^{be}\right)\left(\alpha_\ell - \alpha_\ell^{be}\right)\left(r_m - r_m^{be}\right) p_4^{be}(\mathbf{r},\boldsymbol{\alpha})\,d\boldsymbol{\alpha}\,d\mathbf{r}. \tag{4.125}$$

The lowest approximation for $\mu_3^{be}\left(\alpha_k^{be},\alpha_\ell^{be},r_m^{be}\right)$ will be denoted as $\left[\mu_3^{be}\left(\alpha_k^{be},\alpha_\ell^{be},r_m^{be}\right)\right]_1$ and is determined by using the corresponding expressions of the best-estimate responses and parameters, to obtain the following result:

$$\left[\mu_3^{be}\left(\alpha_k^{be},\alpha_\ell^{be},r_m^{be}\right)\right]_1 = \int_D \left(\delta\alpha_k + \beta_k\right)\left(\delta\alpha_\ell + \beta_\ell\right)\left(\delta r_m - \rho_m\right) p_4^{be}(\mathbf{r},\boldsymbol{\alpha})\,d\boldsymbol{\alpha}\,d\mathbf{r} = -\rho_m\left(c_{k\ell}^\alpha + \beta_k\beta_\ell\right). \tag{4.126}$$

4. The best-estimate predicted triple correlations among three best-estimate calibrated model parameters, $\alpha_k^{be}, \alpha_\ell^{be}, \alpha_m^{be}$, are denoted as $\mu_3^{be}\left(\alpha_k^{be}, \alpha_\ell^{be}, \alpha_m^{be}\right)$ for $k, \ell, m = 1, \dots, TP$, and are defined as follows:

$$\mu_3^{be}\left(\alpha_k^{be}, \alpha_\ell^{be}, \alpha_m^{be}\right) \triangleq \int_D \left(\alpha_k - \alpha_k^{be}\right)\left(\alpha_\ell - \alpha_\ell^{be}\right)\left(\alpha_m - \alpha_m^{be}\right) p_4^{be}(\mathbf{r}, \boldsymbol{\alpha})\, d\boldsymbol{\alpha}\, d\mathbf{r}. \tag{4.127}$$

The lowest approximation for $\mu_3^{be}\left(\alpha_k^{be}, \alpha_\ell^{be}, \alpha_m^{be}\right)$ will be denoted as $\left[\mu_3^{be}\left(\alpha_k^{be}, \alpha_\ell^{be}, \alpha_m^{be}\right)\right]_1$ and is determined by using the corresponding expression of the best-estimate calibrated parameters, to obtain the following result:

$$\begin{aligned}\mu_3^{be}\left(\alpha_k^{be}, \alpha_\ell^{be}, \alpha_m^{be}\right) &= \int_D (\delta\alpha_k + \beta_k)(\delta\alpha_\ell + \beta_\ell)(\delta\alpha_m + \beta_m) p_4^{be}(\mathbf{r}, \boldsymbol{\alpha})\, d\boldsymbol{\alpha}\, d\mathbf{r} \\ &= t_{k\ell m}^{\alpha} + \beta_\ell c_{km}^{\alpha} + \beta_m c_{k\ell}^{\alpha} + \beta_k c_{\ell m}^{\alpha} + \beta_k \beta_\ell \beta_m.\end{aligned} \tag{4.128}$$

4.3.7 Best-Estimate 4th-Order Expression of the Quadruple Correlations among Predicted Responses and Calibrated Model Parameters

Five types of quadruple correlations can occur among the predicted responses and model parameters. The 4th-order expressions corresponding to each of these five types of quadruple correlations are provided.

1. The best-estimate predicted quadruple correlations among four best-estimate predicted responses, $r_k^{be}, r_\ell^{be}, r_m^{be}, r_n^{be}$, are denoted as $\mu_4^{be}\left(r_k^{be}, r_\ell^{be}, r_m^{be}, r_n^{be}\right)$ for $k, \ell, m, n = 1, \dots, TR$, and are defined as follows:

$$\mu_4^{be}\left(r_k^{be}, r_\ell^{be}, r_m^{be}, r_n^{be}\right) \triangleq \int_D \left(r_k - r_k^{be}\right)\left(r_\ell - r_\ell^{be}\right)\left(r_m - r_m^{be}\right)\left(r_n - r_n^{be}\right) p_4^{be}(\mathbf{r}, \boldsymbol{\alpha})\, d\boldsymbol{\alpha}\, d\mathbf{r}. \tag{4.129}$$

The lowest approximation for $\mu_4^{be}\left(r_k^{be}, r_\ell^{be}, r_m^{be}, r_n^{be}\right)$ will be denoted as $\left[\mu_4^{be}\left(r_k^{be}, r_\ell^{be}, r_m^{be}, r_n^{be}\right)\right]_1$ and is determined by using the corresponding expressions of the best-estimate responses, to obtain the following result:

$$\begin{aligned}&\left[\mu_4^{be}\left(r_k^{be}, r_\ell^{be}, r_m^{be}, r_n^{be}\right)\right]_1 \\ &= \int_D (\delta r_k - \rho_k)(\delta r_\ell - \rho_\ell)(\delta r_m - \rho_m)(\delta r_n - \rho_n) p_4^{be}(\mathbf{r}, \boldsymbol{\alpha})\, d\boldsymbol{\alpha}\, d\mathbf{r} \\ &= q_{k\ell mn}^{e} - \rho_\ell t_{kmn}^{e} - \rho_m t_{k\ell n}^{e} - \rho_n t_{k\ell m}^{e} - \rho_k t_{\ell mn}^{e} + \rho_\ell \rho_m c_{kn}^{e} + \rho_k \rho_\ell c_{mn}^{e} \\ &\quad + \rho_k \rho_m c_{\ell n}^{e} + \rho_n \rho_\ell c_{km}^{e} + \rho_n \rho_m c_{k\ell}^{e} + \rho_k \rho_n c_{\ell m}^{e} + \rho_k \rho_\ell \rho_m \rho_n.\end{aligned} \tag{4.130}$$

2. The best-estimate predicted quadruple correlations among a best-estimate calibrated model parameter, α_k^{be}, and three best-estimate predicted

responses, $r_\ell^{be}, r_m^{be}, r_n^{be}$, are denoted as $\mu_4^{be}\left(\alpha_k^{be}, r_\ell^{be}, r_m^{be}, r_n^{be}\right)$ for $k = 1,...,TP$ and $\ell, m, n = 1,...,TR$, and are defined as follows:

$$\mu_4^{be}\left(\alpha_k^{be}, r_\ell^{be}, r_m^{be}, r_n^{be}\right) \triangleq \int_D \left(\alpha_k - \alpha_k^{be}\right)\left(r_\ell - r_\ell^{be}\right)\left(r_m - r_m^{be}\right)\left(r_n - r_n^{be}\right) p_4^{be}\left(\mathbf{r}, \boldsymbol{\alpha}\right) d\boldsymbol{\alpha}\, d\mathbf{r}. \tag{4.131}$$

The lowest approximation for $\mu_4^{be}\left(\alpha_k^{be}, r_\ell^{be}, r_m^{be}, r_n^{be}\right)$ will be denoted as $\left[\mu_4^{be}\left(\alpha_k^{be}, r_\ell^{be}, r_m^{be}, r_n^{be}\right)\right]_1$ and is determined by using the corresponding expressions of the best-estimate responses and parameters, to obtain the following result:

$$\begin{aligned}
&\left[\mu_4^{be}\left(\alpha_k^{be}, r_\ell^{be}, r_m^{be}, r_n^{be}\right)\right]_1 \\
&= \int_D \left(\delta\alpha_k + \beta_k\right)\left(\delta r_n - \rho_n\right)\left(\delta r_\ell - \rho_\ell\right)\left(\delta r_m - \rho_m\right) p_4^{be}\left(\mathbf{r}, \boldsymbol{\alpha}\right) d\boldsymbol{\alpha}\, d\mathbf{r} \\
&= -\rho_\ell t_{kmn}^e - \rho_m t_{k\ell n}^e + \beta_k t_{\ell mn}^e + c_{kn}^e \rho_\ell \rho_m + \beta_k \\
&\left(\rho_n c_{\ell m}^e - \rho_\ell c_{mn}^e - \rho_m c_{\ell n}^e - \rho_n \rho_\ell \rho_m\right).
\end{aligned} \tag{4.132}$$

3. The best-estimate predicted quadruple correlations among two best-estimate calibrated model parameters, $\alpha_k^{be}, \alpha_\ell^{be}$, and two best-estimate predicted responses, r_m^{be} and r_n^{be}, are denoted as $\mu_4^{be}\left(\alpha_k^{be}, \alpha_n^{be}, r_\ell^{be}, r_m^{be}\right)$ for $k, \ell = 1,...,TP$ and $m, n = 1,...,TR$, and are defined as follows:

$$\mu_4^{be}\left(\alpha_k^{be}, \alpha_n^{be}, r_\ell^{be}, r_m^{be}\right) \triangleq \int_D \left(\alpha_k - \alpha_k^{be}\right)\left(\alpha_\ell - \alpha_\ell^{be}\right)\left(r_m - r_m^{be}\right)\left(r_n - r_n^{be}\right) p_4^{be}\left(\mathbf{r}, \boldsymbol{\alpha}\right) d\boldsymbol{\alpha}\, d\mathbf{r}. \tag{4.133}$$

The lowest approximation for $\mu_4^{be}\left(\alpha_k^{be}, \alpha_\ell^{be}, r_m^{be}, r_n^{be}\right)$ will be denoted as $\left[\mu_4^{be}\left(\alpha_k^{be}, \alpha_\ell^{be}, r_m^{be}, r_n^{be}\right)\right]_1$ and is determined by using the corresponding expressions of the best-estimate responses and parameters, to obtain the following result:

$$\begin{aligned}
&\left[\mu_4^{be}\left(\alpha_k^{be}, \alpha_\ell^{be}, r_m^{be}, r_n^{be}\right)\right]_1 \\
&= \int_D \left(\delta\alpha_k + \beta_k\right)\left(\delta\alpha_n + \beta_n\right)\left(\delta r_\ell - \rho_\ell\right)\left(\delta r_m - \rho_m\right) p_4^{be}\left(\mathbf{r}, \boldsymbol{\alpha}\right) d\boldsymbol{\alpha}\, d\mathbf{r} \\
&= c_{kn}^e \rho_\ell \rho_m + \beta_k \beta_n \left(c_{\ell m}^e + \rho_\ell \rho_m\right).
\end{aligned} \tag{4.134}$$

4. The best-estimate predicted quadruple correlations among three best-estimate calibrated model parameters, $\alpha_k^{be}, \alpha_\ell^{be}$, α_m^{be}, and one best-estimate predicted response, r_n^{be}, are denoted as $\mu_4^{be}\left(\alpha_k^{be}, \alpha_\ell^{be}, \alpha_m^{be}, r_n^{be}\right)$ for $k, \ell, m = 1,...,TP$ and $n = 1,...,TR$, and are defined as follows:

$$\mu_4^{be}\left(\alpha_k^{be},\alpha_\ell^{be},\alpha_m^{be},r_n^{be}\right)$$
$$\triangleq \int_D \left(\alpha_k-\alpha_k^{be}\right)\left(\alpha_\ell-\alpha_\ell^{be}\right)\left(\alpha_m-\alpha_m^{be}\right)\left(r_n-r_n^{be}\right)p_4^{be}(\mathbf{r},\boldsymbol{\alpha})\,d\boldsymbol{\alpha}\,d\mathbf{r}. \tag{4.135}$$

The lowest approximation for $\mu_4^{be}\left(\alpha_k^{be},\alpha_\ell^{be},\alpha_m^{be},r_n^{be}\right)$ will be denoted as $\left[\mu_4^{be}\left(\alpha_k^{be},\alpha_\ell^{be},\alpha_m^{be},r_n^{be}\right)\right]_1$ and is determined by using the corresponding expressions of the best-estimate responses and parameters, to obtain the following result:

$$\mu_4^{be}\left(\alpha_k^{be},\alpha_\ell^{be},\alpha_m^{be},r_n^{be}\right)$$
$$=\int_D (\delta\alpha_k+\beta_k)(\delta\alpha_\ell+\beta_\ell)(\delta\alpha_m+\beta_m)(\delta r_n-\rho_n)\,p_4^{be}(\mathbf{r},\boldsymbol{\alpha})\,d\boldsymbol{\alpha}\,d\mathbf{r} \tag{4.136}$$
$$=\rho_n\left(c_{km}^{\alpha}\beta_\ell-\beta_k c_{\ell m}^{\alpha}-\beta_m c_{k\ell}^{\alpha}+\beta_k\beta_m\beta_\ell\right).$$

5. The best-estimate predicted quadruple correlations among four best-estimate calibrated model parameters, $\alpha_k^{be},\alpha_\ell^{be},\ \alpha_m^{be},\ \alpha_n^{be}$, are denoted as $\mu_4^{be}\left(\alpha_k^{be},\alpha_\ell^{be},\alpha_m^{be},\alpha_n^{be}\right)$ for $k,\ell,m,n=1,...,TP$ and are defined as follows:

$$\mu_4^{be}\left(\alpha_k^{be},\alpha_\ell^{be},\alpha_m^{be},\alpha_n^{be}\right)$$
$$\triangleq \int_D \left(\alpha_k-\alpha_k^{be}\right)\left(\alpha_\ell-\alpha_\ell^{be}\right)\left(\alpha_m-\alpha_m^{be}\right)\left(\alpha_n-\alpha_n^{be}\right)p_4^{be}(\mathbf{r},\boldsymbol{\alpha})\,d\boldsymbol{\alpha}\,d\mathbf{r}. \tag{4.137}$$

The lowest approximation for $\mu_4^{be}\left(\alpha_k^{be},\alpha_\ell^{be},\alpha_m^{be},\alpha_n^{be}\right)$ will be denoted as $\left[\mu_4^{be}\left(\alpha_k^{be},\alpha_\ell^{be},\alpha_m^{be},\alpha_n^{be}\right)\right]_1$ and is determined by using the corresponding expressions of the best-estimate parameters, to obtain the following result:

$$\left[\mu_4^{be}\left(\alpha_k^{be},\alpha_\ell^{be},\alpha_m^{be},\alpha_n^{be}\right)\right]_1$$
$$=\int_D (\delta\alpha_k+\beta_k)(\delta\alpha_\ell+\beta_\ell)(\delta\alpha_m+\beta_m)(\delta\alpha_n+\beta_n)\,p_4^{be}(\mathbf{r},\boldsymbol{\alpha})\,d\boldsymbol{\alpha}\,d\mathbf{r} \tag{4.138}$$
$$=q_{k\ell mn}^{\alpha}+\beta_\ell t_{kmn}^{\alpha}+\beta_m t_{k\ell n}^{\alpha}+\beta_k t_{\ell mn}^{\alpha}+\beta_n t_{k\ell m}^{\alpha}+\beta_k\beta_\ell c_{mn}^{\alpha}+\beta_k\beta_m c_{\ell n}^{\alpha}$$
$$+\beta_k\beta_n c_{\ell m}^{\alpha}+\beta_\ell\beta_m c_{kn}^{\alpha}+\beta_\ell\beta_n c_{km}^{\alpha}+\beta_m\beta_n c_{k\ell}^{\alpha}+\beta_k\beta_\ell\beta_m\beta_n.$$

4.3.8 Indicator of Consistency among Parameters and Responses

It is evident from the expressions obtained in this section, compare Eqs. (4.100)–(4.138), that the triple and quadruple correlations among the best-estimate predicted responses and calibrated model parameters do not vanish in general. Therefore, the 4th-order posterior distribution $p_4^{be}(\mathbf{r},\boldsymbol{\alpha})$ of the predicted responses and calibrated model parameters is *not* a multivariate normal (Gaussian) distribution. However, if all of the a priori triple and quadruple correlations among responses and model parameters are neglected, then all of the posterior results obtained in this section reduce to the results predicted by the 2nd-BERRU-PM (2nd-order best-estimate results with reduced uncertainties – predictive modeling) methodology presented in Chapter 1. In this case, the quantity $Q(\boldsymbol{\alpha},\mathbf{r})$, which was defined in Eq. (4.60), becomes a quadratic form, which was shown in Chapter 1 to take on the following minimum value, denoted as $Q_{\min}^{(0)}$ at the 2nd-order saddle-point $\left(\mathbf{r}_s^{(0)},\boldsymbol{\alpha}_s^{(0)}\right)$:

$$Q_{\min}^{(0)} = \begin{pmatrix} \mathbf{r}^{(0)} - \mathbf{E}_c(\mathbf{r}) \\ \boldsymbol{\alpha}^{(0)} - \boldsymbol{\alpha}^0 \end{pmatrix}^{\dagger} \begin{pmatrix} \mathbf{C}_{rr}^c & \mathbf{C}_{r\alpha}^c \\ \mathbf{C}_{\alpha r}^c & \mathbf{C}_{\alpha\alpha} \end{pmatrix}^{-1} \begin{pmatrix} \mathbf{r}^{(0)} - \mathbf{E}_c(\mathbf{r}) \\ \boldsymbol{\alpha}^{(0)} - \boldsymbol{\alpha}^0 \end{pmatrix}$$
$$+\left(\mathbf{r}^{(0)} - \mathbf{r}^e\right)^{\dagger}\left(\mathbf{C}_{rr}^e\right)^{-1}\left(\mathbf{r}^{(0)} - \mathbf{r}^e\right) = \left[\mathbf{E}_c(\mathbf{r}) - \mathbf{r}^e\right]^{\dagger}\left(\mathbf{C}_{rr}^e + \mathbf{C}_{rr}^c\right)^{-1}\left[\mathbf{E}_c(\mathbf{r}) - \mathbf{r}^e\right]. \tag{4.139}$$

As the expression obtained in Equation (4.139) indicates, the quantity $Q_{\min}^{(0)}$ represents the square of the length of the vector $\left[\mathbf{E}_c(\mathbf{r}) - \mathbf{r}^e\right]$ measuring (in the corresponding metric) the deviations between the experimental and nominally computed responses. The quantity $Q_{\min}^{(0)}$ is independent of calibrating (or adjusting) the original data, so it can be evaluated directly from the given data (i.e., model parameters and computed and measured responses, together with their original uncertainties) after having computed the matrix $\left(\mathbf{C}_{rr}^e + \mathbf{C}_{rr}^c\right)^{-1}$. As the dimension of the vector $\left[\mathbf{E}_c(\mathbf{r}) - \mathbf{r}^e\right]$ indicates, the number of degrees of freedom characteristic of the calibration under consideration is equal to the number *TR* of experimental responses. The quantity $Q_{\min}^{(0)}$ plays the role of a "χ^2-like consistency indicator," which can be used in the course of quantitative model validation since it quantifies the degree of agreement between the computed and the experimentally measured responses, before actually combining the computational and experimental information. Agreement between experimental and computational results is indicated when $Q_{\min}^{(0)} \simeq 1$ (per degree of freedom); values of $Q_{\min}^{(0)}$ that differ greatly from unity indicate "inconsistency" (i.e., lack of validation) or perhaps even gross errors. In such cases, the individual contributions to $Q_{\min}^{(0)}$ must be examined and the outliers (i.e., very large or very small individual contributions) need to be investigated as possible sources of inconsistencies that could invalidate either parts of the model or parts of the data (or both).

The results provided in Eqs. (4.100)–(4.138) highlights the fact that they can be utilized both for "forward/direct predictive modeling" and for "inverse predictive modeling." The "forward" or "direct problem" solves the "parameter-to-output" mapping that describes the "cause-to-effect" relationship in the physical process being modeled. The "inverse problem" attempts to solve the "output-to-parameters" mapping. Since the framework of 4th-BERRU-PM methodology comprises the combined phase-space of parameters and responses, it can be used for solving both forward/direct and inverse problems. The solution of the "forward/direct problem" is provided by the expression for the predicted best-estimate responses together with the corresponding reduced predicted uncertainties. Conversely, the solution of the "inverse problem" is provided by the expressions for the predicted best-estimate calibrated parameters and their reduced predicted uncertainties.

4.4 CONCLUSIONS

This chapter has presented the 4th-order predictive modeling methodology designated by the acronym *4th-BERRU-PM* methodology, which uses the MaxEnt principle to incorporate 4th-order moments (means, covariances, skewness, kurtosis) of model parameters, computed and measured model responses, as well as 4th-order (and higher-) sensitivities of computed model responses to model parameters. The acronym *4th-BERRU-PM*, which designates this new methodology, stands for "4th-order best-estimate results with reduced uncertainties." The results predicted by the *4th-BERRU-PM* incorporate, as particular cases, the results previously predicted by the 1st-order BERRU-PM methodology (Cacuci, 2019), the 2nd-order predictive modeling methodology *2nd-BERRU-PM* (presented in Chapter 1) and vastly generalizes the results produced by extant inverse methods, data adjustment, and data assimilation procedures, all of which rely on the minimization of user-defined least-squares-type functionals for estimating the perceived discrepancies between measured and computed results/responses. The key novel features of the 4th-BRERRU-PM methodology presented in this chapter include the following:

1. The mean values, the 2nd-, 3rd-, and 4th-order correlations among the computed responses, which are incorporated within the 4th-BERRU-PM methodology, include high-order (as high as needed) response sensitivities to model parameters, thus generalizing all of the previous formulas of this type found in data adjustment/assimilation procedures published to date. These arbitrarily high-order sensitivities can be computed most efficiently by applying the high-order adjoint methods developed by Cacuci (2022, 2023a).
2. The 4th-BERRU-PM methodology predicts posterior parameter-response correlations that are not obtainable by any extant data assimilation methods.
3. The 4th-BERRU-PM methodology enjoys unrivaled computational efficiency since the most intensive computation arises when computing the matrix $\left(\mathbf{C}_e^{rr} + \mathbf{C}_c^{rr}\right)^{-1}$, which entails the inversion of a matrix of size $TR \times TR$, where TR denotes the number of distinct computed (or measured) responses. This is computationally very advantageous since the number of responses is much smaller than the number of model parameters, in practice.
4. The 4th-BERRU-PM methodology is the first and thus far the only predictive modeling methodology which incorporates triple and quadruple correlations for model parameters and experimentally measured responses.
5. The 4th-BERRU-PM methodology is thus far the only methodology that yields closed-form expressions for computing the predicted best-estimates triple and quadruple correlations among the best-estimate predicted responses and calibrated model parameters.

APPENDIX 4.A AUXILIARY COMPUTATIONS FOR CONSTRUCTING THE 4TH-ORDER MAXIMUM ENTROPY DISTRIBUTION FOR MODEL PARAMETERS AND RESPONSES

It follows from Eqs. (4.12)–(4.26) that the functions $\Psi_k^{(1)}(\mathbf{a})$, $X_k^{(1)}(\mathbf{a})$, for $k=1,...,TR$, and $\Psi_i^{(2)}(\mathbf{a})$, $X_i^{(2)}(\mathbf{a})$ for $i=1,...,TP$ can be obtained by differentiating the function $\Phi(\mathbf{a})$ with respect to the components of the Lagrange multiplier $\mathbf{a} \triangleq \left(\mathbf{a}^{(1)},\mathbf{a}^{(2)}\right)^{\dagger}$, the definition of which is reproduced:

$$\Phi(\mathbf{a}) \triangleq \int_D \exp\left(-\mathbf{a}^{\dagger}\mathbf{u} - \frac{1}{2}\mathbf{u}^{\dagger}\mathbf{B}\mathbf{u}\right)d\mathbf{u} = K_c \exp\left(\frac{1}{2}\mathbf{a}^{\dagger}\mathbf{B}^{-1}\mathbf{a}\right) = K_c \exp\left[\eta(\mathbf{a})\right], \quad (4.140)$$

$$K_c \triangleq \frac{(2\pi)^{(TR+TP)/2}}{\sqrt{Det(\mathbf{B})}} = (2\pi)^{(TR+TP)/2}\left[Det(\mathbf{C}_c)\right]^{1/2}; \quad (4.141)$$

$$\eta(\mathbf{a}) \triangleq \frac{1}{2}\mathbf{a}^{\dagger}\mathbf{C}_c\mathbf{a};\ \mathbf{C}_c \triangleq \begin{pmatrix} \mathbf{C}_{rr}^c & \mathbf{C}_{r\alpha}^c \\ \mathbf{C}_{\alpha r}^c & \mathbf{C}_{\alpha\alpha} \end{pmatrix}. \quad (4.142)$$

The quantity $\eta(\mathbf{a})$, which appears in the exponent of the quantity $\Phi(\mathbf{a})$ on the right-most side of Eq. (4.140), has the following expression:

$$\begin{aligned}\eta(\mathbf{a}) &\triangleq \frac{1}{2}\mathbf{a}^{\dagger}\mathbf{C}_c\mathbf{a} = \frac{1}{2}\left(\mathbf{a}^{(1)},\mathbf{a}^{(2)}\right)^{\dagger}\begin{pmatrix} \mathbf{C}_{rr}^c & \mathbf{C}_{r\alpha}^c \\ \mathbf{C}_{\alpha r}^c & \mathbf{C}_{\alpha\alpha} \end{pmatrix}\begin{pmatrix} \mathbf{a}^{(1)} \\ \mathbf{a}^{(2)} \end{pmatrix} \\ &= \frac{1}{2}\left(\mathbf{a}^{(1)}\right)^{\dagger}\mathbf{C}_{rr}^c\mathbf{a}^{(1)} + \frac{1}{2}\left(\mathbf{a}^{(1)}\right)^{\dagger}\mathbf{C}_{r\alpha}^c\mathbf{a}^{(2)} \\ &+ \frac{1}{2}\left(\mathbf{a}^{(2)}\right)^{\dagger}\mathbf{C}_{\alpha r}^c\mathbf{a}^{(1)} + \frac{1}{2}\left(\mathbf{a}^{(2)}\right)^{\dagger}\mathbf{C}_{\alpha\alpha}\mathbf{a}^{(2)} \\ &= \frac{1}{2}\sum_{i=1}^{TR}\sum_{j=1}^{TR} a_i^{(1)}a_j^{(1)}\mathrm{cov}(r_i,r_j) + \sum_{i=1}^{TR}\sum_{j=1}^{TP} a_i^{(1)}a_j^{(2)}\mathrm{cor}(\alpha_i,r_j) \\ &+ \frac{1}{2}\sum_{i=1}^{TP}\sum_{j=1}^{TP} a_i^{(2)}a_j^{(2)}\mathrm{cov}(\alpha_i,\alpha_j).\end{aligned} \quad (4.143)$$

Differentiating Eq. (4.140) with respect to a Lagrange multiplier $a_k^{(1)}$ yields the following relation, for $k=1,...,TR$:

$$\frac{\partial\Phi(\mathbf{a})}{\partial a_k^{(1)}} = -\int_D (\delta r_k)\exp\left(-\mathbf{a}^{\dagger}\mathbf{u} - \frac{1}{2}\mathbf{u}^{\dagger}\mathbf{B}\mathbf{u}\right)d\mathbf{u} = K_c\Phi_{k,1}^{(1)}(\mathbf{a})\exp\left[\eta(\mathbf{a})\right], \quad (4.144)$$

where, for each $k=1,...,TR$, the quantity $\Phi_{k,1}^{(1)}$ is defined as follows:

$$\Phi_{k,1}^{(1)}(\mathbf{a}) \triangleq \frac{\partial\eta(\mathbf{a})}{\partial a_k^{(1)}} = \sum_{j=1}^{TR}\mathrm{cov}(r_j,r_k)a_j^{(1)} + \sum_{j=1}^{TP}\mathrm{cor}(\alpha_j,r_k)a_j^{(2)};\ k=1,...,TR. \quad (4.145)$$

Differentiating the expression in Eq. (4.144) with respect to the same Lagrange multiplier $a_k^{(1)}$ yields the following relation for the 2nd-order unmixed derivatives of $\Phi(\mathbf{a})$ with respect to $a_k^{(1)}$ for $k = 1,...,TR$:

$$\frac{\partial^2 \Phi(\mathbf{a})}{\partial a_k^{(1)} \partial a_k^{(1)}} = \int_D (\delta r_k)^2 \exp\left(-\mathbf{a}^\dagger \mathbf{u} - \frac{1}{2}\mathbf{u}^\dagger \mathbf{B}\mathbf{u}\right) d\mathbf{u} = K_c \Phi_{k,1}^{(2)} \exp[\eta(\mathbf{a})], \tag{4.146}$$

where, for each $k = 1,...,TR$, the quantity $\Phi_{k,1}^{(2)}$ is defined as follows:

$$\Phi_{k,1}^{(2)} \triangleq \frac{\partial \Phi_{k,1}^{(1)}(\mathbf{a})}{\partial a_k^{(1)}} + \Phi_{k,1}^{(1)}(\mathbf{a}) \frac{\partial \eta(\mathbf{a})}{\partial a_k^{(1)}} = \text{var}(r_k, r_k) + \left[\Phi_{k,1}^{(1)}(\mathbf{a})\right]^2; \; k = 1,...,TR. \tag{4.147}$$

Differentiating the expressions in Eq. (4.146) with respect to the same Lagrange multiplier $a_k^{(1)}$ yields the following relation for the 3rd-order unmixed derivatives of $\Phi(\mathbf{a})$ with respect to $a_k^{(1)}$, for $k = 1,...,TR$:

$$\begin{aligned} \frac{\partial^3 \Phi(\mathbf{a})}{\partial a_k^{(1)} \partial a_k^{(1)} \partial a_k^{(1)}} &= -\int_D (\delta r_k)^3 \exp\left(-\mathbf{a}^\dagger \mathbf{u} - \frac{1}{2}\mathbf{u}^\dagger \mathbf{B}\mathbf{u}\right) d\mathbf{u} \\ &= \Phi_{k,1}^{(3)} K_c \exp[\eta(\mathbf{a})] = -\Psi_k^{(1)}(\mathbf{a}). \end{aligned} \tag{4.148}$$

where:

$$\begin{aligned} \Phi_{k,1}^{(3)} &\triangleq \frac{\partial \Phi_{k,1}^{(2)}(\mathbf{a})}{\partial a_k^{(1)}} + \Phi_{k,1}^{(2)}(\mathbf{a}) \frac{\partial \eta(\mathbf{a})}{\partial a_k^{(1)}} = 2\Phi_{k,1}^{(1)}(\mathbf{a}) \frac{\partial \Phi_{k,1}^{(1)}(\mathbf{a})}{\partial a_k^{(1)}} + \Phi_{k,1}^{(2)}(\mathbf{a}) \Phi_{k,1}^{(1)}(\mathbf{a}) \\ &= 3\Phi_{k,1}^{(1)}(\mathbf{a}) \text{var}(r_k, r_k) + \left[\Phi_{k,1}^{(1)}(\mathbf{a})\right]^3; \; k = 1,...,TR. \end{aligned} \tag{4.149}$$

Differentiating the expressions in Eq. (4.148) with respect to the same Lagrange multiplier $a_k^{(1)}$ yields the following relation for the 4th-order unmixed derivatives of $\Phi(\mathbf{a})$ with respect to $a_k^{(1)}$, for $k = 1,...,TR$:

$$\begin{aligned} \frac{\partial^4 \Phi(\mathbf{a})}{\partial a_k^{(1)} \partial a_k^{(1)} \partial a_k^{(1)} \partial a_k^{(1)}} &= \int_D (\delta r_k)^4 \exp\left(-\mathbf{a}^\dagger \mathbf{u} - \frac{1}{2}\mathbf{u}^\dagger \mathbf{B}\mathbf{u}\right) d\mathbf{u} \\ &= \Phi_{k,1}^{(4)} K_c \exp[\eta(\mathbf{a})] = \mathrm{X}_k^{(1)}(\mathbf{a}); \quad k = 1,...,TR; \end{aligned} \tag{4.150}$$

where:

$$\begin{aligned} \Phi_{k,1}^{(4)} &\triangleq \frac{\partial \Phi_{k,1}^{(3)}(\mathbf{a})}{\partial a_k^{(1)}} + \Phi_{k,1}^{(3)}(\mathbf{a}) \frac{\partial \eta(\mathbf{a})}{\partial a_k^{(1)}} = 3\,\text{var}(r_k, r_k) \frac{\partial \Phi_{k,1}^{(1)}(\mathbf{a})}{\partial a_k^{(1)}} + 3\left[\Phi_{k,1}^{(1)}(\mathbf{a})\right]^2 \frac{\partial \Phi_{k,1}^{(1)}(\mathbf{a})}{\partial a_k^{(1)}} \\ &+ \Phi_{k,1}^{(3)}(\mathbf{a}) \Phi_{k,1}^{(1)}(\mathbf{a}) = 3\left\{\text{var}(r_k, r_k) + \left[\Phi_{k,1}^{(1)}(\mathbf{a})\right]^2\right\} \text{var}(r_k, r_k) \end{aligned} \tag{4.151}$$

$$+\left\{3\Phi_{k,1}^{(1)}(\mathbf{a})\operatorname{var}(r_k,r_k)+\left[\Phi_{k,1}^{(1)}(\mathbf{a})\right]^3\right\}\Phi_{k,1}^{(1)}(\mathbf{a})$$
$$=3\left[\operatorname{var}(r_k,r_k)\right]^2+6\left[\Phi_{k,1}^{(1)}(\mathbf{a})\right]^2\operatorname{var}(r_k,r_k)+\left[\Phi_{k,1}^{(1)}(\mathbf{a})\right]^4;\ k=1,...,TR.$$

Differentiating Eq. (4.140) with respect to a Lagrange multiplier $a_i^{(2)}$ yields the following relation for the 1st-order derivatives of $\Phi(\mathbf{a})$ with respect to $a_i^{(2)}$ for $i=1,...,TP$:

$$\frac{\partial\Phi(\mathbf{a})}{\partial a_i^{(2)}}=-\int_D(\delta\alpha_i)\exp\left(-\mathbf{a}^\dagger\mathbf{u}-\frac{1}{2}\mathbf{u}^\dagger\mathbf{B}\mathbf{u}\right)d\mathbf{u}=K_c\Phi_{i,2}^{(1)}\exp\left[\eta(\mathbf{a})\right], \tag{4.152}$$

where:

$$\Phi_{i,2}^{(1)}\triangleq\frac{\partial\eta(\mathbf{a})}{\partial a_i^{(2)}}=\sum_{j=1}^{TP}a_j^{(1)}\operatorname{cor}(\alpha_i,r_j)+\sum_{j=1}^{TP}a_j^{(2)}\operatorname{cov}(\alpha_i,\alpha_j);\quad i=1,...,TP. \tag{4.153}$$

Differentiating the expression in Eq. (4.152) with respect to the same Lagrange multiplier $a_i^{(2)}$ yields the following relation for the 2nd-order unmixed derivatives of $\Phi(\mathbf{a})$ with respect to $a_i^{(2)}$ for $i=1,...,TP$:

$$\frac{\partial^2\Phi(\mathbf{a})}{\partial a_i^{(2)}\partial a_i^{(2)}}=\int_D(\delta\alpha_i)^2\exp\left(-\mathbf{a}^\dagger\mathbf{u}-\frac{1}{2}\mathbf{u}^\dagger\mathbf{B}\mathbf{u}\right)d\mathbf{u}=K_c\Phi_{i,2}^{(2)}\exp\left[\eta(\mathbf{a})\right], \tag{4.154}$$

where, for each $i=1,...,TP$, the quantity $\Phi_{i,2}^{(2)}$ is defined as follows:

$$\Phi_{i,2}^{(2)}\triangleq\frac{\partial\Phi_{i,2}^{(1)}(\mathbf{a})}{\partial a_i^{(2)}}+\Phi_{i,2}^{(1)}(\mathbf{a})\frac{\partial\eta(\mathbf{a})}{\partial a_i^{(2)}}=\operatorname{var}(\alpha_i,\alpha_i)+\left[\Phi_{i,2}^{(1)}(\mathbf{a})\right]^2;\ i=1,...,TP. \tag{4.155}$$

Differentiating the expression in Eq. (4.154) with respect to the same Lagrange multiplier $a_i^{(2)}$ yields the following relation for the 3rd-order unmixed derivatives of $\Phi(\mathbf{a})$ with respect to $a_i^{(2)}$ for $i=1,...,TP$:

$$\begin{aligned}\frac{\partial^3\Phi(\mathbf{a})}{\partial a_i^{(2)}\partial a_i^{(2)}\partial a_i^{(2)}}&=-\int_D(\delta\alpha_i)^3\exp\left(-\mathbf{a}^\dagger\mathbf{u}-\frac{1}{2}\mathbf{u}^\dagger\mathbf{B}\mathbf{u}\right)d\mathbf{u}\\&=\Phi_{i,2}^{(3)}K_c\exp\left[\eta(\mathbf{a})\right]=-\Psi_i^{(2)}(\mathbf{a});\quad i=1,...,TP.\end{aligned} \tag{4.156}$$

where:

$$\begin{aligned}\Phi_{i,2}^{(3)}&\triangleq\frac{\partial\Phi_{i,2}^{(2)}(\mathbf{a})}{\partial a_2^{(2)}}+\Phi_{i,2}^{(2)}(\mathbf{a})\frac{\partial\eta(\mathbf{a})}{\partial a_i^{(2)}}=2\Phi_{i,2}^{(1)}(\mathbf{a})\frac{\partial\Phi_{i,2}^{(1)}(\mathbf{a})}{\partial a_i^{(2)}}+\Phi_{i,2}^{(2)}(\mathbf{a})\Phi_{i,2}^{(1)}(\mathbf{a})\\&=3\Phi_{i,2}^{(1)}(\mathbf{a})\operatorname{var}(\alpha_i,\alpha_i)+\left[\Phi_{i,2}^{(1)}(\mathbf{a})\right]^3;\quad i=1,...,TP.\end{aligned} \tag{4.157}$$

Differentiating the expressions in Eq. (4.156) with respect to the same Lagrange multiplier $a_i^{(2)}$ yields the following relation for the 4th-order unmixed derivatives of $\Phi(\mathbf{a})$ with respect to $a_i^{(2)}$ for $i = 1, ..., TP$:

$$\frac{\partial^4 \Phi(\mathbf{a})}{\partial a_i^{(2)} \partial a_i^{(2)} \partial a_i^{(2)} \partial a_i^{(2)}} = \int_D (\delta \alpha_i)^4 \exp\left(-\mathbf{a}^\dagger \mathbf{u} - \frac{1}{2} \mathbf{u}^\dagger \mathbf{B} \mathbf{u} \right) d\mathbf{u}$$
$$= \Phi_{i,2}^{(4)} K_c \exp[\eta(\mathbf{a})] = \mathrm{X}_i^{(2)}(\mathbf{a}); \quad i = 1, ..., TP. \tag{4.158}$$

where:

$$\Phi_{i,2}^{(4)} \triangleq \frac{\partial \Phi_{i,2}^{(3)}}{\partial a_i^{(2)}} + \Phi_{i,2}^{(3)}(\mathbf{a}) \frac{\partial \eta(\mathbf{a})}{\partial a_i^{(2)}} = 3[\mathrm{var}(\alpha_i, \alpha_i)]^2$$
$$+ 6\left[\Phi_{i,2}^{(1)}(\mathbf{a})\right]^2 \mathrm{var}(\alpha_i, \alpha_i) + \left[\Phi_{i,2}^{(1)}(\mathbf{a})\right]^4; \quad i = 1, ..., TP. \tag{4.159}$$

5 Application of the 4th-BERRU-PM Methodology to the PERP Reactor Physics Benchmark

5.1 INTRODUCTION

The fundamental importance of the new results provided by the 4th-BERRU-PM methodology will be highlighted in this chapter by applying this predictive modeling methodology to the *P*oly*e*thylene-*R*eflected *P*lutonium (acronym: PERP) OECD/NEA reactor physics benchmark (Valentine, 2006). Section 5.2 of this chapter presents illustrative results predicted by the 4th-BERRU-PM methodology for the best-estimate predicted mean value (1st-order moment) and predicted best-estimate variance 2nd-order moment) for the PERP benchmark's leakage response. The effects of sensitivities on the predicted mean value and variance of the computed model response are illustrated as follows: (i) only the 1st-order sensitivities are considered; (ii) the 1st- and 2nd-order sensitivities are included; (iii) all sensitivities up to and including 3rd-order are included; (iv) all sensitivities up to and including 4th-order are included. The precision of the model parameters will also be investigated, as follows: (i) parameters that are known with "high precision" having uniform standard deviations of 2%; (ii) parameters known with "typical precision" having uniform standard deviations of 5%; and (iii) parameters known with "low precision" having uniform standard deviations of 10%, respectively. The Taylor series that represents the computed response in terms of parameter variations is convergent for parameters having uniform standard deviations of 2% but diverges for parameters having uniform standard deviations of 5% or larger. The impact of the precision (low vs. standard) of the measured response will also be investigated.

Section 5.3 presents illustrative results predicted by the 4th-BERRU-PM methodology for the best-estimate predicted mean value (first-order moment) and predicted best-estimate variance (2nd-order moment) for the PERP benchmark's group-averaged total microscopic cross-sections; these cross-sections are the "model parameters" considered for illustrating the model calibration capabilities of the 4th-BERRU-PM methodology. The impact of the precision ("high" vs. "typical") of these parameters will be investigated, along with the impact of the sensitivities of

DOI: 10.1201/9781003478119-5

various orders. Section 5.4 presents illustrative results predicted by the 4th-BERRU-PM methodology for the best-estimate predicted posterior correlations between predicted responses and calibrated model parameters. The impact of the order of the sensitivities as well as the precision of the model parameters is illustrated. Section 5.5 presents illustrative results predicted by the 4th-BERRU-PM methodology for the best-estimate predicted covariance matrix of calibrated model parameters, investigating the impact of the order of the sensitivities retained as well as the precision of the model parameters. Section 5.6 presents illustrative results predicted by the 4th-BERRU-PM methodology for the best-estimate skewness of predicted response and calibrated model parameters, while Section 5.7 presents illustrative results predicted by the 4th-BERRU-PM methodology for the best-estimate kurtosis of the predicted response and calibrated model parameters. The results presented in these Sections illustrate the impact of the order of the sensitivities retained as well as the precision of the parameters on these high-order moments of the posterior predicted distribution of best-estimate responses and parameters.

The concluding remarks presented in Section 5.8 underscore the unique capabilities of the 4th-BERRU-PM methodology, which incorporates, as particular cases, the results previously predicted by the 2nd-BERRU-PM methodology, and significantly generalizes the results produced by extant data adjustment (SCALE, 2009; Venard et al., 2009) and data assimilation methodologies (Rabier, 2005; Lewis et al., 2006; Lahoz et al., 2010; Práger and Kelemen, 2013; Cacuci et al., 2014).

5.2 4TH-BERRU-PM PREDICTED BEST-ESTIMATE POSTERIOR MEAN AND VARIANCE FOR THE PERP LEAKAGE RESPONSE

This section presents illustrative results predicted by the 4th-BERRU-PM methodology for the best-estimate predicted mean value (1st-order moment) and predicted best-estimate variance (2nd-order moment) for the PERP benchmark's leakage response. For the specific characteristics of the PERP benchmark's selected parameters (i.e., uncorrelated and normally distributed cross-sections, initially uncorrelated parameters and measured response) the mathematical expressions produced by the 4th-BERRU-PM methodology for the best-estimate predicted mean value and variance for the leakage response coincide with the respective expressions provided by the 2nd-BERRU-PM methodology (as presented in Chapter 2), since the higher-order terms considered in the 4th-BERRU-PM formulas for the response mean value and variance happen to vanish for the specific PERP leakage response. For convenient referencing, these expressions are provided. Using the index $i = 1,...,4$ to denote the highest-order of the retained sensitivities, the expression of the best-estimate predicted mean value, $L^{(be,i)}$, has the following form:

$$L^{(be,i)} = L^{e} + \left[SD^{(e)}\right]^{2}\left\{\left[SD^{(e)}\right]^{2} + \left[\text{var}(L)\right]^{(i,U,N)}\right\}^{-1}\left[E^{(i)}(L^{c}) - L^{e}\right],\ i = 1,...,4, \quad (5.1)$$

where the superscript "U" denotes "uncorrelated parameters", the superscript "N" denotes "normally distributed parameters" and where the following definitions are used:

1. Only 1st-order sensitivities are retained $(i=1)$:

$$\left[\operatorname{var}(L)\right]^{(1,U,N)} \triangleq \sum_{i=1}^{TP}\left[\frac{\partial L(\boldsymbol{\alpha})}{\partial \alpha_i}\right]^2 \sigma_i^2; \\ SD^{(1)} \triangleq \sqrt{\left[\operatorname{var}(L)\right]^{(1,U,N)}};\; E^{(1)}(L^c)=L(\boldsymbol{\alpha}^0) \tag{5.2}$$

2. The 1st- and 2nd-order sensitivities are retained $(i=2)$:

$$\left[\operatorname{var}(L)\right]^{(2,U,N)} \triangleq \left[\operatorname{var}(L)\right]^{(1,U,N)} + \frac{1}{2}\sum_{i=1}^{TP}\left[\frac{\partial^2 L(\boldsymbol{\alpha})}{\partial \alpha_i \partial \alpha_i}\right]^2 \sigma_i^4; \\ SD^{(2)} \triangleq \sqrt{\left[\operatorname{var}(L)\right]^{(2,U,N)}}; \tag{5.3}$$

$$E^{(2)}(L^c)=L(\boldsymbol{\alpha}^0)+\frac{1}{2}\sum_{i=1}^{TP}\frac{\partial^2 L(\boldsymbol{\alpha})}{(\partial \alpha_i)^2}\sigma_i^2. \tag{5.4}$$

3. The 1st-, 2nd-, and 3rd-order sensitivities are retained $(i=3)$:

$$\left[\operatorname{var}(L)\right]^{(3,U,N)} \triangleq \left[\operatorname{var}(L)\right]^{(2,U,N)} + \sum_{i=1}^{TP}\left[\frac{\partial^3 L(\boldsymbol{\alpha})}{(\partial \alpha_i)^3}\frac{\partial L(\boldsymbol{\alpha})}{\partial \alpha_i}\right]\sigma_i^4 \\ + \frac{15}{36}\sum_{i=1}^{TP}\left[\frac{\partial^3 L(\boldsymbol{\alpha})}{(\partial \alpha_i)^3_i}\right]^2 \sigma_i^6;\; SD^{(3)} \triangleq \sqrt{\left[\operatorname{var}(L)\right]^{(3,U,N)}}; \tag{5.5}$$

$$E^{(3)}(L^c)=L(\boldsymbol{\alpha}^0)+\frac{1}{2}\sum_{i=1}^{TP}\frac{\partial^2 L(\boldsymbol{\alpha})}{(\partial \alpha_i)^2}\sigma_i^2. \tag{5.6}$$

4. The 1st-, 2nd-, 3rd-, and 4th-order sensitivities are retained $(i=4)$:

$$\left[\operatorname{var}(L)\right]^{(4,U,N)} \triangleq \left[\operatorname{var}(L)\right]^{(3,U,N)} + \frac{1}{2}\sum_{i=1}^{TP}\left[\frac{\partial^4 L(\boldsymbol{\alpha})}{(\partial \alpha_i)^4}\frac{\partial^2 L(\boldsymbol{\alpha})}{(\partial \alpha_i)^2}\right]\sigma_i^6; \\ SD^{(4)} \triangleq \sqrt{\left[\operatorname{var}(L)\right]^{(4,U,N)}}; \tag{5.7}$$

$$E^{(4)}(L^c)=L(\boldsymbol{\alpha}^0)+\frac{1}{2}\sum_{i=1}^{TP}\frac{\partial^2 L(\boldsymbol{\alpha})}{(\partial \alpha_i)^2}\sigma_i^2+\frac{1}{8}\sum_{i=1}^{TP}\frac{\partial^4 L(\boldsymbol{\alpha})}{(\partial \alpha_i)^4}\sigma_i^4. \tag{5.8}$$

Using the index $i=1,2,3,4,$ to denote the highest order of the retained sensitivities, the expression of the predicted best-estimate variance of the predicted best-estimate leakage response, which will be denoted as $SD^{(be,i)}$, will have the following expression:

$$Var^{(be,i)} = \left[SD^{(e)}\right]^2 - \left[SD^{(e)}\right]^2 \left\{\left[SD^{(e)}\right]^2 + \left[\text{var}(L)\right]^{(i,U,N)}\right\}^{-1} \left[SD^{(e)}\right]^2,$$
$$SD^{(be,i)} \triangleq \sqrt{Var^{(be,i)}}, \quad i = 1,2,3,4. \tag{5.9}$$

5.2.1 Low Precision (Standard Deviation=10%) Measurement of the Leakage Response

This Subsection presents the results obtained by applying the 4th-BERRU-PM methodology to the PERP benchmark when several relatively imprecise experimental measurements of the benchmark's leakage response have been performed, having a mean value $L^e = 1.94 \times 10^6$ neutrons/sec, and a relatively large standard deviation $SD^{(e)} = 10\%$. These values were chosen in order to examine the effects of applying the 4th-BERRU-PM methodology to a case when the mean value of the measurements is about "one experimental standard deviation" away from the computed values of the leakage response (since $L^e = L^c \times 110\%$).

Three cases will be considered regarding the precision of the model parameters, namely:

1. Parameters having uniform standard deviations of 2%, which are expected to provide physically meaningful results, within the radius of convergence of the Taylor series expansion of the leakage response with respect to the PERP total cross-sections (parameters);
2. Parameters having uniform standard deviations of 5%, which are expected to illustrate the results still obtainable not far outside the radius of convergence of the Taylor series expansion of the leakage response;
3. Parameters having uniform standard deviations of 10%, which are expected to illustrate the results still obtainable outside the radius of convergence of the Taylor series expansion of the leakage response.

The effects of the order of the sensitivities retained in the expressions of the best-estimate predicted mean value and standard deviation of the leakage response will also investigated, as follows: (i) $i = 1$: only the 1st-order sensitivities are considered; (ii) $i = 2$: the 1st- and 2nd-order sensitivities are included; (iii) $i = 3$: all sensitivities up to and including 3rd-order are included. As has been seen in Chapter 2, the inclusion of the 4th-order sensitivities does not change the conclusions that can be drawn after having included just the 1st-, 2nd-, and 3rd-order sensitivities; hence the 4th-order sensitivities are not included in the numerical results presented in this section.

5.2.1.1 "High Precision" Parameters (Relative Standard Deviations *SD* = 2%)

The results of applying Eqs. (5.1)–(5.9) to the PERP benchmark, in conjunction with the experimental information previously mentioned (i.e., $L^e = 1.94 \times 10^6$ with $SD^{(e)} = 10\%$), are illustrated in Figures 5.1–5.3, when considering that the

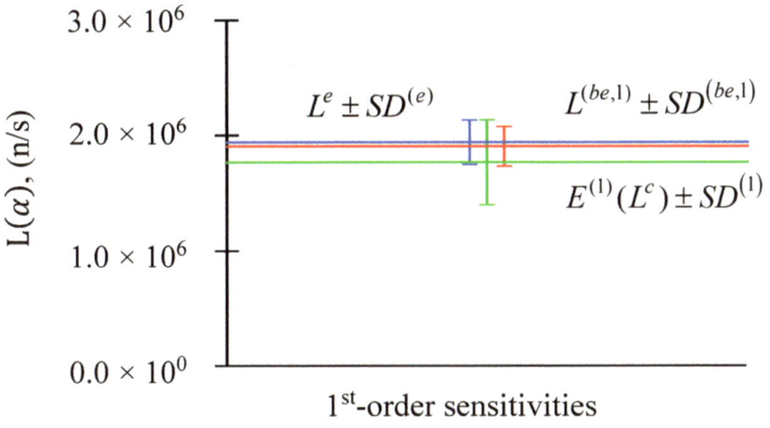

FIGURE 5.1 Parameter relative standard deviations $SD = 2\%$: (i) experimentally measured response and its standard deviation $L^e \pm SD^{(e)}$ (blue); (ii) expected value of computed response and its standard deviation $E^{(1)}(L^c) \pm SD^{(1)}$ (green) including only 1st-order sensitivities; (iii) best-estimate predicted leakage response and its predicted best-estimate standard deviation $L^{(be,1)} \pm SD^{(be,1)}$ (red) including only 1st-order sensitivities.

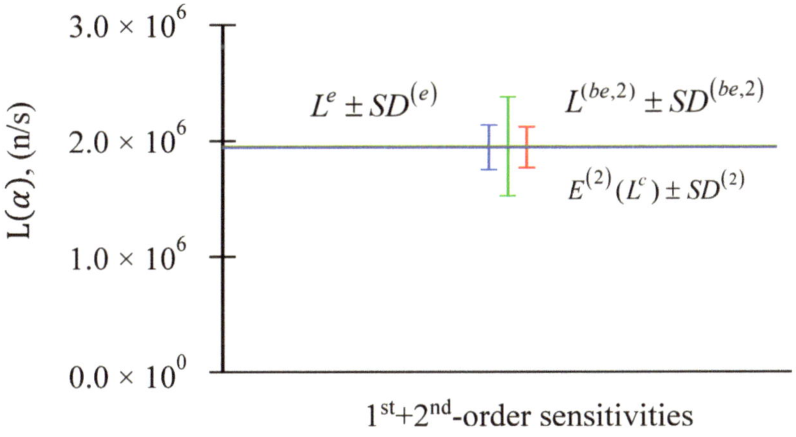

FIGURE 5.2 Parameter relative standard deviations $SD = 2\%$: (i) experimentally measured response and its standard deviation $L^e \pm SD^{(e)}$ (blue); (ii) expected value of computed response and its standard deviation $E^{(2)}(L^c) \pm SD^{(2)}$ (green) including 1st- and 2nd-order sensitivities; (iii) best-estimate predicted leakage response and its predicted best-estimate standard deviation $L^{(be,2)} \pm SD^{(be,2)}$ (red) including 1st- and 2nd-order sensitivities.

parameters have been measured precisely, having uniform relative standard deviations of 2%. The results presented in Figure 5.1 include only the 1st-order sensitivities of the leakage response with respect to the model parameters (total microscopic group cross-sections). The results presented in Figure 5.2 include the 1st- and 2nd-order sensitivities. The results depicted in Figure 5.3 include 1st- through

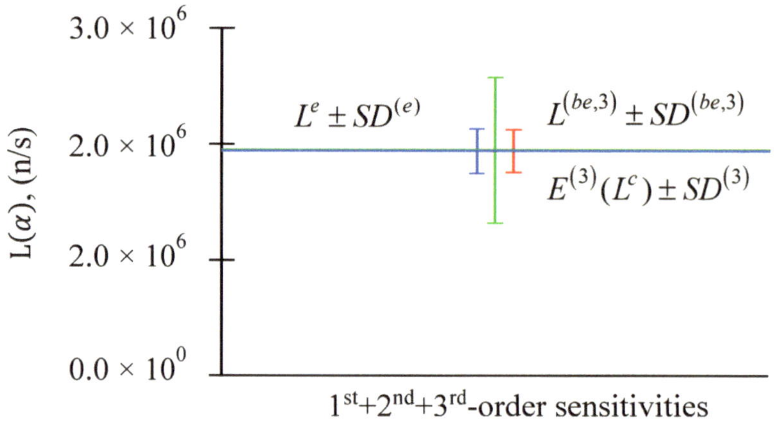

FIGURE 5.3 Parameter relative standard deviations $SD = 2\%$: (i) experimentally measured response and its standard deviation $L^e \pm SD^{(e)}$ (blue); (ii) expected value of computed response and its standard deviation $E^{(3)}(L^c) \pm SD^{(3)}$ (green) including 1st-, 2nd-, and 3rd-order sensitivities; (iii) best-estimate predicted leakage response and its predicted best-estimate standard deviation $L^{(be,3)} \pm SD^{(be,3)}$ (red) including 1st-, 2nd-, and 3rd-order sensitivities.

TABLE 5.1

Results for $E^{(i)}(L^c) \pm SD^{(i)}$ and $L^{(be,i)} \pm SD^{(be,i)}$, $i = 1,2,3$; Response Measurement $L^e = 1.94 \times 10^6$ N/S with $SD^{(e)} = 10\%$; Parameters $SD = 2\%$

Order of Sensitivities Considered	Values of $L^e \pm SD^{(e)}$ (Neutrons/sec)	Values of $E^{(i)}(L^c) \pm SD^{(i)}$ (Neutrons/sec)	Values of $L^{(be,i)} \pm SD^{(be,i)}$ (Neutrons/sec)
1st-order ($i = 1$)	$1.941 \times 10^6 \pm 1.941 \times 10^5$	$1.765 \times 10^6 \pm 3.698 \times 10^5$	$1.903 \times 10^6 \pm 1.719 \times 10^5$
1st+2nd-order ($i = 2$)		$1.949 \times 10^6 \pm 4.276 \times 10^5$	$1.943 \times 10^6 \pm 1.768 \times 10^5$
1st+2nd+3rd-order ($i = 3$)		$1.949 \times 10^6 \pm 6.288 \times 10^5$	$1.942 \times 10^6 \pm 1.855 \times 10^5$

3rd-order sensitivities. The numerical values of the results depicted in Figures 5.1–5.3 are tabulated in Table 5.1.

The results presented in Figures 5.1–5.3 and Table 5.1 indicate that the application of the BERRU-PM methodology yields a best-estimate value for the predicted leakage response which is in between the experimentally measured and computed values of this response (but closer to the more precisely known experimental value), with an accompanying predicted standard deviation which is smaller than either the measured or the computed standard deviations. Since the standard deviations (of 2%) for the model parameters are within the radius of convergence of the Taylor series that represents the computed response in terms of parameter deviations, the contributions stemming from the higher-order sensitivities are expected to diminish with increasing order. This expectation is confirmed by the results shown in Figures 5.1–5.3 and Table 5.1, which indicate that the impact of the 2nd- and 3rd-order sensitivities on the best-estimate predicted leakage response and accompanying predicted standard deviation are small.

5.2.1.2 "Typical Precision" Parameters (Relative Standard Deviations $SD = 5\%$)

Figures 5.4–5.6 depict the results of applying Eqs. (5.1)–(5.9) to the PERP benchmark, in conjunction with the same experimental information as mentioned previously (i.e., $L^e = 1.94 \times 10^6$ with $SD^{(e)} = 10\%$), but considering that the parameters have been measured with a precision which is typical of cross-section measurements, having uniform relative standard deviations of 5%. The results presented in Figure 5.4 include only the 1st-order sensitivities of the leakage response with respect to the model parameters (total microscopic group cross-sections). The results presented in Figure 5.5 include the 1st- and 2nd-order sensitivities. The results depicted in Figure 5.6 include 1st- through 3rd-order sensitivities. The numerical values of the results depicted in Figures 5.4–5.6 are tabulated in Table 5.2.

As indicated by the results shown in Figures 5.4–5.6 and Table 5.2, the contributions stemming from the higher-order sensitivities to the quantities $E^{(i)}(L^c) \pm SD^{(i)}$ increase with increasing order of sensitivities retained, thus confirming the expectation that the parameter standard deviations are sufficiently large to cause divergence of the Taylor series that are used to represent the expected value and the standard deviation of the computed response in terms of parameter standard deviations.

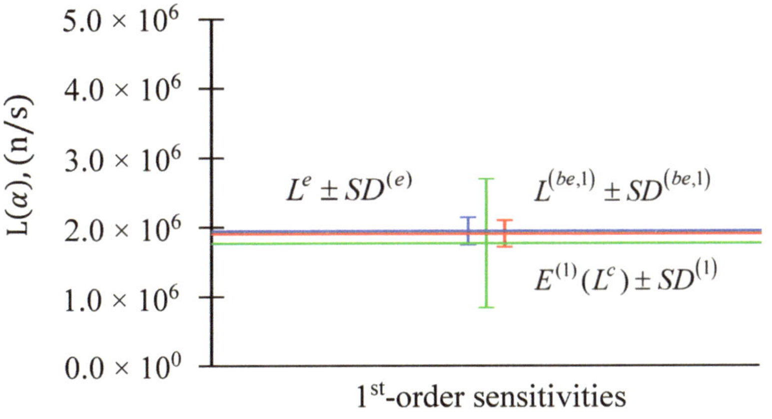

FIGURE 5.4 Parameter relative standard deviations $SD = 5\%$: (i) experimentally measured response and its standard deviation $L^e \pm SD^{(e)}$ (blue); (ii) expected value of computed response and its standard deviation $E^{(1)}(L^c) \pm SD^{(1)}$ (green) including only 1st-order sensitivities; (iii) best-estimate predicted leakage response and its predicted best-estimate standard deviation $L^{(be,1)} \pm SD^{(be,1)}$ (red) including only 1st-order sensitivities.

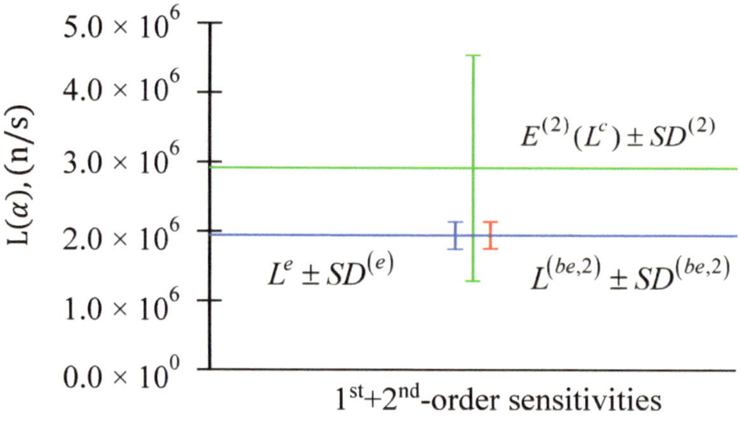

FIGURE 5.5 Parameter relative standard deviations $SD = 5\%$: (i) experimentally measured response and its standard deviation $L^e \pm SD^{(e)}$ (blue); (ii) expected value of computed response and its standard deviation $E^{(2)}(L^c) \pm SD^{(2)}$ (green) including 1st- and 2nd-order sensitivities; (iii) best-estimate predicted leakage response and its predicted best-estimate standard deviation $L^{(be,2)} \pm SD^{(be,2)}$ (red) including 1st- and 2nd-order sensitivities.

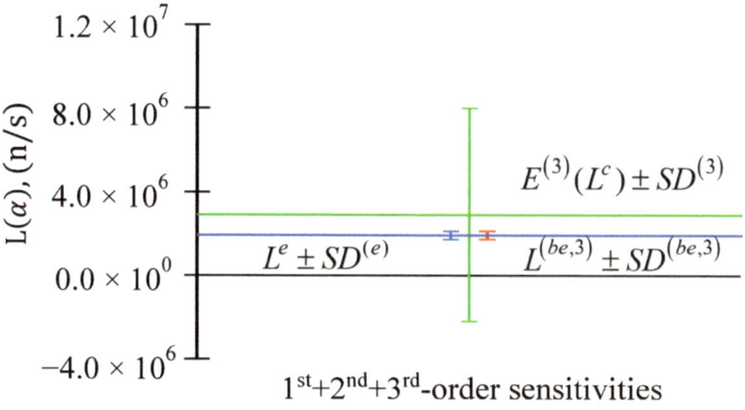

FIGURE 5.6 Parameter relative standard deviations $SD = 5\%$: (i) experimentally measured response and its standard deviation $L^e \pm SD^{(e)}$ (blue); (ii) expected value of computed response and its standard deviation $E^{(3)}(L^c) \pm SD^{(3)}$ (green) including 1st-, 2nd-, and 3rd-order sensitivities; (iii) best-estimate predicted leakage response and its predicted best-estimate standard deviation $L^{(be,3)} \pm SD^{(be,3)}$ (red) including 1st-, 2nd-, and 3rd-order sensitivities.

Nevertheless, the application of the BERRU-PM methodology yields a best-estimate value for the predicted leakage response which falls in between the experimentally measured and computed values of this response (but closer to the more precisely known experimental value), with an accompanying predicted standard deviation which is smaller than either the measured or the computed standard deviations.

TABLE 5.2

Results for $E^{(i)}(L^c) \pm SD^{(i)}$ and $L^{(be,i)} \pm SD^{(be,i)}$, $i = 1,2,3$; Response Measurement $L^e = 1.94 \times 10^6$ N/S with $SD^{(e)} = 10\%$; Parameters $SD = 5\%$

Order of Sensitivities Considered	Values of $L^e \pm SD^{(e)}$ (Neutrons/sec)	Values of $E^{(i)}(L^c) \pm SD^{(i)}$ (Neutrons/sec)	Values of $L^{(be,i)} \pm SD^{(be,i)}$ (Neutrons/sec)
1st-order ($i = 1$)	$1.941 \times 10^6 \pm 1.941 \times 10^5$	$1.765 \times 10^6 \pm 9.246 \times 10^5$	$1.934 \times 10^6 \pm 1.899 \times 10^5$
1st+2nd-order ($i = 2$)		$2.914 \times 10^6 \pm 1.629 \times 10^6$	$1.955 \times 10^6 \pm 1.928 \times 10^5$
1st+2nd+3rd-order ($i = 3$)		$2.914 \times 10^6 \pm 5.103 \times 10^6$	$1.943 \times 10^6 \pm 1.939 \times 10^5$

These results indicate that the application of the 4th-BERRU-PM methodology can produce physically useful/meaningful results even when the Taylor series expansion of the response in terms of parameter standard deviations diverges slightly.

5.2.1.3 "Low Precision" Parameters (Standard Deviations $SD = 10\%$)

Figures 5.7–5.9 depict the results of applying Eqs. (5.1)–(5.9) to the PERP benchmark, in conjunction with the same experimental information as previously mentioned (i.e., $L^e = 1.94 \times 10^6$ with $SD^{(e)} = 10\%$), but considering that the parameters have been measured with low precision, having uniform relative standard deviations of 10%. The results presented in Figure 5.7 include only the 1st-order sensitivities of the leakage response with respect to the model parameters (total microscopic group

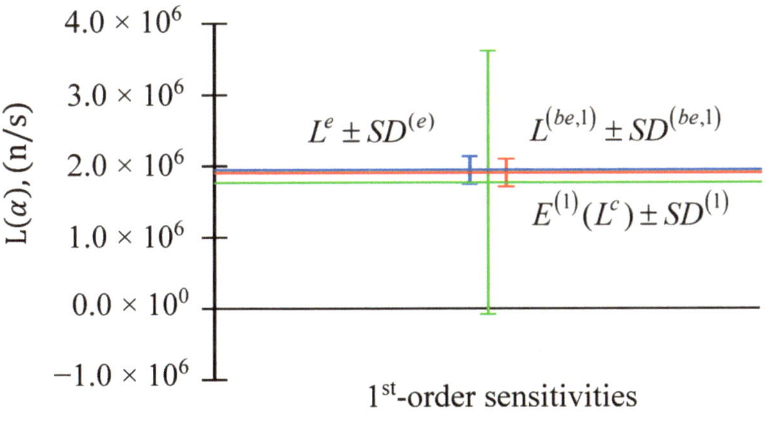

FIGURE 5.7 Comparison of $E^{(i)}(L^c) \pm SD^{(i)}$ (in green), $L^{(be,i)} \pm SD^{(be,i)}$ (in red), and $L^e \pm SD^{(e)}$ (in blue), for response measurement for response measurement $L^e = 1.94 \times 10^6$ n/s, $SD^{(e)} = 10\%$, and parameters $SD = 10\%$. Only the 1st-order ($i = 1$) sensitivities are considered.

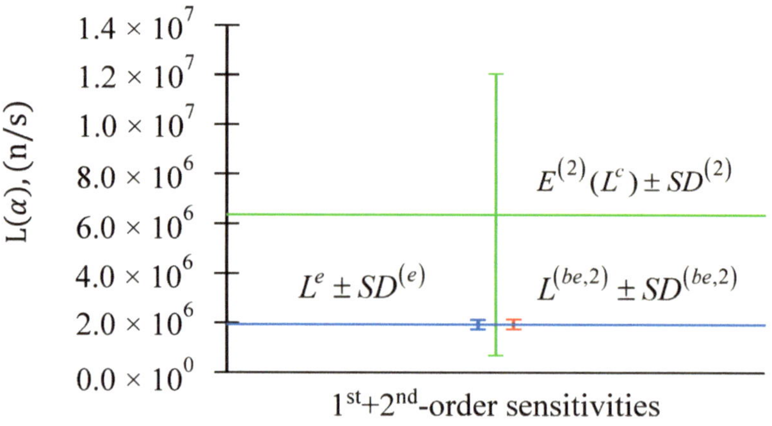

FIGURE 5.8 Comparison of $E^{(i)}(L^c) \pm SD^{(i)}$ (in green), $L^{(be,i)} \pm SD^{(be,i)}$ (in red), and $L^e \pm SD^{(e)}$ (in blue), for response measurement for response measurement $L^e = 1.94 \times 10^6$ n/s, $SD^{(e)} = 10\%$, and parameters $SD = 10\%$. The 1st+2nd-order ($i = 2$) sensitivities are included.

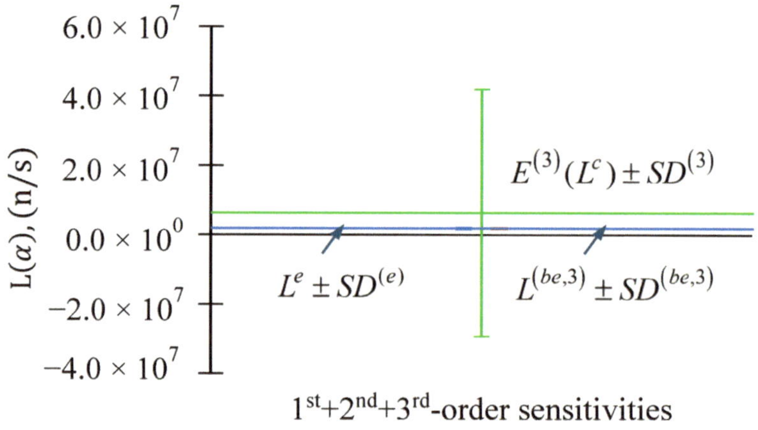

FIGURE 5.9 Comparison of $E^{(i)}(L^c) \pm SD^{(i)}$ (in green), $L^{(be,i)} \pm SD^{(be,i)}$ (in red), and $L^e \pm SD^{(e)}$ (in blue), for response measurement for response measurement $L^e = 1.94 \times 10^6$ n/s, $SD^{(e)} = 10\%$, and parameters $SD = 10\%$. The 1st+2nd+3rd-order ($i = 3$) sensitivities are included.

cross-sections). The results presented in Figure 5.8 include the 1st- and 2nd-order sensitivities. The results depicted in Figure 5.9 include 1st- through 3rd-order sensitivities. The numerical values of the results depicted in Figures 5.7–5.9 are tabulated in Table 5.3.

The results presented in Figures 5.7–5.9 and Table 5.3 confirm the expectation that parameter standard deviations of 10% cause massive divergence of the Taylor

TABLE 5.3

Results for $E^{(i)}(L^c) \pm SD^{(i)}$ and $L^{(be,i)} \pm SD^{(be,i)}$, $i = 1,2,3$; Response Measurement $L^e = 1.94 \times 10^6$ N/S with $SD^{(e)} = 10\%$; Parameters $SD = 10\%$

Order of Sensitivities Considered	Values of $L^e \pm SD^{(e)}$ (Neutrons/sec)	Values of $E^{(i)}(L^c) \pm SD^{(i)}$ (Neutrons/sec)	Values of $L^{(be,i)} \pm SD^{(be,i)}$ (Neutrons/sec)
1st-order ($i = 1$)	$1.941 \times 10^6 \pm 1.941 \times 10^5$	$1.765 \times 10^6 \pm 1.849 \times 10^6$	$1.939 \times 10^6 \pm 1.931 \times 10^5$
1st+2nd-order ($i = 2$)		$6.363 \times 10^6 \pm 5.675 \times 10^6$	$1.946 \times 10^6 \pm 1.940 \times 10^5$
1st+2nd+3rd-order ($i = 3$)		$6.363 \times 10^6 \pm 3.561 \times 10^7$	$1.941 \times 10^6 \pm 1.941 \times 10^5$

series that underlies the expressions used for computing the expected value and standard deviation of the computed response in terms of parameter standard deviations. Nevertheless, as indicated by the last column in Table 5.3, the application of the BERRU-PM methodology yields a best-estimate value and reduced predicted standard deviation for the predicted leakage response, which is close to the more precisely known experimental values. These results indicate that the application of the 4th-BERRU-PM methodology predicts results that remain physically useful/meaningful even when the parameters uncertainties (measured by the respective standard deviations) are sufficiently large to cause indisputable divergence of the Taylor series expansion of the response in terms of parameter deviations. However, since parameter standard deviations of 5% already cause the Taylor series of the response to diverge, and since microscopic cross-sections are generally known with a precision of 5% or better, only parameter standard deviations of 2% and 5% will, henceforth, be considered for the subsequent illustrative applications of the 4th-BERRU-PM methodology to the PERP benchmark.

5.2.2 Typical Precision (Standard Deviation = 5%) Measurement of the Leakage Response

The results to be presented in this Subsection will illustrate the effects of increasing the precision of the response measurement, by keeping the same mean value for the leakage response as in Subsection 5.2.1 but lowering the standard deviation of the measured response from 10% (considered in Subsection 5.2.1) to 5%, which is typical of the measurement precision of such responses.

Two cases will be considered to illustrate the impact of the model parameters' standard deviations, namely:

1. Parameters measured with "high precision," having uniform standard deviations of 2%, which are within the radius of convergence of the Taylor series of the leakage response; and

2. Parameters measured with "typical precision," having uniform standard deviations of 5%, which will illustrate predictions still obtainable not far outside the radius of convergence of the Taylor series of the leakage response.

5.2.2.1 "High Precision" Parameters (Relative Standard Deviations $SD = 2\%$)

The results of applying Eqs. (5.1)–(5.9) to the PERP benchmark, in conjunction with the experimental information mentioned previously (i.e., $L^e = 1.94 \times 10^6$ with $SD^{(e)} = 5\%$) are tabulated in Table 5.4 and illustrated in Figures 5.10–12, considering that the parameters have been measured with "high" precision, having uniform relative standard deviations of 2%. The results presented in Figure 5.10 include only the 1st-order sensitivities of the leakage response with respect to the model parameters (total microscopic group cross-sections). The results presented in Figure 5.11 include the 1st- and 2nd-order sensitivities. The results depicted in Figure 5.12 include 1st- through 3rd-order sensitivities.

Since the standard deviations (of 2%) for the model parameters are within the radius of convergence of the Taylor series that represents the computed response in terms of parameter deviations, the contributions stemming from the higher-order sensitivities are expected to diminish with increasing order. This expectation is confirmed by the results shown in Figures 5.10–5.12 and Table 5.4, which indicate that the impact of the 2nd- and higher-order sensitivities on the best-estimate predicted leakage response and accompanying predicted standard deviation are small. Also as expected, the results presented in Figures 5.10–5.12 and Table 5.4 indicate that

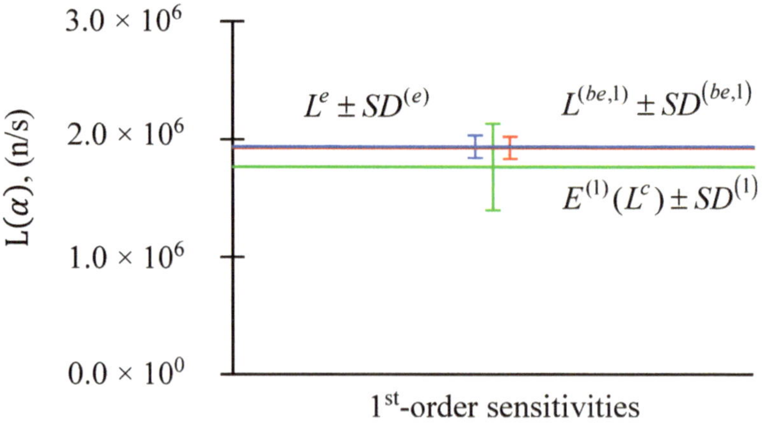

FIGURE 5.10 Parameter relative standard deviations $SD = 2\%$: (i) experimentally measured response and its standard deviation $L^e \pm SD^{(e)}$ (blue); (ii) expected value of computed response and its standard deviation $E^{(1)}(L^c) \pm SD^{(1)}$ (green) including only 1st-order sensitivities; (iii) best-estimate predicted leakage response and its predicted best-estimate standard deviation $L^{(be,1)} \pm SD^{(be,1)}$ (red) including only 1st-order sensitivities.

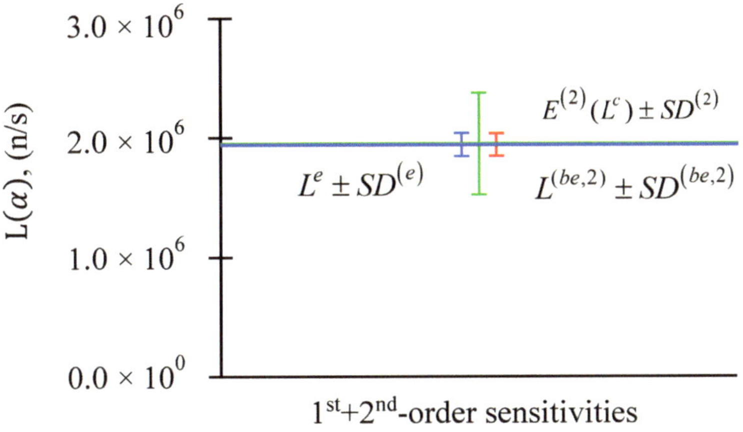

FIGURE 5.11 Parameter relative standard deviations $SD = 2\%$: (i) experimentally measured response and its standard deviation $L^e \pm SD^{(e)}$ (blue); (ii) expected value of computed response and its standard deviation $E^{(2)}(L^c) \pm SD^{(2)}$ (green) including 1st- and 2nd-order sensitivities; (iii) best-estimate predicted leakage response and its predicted best-estimate standard deviation $L^{(be,2)} \pm SD^{(be,2)}$ (red) including 1st- and 2nd-order sensitivities.

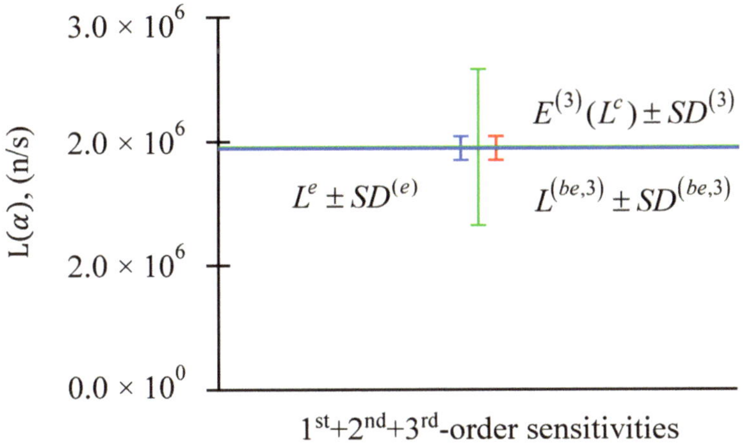

FIGURE 5.12 Parameter relative standard deviations $SD = 2\%$: (i) experimentally measured response and its standard deviation $L^e \pm SD^{(e)}$ (blue); (ii) expected value of computed response and its standard deviation $E^{(3)}(L^c) \pm SD^{(3)}$ (green) including 1st-, 2nd-, and 3rd-order sensitivities; (iii) best-estimate predicted leakage response and its predicted best-estimate standard deviation $L^{(be,3)} \pm SD^{(be,3)}$ (red) including 1st-, 2nd-, and 3rd-order sensitivities.

the application of the BERRU-PM methodology yields a best-estimate value for the predicted leakage response which is in between the experimentally measured and computed values of this response but closer to the more precisely known experimental value.

TABLE 5.4
Results for $E^{(i)}(L^c) \pm SD^{(i)}$ and $L^{(be,i)} \pm SD^{(be,i)}$, $i = 1,\ldots,4$, for Response Measurement $L^e = 1.941\times10^6$ N/S, $SD^{(e)} = 5\%$, and Parameters $SD = 2\%$

Order of Sensitivities Considered	Values of $L^e \pm SD^{(e)}$ (Neutrons/sec)	Values of $E^{(i)}(L^c) \pm SD^{(i)}$ (Neutrons/sec)	Values of $L^{(be,i)} \pm SD^{(be,i)}$ (Neutrons/sec)
1st-order ($i=1$)	$1.941 \times 10^6 \pm 9.706 \times 10^4$	$1.765 \times 10^6 \pm 3.698 \times 10^5$	$1.930 \times 10^6 \pm 9.388 \times 10^4$
1st+2nd-order ($i=2$)		$1.949 \times 10^6 \pm 4.276 \times 10^5$	$1.942 \times 10^6 \pm 9.466 \times 10^4$
1st+2nd+3rd-order ($i=3$)		$1.949 \times 10^6 \pm 6.288 \times 10^5$	$1.942 \times 10^6 \pm 9.593 \times 10^4$

5.2.2.2 "Typical Precision" Parameters (Relative Standard Deviations $SD = 5\%$)

The numerical results tabulated in Table 5.5 and depicted in Figures 5.13–5.15 have been obtained by applying Eqs. (5.1)–(5.9) to the PERP benchmark, in conjunction with the same experimental information as previously mentioned (i.e., $L^e = 1.94\times10^6$ with $SD^{(e)} = 5\%$), but considering that the parameters have been measured with a precision which is typical of cross-section measurements, having uniform relative standard deviations of 5%. The results presented in Figure 5.13 include only the 1st-order sensitivities of the leakage response with respect to the

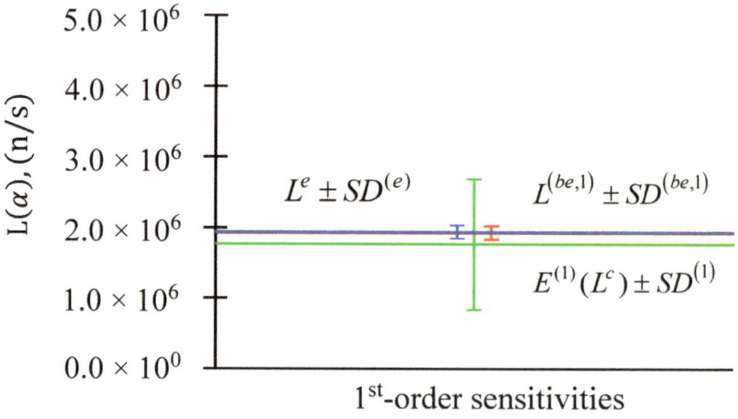

FIGURE 5.13 Parameter relative standard deviations $SD = 5\%$: (i) experimentally measured response and its standard deviation $L^e \pm SD^{(e)}$ (blue); (ii) expected value of computed response and its standard deviation $E^{(1)}(L^c) \pm SD^{(1)}$ (green) including only 1st-order sensitivities; (iii) best-estimate predicted leakage response and its predicted best-estimate standard deviation $L^{(be,1)} \pm SD^{(be,1)}$ (red) including only 1st-order sensitivities.

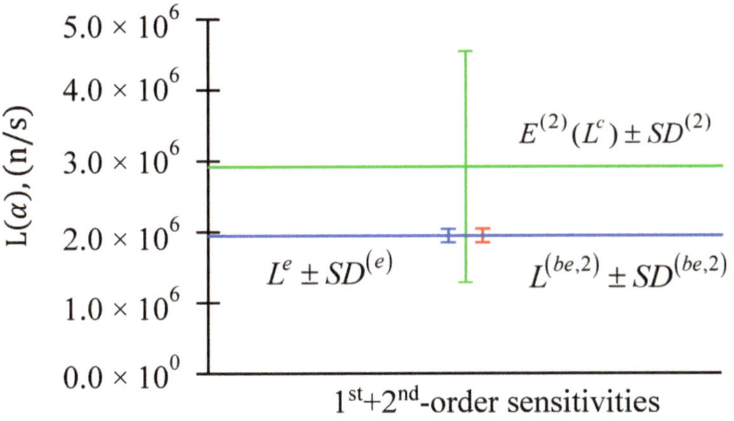

FIGURE 5.14 Parameter relative standard deviations $SD = 5\%$: (i) experimentally measured response and its standard deviation $L^e \pm SD^{(e)}$ (blue); (ii) expected value of computed response and its standard deviation $E^{(2)}(L^c) \pm SD^{(2)}$ (green) including 1st- and 2nd-order sensitivities; (iii) best-estimate predicted leakage response and its predicted best-estimate standard deviation $L^{(be,2)} \pm SD^{(be,2)}$ (red) including 1st- and 2nd-order sensitivities.

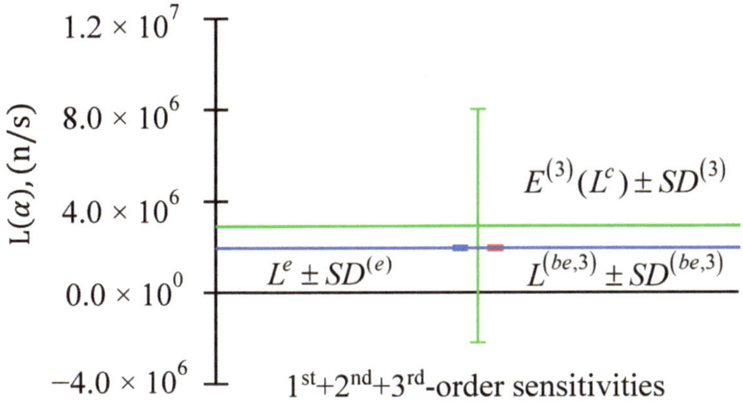

FIGURE 5.15 Parameter relative standard deviations $SD = 5\%$: (i) experimentally measured response and its standard deviation $L^e \pm SD^{(e)}$ (blue); (ii) expected value of computed response and its standard deviation $E^{(3)}(L^c) \pm SD^{(3)}$ (green) including 1st-, 2nd-, and 3rd-order sensitivities; (iii) best-estimate predicted leakage response and its predicted best-estimate standard deviation $L^{(be,3)} \pm SD^{(be,3)}$ (red) including 1st-, 2nd-, and 3rd-order sensitivities.

model parameters (total microscopic group cross-sections). The results presented in Figure 5.14 include the 1st- and 2nd-order sensitivities. The results depicted in Figure 5.15 include 1st- through 3rd-order sensitivities.

As indicated by the results presented in Table 5.5 and depicted in Figures 5.13–5.15, the contributions stemming from the higher-order sensitivities to the quantities

TABLE 5.5
Results for $E^{(i)}(L^c) \pm SD^{(i)}$ and $L^{(be,i)} \pm SD^{(be,i)}$, $i = 1,2,3$; Response Measurement $L^e = 1.941 \times 10^6$ N/S with $SD^{(e)} = 5\%$; Parameters $SD = 5\%$

Order of Sensitivities Considered	Values of $L^e \pm SD^{(e)}$ (Neutrons/sec)	Values of $E^{(i)}(L^c) \pm SD^{(i)}$ (Neutrons/sec)	Values of $L^{(be,i)} \pm SD^{(be,i)}$ (Neutrons/sec)
1st-order ($i=1$)	$1.941 \times 10^6 \pm 9.706 \times 10^4$	$1.765 \times 10^6 \pm 9.246 \times 10^5$	$1.939 \times 10^6 \pm 9.653 \times 10^4$
1st+2nd-order ($i=2$)		$2.914 \times 10^6 \pm 1.629 \times 10^6$	$1.945 \times 10^6 \pm 9.689 \times 10^4$
1st+2nd+3rd-order ($i=3$)		$2.914 \times 10^6 \pm 5.103 \times 10^6$	$1.942 \times 10^6 \pm 9.705 \times 10^4$

$E^{(i)}(L^c) \pm SD^{(i)}$ increase with increasing order of sensitivities retained, thus confirming the expectation that the parameter standard deviations are sufficiently large to cause divergence of the Taylor series that is used to represent the expected value and the standard deviation of the computed response in terms of parameter standard deviations. Nevertheless, the application of the BERRU-PM methodology yields a best-estimate value for the predicted leakage response which falls in between the experimentally measured and computed values of this response (but closer to the more precisely known experimental value), with an accompanying predicted standard deviation which is smaller than either the measured or the computed standard deviations. These results indicate that the application of the 4th-BERRU-PM methodology can produce physically useful/meaningful results even when the Taylor series expansion of the response in terms of parameter standard deviations diverges slightly.

The only difference between the results presented in Table 5.1 and those presented in Table 6.4 is the precision of the measured response: the results presented in Table 5.1 are for response measurement standard deviation SD = 10%, whereas the results presented in Table 5.1 are for a more precise response measurement, with standard deviation SD = 5%. Comparing the corresponding results presented in the rightmost columns of these tables reveals that the predicted best-estimate mean values are quite close to the respective measured values, while the predicted standard deviations are clearly smaller than the respective computed values and are smaller by ca. 1% than the corresponding measured values. The same conclusion follows when comparing the corresponding results presented in the rightmost columns of Tables 5.2 and 5.5, respectively, for parameter standard deviations of 5%.

5.3 4TH-BERRU-PM BEST-ESTIMATE POSTERIOR MEAN VALUES OF CALIBRATED MODEL PARAMETERS

The best-estimate 4th-order expression of the mean values of the predicted calibrated model parameters, denoted as $\boldsymbol{\alpha}^{be} \triangleq \left(\alpha_1^{be}, \ldots, \alpha_{TP}^{be}\right)^\dagger$, has the following expression:

$$\boldsymbol{\alpha}^{be} \triangleq \int_D \boldsymbol{\alpha} p_4^{be}(\mathbf{r},\boldsymbol{\alpha})\, d\boldsymbol{\alpha}\, d\mathbf{r} = \boldsymbol{\alpha}_s^{(0)} + \mathbf{p}_2\left(\mathbf{r}_s^{(0)},\boldsymbol{\alpha}_s^{(0)}\right), \tag{5.10}$$

where

$$\begin{aligned} \mathbf{p}_2\left(\mathbf{r}_s^{(0)},\boldsymbol{\alpha}_s^{(0)}\right) &\triangleq \mathbf{C}_{\alpha r}^{c}\left[\mathbf{I}_r - \left(\mathbf{C}^e + \mathbf{C}_{rr}^{c}\right)^{-1}\mathbf{C}_{rr}^{c}\right]\mathbf{v}\left(\mathbf{r}_s^{(0)}\right) \\ &+ \left[\mathbf{C}_{\alpha\alpha} - \mathbf{C}_{\alpha r}^{c}\left(\mathbf{C}^e + \mathbf{C}_{rr}^{c}\right)^{-1}\mathbf{C}_{r\alpha}^{c}\right]\mathbf{w}\left(\boldsymbol{\alpha}_s^{(0)}\right). \end{aligned} \tag{5.11}$$

For the particular information available for the PERP reactor physics benchmark, the higher-order terms represented by the quantity $\mathbf{p}_2\left(\mathbf{r}_s^{(0)},\boldsymbol{\alpha}_s^{(0)}\right)$ are expected to be negligible by comparison to the leading term. Therefore, the expression of the best-estimate mean values of the predicted calibrated model parameters which will be denoted as $\boldsymbol{\alpha}^{(be,i)} \triangleq \left(\alpha_1^{(be,i)},...,\alpha_{TP}^{(be,i)}\right)^{\dagger}$, will reduce to the following form:

$$\begin{aligned} &\boldsymbol{\alpha}^{(be,i)} = \boldsymbol{\alpha}^{0} - \mathbf{C}_{\alpha r}^{(c,i)}\left\{\left[SD^{(e)}\right]^2 + \left[\mathrm{var}(L)\right]^{(i,U,N)}\right\}^{-1}\left[E^{(i)}(L^c) - L^e\right], \\ &i = 1,2,3,4; \end{aligned} \tag{5.12}$$

where the index $i = 1,2,3,4,$ indicates (as before) the highest-order of retained sensitivities.

All of the numerical results presented in this section for the best-estimate predicted calibrated parameter values are in units of "barn", where 1 barn = 10^{-24} cm^2.

5.3.1 "High" Precision Parameters (Relative Standard Deviations $SD = 2\%$)

Tables 5.6 and 5.7 present the results obtained using Eq. (5.12) for the best-estimate predicted values for the top ten calibrated model parameters (ranked in the order of the magnitude of their 1st-order relative sensitivities). The results presented in Tables 5.6 and 5.7 are for a nominal value of the measured response is $L^e = 1.94 \times 10^6$ n/s, and for parameters that are measured with "high" precision, having uniform relative standard deviations $SD = 2\%$. The difference between the results presented in these Tables stems solely from the assumed precision of the measured response: the results in Table 5.6 are for a "low" precision measured response, having a measured relative standard deviation $SD^{(e)} = 10\%$, while the results in Table 5.7 are for a measured response having a "typical" precision with a relative standard deviation $SD^{(e)} = 5\%$.

The results presented in Tables 5.6 and 5.7 indicate that the calibration (i.e., adjustment) of each of the respective parameter's nominal value is negligible for the precisely measured (2%) parameters; the precision of the measured response is inconsequential, as is the incorporation of sensitivities higher than 2nd-order.

TABLE 5.6

Best-Estimate Parameter Values $\alpha^{(be,i)}$, $i = 1,2,3,4$; Measured Response: $L^e = 1.94 \times 10^6$ N/S, with $SD^{(e)} = 10\%$; Parameter $SD = 2\%$

Ranking	Parameter $\sigma_{t,i}^{g}$	Nominal value α^0	$\alpha^{(be,1)}$	$\alpha^{(be,2)}$	$\alpha^{(be,3)}$	$\alpha^{(be,4)}$
1	$\sigma_{t,i=6}^{g=30}$	29.542	29.346	29.550	29.549	29.604
2	$\sigma_{t,i=1}^{g=12}$	6.971	6.965	6.972	6.971	6.973
3	$\sigma_{t,i=6}^{g=17}$	16.182	16.169	16.183	16.183	16.185
4	$\sigma_{t,i=6}^{g=16}$	12.279	12.268	12.279	12.279	12.281
5	$\sigma_{t,i=1}^{g=13}$	7.928	7.921	7.928	7.928	7.929
6	$\sigma_{t,i=6}^{g=18}$	18.571	18.556	18.571	18.571	18.574
7	$\sigma_{t,i=6}^{g=19}$	19.697	19.681	19.697	19.697	19.700
8	$\sigma_{t,i=6}^{g=20}$	20.158	20.143	20.158	20.158	20.160
9	$\sigma_{t,i=6}^{g=21}$	20.334	20.320	20.335	20.334	20.337
10	$\sigma_{t,i=1}^{g=14}$	9.186	9.179	9.186	9.186	9.187

TABLE 5.7

Best-Estimate Parameter Values $\alpha^{(be,i)}$, $i = 1,2,3,4$; Measured Response: $L^e = 1.94 \times 10^6$ N/S, with $SD^{(e)} = 5\%$; Parameter $SD = 2\%$

Ranking	Parameter $\sigma_{t,i}^{g}$	Nominal value α^0	$\alpha^{(be,1)}$	$\alpha^{(be,2)}$	$\alpha^{(be,3)}$	$\alpha^{(be,4)}$
1	$\sigma_{t,i=6}^{g=30}$	29.542	29.308	29.551	29.549	29.607
2	$\sigma_{t,i=1}^{g=12}$	6.971	6.963	6.972	6.971	6.973
3	$\sigma_{t,i=6}^{g=17}$	16.182	16.166	16.183	16.183	16.185
4	$\sigma_{t,i=6}^{g=16}$	12.279	12.266	12.279	12.279	12.281
5	$\sigma_{t,i=1}^{g=13}$	7.928	7.920	7.928	7.928	7.929
6	$\sigma_{t,i=6}^{g=18}$	18.571	18.553	18.571	18.571	18.574
7	$\sigma_{t,i=6}^{g=19}$	19.697	19.678	19.697	19.697	19.700
8	$\sigma_{t,i=6}^{g=20}$	20.158	20.140	20.158	20.158	20.160
9	$\sigma_{t,i=6}^{g=21}$	20.334	20.317	20.335	20.334	20.337
10	$\sigma_{t,i=1}^{g=14}$	9.186	9.178	9.186	9.186	9.187

5.3.2 "Typical" Precision Parameters (Relative Standard Deviations $SD = 5\%$)

Tables 5.8 and 5.9 present the results obtained using Eq. (5.12) for the best-estimate predicted values for the top ten calibrated model parameters (ranked in the order of the magnitude of their 1st-order relative sensitivities), for a nominal value of the measured response is $L^e = 1.94 \times 10^6$ n/s, and for parameters that are measured with "typical" precision, having uniform relative standard deviations $SD = 5\%$. The difference between the results presented in these tables stems solely from the assumed precision of the measured response: the results in Table 5.8 are for a "low" precision measured response, having a measured relative standard deviation $SD^{(e)} = 10\%$, while the results in Table 5.9 are for a measured response having a "typical" precision with a relative standard deviation $SD^{(e)} = 5\%$. The results presented in Tables 5.8 and 5.9 indicate that the calibration (i.e., adjustment) of each of the respective parameter's nominal value is minimal for the parameters having "typical" relative standard deviations of 5%. As was the case for the results presented in Tables 5.6 and 5.7, the precision of the measured response and the incorporation of sensitivities higher than 2nd-order are inconsequential for the results presented in Tables 5.8 and 5.9.

TABLE 5.8

Best-Estimate Parameter Values $\alpha^{(be,i)}$, $i = 1,2,3,4$; Measured Response: $L^e = 1.94 \times 10^6$ N/S, with $SD^{(e)} = 10\%$; Parameter $SD = 5\%$

Ranking	Parameter $\sigma_{t,i}^{g}$	Nominal value α^0	$\alpha^{(be,1)}$	$\alpha^{(be,2)}$	$\alpha^{(be,3)}$	$\alpha^{(be,4)}$
1	$\sigma_{t,i=6}^{g=30}$	29.542	29.302	29.985	29.770	30.070
2	$\sigma_{t,i=1}^{g=12}$	6.971	6.963	6.986	6.973	6.975
3	$\sigma_{t,i=6}^{g=17}$	16.182	16.166	16.213	16.186	16.190
4	$\sigma_{t,i=6}^{g=16}$	12.279	12.266	12.301	12.281	12.284
5	$\sigma_{t,i=1}^{g=13}$	7.928	7.920	7.942	7.929	7.931
6	$\sigma_{t,i=6}^{g=18}$	18.571	18.552	18.604	18.574	18.579
7	$\sigma_{t,i=6}^{g=19}$	19.697	19.678	19.731	19.700	19.705
8	$\sigma_{t,i=6}^{g=20}$	20.158	20.139	20.191	20.161	20.166
9	$\sigma_{t,i=6}^{g=21}$	20.334	20.317	20.366	20.338	20.342
10	$\sigma_{t,i=1}^{g=14}$	9.186	9.178	9.199	9.187	9.189

TABLE 5.9

Best-Estimate Parameter Values $\alpha^{(be,i)}$, $i = 1,2,3,4$; Measured Response: $L^e = 1.94\times10^6$ N/S, with $SD^{(e)} = 5\%$; Parameter $SD = 5\%$

Ranking	Parameter $\sigma^g_{t,i}$	Nominal value α^0	$\alpha^{(be,1)}$	$\alpha^{(be,2)}$	$\alpha^{(be,3)}$	$\alpha^{(be,4)}$
1	$\sigma^{g=30}_{t,i=6}$	29.542	29.294	29.989	29.770	30.070
2	$\sigma^{g=12}_{t,i=1}$	6.971	6.963	6.986	6.973	6.975
3	$\sigma^{g=17}_{t,i=6}$	16.182	16.165	16.213	16.186	16.190
4	$\sigma^{g=16}_{t,i=6}$	12.279	12.266	12.302	12.281	12.284
5	$\sigma^{g=13}_{t,i=1}$	7.928	7.919	7.942	7.929	7.931
6	$\sigma^{g=18}_{t,i=6}$	18.571	18.552	18.605	18.574	18.579
7	$\sigma^{g=19}_{t,i=6}$	19.697	19.677	19.731	19.700	19.705
8	$\sigma^{g=20}_{t,i=6}$	20.158	20.139	20.191	20.161	20.166
9	$\sigma^{g=21}_{t,i=6}$	20.334	20.316	20.366	20.338	20.342
10	$\sigma^{g=14}_{t,i=1}$	9.186	9.178	9.200	9.187	9.189

5.4 4TH-BERRU-PM BEST-ESTIMATE POSTERIOR CORRELATIONS BETWEEN PREDICTED RESPONSES AND CALIBRATED MODEL PARAMETERS

The predicted best-estimate correlation matrix of the predicted best-estimate responses and calibrated model parameters, denoted as $\mathbf{C}^{be}_{\alpha r}$, is defined as follows:

$$\begin{aligned}\mathbf{C}^{be}_{\alpha r} &\triangleq \int_D \left(\boldsymbol{\alpha}-\boldsymbol{\alpha}^{be}\right)\left(\mathbf{r}-\mathbf{r}^{be}\right)^{\dagger} p_4^{be}\left(\mathbf{r},\boldsymbol{\alpha}\right) d\boldsymbol{\alpha}\, d\mathbf{r} \\ &\cong \mathbf{C}^{e}_{\alpha r} - \mathbf{C}^{c}_{\alpha r}\left(\mathbf{C}^{e}_{rr}+\mathbf{C}^{c}_{rr}\right)^{-1}\mathbf{C}^{e}_{rr} + HOT,\end{aligned} \tag{5.13}$$

where $\mathbf{C}^{e}_{\alpha r} \triangleq \int \left(\boldsymbol{\alpha}-\boldsymbol{\alpha}^{0}\right)\left(\mathbf{r}-\mathbf{r}^{e}\right)^{\dagger} p_4^{be}\left(\mathbf{r},\boldsymbol{\alpha}\right) d\boldsymbol{\alpha}\, d\mathbf{r}$ denotes the correlation matrix between the model parameters and the computed responses, and where "*HOT*" denotes "higher-order terms" involving products of triple and quadruple correlations, which are expected to be negligible by comparison to the leading term. For the particular information available for the PERP reactor physics benchmark, the model parameters are not correlated to the experimentally measured response so that $\mathbf{C}^{e}_{\alpha r} \equiv \mathbf{0}$. Furthermore, the higher-order terms are expected to be negligible by comparison to the leading term, so that $\mathbf{C}^{(be,i)}_{\alpha r}$ becomes a $TP\times 1$-dimensional vector which will have one of the expressions provided here, differing from one another depending on the highest-order of retained sensitivities:

$$\mathbf{C}^{(be,i)}_{\alpha r} = -\mathbf{C}^{(c,i)}_{\alpha r}\left\{\left[SD^{(e)}\right]^2 + \left[\mathrm{var}\left(L\right)\right]^{(i,U,N)}\right\}^{-1}\left[SD^{(e)}\right]^2;\ j=1,\ldots,TP;\ i=1,2,3,4. \tag{5.14}$$

In Eq. (5.14), the correlation matrix $\mathbf{C}_{\alpha r}^{(c,i)} \triangleq \left[\operatorname{cor}\left(\alpha_j, L\right)\right]_{TP\times 1}$ between parameters and the computed leakage response reduces to a column vector (since there is a single response) having TP components of the following form:

$$\operatorname{cor}^{(c,i)}\left(\alpha_j, L\right) = \left\{\frac{\partial L(\boldsymbol{\alpha})}{\partial \alpha_j}\right\}_{\boldsymbol{\alpha}^0} c_j^{\alpha}; \quad j = 1,\ldots,TP; \quad i = 1,2; \tag{5.15}$$

$$\operatorname{cor}^{(c,i)}\left(\alpha_j, L\right) = \left\{\frac{\partial L(\boldsymbol{\alpha})}{\partial \alpha_j}\right\}_{\boldsymbol{\alpha}^0} c_j^{\alpha} + \frac{1}{2}\left\{\frac{\partial^3 L(\boldsymbol{\alpha})}{\left(\partial \alpha_j\right)^3}\right\}_{\boldsymbol{\alpha}^0} \left(c_j^{\alpha}\right)^2; \tag{5.16}$$

$$j = 1,\ldots,TP; \quad i = 3,4.$$

The components of the vectors $\mathbf{C}_{\alpha r}^{(be,i)}$, $i = 1,\ldots,4$ of best-estimate correlations among the benchmark's leakage response and parameters are computed using the expression provided in Eq. (5.14). The impact of the parameters' precision and of the sensitivities of various orders will be illustrated in Subsection 5.4.1 (for "high" precision parameters having uniform relative standard deviations of 2%) and, respectively, Subsection 5.4.2 (for "typical" precision parameters having uniform relative standard deviations of 5%). All of the numerical results presented in this section for the components of the vectors $\mathbf{C}_{\alpha r}^{(be,i)}$, $i = 1,\ldots,4$ are in units of "[parameter units] × [response units]", in other words, [barns × neutrons/sec], where 1 barn = 10^{-24} cm^2.

All of the results presented in this section are for a "low" precision measured response, having the nominal value $L^e = 1.94\times 10^6$ n/s, and a measured relative standard deviation $SD^{(e)} = 10\%$. Decreasing the relative standard deviation of the measured response to $SD^{(e)} = 5\%$ had no consequential numerical effects on the components of the vectors $\mathbf{C}_{\alpha r}^{(be,i)}$, $i = 1,\ldots,4$.

5.4.1 "High" Precision Parameters (Relative Standard Deviations $SD = 2\%$)

When only the 1st- and 2nd-order sensitivities are considered, the numerical results for the components of the vectors $\mathbf{C}_{\alpha r}^{(c,i)}$, $i = 1,2$, are computed using the expression provided in Eq. (5.15). The numerical results thus obtained for the components of the vector $\mathbf{C}_{\alpha r}^{(be,1)}$ and, respectively, $\mathbf{C}_{\alpha r}^{(be,2)}$, are presented in Tables 5.10 and 5.11, respectively.

Comparing the results presented in Tables 5.10 and 5.11 indicates the following conclusions:

1. Even though the original correlations $\mathbf{C}_{\alpha r}^{(c,1)} = \mathbf{C}_{\alpha r}^{(c,2)}$ between the model parameters and the computed leakage response were negative, the best-estimate correlations are positive, in other words, the components of $\mathbf{C}_{\alpha r}^{(be,1)}$ and $\mathbf{C}_{\alpha r}^{(be,2)}$ are all positive.

TABLE 5.10

Results for the Components of $\mathbf{C}_{\alpha r}^{(be,1)}$ when Only the 1st-Order Sensitivities for the Top Ten Parameters Are Considered; Parameters Relative Standard Deviations: *SD* = 2%

Ranking of parameter	1	2	3	4	5	6	7	8	9	10
Parameter $\sigma_{t,i}^{g}$	$\sigma_{t,i=6}^{g=30}$	$\sigma_{t,i=1}^{g=12}$	$\sigma_{t,i=6}^{g=17}$	$\sigma_{t,i=6}^{g=16}$	$\sigma_{t,i=1}^{g=13}$	$\sigma_{t,i=6}^{g=18}$	$\sigma_{t,i=6}^{g=19}$	$\sigma_{t,i=6}^{g=20}$	$\sigma_{t,i=6}^{g=21}$	$\sigma_{t,i=1}^{g=14}$
$\mathbf{C}_{\alpha r}^{(c,1)}$	−195330	-6496	−13396	−10092	-6458	−14959	−15207	−14692	−13912	-6176
$\mathbf{C}_{\alpha r}^{(be,1)}$	42191	1403	2894	2180	1395	3231	3285	3173	3005	1334

TABLE 5.11

Results for the Components of $\mathbf{C}_{\alpha r}^{(be,2)}$ when the 1st- and 2nd-Order Sensitivities for the Top Ten Parameters Are Considered; Parameters Relative Standard Deviations: *SD* = 2%

Ranking of parameter	1	2	3	4	5	6	7	8	9	10
Parameter $\sigma_{t,i}^{g}$	$\sigma_{t,i=6}^{g=30}$	$\sigma_{t,i=1}^{g=12}$	$\sigma_{t,i=6}^{g=17}$	$\sigma_{t,i=6}^{g=16}$	$\sigma_{t,i=1}^{g=13}$	$\sigma_{t,i=6}^{g=18}$	$\sigma_{t,i=6}^{g=19}$	$\sigma_{t,i=6}^{g=20}$	$\sigma_{t,i=6}^{g=21}$	$\sigma_{t,i=1}^{g=14}$
$\mathbf{C}_{\alpha r}^{(c,2)}$	−195330	-6496	−13396	−10092	-6458	−14959	−15207	−14692	−13912	-6176
$\mathbf{C}_{\alpha r}^{(be,2)}$	33379	1110	2289	1725	1104	2556	2599	2511	2377	1055

2. The absolute values of the components of $\mathbf{C}_{\alpha r}^{(be,1)}$ and $\mathbf{C}_{\alpha r}^{(be,2)}$ are all considerably smaller than the absolute values of the original correlations $\mathbf{C}_{\alpha r}^{(c,1)} = \mathbf{C}_{\alpha r}^{(c,2)}$.
3. The values of the components of $\mathbf{C}_{\alpha r}^{(be,1)}$ are larger than the corresponding components of $\mathbf{C}_{\alpha r}^{(be,2)}$, which indicates that the inclusion of the 2nd-order sensitivities reduces the best-estimate correlations between the best-estimate parameters and the predicted best-estimate leakage response.

When the 3rd- and, respectively, 4th-order sensitivities are also included in addition to the 1st- and 2nd-order sensitivities, the numerical results for the components of the vectors $\mathbf{C}_{\alpha r}^{(c,i)}$, $i = 3,4$, are computed using the expression provided in Eq. (5.16). The numerical results thus obtained for the components of the vector $\mathbf{C}_{\alpha r}^{(be,3)}$ and, respectively, $\mathbf{C}_{\alpha r}^{(be,4)}$, are presented in Tables 5.12 and 5.13, respectively.

Comparing the results presented in Tables 5.10–5.13 indicates the following conclusions, in addition to the conclusions drawn from the results presented in Tables 5.10 and 5.11:

TABLE 5.12

Results for the Components of $\mathbf{C}_{\alpha r}^{(be,3)}$ when Sensitivities up to 3rd-Order Are Considered for the Top Ten Parameters; Parameters Relative Standard Deviations: *SD* = 2%

Ranking of parameter	1	2	3	4	5	6	7	8	9	10
Parameter $\sigma_{t,i}^{g}$	$\sigma_{t,i=6}^{g=30}$	$\sigma_{t,i=1}^{g=12}$	$\sigma_{t,i=6}^{g=17}$	$\sigma_{t,i=6}^{g=16}$	$\sigma_{t,i=1}^{g=13}$	$\sigma_{t,i=6}^{g=18}$	$\sigma_{t,i=6}^{g=19}$	$\sigma_{t,i=6}^{g=20}$	$\sigma_{t,i=6}^{g=21}$	$\sigma_{t,i=1}^{g=14}$
$\mathbf{C}_{\alpha r}^{(c,3)}$	−319430	−6519	−13464	−10136	−6481	−15039	−15288	−14767	−13981	−6199
$\mathbf{C}_{\alpha r}^{(be,3)}$	27793	567	1171	882	564	1309	1330	1285	1216	539

TABLE 5.13

Results for the Components of $\mathbf{C}_{\alpha r}^{(be,4)}$ when Sensitivities up to 4th-Order Are Considered for the Top Ten Parameters; Parameters Relative Standard Deviations: *SD* = 2%

Ranking of parameter	1	2	3	4	5	6	7	8	9	10
Parameter $\sigma_{t,i}^{g}$	$\sigma_{t,i=6}^{g=30}$	$\sigma_{t,i=1}^{g=12}$	$\sigma_{t,i=6}^{g=17}$	$\sigma_{t,i=6}^{g=16}$	$\sigma_{t,i=1}^{g=13}$	$\sigma_{t,i=6}^{g=18}$	$\sigma_{t,i=6}^{g=19}$	$\sigma_{t,i=6}^{g=20}$	$\sigma_{t,i=6}^{g=21}$	$\sigma_{t,i=1}^{g=14}$
$\mathbf{C}_{\alpha r}^{(c,4)}$	−319430	−6519	−13464	−10136	−6481	−15039	−15288	−14767	−13981	−6199
$\mathbf{C}_{\alpha r}^{(be,4)}$	21889	447	923	695	444	1031	1048	1012	958	425

1. All of the components of the original correlations $\mathbf{C}_{\alpha r}^{(c,3)} = \mathbf{C}_{\alpha r}^{(c,4)}$ and $\mathbf{C}_{\alpha r}^{(c,1)} = \mathbf{C}_{\alpha r}^{(c,2)}$ between the model parameters and the computed leakage response are negative. Furthermore, including the 3rd- and 4th-order sensitivities causes the absolute values of the components of $\mathbf{C}_{\alpha r}^{(c,3)} = \mathbf{C}_{\alpha r}^{(c,4)}$ to be larger than the corresponding components of the vector $\mathbf{C}_{\alpha r}^{(c,1)} = \mathbf{C}_{\alpha r}^{(c,2)}$.
2. The components of the vectors $\mathbf{C}_{\alpha r}^{(be,i)}$, $i = 1,...,4$ representing the best-estimate correlations among the benchmark's best-estimate leakage response and best-estimate parameters, are all positive.
3. The values of the components of $\mathbf{C}_{\alpha r}^{(be,1)}$ are larger than the corresponding components of $\mathbf{C}_{\alpha r}^{(be,2)}$, which in turn are larger than the components of $\mathbf{C}_{\alpha r}^{(be,3)}$, which in turn are larger than the components of $\mathbf{C}_{\alpha r}^{(be,4)}$. This sequence of inequalities indicates that the inclusion of the successively higher-order sensitivities reduces successively the corresponding best-estimate correlations between the best-estimate parameters and the predicted best-estimate leakage response.

5.4.2 "Typical Precision" Parameters (Relative Standard Deviations $SD = 5\%$)

When the parameters are considered to be characterized by uniform relative standard deviations of 5%, the numerical results obtained for the components of the vectors $\mathbf{C}_{\alpha r}^{(be,i)}$, $i = 1,...,4$ of best-estimate correlations among the benchmark's best-estimate leakage response are presented in Tables 5.14–5.17, respectively. As before, when only the 1st- and 2nd-order sensitivities are considered, the numerical results

TABLE 5.14

Results for the Components of $\mathbf{C}_{\alpha r}^{(be,1)}$ when Only the 1st-Order Sensitivities for the Top Ten Parameters Are Considered; Parameters Relative Standard Deviations: $SD = 5\%$

Ranking of parameter	1	2	3	4	5	6	7	8	9	10
Parameter $\sigma_{t,i}^{g}$	$\sigma_{t,i=6}^{g=30}$	$\sigma_{t,i=1}^{g=12}$	$\sigma_{t,i=6}^{g=17}$	$\sigma_{t,i=6}^{g=16}$	$\sigma_{t,i=1}^{g=13}$	$\sigma_{t,i=6}^{g=18}$	$\sigma_{t,i=6}^{g=19}$	$\sigma_{t,i=6}^{g=20}$	$\sigma_{t,i=6}^{g=21}$	$\sigma_{t,i=1}^{g=14}$
$C_{\alpha r}^{(c,1)}$	−1220812	−40598	−83724	−63075	−40362	−93494	−95043	−91823	−86952	−38598
$C_{\alpha r}^{(be,1)}$	51544	1714	3535	2663	1704	3947	4013	3877	3671	1630

TABLE 5.15

Results for the Components of $\mathbf{C}_{\alpha r}^{(be,2)}$ when the 1st- and 2nd-Order Sensitivities for the Top Ten Parameters Are Considered; Parameters Relative Standard Deviations: $SD = 5\%$

Ranking of parameter	1	2	3	4	5	6	7	8	9	10
Parameter $\sigma_{t,i}^{g}$	$\sigma_{t,i=6}^{g=30}$	$\sigma_{t,i=1}^{g=12}$	$\sigma_{t,i=6}^{g=17}$	$\sigma_{t,i=6}^{g=16}$	$\sigma_{t,i=1}^{g=13}$	$\sigma_{t,i=6}^{g=18}$	$\sigma_{t,i=6}^{g=19}$	$\sigma_{t,i=6}^{g=20}$	$\sigma_{t,i=6}^{g=21}$	$\sigma_{t,i=1}^{g=14}$
$C_{\alpha r}^{(c,2)}$	−1220812	−40598	−83724	−63075	−40362	−93494	−95043	−91823	−86952	−38598
$C_{\alpha r}^{(be,2)}$	17091	568	1172	883	565	1309	1331	1285	1217	540

TABLE 5.16

Results for the Components of $\mathbf{C}_{\alpha r}^{(be,3)}$ when Sensitivities up to 3rd-Order Are Considered for the Top Ten Parameters; Parameters Relative Standard Deviations: $SD = 5\%$

Ranking of parameter	1	2	3	4	5	6	7	8	9	10
Parameter $\sigma_{t,i}^{g}$	$\sigma_{t,i=6}^{g=30}$	$\sigma_{t,i=1}^{g=12}$	$\sigma_{t,i=6}^{g=17}$	$\sigma_{t,i=6}^{g=16}$	$\sigma_{t,i=1}^{g=13}$	$\sigma_{t,i=6}^{g=18}$	$\sigma_{t,i=6}^{g=19}$	$\sigma_{t,i=6}^{g=20}$	$\sigma_{t,i=6}^{g=21}$	$\sigma_{t,i=1}^{g=14}$
$C_{\alpha r}^{(c,3)}$	−6068475	−41512	−86401	−64793	−41281	−96626	−98217	−94783	−89638	−39528
$C_{\alpha r}^{(be,3)}$	8771	60	125	94	60	140	142	137	130	57

TABLE 5.17

Results for the Components of $\mathbf{C}_{\alpha r}^{(be,4)}$ when Sensitivities up to 4th-Order Are Considered for the Top Ten Parameters; Parameters Relative Standard Deviations: *SD* = 5%

Ranking of parameter	1	2	3	4	5	6	7	8	9	10
Parameter $\sigma_{t,i}^{g}$	$\sigma_{t,i=6}^{g=30}$	$\sigma_{t,i=1}^{g=12}$	$\sigma_{t,i=6}^{g=17}$	$\sigma_{t,i=6}^{g=16}$	$\sigma_{t,i=1}^{g=13}$	$\sigma_{t,i=6}^{g=18}$	$\sigma_{t,i=6}^{g=19}$	$\sigma_{t,i=6}^{g=20}$	$\sigma_{t,i=6}^{g=21}$	$\sigma_{t,i=1}^{g=14}$
$\mathbf{C}_{\alpha r}^{(c,4)}$	−6068475	−41512	−86401	−64793	−41281	−96626	−98217	−94783	−89638	−39528
$\mathbf{C}_{\alpha r}^{(be,4)}$	4189	29	60	45	28	67	68	65	62	27

for the components of the vectors $\mathbf{C}_{\alpha r}^{(c,i)}$, $i=1,2$ are computed using the expression provided in Eq. (5.15). Similarly, when the 3rd- and 4th-order sensitivities are also included, the numerical results for the components of the vectors $\mathbf{C}_{\alpha r}^{(c,i)}$, $i=3,4$ are computed using the expression provided in Eq. (5.16).

Comparing the results presented in Tables 5.14–5.17 with the corresponding results presented in Tables 5.10–5.13 reveals that the larger parameters' standard deviations ($SD=5\%$) accentuate the characteristics displayed by the corresponding results for the smaller parameter standard deviations ($SD=2\%$), as follows:

1. Comparing the results in Table 5.14 with the results presented in Table 5.10 reveals that the components of the correlations $\mathbf{C}_{\alpha r}^{(c,1)}$ for $SD=5\%$ are considerably larger in absolute values than the components of the correlations $\mathbf{C}_{\alpha r}^{(c,1)}$ for $SD=2\%$. Conversely (and consequently), the components of the best-estimate correlations $\mathbf{C}_{\alpha r}^{(be,1)}$ for $SD=5\%$ are considerably larger in absolute values than the components of the correlations $\mathbf{C}_{\alpha r}^{(be,1)}$ for $SD=2\%$. Similar conclusion can be drawn by: (i) comparing the results in Table 5.15 with the results presented in Table 5.11; (ii) comparing the results in Table 5.16 with the results presented in Table 5.12; and (iii) comparing the results in Table 5.17 with the results presented in Table 5.13.
2. All of the conclusions previously drawn from the results obtained for the smaller parameter standard deviations ($SD=2\%$) also hold for the larger parameters' standard deviations ($SD=5\%$), so they will not be repeated here, other than to note that these conclusions are even more evident/accentuated by the contrast between the original values for the correlations $\mathbf{C}_{\alpha r}^{(c,i)}$ (between the model parameters and computed leakage response) and the predicted best-estimate values for the posterior best-estimate correlations, $\mathbf{C}_{\alpha r}^{(be,i)}$, between the calibrated best-estimate parameters and the best-estimate predicted response. In particular, the starkest contrasts among the original and predicted posterior values for the parameter-response correlations are evidenced by the results presented in Table 5.17: the original components of the vector $\mathbf{C}_{\alpha r}^{(c,4)}$ all have large *negative* values of the order of $O\left(10^5-10^7\right)$, while the posterior best-estimate correlations $\mathbf{C}_{\alpha r}^{(be,4)}$ have *positive* values of the order of $O\left(10-10^3\right)$.

The overall conclusion that follows from the results presented in Tables 5.10–5.17 is that the larger the parameter standard deviations and the higher the order of retained sensitivities, the larger the negative values of the original correlations among the model parameters and the leakage response, and the smaller the positive values of the best-estimate predicted posterior correlations among the best-estimate posterior calibrated model parameters and the predicted posterior leakage response.

5.5 4TH-BERRU-PM BEST-ESTIMATE POSTERIOR COVARIANCE MATRIX OF CALIBRATED MODEL PARAMETERS

The predicted best-estimate covariance matrix of the predicted best-estimate calibrated model parameters is denoted as $\mathbf{C}_{\alpha\alpha}^{be}$ and has the following expression:

$$\begin{aligned}\mathbf{C}_{\alpha\alpha}^{be} &\triangleq \int_D \left(\boldsymbol{\alpha}-\boldsymbol{\alpha}^{be}\right)\left(\boldsymbol{\alpha}-\boldsymbol{\alpha}^{be}\right)^{\dagger} p_4^{be}\left(\mathbf{r},\boldsymbol{\alpha}\right) d\boldsymbol{\alpha}\, d\mathbf{r} \\ &= \mathbf{C}_{\alpha\alpha} - \mathbf{C}_{\alpha r}^{c}\left(\mathbf{C}_{rr}^{e}+\mathbf{C}_{rr}^{c}\right)^{-1}\mathbf{C}_{r\alpha}^{c} + O\left[\left(t_i^{\alpha}\right)^2\right] + O\left[\left(q_i^{\alpha}\right)^2\right].\end{aligned} \tag{5.17}$$

For the particular information available for the PERP reactor physics benchmark, the 2nd-order terms involving the "squared triple correlations" and "squared quadruple correlations" shown in Eq. (5.17) arise from terms involving the quantity $\mathbf{p}_2\left(\mathbf{r}_s^{(0)},\boldsymbol{\alpha}_s^{(0)}\right)$ and are expected to be negligible by comparison to the leading term. Not explicitly shown in Eq. (5.17) are terms involving products of triple correlations among parameters and responses, which also stem from the quantity $\mathbf{p}_2\left(\mathbf{r}_s^{(0)},\boldsymbol{\alpha}_s^{(0)}\right)$ and which are also expected to be negligible by comparison to the leading term. Depending on the highest-order retained sensitivities, the predicted best-estimate covariance matrix of the best-estimate calibrated model parameters, denoted as $\mathbf{C}_{\alpha\alpha}^{(be,i)}$, $i=1,2,3,4$, will have one of the expressions provided:

$$\mathbf{C}_{\alpha\alpha}^{(be,i)} = \mathbf{C}_{\alpha\alpha} - \mathbf{C}_{\alpha r}^{(c,i)}\left\{\left[SD^{(e)}\right]^2 + \left[\mathrm{var}(L)\right]^{(i,U,N)}\right\}^{-1}\left[\mathbf{C}_{\alpha r}^{(c,i)}\right]^{\dagger};\ i=1,2,3,4. \tag{5.18}$$

The components of the predicted posterior covariance matrix $\mathbf{C}_{\alpha\alpha}^{(be,i)}$, $i=1,2,3,4$ of the best-estimate calibrated parameters are computed by using Eq. (5.18). All of the numerical results presented in this section for the components of the best-estimate parameter covariance matrices $\mathbf{C}_{\alpha\alpha}^{(be,i)}$, $i=1,...,4$, are in units of "$\left[barn\right]^2$," in other words, $\left[10^{-24}\,cm^2\right]^2$.

5.5.1 "HIGH" PRECISION PARAMETERS (RELATIVE STANDARD DEVIATIONS $SD = 2\%$)

The diagonal matrix of apriori variances for the top ten model parameters (ranked in the order of the magnitude of their 1st-order relative sensitivities), which are considered to be uncorrelated, normally distributed, and having uniform relative standard deviations $SD = 2\%$, is presented in Table 5.18.

The results presented in Tables 5.19–5.22 were obtained by considering a “low” precision measured response, having the nominal value $L^e = 1.94 \times 10^6$ n/s, and a relative standard deviation $SD^{(e)} = 10\%$.

The results presented in Tables 5.19–5.22 indicate that the variance of each of the calibrated predicted parameters has been significantly reduced by comparison to its original value. However, this reduction is not monotonic but oscillates with decreasing amplitudes as sensitivities of increasingly higher-order are considered. Furthermore, the originally uncorrelated parameters have become negatively correlated after the application of the 4th-BERRU-PM methodology. The impact of sensitivities higher than 1st-order is small.

To investigate the effects of the precision of the measured response, a set of computations similar to the ones presented n Table 5.19–5.22 was performed by considering a “typical” precision measured response, having the nominal value $L^e = 1.94 \times 10^6$ n/s, and a relative standard deviation $SD^{(e)} = 5\%$. The results obtained for this “more precise” measurement are presented in Table 5.23 when only the 1st-order sensitivities are considered, and in Table 5.24, when considering the sensitivities of 1st-order through 4th-order. Comparing the results presented in Table 5.23 with the results presented in Table 5.19 indicates that these results are very similar to each other, the same conclusion can be drawn by intercomparing the results presented in Table 5.24 with those presented in Table 5.22. It can, therefore, be concluded that the precision of the measured leakage response has no impact on the calibration of the benchmark's parameters (microscopic group-averaged total cross-sections).

TABLE 5.18
Nominal Values of the Components of $\mathbf{C}_{\alpha\alpha}$; $SD = 2\%$

	$\sigma_{t,i=6}^{g=30}$	$\sigma_{t,i=1}^{g=12}$	$\sigma_{t,i=6}^{g=17}$	$\sigma_{t,i=6}^{g=16}$	$\sigma_{t,i=1}^{g=13}$	$\sigma_{t,i=6}^{g=18}$	$\sigma_{t,i=6}^{g=19}$	$\sigma_{t,i=6}^{g=20}$	$\sigma_{t,i=6}^{g=21}$	$\sigma_{t,i=1}^{g=14}$
$\sigma_{t,i=6}^{g=30}$	0.3491	0.000	0.0000	0.0000	0.0000	0.0000	0.0000	0.0000	0.0000	0.0000
$\sigma_{t,i=1}^{g=12}$	0.0000	0.0194	0.0000	0.0000	0.0000	0.0000	0.0000	0.0000	0.0000	0.0000
$\sigma_{t,i=6}^{g=17}$	0.0000	0.0000	0.1047	0.0000	0.0000	0.0000	0.0000	0.0000	0.0000	0.0000
$\sigma_{t,i=6}^{g=16}$	0.0000	0.0000	0.0000	0.0603	0.0000	0.0000	0.0000	0.0000	0.0000	0.0000
$\sigma_{t,i=1}^{g=13}$	0.0000	0.0000	0.0000	0.0000	0.0251	0.0000	0.0000	0.0000	0.0000	0.0000
$\sigma_{t,i=6}^{g=18}$	0.0000	0.0000	0.0000	0.0000	0.0000	0.1379	0.0000	0.0000	0.0000	0.0000
$\sigma_{t,i=6}^{g=19}$	0.0000	0.0000	0.0000	0.0000	0.0000	0.0000	0.1552	0.0000	0.0000	0.0000
$\sigma_{t,i=6}^{g=20}$	0.0000	0.0000	0.0000	0.0000	0.0000	0.0000	0.0000	0.1625	0.0000	0.0000
$\sigma_{t,i=6}^{g=21}$	0.0000	0.0000	0.0000	0.0000	0.0000	0.0000	0.0000	0.0000	0.1654	0.0000
$\sigma_{t,i=1}^{g=14}$	0.0000	0.0000	0.0000	0.0000	0.0000	0.0000	0.0000	0.0000	0.0000	0.0337

TABLE 5.19
Results for $\mathbf{C}_{\alpha\alpha}^{(be,1)}$; Only the 1st-Order Sensitivities Are Considered; $SD = 2\%$; Measured Response $SD^{(e)} = 10\%$

	$\sigma_{t,i=6}^{g=30}$	$\sigma_{t,i=1}^{g=12}$	$\sigma_{t,i=6}^{g=17}$	$\sigma_{t,i=6}^{g=16}$	$\sigma_{t,i=1}^{g=13}$	$\sigma_{t,i=6}^{g=18}$	$\sigma_{t,i=6}^{g=19}$	$\sigma_{t,i=6}^{g=20}$	$\sigma_{t,i=6}^{g=21}$	$\sigma_{t,i=1}^{g=14}$
$\sigma_{t,i=6}^{g=30}$	0.1304	−0.0073	−0.0150	−0.0113	−0.0072	−0.0167	−0.0170	−0.0164	−0.0156	−0.0069
$\sigma_{t,i=1}^{g=12}$	−0.0073	0.0192	−0.0005	−0.0004	−0.0002	−0.0006	−0.0006	−0.0005	−0.0005	−0.0002
$\sigma_{t,i=6}^{g=17}$	−0.0150	−0.0005	0.1037	−0.0008	−0.0005	−0.0011	−0.0012	−0.0011	−0.0011	−0.0005
$\sigma_{t,i=6}^{g=16}$	−0.0113	−0.0004	−0.0008	0.0597	−0.0004	−0.0009	−0.0009	−0.0008	−0.0008	−0.0004
$\sigma_{t,i=1}^{g=13}$	−0.0072	−0.0002	−0.0005	−0.0004	0.0249	−0.0006	−0.0006	−0.0005	−0.0005	−0.0002
$\sigma_{t,i=6}^{g=18}$	−0.0167	−0.0006	−0.0011	−0.0009	−0.0006	0.1367	−0.0013	−0.0013	−0.0012	−0.0005
$\sigma_{t,i=6}^{g=19}$	−0.0170	−0.0006	−0.0012	−0.0009	−0.0006	−0.0013	0.1539	−0.0013	−0.0012	−0.0005
$\sigma_{t,i=6}^{g=20}$	−0.0164	−0.0005	−0.0011	−0.0008	−0.0005	−0.0013	−0.0013	0.1613	−0.0012	−0.0005
$\sigma_{t,i=6}^{g=21}$	−0.0156	−0.0005	−0.0011	−0.0008	−0.0005	−0.0012	−0.0012	−0.0012	0.1643	−0.0005
$\sigma_{t,i=1}^{g=14}$	−0.0069	−0.0002	−0.0005	−0.0004	−0.0002	−0.0005	−0.0005	−0.0005	−0.0005	0.0335

TABLE 5.20
Results for $\mathbf{C}_{\alpha\alpha}^{(be,2)}$; 1st- and 2nd-Order Sensitivities Are Considered; $SD = 2\%$; Measured Response $SD^{(e)} = 10\%$

	$\sigma_{t,i=6}^{g=30}$	$\sigma_{t,i=1}^{g=12}$	$\sigma_{t,i=6}^{g=17}$	$\sigma_{t,i=6}^{g=16}$	$\sigma_{t,i=1}^{g=13}$	$\sigma_{t,i=6}^{g=18}$	$\sigma_{t,i=6}^{g=19}$	$\sigma_{t,i=6}^{g=20}$	$\sigma_{t,i=6}^{g=21}$	$\sigma_{t,i=1}^{g=14}$
$\sigma_{t,i=6}^{g=30}$	0.1761	−0.0058	−0.0119	−0.0089	−0.0057	−0.0132	−0.0135	−0.0130	−0.0123	−0.0055
$\sigma_{t,i=1}^{g=12}$	−0.0058	0.0192	−0.0004	−0.0003	−0.0002	−0.0004	−0.0004	−0.0004	−0.0004	−0.0002
$\sigma_{t,i=6}^{g=17}$	−0.0119	−0.0004	0.1039	−0.0006	−0.0004	−0.0009	−0.0009	−0.0009	−0.0008	−0.0004
$\sigma_{t,i=6}^{g=16}$	−0.0089	−0.0003	−0.0006	0.0598	−0.0003	−0.0007	−0.0007	−0.0007	−0.0006	−0.0003
$\sigma_{t,i=1}^{g=13}$	−0.0057	−0.0002	−0.0004	−0.0003	0.0249	−0.0004	−0.0004	−0.0004	−0.0004	−0.0002
$\sigma_{t,i=6}^{g=18}$	−0.0132	−0.0004	−0.0009	−0.0007	−0.0004	0.1369	−0.0010	−0.0010	−0.0009	−0.0004
$\sigma_{t,i=6}^{g=19}$	−0.0135	−0.0004	−0.0009	−0.0007	−0.0004	−0.0010	0.1541	−0.0010	−0.0010	−0.0004
$\sigma_{t,i=6}^{g=20}$	−0.0130	−0.0004	−0.0009	−0.0007	−0.0004	−0.0010	−0.0010	0.1616	−0.0009	−0.0004
$\sigma_{t,i=6}^{g=21}$	−0.0123	−0.0004	−0.0008	−0.0006	−0.0004	−0.0009	−0.0010	−0.0009	0.1645	−0.0004
$\sigma_{t,i=1}^{g=14}$	−0.0055	−0.0002	−0.0004	−0.0003	−0.0002	−0.0004	−0.0004	−0.0004	−0.0004	0.0336

TABLE 5.21

Results for $C_{\alpha\alpha}^{(be,3)}$; 1st-, 2nd-, and 3rd-Order Sensitivities Are Considered; $SD = 2\%$; Measured Response $SD^{(e)} = 10\%$

	$\sigma_{t,i=6}^{g=30}$	$\sigma_{t,i=1}^{g=12}$	$\sigma_{t,i=6}^{g=17}$	$\sigma_{t,i=6}^{g=16}$	$\sigma_{t,i=1}^{g=13}$	$\sigma_{t,i=6}^{g=18}$	$\sigma_{t,i=6}^{g=19}$	$\sigma_{t,i=6}^{g=20}$	$\sigma_{t,i=6}^{g=21}$	$\sigma_{t,i=1}^{g=14}$
$\sigma_{t,i=6}^{g=30}$	0.1135	−0.0048	−0.0099	−0.0075	−0.0048	−0.0111	−0.0113	−0.0109	−0.0103	−0.0046
$\sigma_{t,i=1}^{g=12}$	−0.0048	0.0193	−0.0002	−0.0002	−0.0001	−0.0002	−0.0002	−0.0002	−0.0002	−0.0001
$\sigma_{t,i=6}^{g=17}$	−0.0099	−0.0002	0.1043	−0.0003	−0.0002	−0.0005	−0.0005	−0.0005	−0.0004	−0.0002
$\sigma_{t,i=6}^{g=16}$	−0.0075	−0.0002	−0.0003	0.0601	−0.0002	−0.0004	−0.0004	−0.0003	−0.0003	−0.0001
$\sigma_{t,i=1}^{g=13}$	−0.0048	−0.0001	−0.0002	−0.0002	0.0250	−0.0002	−0.0002	−0.0002	−0.0002	−0.0001
$\sigma_{t,i=6}^{g=18}$	−0.0111	−0.0002	−0.0005	−0.0004	−0.0002	0.1374	−0.0005	−0.0005	−0.0005	−0.0002
$\sigma_{t,i=6}^{g=19}$	−0.0113	−0.0002	−0.0005	−0.0004	−0.0002	−0.0005	0.1546	−0.0005	−0.0005	−0.0002
$\sigma_{t,i=6}^{g=20}$	−0.0109	−0.0002	−0.0005	−0.0003	−0.0002	−0.0005	−0.0005	0.1620	−0.0005	−0.0002
$\sigma_{t,i=6}^{g=21}$	−0.0103	−0.0002	−0.0004	−0.0003	−0.0002	−0.0005	−0.0005	−0.0005	0.1649	−0.0002
$\sigma_{t,i=1}^{g=14}$	−0.0046	−0.0001	−0.0002	−0.0001	−0.0001	−0.0002	−0.0002	−0.0002	−0.0002	0.0337

TABLE 5.22

Results for $C_{\alpha\alpha}^{(be,4)}$; 1st- through 4th-Order Sensitivities Are Considered; $SD = 2\%$; Measured Response $SD^{(e)} = 10\%$

	$\sigma_{t,i=6}^{g=30}$	$\sigma_{t,i=1}^{g=12}$	$\sigma_{t,i=6}^{g=17}$	$\sigma_{t,i=6}^{g=16}$	$\sigma_{t,i=1}^{g=13}$	$\sigma_{t,i=6}^{g=18}$	$\sigma_{t,i=6}^{g=19}$	$\sigma_{t,i=6}^{g=20}$	$\sigma_{t,i=6}^{g=21}$	$\sigma_{t,i=1}^{g=14}$
$\sigma_{t,i=6}^{g=30}$	0.1636	−0.0038	−0.0078	−0.0059	−0.0038	−0.0087	−0.0089	−0.0086	−0.0081	−0.0036
$\sigma_{t,i=1}^{g=12}$	−0.0038	0.0194	−0.0002	−0.0001	−0.0001	−0.0002	−0.0002	−0.0002	−0.0002	−0.0001
$\sigma_{t,i=6}^{g=17}$	−0.0078	−0.0002	0.1044	−0.0002	−0.0002	−0.0004	−0.0004	−0.0004	−0.0003	−0.0002
$\sigma_{t,i=6}^{g=16}$	−0.0059	−0.0001	−0.0002	0.0601	−0.0001	−0.0003	−0.0003	−0.0003	−0.0003	−0.0001
$\sigma_{t,i=1}^{g=13}$	−0.0038	−0.0001	−0.0002	−0.0001	0.0251	−0.0002	−0.0002	−0.0002	−0.0002	−0.0001
$\sigma_{t,i=6}^{g=18}$	−0.0087	−0.0002	−0.0004	−0.0003	−0.0002	0.1375	−0.0004	−0.0004	−0.0004	−0.0002
$\sigma_{t,i=6}^{g=19}$	−0.0089	−0.0002	−0.0004	−0.0003	−0.0002	−0.0004	0.1548	−0.0004	−0.0004	−0.0002
$\sigma_{t,i=6}^{g=20}$	−0.0086	−0.0002	−0.0004	−0.0003	−0.0002	−0.0004	−0.0004	0.1621	−0.0004	−0.0002
$\sigma_{t,i=6}^{g=21}$	−0.0081	−0.0002	−0.0003	−0.0003	−0.0002	−0.0004	−0.0004	−0.0004	0.1650	−0.0002
$\sigma_{t,i=1}^{g=14}$	−0.0036	−0.0001	−0.0002	−0.0001	−0.0001	−0.0002	−0.0002	−0.0002	−0.0002	0.0337

TABLE 5.23
Results for $\mathbf{C}_{\alpha\alpha}^{(be,1)}$; Only the 1st-Order Sensitivities Are Considered; $SD = 2\%$; Measured Response $SD^{(e)} = 5\%$

	$\sigma_{t,i=6}^{g=30}$	$\sigma_{t,i=1}^{g=12}$	$\sigma_{t,i=6}^{g=17}$	$\sigma_{t,i=6}^{g=16}$	$\sigma_{t,i=1}^{g=13}$	$\sigma_{t,i=6}^{g=18}$	$\sigma_{t,i=6}^{g=19}$	$\sigma_{t,i=6}^{g=20}$	$\sigma_{t,i=6}^{g=21}$	$\sigma_{t,i=1}^{g=14}$
$\sigma_{t,i=6}^{g=30}$	0.0882	−0.0087	−0.0179	−0.0135	−0.0086	−0.0200	−0.0203	−0.0196	−0.0186	−0.0083
$\sigma_{t,i=1}^{g=12}$	−0.0087	0.0192	−0.0006	−0.0004	−0.0003	−0.0007	−0.0007	−0.0007	−0.0006	−0.0003
$\sigma_{t,i=6}^{g=17}$	−0.0179	−0.0006	0.1035	−0.0009	−0.0006	−0.0014	−0.0014	−0.0013	−0.0013	−0.0006
$\sigma_{t,i=6}^{g=16}$	−0.0135	−0.0004	−0.0009	0.0596	−0.0004	−0.0010	−0.0010	−0.0010	−0.0010	−0.0004
$\sigma_{t,i=1}^{g=13}$	−0.0086	−0.0003	−0.0006	−0.0004	0.0249	−0.0007	−0.0007	−0.0006	−0.0006	−0.0003
$\sigma_{t,i=6}^{g=18}$	−0.0200	−0.0007	−0.0014	−0.0010	−0.0007	0.1364	−0.0016	−0.0015	−0.0014	−0.0006
$\sigma_{t,i=6}^{g=19}$	−0.0203	−0.0007	−0.0014	−0.0010	−0.0007	−0.0016	0.1536	−0.0015	−0.0014	−0.0006
$\sigma_{t,i=6}^{g=20}$	−0.0196	−0.0007	−0.0013	−0.0010	−0.0006	−0.0015	−0.0015	0.1611	−0.0014	−0.0006
$\sigma_{t,i=6}^{g=21}$	−0.0186	−0.0006	−0.0013	−0.0010	−0.0006	−0.0014	−0.0014	−0.0014	0.1641	−0.0006
$\sigma_{t,i=1}^{g=14}$	−0.0083	−0.0003	−0.0006	−0.0004	−0.0003	−0.0006	−0.0006	−0.0006	−0.0006	0.0335

TABLE 5.24
Results for $\mathbf{C}_{\alpha\alpha}^{(be,4)}$; 1st- through 4th-Order Sensitivities Are Considered; $SD = 2\%$; Measured Response $SD^{(e)} = 5\%$

	$\sigma_{t,i=6}^{g=30}$	$\sigma_{t,i=1}^{g=12}$	$\sigma_{t,i=6}^{g=17}$	$\sigma_{t,i=6}^{g=16}$	$\sigma_{t,i=1}^{g=13}$	$\sigma_{t,i=6}^{g=18}$	$\sigma_{t,i=6}^{g=19}$	$\sigma_{t,i=6}^{g=20}$	$\sigma_{t,i=6}^{g=21}$	$\sigma_{t,i=1}^{g=14}$
$\sigma_{t,i=6}^{g=30}$	0.1535	−0.0040	−0.0082	−0.0062	−0.0040	−0.0092	−0.0094	−0.0090	−0.0086	−0.0038
$\sigma_{t,i=1}^{g=12}$	−0.0040	0.0194	−0.0002	−0.0001	−0.0001	−0.0002	−0.0002	−0.0002	−0.0002	−0.0001
$\sigma_{t,i=6}^{g=17}$	−0.0082	−0.0002	0.1044	−0.0003	−0.0002	−0.0004	−0.0004	−0.0004	−0.0004	−0.0002
$\sigma_{t,i=6}^{g=16}$	−0.0062	−0.0001	−0.0003	0.0601	−0.0001	−0.0003	−0.0003	−0.0003	−0.0003	−0.0001
$\sigma_{t,i=1}^{g=13}$	−0.0040	−0.0001	−0.0002	−0.0001	0.0251	−0.0002	−0.0002	−0.0002	−0.0002	−0.0001
$\sigma_{t,i=6}^{g=18}$	−0.0092	−0.0002	−0.0004	−0.0003	−0.0002	0.1375	−0.0004	−0.0004	−0.0004	−0.0002
$\sigma_{t,i=6}^{g=19}$	−0.0094	−0.0002	−0.0004	−0.0003	−0.0002	−0.0004	0.1547	−0.0004	−0.0004	−0.0002
$\sigma_{t,i=6}^{g=20}$	−0.0090	−0.0002	−0.0004	−0.0003	−0.0002	−0.0004	−0.0004	0.1621	−0.0004	−0.0002
$\sigma_{t,i=6}^{g=21}$	−0.0086	−0.0002	−0.0004	−0.0003	−0.0002	−0.0004	−0.0004	−0.0004	0.1650	−0.0002
$\sigma_{t,i=1}^{g=14}$	−0.0038	−0.0001	−0.0002	−0.0001	−0.0001	−0.0002	−0.0002	−0.0002	−0.0002	0.0337

5.5.2 "Typical" Precision Parameters (Relative Standard Deviations $SD = 5\%$)

The diagonal matrix of apriori parameter variances for the top ten model parameters (ranked in the order of the magnitude of their 1st-order relative sensitivities), which are considered to be uncorrelated, normally distributed, and having uniform relative standard deviations $SD = 5\%$, is presented in Table 5.25.

Table 5.26 presents the results for the predicted values for the components of the posterior matrix $\mathbf{C}_{\alpha\alpha}^{(be,1)}$, while Table 5.27 presents the results for the predicted values for the components of the posterior matrix $\mathbf{C}_{\alpha\alpha}^{(be,4)}$. The results presented in these tables are similar to the results presented in Tables 5.19 and 5.22, respectively, indicating that: (i) the variance of each of the calibrated predicted parameters is reduced by comparison to its original value; this reduction is not monotonic but oscillates with decreasing amplitudes as sensitivities of increasingly higher-order are considered.; (ii) the originally uncorrelated parameters have become negatively correlated after the application of the 4th-BERRU-PM methodology; and (iii) the impact of sensitivities higher than 1st-order are small.

To investigate the effects of the precision of the measured response, computations similar to the ones presented in Tables 5.26 and 5.27 were performed by considering a "typical" precision measured response, having the nominal value $L^e = 1.94 \times 10^6$ n/s, and a relative standard deviation $SD^{(e)} = 5\%$ (rather than $SD^{(e)} = 10\%$). The results obtained for this "more precise" measurement are presented in Table 5.28

TABLE 5.25
Nominal Values of the Components of $\mathbf{C}_{\alpha\alpha}$; $SD = 5\%$

	$\sigma_{t,i=6}^{g=30}$	$\sigma_{t,i=1}^{g=12}$	$\sigma_{t,i=6}^{g=17}$	$\sigma_{t,i=6}^{g=16}$	$\sigma_{t,i=1}^{g=13}$	$\sigma_{t,i=6}^{g=18}$	$\sigma_{t,i=6}^{g=19}$	$\sigma_{t,i=6}^{g=20}$	$\sigma_{t,i=6}^{g=21}$	$\sigma_{t,i=1}^{g=14}$
$\sigma_{t,i=6}^{g=30}$	2.1820	0.000	0.0000	0.0000	0.0000	0.0000	0.0000	0.0000	0.0000	0.0000
$\sigma_{t,i=1}^{g=12}$	0.0000	0.1215	0.0000	0.0000	0.0000	0.0000	0.0000	0.0000	0.0000	0.0000
$\sigma_{t,i=6}^{g=17}$	0.0000	0.0000	0.6547	0.0000	0.0000	0.0000	0.0000	0.0000	0.0000	0.0000
$\sigma_{t,i=6}^{g=16}$	0.0000	0.0000	0.0000	0.3769	0.0000	0.0000	0.0000	0.0000	0.0000	0.0000
$\sigma_{t,i=1}^{g=13}$	0.0000	0.0000	0.0000	0.0000	0.1571	0.0000	0.0000	0.0000	0.0000	0.0000
$\sigma_{t,i=6}^{g=18}$	0.0000	0.0000	0.0000	0.0000	0.0000	0.8622	0.0000	0.0000	0.0000	0.0000
$\sigma_{t,i=6}^{g=19}$	0.0000	0.0000	0.0000	0.0000	0.0000	0.0000	0.9699	0.0000	0.0000	0.0000
$\sigma_{t,i=6}^{g=20}$	0.0000	0.0000	0.0000	0.0000	0.0000	0.0000	0.0000	1.0158	0.0000	0.0000
$\sigma_{t,i=6}^{g=21}$	0.0000	0.0000	0.0000	0.0000	0.0000	0.0000	0.0000	0.0000	1.0337	0.0000
$\sigma_{t,i=1}^{g=14}$	0.0000	0.0000	0.0000	0.0000	0.0000	0.0000	0.0000	0.0000	0.0000	0.2109

TABLE 5.26

Results for $C_{\alpha\alpha}^{(be,1)}$; Only the 1st-Order Sensitivities Are Considered; $SD = 5\%$; Measured Response $SD^{(e)} = 10\%$

	$\sigma_{t,i=6}^{g=30}$	$\sigma_{t,i=1}^{g=12}$	$\sigma_{t,i=6}^{g=17}$	$\sigma_{t,i=6}^{g=16}$	$\sigma_{t,i=1}^{g=13}$	$\sigma_{t,i=6}^{g=18}$	$\sigma_{t,i=6}^{g=19}$	$\sigma_{t,i=6}^{g=20}$	$\sigma_{t,i=6}^{g=21}$	$\sigma_{t,i=1}^{g=14}$
$\sigma_{t,i=6}^{g=30}$	0.5123	−0.0555	−0.1145	−0.0863	−0.0552	−0.1279	−0.1300	−0.1256	−0.1189	−0.0528
$\sigma_{t,i=1}^{g=12}$	−0.0555	0.1197	−0.0038	−0.0029	−0.0018	−0.0043	−0.0043	−0.0042	−0.0040	−0.0018
$\sigma_{t,i=6}^{g=17}$	−0.1145	−0.0038	0.6468	−0.0059	−0.0038	−0.0088	−0.0089	−0.0086	−0.0082	−0.0036
$\sigma_{t,i=6}^{g=16}$	−0.0863	−0.0029	−0.0059	0.3725	−0.0029	−0.0066	−0.0067	−0.0065	−0.0061	−0.0027
$\sigma_{t,i=1}^{g=13}$	−0.0552	−0.0018	−0.0038	−0.0029	0.1553	−0.0042	−0.0043	−0.0042	−0.0039	−0.0017
$\sigma_{t,i=6}^{g=18}$	−0.1279	−0.0043	−0.0088	−0.0066	−0.0042	0.8524	−0.0100	−0.0096	−0.0091	−0.0040
$\sigma_{t,i=6}^{g=19}$	−0.1300	−0.0043	−0.0089	−0.0067	−0.0043	−0.0100	0.9598	−0.0098	−0.0093	−0.0041
$\sigma_{t,i=6}^{g=20}$	−0.1256	−0.0042	−0.0086	−0.0065	−0.0042	−0.0096	−0.0098	1.0064	−0.0089	−0.0040
$\sigma_{t,i=6}^{g=21}$	−0.1189	−0.0040	−0.0082	−0.0061	−0.0039	−0.0091	−0.0093	−0.0089	1.0252	−0.0038
$\sigma_{t,i=1}^{g=14}$	−0.0528	−0.0018	−0.0036	−0.0027	−0.0017	−0.0040	−0.0041	−0.0040	−0.0038	0.2093

TABLE 5.27

Results for $C_{\alpha\alpha}^{(be,4)}$; 1st- through 4th-Order Sensitivities Are Considered; $SD = 5\%$; Measured Response $SD^{(e)} = 10\%$

	$\sigma_{t,i=6}^{g=30}$	$\sigma_{t,i=1}^{g=12}$	$\sigma_{t,i=6}^{g=17}$	$\sigma_{t,i=6}^{g=16}$	$\sigma_{t,i=1}^{g=13}$	$\sigma_{t,i=6}^{g=18}$	$\sigma_{t,i=6}^{g=19}$	$\sigma_{t,i=6}^{g=20}$	$\sigma_{t,i=6}^{g=21}$	$\sigma_{t,i=1}^{g=14}$
$\sigma_{t,i=6}^{g=30}$	1.5075	−0.0046	−0.0096	−0.0072	−0.0046	−0.0107	−0.0109	−0.0105	−0.0100	−0.0044
$\sigma_{t,i=1}^{g=12}$	−0.0046	0.1215	−0.0001	−0.0001	−0.0001	−0.0001	−0.0001	−0.0001	−0.0001	−0.0001
$\sigma_{t,i=6}^{g=17}$	−0.0096	−0.0001	0.6545	−0.0001	−0.0001	−0.0002	−0.0002	−0.0002	−0.0001	−0.0001
$\sigma_{t,i=6}^{g=16}$	−0.0072	−0.0001	−0.0001	0.3768	−0.0001	−0.0001	−0.0001	−0.0001	−0.0001	−0.0001
$\sigma_{t,i=1}^{g=13}$	−0.0046	−0.0001	−0.0001	−0.0001	0.1571	−0.0001	−0.0001	−0.0001	−0.0001	−0.0001
$\sigma_{t,i=6}^{g=18}$	−0.0107	−0.0001	−0.0002	−0.0001	−0.0001	0.8620	−0.0002	−0.0002	−0.0002	−0.0001
$\sigma_{t,i=6}^{g=19}$	−0.0109	−0.0001	−0.0002	−0.0001	−0.0001	−0.0002	0.9697	−0.0002	−0.0002	−0.0001
$\sigma_{t,i=6}^{g=20}$	−0.0105	−0.0001	−0.0002	−0.0001	−0.0001	−0.0002	−0.0002	1.0156	−0.0002	−0.0001
$\sigma_{t,i=6}^{g=21}$	−0.0100	−0.0001	−0.0001	−0.0001	−0.0001	−0.0002	−0.0002	−0.0002	1.0336	−0.0001
$\sigma_{t,i=1}^{g=14}$	−0.0044	−0.0001	−0.0001	−0.0001	−0.0001	−0.0001	−0.0001	−0.0001	−0.0001	0.2109

TABLE 5.28
Results for $C_{\alpha\alpha}^{(be,1)}$; Only the 1st-Order Sensitivities Are Considered; $SD = 5\%$; Measured Response $SD^{(e)} = 5\%$

	$\sigma_{t,i=6}^{g=30}$	$\sigma_{t,i=1}^{g=12}$	$\sigma_{t,i=6}^{g=17}$	$\sigma_{t,i=6}^{g=16}$	$\sigma_{t,i=1}^{g=13}$	$\sigma_{t,i=6}^{g=18}$	$\sigma_{t,i=6}^{g=19}$	$\sigma_{t,i=6}^{g=20}$	$\sigma_{t,i=6}^{g=21}$	$\sigma_{t,i=1}^{g=14}$
$\sigma_{t,i=6}^{g=30}$	0.4577	−0.0573	−0.1183	−0.0891	−0.0570	−0.1321	−0.1342	−0.1297	−0.1228	−0.0545
$\sigma_{t,i=1}^{g=12}$	−0.0573	0.1196	−0.0039	−0.0030	−0.0019	−0.0044	−0.0045	−0.0043	−0.0041	−0.0018
$\sigma_{t,i=6}^{g=17}$	−0.1183	−0.0039	0.6466	−0.0061	−0.0039	−0.0091	−0.0092	−0.0089	−0.0084	−0.0037
$\sigma_{t,i=6}^{g=16}$	−0.0891	−0.0030	−0.0061	0.3723	−0.0029	−0.0068	−0.0069	−0.0067	−0.0063	−0.0028
$\sigma_{t,i=1}^{g=13}$	−0.0570	−0.0019	−0.0039	−0.0029	0.1552	−0.0044	−0.0044	−0.0043	−0.0041	−0.0018
$\sigma_{t,i=6}^{g=18}$	−0.1321	−0.0044	−0.0091	−0.0068	−0.0044	0.8521	−0.0103	−0.0099	−0.0094	−0.0042
$\sigma_{t,i=6}^{g=19}$	−0.1342	−0.0045	−0.0092	−0.0069	−0.0044	−0.0103	0.9594	−0.0101	−0.0096	−0.0042
$\sigma_{t,i=6}^{g=20}$	−0.1297	−0.0043	−0.0089	−0.0067	−0.0043	−0.0099	−0.0101	1.0061	−0.0092	−0.0041
$\sigma_{t,i=6}^{g=21}$	−0.1228	−0.0041	−0.0084	−0.0063	−0.0041	−0.0094	−0.0096	−0.0092	1.0250	−0.0039
$\sigma_{t,i=1}^{g=14}$	−0.0545	−0.0018	−0.0037	−0.0028	−0.0018	−0.0042	−0.0042	−0.0041	−0.0039	0.2092

TABLE 5.29
Results for $C_{\alpha\alpha}^{(be,4)}$; 1st- through 4th-Order Sensitivities Are Considered; $SD = 5\%$; Measured Response $SD^{(e)} = 5\%$

	$\sigma_{t,i=6}^{g=30}$	$\sigma_{t,i=1}^{g=12}$	$\sigma_{t,i=6}^{g=17}$	$\sigma_{t,i=6}^{g=16}$	$\sigma_{t,i=1}^{g=13}$	$\sigma_{t,i=6}^{g=18}$	$\sigma_{t,i=6}^{g=19}$	$\sigma_{t,i=6}^{g=20}$	$\sigma_{t,i=6}^{g=21}$	$\sigma_{t,i=1}^{g=14}$
$\sigma_{t,i=6}^{g=30}$	1.5071	−0.0046	−0.0096	−0.0072	−0.0046	−0.0107	−0.0109	−0.0105	−0.0100	−0.0044
$\sigma_{t,i=1}^{g=12}$	−0.0046	0.1215	−0.0001	0.0000	0.0000	−0.0001	−0.0001	−0.0001	−0.0001	0.0000
$\sigma_{t,i=6}^{g=17}$	−0.0096	−0.0001	0.6545	−0.0001	−0.0001	−0.0002	−0.0002	−0.0002	−0.0001	−0.0001
$\sigma_{t,i=6}^{g=16}$	−0.0072	0.0000	−0.0001	0.3768	0.0000	−0.0001	−0.0001	−0.0001	−0.0001	0.0000
$\sigma_{t,i=1}^{g=13}$	−0.0046	0.0000	−0.0001	0.0000	0.1571	−0.0001	−0.0001	−0.0001	−0.0001	0.0000
$\sigma_{t,i=6}^{g=18}$	−0.0107	−0.0001	−0.0002	−0.0001	−0.0001	0.8620	−0.0002	−0.0002	−0.0002	−0.0001
$\sigma_{t,i=6}^{g=19}$	−0.0109	−0.0001	−0.0002	−0.0001	−0.0001	−0.0002	0.9697	−0.0002	−0.0002	−0.0001
$\sigma_{t,i=6}^{g=20}$	−0.0105	−0.0001	−0.0002	−0.0001	−0.0001	−0.0002	−0.0002	1.0156	−0.0002	−0.0001
$\sigma_{t,i=6}^{g=21}$	−0.0100	−0.0001	−0.0001	−0.0001	−0.0001	−0.0002	−0.0002	−0.0002	1.0336	−0.0001
$\sigma_{t,i=1}^{g=14}$	−0.0044	0.0000	−0.0001	0.0000	0.0000	−0.0001	−0.0001	−0.0001	−0.0001	0.2109

(below) when only the 1st-order sensitivities are considered and in Table 5.29 when considering the sensitivities of 1st-order through 4th-order. Comparing the results presented in Table 5.28 with the results presented in Table 5.26 indicates that these results are very similar to each other, the same conclusion can be drawn by inter-comparing the results presented in Table 5.29 with those presented in Table 5.27. It can, therefore, be concluded that the precision of the measured leakage response has no impact on the calibration of the benchmark's parameters (microscopic group-averaged total cross-sections).

5.5.3 "Low" Precision Parameters (Relative Standard Deviations $SD = 10\%$); "Low" Precision Measurement ($SD^{(e)} = 10\%$)

Recall that the Taylor series expansion of the leakage response in terms of parameter variations is divergent for parameter standard deviations of 5% and larger. Nevertheless, it is of interest to show that the application of the 4th-BERRU-PM methodology reduces the standard deviations of the predicted calibrated parameters even in such rather extreme situations even if only the 1st-order sensitivities are considered. In this vein, consider that the uncorrelated, normally distributed, parameters (total cross-sections) have uniform relative standard deviations $SD = 10\%$ as presented in Table 5.30. The application of the 4th-BERRU-PM methodology, in other words, using Eq. (5.18) with a measured response with the same nominal value and standard deviation as for the results presented in Table 5.19 (i.e., using $SD^{(e)} = 10\%$ and incorporating only the 1st-order sensitivities) yields the results presented in Table 5.31 for the components of the posterior matrices $\mathbf{C}_{\alpha\alpha}^{(be,1)}$. Remarkably, the

TABLE 5.30
Nominal Values of the Components of $\mathbf{C}_{\alpha\alpha}$; $SD = 10\%$

	$\sigma_{t,i=6}^{g=30}$	$\sigma_{t,i=1}^{g=12}$	$\sigma_{t,i=6}^{g=17}$	$\sigma_{t,i=6}^{g=16}$	$\sigma_{t,i=1}^{g=13}$	$\sigma_{t,i=6}^{g=18}$	$\sigma_{t,i=6}^{g=19}$	$\sigma_{t,i=6}^{g=20}$	$\sigma_{t,i=6}^{g=21}$	$\sigma_{t,i=1}^{g=14}$
$\sigma_{t,i=6}^{g=30}$	8.7280	0.000	0.0000	0.0000	0.0000	0.0000	0.0000	0.0000	0.0000	0.0000
$\sigma_{t,i=1}^{g=12}$	0.0000	0.4860	0.0000	0.0000	0.0000	0.0000	0.0000	0.0000	0.0000	0.0000
$\sigma_{t,i=6}^{g=17}$	0.0000	0.0000	2.6187	0.0000	0.0000	0.0000	0.0000	0.0000	0.0000	0.0000
$\sigma_{t,i=6}^{g=16}$	0.0000	0.0000	0.0000	1.5076	0.0000	0.0000	0.0000	0.0000	0.0000	0.0000
$\sigma_{t,i=1}^{g=13}$	0.0000	0.0000	0.0000	0.0000	0.6285	0.0000	0.0000	0.0000	0.0000	0.0000
$\sigma_{t,i=6}^{g=18}$	0.0000	0.0000	0.0000	0.0000	0.0000	3.4487	0.0000	0.0000	0.0000	0.0000
$\sigma_{t,i=6}^{g=19}$	0.0000	0.0000	0.0000	0.0000	0.0000	0.0000	3.8796	0.0000	0.0000	0.0000
$\sigma_{t,i=6}^{g=20}$	0.0000	0.0000	0.0000	0.0000	0.0000	0.0000	0.0000	4.0633	0.0000	0.0000
$\sigma_{t,i=6}^{g=21}$	0.0000	0.0000	0.0000	0.0000	0.0000	0.0000	0.0000	0.0000	4.1348	0.0000
$\sigma_{t,i=1}^{g=14}$	0.0000	0.0000	0.0000	0.0000	0.0000	0.0000	0.0000	0.0000	0.0000	0.8437

results presented in Table 5.31 are similar to the results presented in Table 5.19, which were obtained for small (2%) parameter standard deviations, indicating that the variance of each of the calibrated predicted parameters has been reduced by comparison to its original value, and the originally uncorrelated parameters have become negatively correlated after the application of the 4th-BERRU-PM methodology.

5.6 4TH-BERRU-PM BEST-ESTIMATE POSTERIOR SKEWNESS OF PREDICTED RESPONSES AND CALIBRATED MODEL PARAMETERS

This section will present illustrative numerical results for the skewness of the PERP benchmark model's leakage response as well as the skewness of the three most important model parameters (total cross-sections), as predicted by the 4th-BERRU-PM methodology. It has been shown in Chapter 4 that, in the lowest approximation provided by the 4th-BERRU-PM methodology, the best-estimate predicted triple correlation among three best-estimate predicted responses, $r_k^{be}, r_\ell^{be}, r_m^{be}$, denoted as $\mu_3^{be}\left(r_k^{be}, r_\ell^{be}, r_m^{be}\right)$, has the following expression for $k,\ell,m = 1,...,TR$:

$$\begin{aligned}\mu_3^{be}\left(r_k^{be}, r_\ell^{be}, r_m^{be}\right) &\triangleq \int_D \left(r_k - r_k^{be}\right)\left(r_\ell - r_\ell^{be}\right)\left(r_m - r_m^{be}\right) p_4^{be}\left(\mathbf{r}, \boldsymbol{\alpha}\right) d\boldsymbol{\alpha}\, d\mathbf{r} \\ &= t_{ijk}^{e} - \rho_\ell c_{km}^{e} - \rho_m c_{k\ell}^{e} - \rho_k c_{\ell m}^{e} - \rho_k \rho_\ell \rho_m; \quad k,\ell,m = 1,...,TR;\end{aligned} \tag{5.19}$$

TABLE 5.31

Results for $\mathbf{C}_{\alpha\alpha}^{(be,1)}$; Only the 1st-Order Sensitivities Are Considered; $SD = 10\%$; Measured Response $SD^{(e)} = 10\%$

	$\sigma_{t,i=6}^{g=30}$	$\sigma_{t,i=1}^{g=12}$	$\sigma_{t,i=6}^{g=17}$	$\sigma_{t,i=6}^{g=16}$	$\sigma_{t,i=1}^{g=13}$	$\sigma_{t,i=6}^{g=18}$	$\sigma_{t,i=6}^{g=19}$	$\sigma_{t,i=6}^{g=20}$	$\sigma_{t,i=6}^{g=21}$	$\sigma_{t,i=1}^{g=14}$
$\sigma_{t,i=6}^{g=30}$	1.8307	−0.2294	−0.4730	−0.3564	−0.2280	−0.5282	−0.5370	−0.5188	−0.4913	−0.2181
$\sigma_{t,i=1}^{g=12}$	−0.2294	0.4784	−0.0157	−0.0119	−0.0076	−0.0176	−0.0179	−0.0173	−0.0163	−0.0073
$\sigma_{t,i=6}^{g=17}$	−0.4730	−0.0157	2.5862	−0.0244	−0.0156	−0.0362	−0.0368	−0.0356	−0.0337	−0.0150
$\sigma_{t,i=6}^{g=16}$	−0.3564	−0.0119	−0.0244	1.4892	−0.0118	−0.0273	−0.0277	−0.0268	−0.0254	−0.0113
$\sigma_{t,i=1}^{g=13}$	−0.2280	−0.0076	−0.0156	−0.0118	0.6209	−0.0175	−0.0178	−0.0172	−0.0162	−0.0072
$\sigma_{t,i=6}^{g=18}$	−0.5282	−0.0176	−0.0362	−0.0273	−0.0175	3.4083	−0.0411	−0.0397	−0.0376	−0.0167
$\sigma_{t,i=6}^{g=19}$	−0.5370	−0.0179	−0.0368	−0.0277	−0.0178	−0.0411	3.8378	−0.0404	−0.0382	−0.0170
$\sigma_{t,i=6}^{g=20}$	−0.5188	−0.0173	−0.0356	−0.0268	−0.0172	−0.0397	−0.0404	4.0242	−0.0369	−0.0164
$\sigma_{t,i=6}^{g=21}$	−0.4913	−0.0163	−0.0337	−0.0254	−0.0162	−0.0376	−0.0382	−0.0369	4.0998	−0.0155
$\sigma_{t,i=1}^{g=14}$	−0.2181	−0.0073	−0.0150	−0.0113	−0.0072	−0.0167	−0.0170	−0.0164	−0.0155	0.8368

$$\rho_k \triangleq \left\{ \mathbf{C}^e \left(\mathbf{C}^e + \mathbf{C}^c_{rr} \right)^{-1} \left[\mathbf{E}_c(\mathbf{r}) - \mathbf{r}^e \right] + \mathbf{p}_1 \left(\mathbf{r}_s^{(0)}, \boldsymbol{\alpha}_s^{(0)} \right) \right\}_k . \tag{5.20}$$

For the particular information available for the PERP reactor physics benchmark, the experimentally measured skewness of the leakage response is not available so the respective measurement is assumed to follow a normal distribution (implying that its skewness vanishes, i.e., $t_L^e = 0$), while the higher-order terms in Eq. (5.19) are negligible by comparison to the leading term. Thus, the best-estimate skewness of the predicted leakage response, denoted as $\mu_3^{(be,i)}\left(L^{(be,i)}\right)$, $i = 1,2,3,4,$ is a scalar which will have one of the expressions provided, differing from one another depending on the highest-order of retained sensitivities:

$$\mu_3^{(be,i)}\left(L^{(be,i)}\right) = -3\rho^{(i)}\left[SD^{(e)}\right]^2 - \left[\rho^{(i)}\right]^3; \quad i = 1,2,3,4; \tag{5.21}$$

where:

$$\rho^{(i)} = \left[SD^{(e)}\right]^2 \left\{ \left[SD^{(e)}\right]^2 + \left[\mathrm{var}(L)\right]^{(i,U,N)} \right\}^{-1} \left[E^{(i)}(L^c) - L^e \right]. \tag{5.22}$$

The most important triple correlation is the skewness of each individual best-estimate predicted response and calibrated parameters. The skewness of the predicted best-estimate leakage response denoted as $Skew^{(be,i)}\left(L^{(be,i)}\right)$, where the index "$i$" denotes the highest-order of sensitivities retained in the respective computation, is computed using the expression:

$$Skew^{(be,i)}\left(L^{(be,i)}\right) = \mu_3^{(be,i)}\left(L^{(be,i)}\right)\left[SD\left(L^{(be,i)}\right)\right]^{-3}, \quad i = 1,2,3,4, \tag{5.23}$$

where the expression of the triple correlation $\mu_3^{(be,i)}\left(L^{(be,i)}\right)$ is provided in Eq. (5.21), while the expression of the standard deviation $SD\left(L^{(be,i)}\right)$ is provided in one of Eqs. (5.2)–(5.9), depending on the value of the index "i."

It has been shown in Chapter 4 that, in the lowest approximation provided by the 4th-BERRU-PM methodology, the best-estimate predicted triple correlation among three best-estimate calibrated model parameters, α_k^{be}, α_ℓ^{be}, α_m^{be}, denoted as $\mu_3^{be}\left(\alpha_k^{be}, \alpha_\ell^{be}, \alpha_m^{be}\right)$ for $k, \ell, m = 1, ..., TP$, has the following expression:

$$\begin{aligned} \mu_3^{be}\left(\alpha_k^{be}, \alpha_\ell^{be}, \alpha_m^{be}\right) &\triangleq \int_D \left(\alpha_k - \alpha_k^{be}\right)\left(\alpha_\ell - \alpha_\ell^{be}\right)\left(\alpha_m - \alpha_m^{be}\right) p_4^{be}(\mathbf{r}, \boldsymbol{\alpha})\, d\boldsymbol{\alpha}\, d\mathbf{r} \\ &= t_{k\ell m}^{\alpha} + \beta_\ell c_{km}^{\alpha} + \beta_m c_{k\ell}^{\alpha} + \beta_k c_{\ell m}^{\alpha} + \beta_k \beta_\ell \beta_m . \end{aligned} \tag{5.24}$$

For the information available for the PERP reactor physics benchmark, the expression in Eq. (5.24) reduces to the following expression for the best-estimate triple auto-correlation, $\mu_3^{(be,i)}\left(\alpha_k^{(be,i)}\right)$, of a calibrated best-estimate parameter, $\alpha_k^{(be,i)}$:

$$\mu_3^{(be,i)}\left(\alpha_k^{(be,i)}\right)=\beta_k^{(i)}\left[3\operatorname{var}\left(\alpha_k\right)+\left(\beta_k^{(i)}\right)^2\right];\ k=\ell=m;\ k=1,...,TP. \quad (5.25)$$

where:

$$\beta_k^{(i)} \triangleq \left\{\mathbf{C}_{\alpha r}^{(c,i)}\right\}_k \left\{\left[SD^{(e)}\right]^2+\left[\operatorname{var}(L)\right]^{(i,U,N)}\right\}^{-1}\left[E^{(i)}(L^c)-L^e\right]. \quad (5.26)$$

The skewness of a predicted best-estimate calibrated parameter $\alpha_k^{(be,i)}$, $k=1,...,TP$, will be denoted as $Skew^{(be,i)}\left(\alpha_k^{(be,i)}\right)$, where the index "$i$" denotes the highest-order of sensitivities retained in the respective computation, and is computed using the expression:

$$Skew^{(be,i)}\left(\alpha_k^{(be,i)}\right)=\mu_3^{(be,i)}\left(\alpha_k^{(be,i)}\right)\left[SD\left(\alpha_k^{(be,i)}\right)\right]^{-3},\ i=1,2,3,4. \quad (5.27)$$

In Eq. (5.27), the expression of the triple correlation $\mu_3^{(be,i)}\left(\alpha_k^{(be,i)}\right)$ is provided in Eq. (5.25), while the standard deviation $SD\left(\alpha_k^{(be,i)}\right)$ is the square root of the respective diagonal element of $\mathbf{C}_{\alpha\alpha}^{(be,i)}$, which is obtained using the expression provided in Eq. (5.18).

Considering a "*low*" precision measured response, having the nominal value $L^e=1.94\times10^6$ n/s, and a relative standard deviation $SD^{(e)}=10\%$, the numerical results obtained for the skewness $Skew^{(be,i)}\left(L^{(be,i)}\right)$ of the predicted best-estimate leakage response, and for the skewness $Skew^{(be,i)}\left(\alpha_k^{(be,i)}\right)$ for each of the three most important predicted calibrated model parameters are presented in Tables 5.32 and 5.33, respectively.

The results presented in Table 5.32 are for uniform relative standard deviations of 2% for the model parameters, for which the Taylor series expansion of the computed response in terms of parameter deviations is convergent. Since the measurements were considered to be normally distributed, their skewness vanishes. Furthermore, the contribution of the initial skewness of the computed responses enters in the

TABLE 5.32

Best-Estimate Values $\mu_3^{(be,i)}$ and $Skew^{(be,i)}$, $i = 1,2,3,4$; Parameter $SD = 2\%$

Skewness	$i = 1$	$i = 2$	$i = 3$	$i = 4$
$Skew^{(be,i)}\left(L^{(be,i)}\right)$	8.595×10^{-1}	-2.604×10^{-2}	-1.147×10^{-2}	-1.224×10^{-1}
$Skew^{(be,i)}\left[\left(\sigma_{t,i=6}^{g=30}\right)^{(be,i)}\right]$	1.805×10^{0}	-4.713×10^{-2}	-4.888×10^{-2}	-4.508×10^{-1}
$Skew^{(be,i)}\left[\left(\sigma_{t,i=1}^{g=12}\right)^{(be,i)}\right]$	1.424×10^{-1}	-4.741×10^{-3}	-2.417×10^{-3}	-2.654×10^{-2}
$Skew^{(be,i)}\left[\left(\sigma_{t,i=6}^{g=17}\right)^{(be,i)}\right]$	1.263×10^{-1}	-4.207×10^{-3}	-2.149×10^{-3}	-2.361×10^{-2}

TABLE 5.33

Best-Estimate Values $\mu_3^{(be,i)}$ and $Skew^{(be,i)}$, $i = 1,2,3,4$; Parameter $SD = 5\%$

Skewness	$i = 1$	$i = 2$	$i = 3$	$i = 4$
$Skew^{(be,i)}\left(L^{(be,i)}\right)$	1.229×10^{-1}	-2.154×10^{-1}	-2.178×10^{-2}	-5.061×10^{-2}
$Skew^{(be,i)}\left[\left(\sigma_{t,i=6}^{g=30}\right)^{(be,i)}\right]$	1.050×10^{0}	-1.079×10^{0}	-7.916×10^{-1}	-1.366×10^{0}
$Skew^{(be,i)}\left[\left(\sigma_{t,i=1}^{g=12}\right)^{(be,i)}\right]$	6.963×10^{-2}	-1.267×10^{-1}	-1.334×10^{-2}	-3.102×10^{-2}
$Skew^{(be,i)}\left[\left(\sigma_{t,i=6}^{g=17}\right)^{(be,i)}\right]$	6.176×10^{-2}	-1.125×10^{-1}	-1.196×10^{-2}	-2.781×10^{-2}

expression of the higher-order term $\mathbf{p}_1\left(\mathbf{r}_s^{(0)},\boldsymbol{\alpha}_s^{(0)}\right)$ in Eq. (5.20), which was neglected by comparison to the other terms in this equation. Thus, the skewness of the measured response and the skewness of the computed responses do not influence the predicted skewness, within the 2nd-order approximation inherent in the expression of the predicted skewness. When only the 1st-order sensitivities are considered, the small positive number for the predicted skewness in Table 5.32 indicates that the MaxEnt distribution of the predicted best-estimate leakage response is slightly tilted to the right of the distribution's mean value. However, when the 2nd- and higher-order sensitivities are also considered, the MaxEnt distribution of the predicted best-estimate leakage response becomes tilted just slightly to the left of the distribution's mean value, but the respective tilt is negligibly small. These results indicate that results based exclusively on 1st-order sensitivities are unreliable. It is, therefore, paramount to at least consider the 2nd-order sensitivities, in addition to the 1st-order ones. The previous conclusions hold both for the predicted leakage response and for the calibrated predicted parameters. The results presented in Table 5.33 for uniform parameter standard deviations of 5% indicate that the conclusions drawn from the results in Table 5.32 also hold for these larger standard deviations, which are just outside the convergence radius of the Taylor series.

Considering a "*typical*" precision-measured response, having the nominal value $L^e = 1.94\times10^6$ n/s, and a relative standard deviation $SD^{(e)} = 5\%$ (rather than $SD^{(e)} = 10\%$), the numerical results obtained for the skewness $Skew^{(be,i)}\left(L^{(be,i)}\right)$ of the predicted best-estimate leakage response, and for the skewness $Skew^{(be,i)}\left(\alpha_k^{(be,i)}\right)$ for each of the three most important predicted calibrated model parameters are presented in Tables 5.34 and 5.35, respectively.

The results presented in Table 5.34 are for parameter relative standard deviations of 2% for the model parameters, for which the Taylor series expansion of the computed response in terms of parameter deviations is convergent, while the results presented in Table 5.35 are for parameter larger standard deviations (5%), for which this Taylor series diverges. Within the 2nd-order approximation inherent in the expressions of the predicted skewness, namely Eqs. (5.23) and (5.27), the skewness of the

TABLE 5.34

Best-Estimate Values for $Skew^{(be,i)}\left(L^{(be,i)}\right)$ and $Skew^{(be,i)}\left(\alpha_k^{(be,i)}\right)$, $i = 1,2,3,4$; Parameter Relative Standard Deviations $SD = 2\%$

Skewness	$i = 1$	$i = 2$	$i = 3$	$i = 4$
$Skew^{(be,i)}\left(L^{(be,i)}\right)$	3.902×10^{-1}	-1.216×10^{-2}	-5.546×10^{-3}	-5.959×10^{-2}
$Skew^{(be,i)}\left[\left(\sigma_{t,i=6}^{g=30}\right)^{(be,i)}\right]$	2.882×10^{0}	-1.114×10^{-1}	-5.654×10^{-2}	-4.912×10^{-1}
$Skew^{(be,i)}\left[\left(\sigma_{t,i=1}^{g=12}\right)^{(be,i)}\right]$	1.134×10^{-1}	-9.040×10^{-3}	-1.725×10^{-3}	-1.864×10^{-2}
$Skew^{(be,i)}\left[\left(\sigma_{t,i=6}^{g=17}\right)^{(be,i)}\right]$	1.509×10^{-1}	-1.204×10^{-2}	-2.299×10^{-3}	-2.489×10^{-2}

TABLE 5.35

Best-Estimate Values for $Skew^{(be,i)}\left(L^{(be,i)}\right)$ and $Skew^{(be,i)}\left(\alpha_k^{(be,i)}\right)$, $i = 1,2,3,4$; Parameter Relative Standard Deviations $SD = 5\%$

Skewness	$i = 1$	$i = 2$	$i = 3$	$i = 4$
$Skew^{(be,i)}\left(L^{(be,i)}\right)$	6.045×10^{-2}	-1.070×10^{-2}	-1.088×10^{-2}	-2.530×10^{-2}
$Skew^{(be,i)}\left[\left(\sigma_{t,i=6}^{g=30}\right)^{(be,i)}\right]$	1.155×10^{0}	-5.576×10^{-1}	-7.933×10^{-1}	-1.367×10^{0}
$Skew^{(be,i)}\left[\left(\sigma_{t,i=1}^{g=12}\right)^{(be,i)}\right]$	4.795×10^{-2}	-3.425×10^{-2}	-8.902×10^{-3}	-2.069×10^{-2}
$Skew^{(be,i)}\left[\left(\sigma_{t,i=6}^{g=17}\right)^{(be,i)}\right]$	6.379×10^{-2}	-4.557×10^{-2}	-1.197×10^{-2}	-2.783×10^{-2}

measured response and the skewness of the computed responses do not influence the predicted skewness. The results presented in Tables 5.34 and 5.35 display the same trends as the results presented in Tables 5.32 and 5.33, thereby indicating that the precision (i.e., $SD^{(e)} = 5\%$ versus $SD^{(e)} = 10\%$) of the measured response slightly affects the numerical values of the individual response and parameter skewness but does not affect the general trend, which is significantly influenced by the order of the sensitivities retained in the respective computations.

5.7 4TH-BERRU-PM BEST-ESTIMATE POSTERIOR KURTOSIS OF PREDICTED RESPONSES AND CALIBRATED MODEL PARAMETERS

It has been shown in Chapter 4 that the best-estimate predicted quadruple correlation among four best-estimate predicted responses, $r_k^{be}, r_\ell^{be}, r_m^{be}$, r_n^{be}, denoted as $\mu_4^{be}\left(r_k^{be}, r_\ell^{be}, r_m^{be}, r_n^{be}\right)$ for $k, \ell, m, n = 1,...,TR$, has the following expressions:

$$\mu_4^{be}\left(r_k^{be},r_\ell^{be},r_m^{be},r_n^{be}\right)\triangleq\int_D\left(r_k-r_k^{be}\right)\left(r_\ell-r_\ell^{be}\right)\left(r_m-r_m^{be}\right)\left(r_n-r_n^{be}\right)p_4^{be}\left(\mathbf{r},\boldsymbol{\alpha}\right)d\boldsymbol{\alpha}\,d\mathbf{r}$$
$$=q_{k\ell mn}^e-\rho_\ell t_{kmn}^e-\rho_m t_{k\ell n}^e-\rho_n t_{k\ell m}^e-\rho_k t_{\ell mn}^e+\rho_\ell\rho_m c_{kn}^e$$
$$+\rho_k\rho_\ell c_{mn}^e+\rho_k\rho_m c_{\ell n}^e+\rho_n\rho_\ell c_{km}^e+\rho_n\rho_m c_{k\ell}^e+\rho_k\rho_n c_{\ell m}^e+\rho_k\rho_\ell\rho_m\rho_n. \quad (5.28)$$

In particular, the best-estimate kurtosis of the best-estimate predicted leakage response, denoted as $\mu_4^{(be,i)}\left(L^{(be,i)}\right)$, $i=1,2,3,4$, is obtained as a particular case of Eq. (5.28). For the PERP benchmark, the experimentally measured skewness of the leakage response is not available, so the respective measurement is assumed to follow a normal distribution with $q_L^e=3\left[SD^{(e)}\right]^4$. Consequently, the best-estimate kurtosis of the predicted leakage response, $\mu_4^{(be,i)}\left(L^{(be,i)}\right)$, is a scalar which will have one of the expressions provided, differing from one another depending on the highest-order $\left(i=1,2,3,4\right)$ of retained sensitivities:

$$\mu_4^{(be,i)}\left(L^{(be,i)}\right)=q_L^e+6\left[\rho^{(i)}\right]^2\left[SD^{(e)}\right]^2+\left[\rho^{(i)}\right]^4. \quad (5.29)$$

The kurtosis of the predicted best-estimate leakage response denoted as $Kurt^{(be,i)}\left(L^{(be,i)}\right)$, where the index "$i$" denotes the highest-order of sensitivities retained in the respective computation, is computed using the following expression:

$$Kurt^{(be,i)}\left(L^{(be,i)}\right)=\mu_4^{(be,i)}\left(L^{(be,i)}\right)\left[SD\left(L^{(be,i)}\right)\right]^{-4},\; i=1,2,3,4, \quad (5.30)$$

where the expression of the quadruple correlation $\mu_4^{(be,i)}\left(L^{(be,i)}\right)$ is provided in Eq. (5.29), while the expressions of the standard deviation $SD\left(L^{(be,i)}\right)$ are provided in the corresponding Eqs. (5.2)–(5.9).

As has been shown in Chapter 4, the best-estimate predicted quadruple correlation among four best-estimate calibrated model parameters, α_k^{be}, α_ℓ^{be}, α_m^{be}, α_n^{be}, denoted as $\mu_4^{be}\left(\alpha_k^{be},\alpha_\ell^{be},\alpha_m^{be},\alpha_n^{be}\right)$ for $k,\ell,m,n=1,...,TP$, has the following expressions:

$$\mu_4^{be}\left(\alpha_k^{be},\alpha_\ell^{be},\alpha_m^{be},\alpha_n^{be}\right)\triangleq\int_D\left(\alpha_k-\alpha_k^{be}\right)\left(\alpha_\ell-\alpha_\ell^{be}\right)\left(\alpha_m-\alpha_m^{be}\right)\left(\alpha_n-\alpha_n^{be}\right)$$
$$p_4^{be}\left(\mathbf{r},\boldsymbol{\alpha}\right)d\boldsymbol{\alpha}\,d\mathbf{r} \quad (5.31)$$
$$=q_{k\ell mn}^\alpha+\beta_\ell t_{kmn}^\alpha+\beta_m t_{k\ell n}^\alpha+\beta_k t_{\ell mn}^\alpha+\beta_n t_{k\ell m}^\alpha+\beta_k\beta_\ell c_{mn}^\alpha+\beta_k\beta_m c_{\ell n}^\alpha$$
$$+\beta_k\beta_n c_{\ell m}^\alpha+\beta_\ell\beta_m c_{kn}^\alpha+\beta_\ell\beta_n c_{km}^\alpha+\beta_m\beta_n c_{k\ell}^\alpha+\beta_k\beta_\ell\beta_m\beta_n.$$

For the PERP benchmark, the best-estimate kurtosis of a calibrated best-estimate parameter (total cross-section), denoted as α_k^{be}, is provided by the following particular case of Eq. (5.31), for $k=\ell=m=n=1,...,TP;\; i=1,2,3,4$:

$$\mu_4^{(be,i)}\left(\alpha_k^{(be,i)}\right)=3\left[\mathrm{var}\left(\alpha_k\right)\right]^2+\left(\beta_k^{(i)}\right)^4+6\left(\beta_k^{(i)}\right)^2\mathrm{var}\left(\alpha_k\right). \tag{5.32}$$

The kurtosis of a predicted best-estimate calibrated parameter $\alpha_k^{(be,i)}$, $k=1,...,TP$, will be denoted as $Kurt^{(be,i)}\left(\alpha_k^{(be,i)}\right)$, where the index "*i*" denotes the highest-order of sensitivities retained in the respective computation, and is computed using the following expression:

$$Kurt^{(be,i)}\left(\alpha_k^{(be,i)}\right)=\mu_4^{(be,i)}\left(\alpha_k^{(be,i)}\right)\left[SD\left(\alpha_k^{(be,i)}\right)\right]^{-4},\ i=1,2,3,4 \tag{5.33}$$

In Eq. (5.33), the expression of the quadruple correlation $\mu_4^{(be,i)}\left(\alpha_k^{(be,i)}\right)$ is provided in Eq. (5.32) while the standard deviation $SD\left(\alpha_k^{(be,i)}\right)$ is the square root of the respective diagonal element of $\mathbf{C}_{\alpha\alpha}^{(be,i)}$, which is obtained using the expression provided in Eq. (5.18).

Considering a "*low*" precision measured response, having the nominal value $L^e=1.94\times10^6$ n/s, and a relative standard deviation $SD^{(e)}=10\%$, the numerical results obtained for the kurtosis $Kurt^{(be,i)}\left(L^{(be,i)}\right)$ of the predicted best-estimate leakage response, and for the kurtosis $Kurt^{(be,i)}\left(\alpha_k^{(be,i)}\right)$, for each of the three most important calibrated model parameters, are presented in Table 5.36 (for parameter standard deviations of 2%) and Table 5.37 (for parameter standard deviations of 5%).

The results presented in Table 5.36, for which the Taylor series expansion of the computed response in terms of parameter deviations is convergent, indicate that when only the 1st-order sensitivities are considered, the kurtosis of the MaxEnt distribution of the predicted best-estimate leakage response is somewhat "more peaked" than the initial normal distribution. However, when the 2nd- and higher-order sensitivities are also considered, the MaxEnt distribution of the predicted best-estimate leakage response tends towards the normal distribution (for which "kurtosis = 3.00").

TABLE 5.36
Best-Estimate Kurtosis of Predicted Response and Parameters; Parameter *SD* = 2%

Kurtosis	*i* = 1	*i* = 2	*i* = 3	*i* = 4
$Kurt^{(be,i)}\left(L^{(be,i)}\right)$	5.260	4.364	3.599	3.548
$Kurt^{(be,i)}\left[\left(\sigma_{t,i=6}^{g=30}\right)^{(be,i)}\right]$	4.885	3.001	3.002	3.134
$Kurt^{(be,i)}\left[\left(\sigma_{t,i=1}^{g=12}\right)^{(be,i)}\right]$	3.013	3.000	3.000	3.000
$Kurt^{(be,i)}\left[\left(\sigma_{t,i=6}^{g=17}\right)^{(be,i)}\right]$	3.011	3.000	3.000	3.000

TABLE 5.37
Best-Estimate Kurtosis of Predicted Response and Parameters; Parameter $SD = 5\%$

Kurtosis	$i = 1$	$i = 2$	$i = 3$	$i = 4$
$Kurt^{(be,i)}\left(L^{(be,i)}\right)$	3.280	3.116	3.009	3.004
$Kurt^{(be,i)}\left[\left(\sigma_{t,i=6}^{g=30}\right)^{(be,i)}\right]$	3.695	3.732	3.404	4.138
$Kurt^{(be,i)}\left[\left(\sigma_{t,i=1}^{g=12}\right)^{(be,i)}\right]$	3.003	3.011	3.000	3.001
$Kurt^{(be,i)}\left[\left(\sigma_{t,i=6}^{g=17}\right)^{(be,i)}\right]$	3.003	3.008	3.000	3.001

The same trend is also observed for the most important parameters: recall that the parameters were considered to have initially been normally distributed and uncorrelated, which implies that the distribution of each parameter initially had the value "*initial parameter kurtosis = 3*." The MaxEnt distribution of the best-estimate calibrated parameters also remains around the value (= 3) of the corresponding normal distribution. The results presented in Table 5.36 for uniform parameter standard deviations of 5% (for which the Taylor series expansion of the computed response in terms of parameter deviations diverges) indicate that the conclusions drawn from the results in Table 5.35 also hold for these larger standard deviations.

Considering a "*typical*" precision-measured response with the nominal value $L^e = 1.94 \times 10^6$ n/s and a relative standard deviation $SD^{(e)} = 5\%$, the numerical results obtained for the kurtosis $Kurt^{(be,i)}\left(L^{(be,i)}\right)$ of the predicted best-estimate leakage response, and for the kurtosis $Kurt^{(be,i)}\left(\alpha_k^{(be,i)}\right)$; for each of the three most important calibrated model parameters, are presented in Tables 5.38 and 5.38.

The results presented in Table 5.38 are for parameter standard deviations of 2%, for which the Taylor series expansion of the computed response in terms of parameter deviations converges. When only the 1st-order sensitivities are considered, the kurtosis of the MaxEnt distribution of the predicted best-estimate leakage response is somewhat "more peaked" than the initial normal distribution. However, when the 2nd- and higher-order sensitivities are also considered, the MaxEnt distribution of the predicted best-estimate leakage response converges to the normal distribution (for which "kurtosis = 3.00"). The same trend is also observed for the most important parameters: recall that the parameters were considered to have initially been normally distributed and uncorrelated, which implies that the distribution of each parameter initially had the value "*initial parameter kurtosis = 3*." The MaxEnt distribution of the best-estimate calibrated parameters also remains around the value (= 3) of the corresponding normal distribution. The results presented in Table 5.39 for uniform parameter standard deviations of 5% (for which the Taylor series expansion of the computed response in terms of parameter deviations diverges) indicate

TABLE 5.38
Best-Estimate Kurtosis of Predicted Response and Parameters; Parameter $SD = 2\%$

Kurtosis	$i = 1$	$i = 2$	$i = 3$	$i = 4$
$Kurt^{(be,i)}\left(L^{(be,i)}\right)$	3.522	3.317	3.145	3.114
$Kurt^{(be,i)}\left[\left(\sigma_{t,i=6}^{g=30}\right)^{(be,i)}\right]$	7.179	3.008	3.002	3.159
$Kurt^{(be,i)}\left[\left(\sigma_{t,i=1}^{g=12}\right)^{(be,i)}\right]$	3.019	3.000	3.000	3.001
$Kurt^{(be,i)}\left[\left(\sigma_{t,i=6}^{g=17}\right)^{(be,i)}\right]$	3.015	3.000	3.000	3.000

TABLE 5.39
Best-Estimate Kurtosis of Predicted Response and Parameters; Parameter $SD = 5\%$

Kurtosis	$i = 1$	$i = 2$	$i = 3$	$i = 4$
$Kurt^{(be,i)}\left(L^{(be,i)}\right)$	3.069	3.029	3.002	3.001
$Kurt^{(be,i)}\left[\left(\sigma_{t,i=6}^{g=30}\right)^{(be,i)}\right]$	3.833	3.204	3.406	4.140
$Kurt^{(be,i)}\left[\left(\sigma_{t,i=1}^{g=12}\right)^{(be,i)}\right]$	3.003	3.002	3.000	3.001
$Kurt^{(be,i)}\left[\left(\sigma_{t,i=6}^{g=17}\right)^{(be,i)}\right]$	3.003	3.001	3.000	3.001

that the conclusions drawn from the results in Table 5.38 also hold for these larger standard deviations.

It follows form, the results presented in Tables 5.36–5.39 that the standard deviations (i.e., "precision") of the measured response do not influence the predicted skewness of either the best-estimate predicted response distribution or the predicted distributions of the best-estimate model parameters (total cross-sections) for the PERP reactor physics benchmark.

5.8 CONCLUDING REMARKS

This chapter has illustrated quantitatively the end results obtained by applying the 4th-BERRU-PM methodology to the PERP reactor physics benchmark. The result ("response") of interest for the PERP benchmark is the leakage of neutrons through the outer surface of this spherical benchmark. Only the 180 group-averaged microscopic total cross-sections (out of the much larger number of PERP model parameters) were considered in this Chapter for illustrating numerically the characteristic results produced by the fundamental components of the 4th-BERRU-PM methodology, namely the 4th-order "sensitivity analysis (SA)," the 4th-order "uncertainty quantification (UQ)," the 4th-order "data assimilation (DA)" and the 4th-order "model calibration (MC)" outputs of the 4th-BERRU-PM methodology, encompassing the 1st-order through 4th-order moments (predicted mean values, covariances, 3rd- and 4th-order correlations, including skewness and kurtosis of responses and calibrated parameters) of the 4th-order MaxEnt posterior combined distribution of experimentally measured and computed model responses and parameters.

The 4th-BERRU-PM methodology enjoys unrivaled computational efficiency since the only matrix inversion involved is the computation of $\left(\mathbf{C}^{e}+\mathbf{C}_{rr}^{c}\right)^{-1}$, which in the case of a single response (i.e., the benchmark's leakage response) reduces to the trivial reciprocal of a scalar, which for the PERP benchmark is the quantity $\left[SD^{(e)}\right]^{2}+\left[\operatorname{var}(L)\right]_{t}^{(i,U,N)}$. Also, the 4th-BERRU-PM methodology predicted posterior parameter-response correlations are not obtainable by any of the extant data assimilation methods.

The impact of the high-order sensitivities was quantified for "high precision" parameters (2% standard deviations) and "typical precision" parameters (5% standard deviations). Analyzing the best-estimate results with reduced uncertainties for the 1st- through 4th-order moments (mean values, covariance, skewness, and kurtosis) produced by the 4th-BERRU-PM methodology for the PERP benchmark has indicated that, even for systems modeled by linear equations (such as the PERP benchmark), retaining only 1st-order sensitivities is insufficient for reliable predictive modeling (including SA, UQ, DA, MC). At least 2nd-order sensitivities should be retained in order to obtain reliable predictions. For large parameter uncertainties, it is also advisable to apply the 4th-BERRU-PM methodology sequentially and repeatedly, using new experimental information in conjunction with the calibrated model parameters from the previous application of the 4th-BERRU-PM methodology.

6 4th-Order Comprehensive Adjoint Sensitivity Analysis Methodology

Mathematical Framework

6.1 INTRODUCTION

The material presented in this chapter is based on the work of Cacuci (2023a) and summarizes the mathematical underpinnings of the 4th-order comprehensive sensitivity analysis methodology for nonlinear models. This methodology is used for obtaining and subsequently computing, most efficiently, the exact expressions of the 1st-, 2nd-, 3rd, and 4th-order sensitivities of a nonlinear model's response with respect to the model's parameters. This 4th-order sensitivity analysis methodology can also be applied for linear models if the response of interest only depends on the forward state functions but does not simultaneously depend on the linear model's adjoint state function(s). This is the case for the PERP benchmark's leakage response, so the methodology presented in this chapter provides the basic concepts for establishing the specific mathematical framework that enables the computations of the sensitivities (1st- through 4th-order) of the PERP benchmark's leakage response with respect to the benchmark model's parameters, as will be presented in Chapter 7.

6.2 1ST-ORDER COMPREHENSIVE ADJOINT SENSITIVITY ANALYSIS METHODOLOGY

As has been discussed in Chapter 1, a nonlinear physical system can be generally modeled by means of coupled equations which can be represented in operator form as follows:

$$\mathbf{N}\left[\mathbf{u}(\mathbf{x}),\boldsymbol{\alpha}\right]=\mathbf{Q}(\mathbf{x},\boldsymbol{\alpha}),\ \ \mathbf{x}\in\Omega_x(\boldsymbol{\alpha}), \tag{6.1}$$

The boundary and/or initial conditions associated with Eq. (6.1) are represented as follows:

$$\mathbf{B}\left[\mathbf{u}(\mathbf{x});\boldsymbol{\alpha};\mathbf{x}\right]-\mathbf{B}^s(\mathbf{x},\boldsymbol{\alpha})=\mathbf{0},\ \ \mathbf{x}\in\partial\Omega_x(\boldsymbol{\alpha}), \tag{6.2}$$

DOI: 10.1201/9781003478119-6

where the vector $\mathbf{B}^s(\mathbf{x},\boldsymbol{\alpha})$ represents "source terms on the model's boundary."

Solving Eqs. (6.1) and (6.2) at the nominal parameter values, $\boldsymbol{\alpha}^0$, provides the "nominal solution" $\mathbf{u}^0(\mathbf{x})$, in other words, the vectors $\mathbf{u}^0(\mathbf{x})$ and $\boldsymbol{\alpha}^0$ satisfy the following equations:

$$\mathbf{N}\left[\mathbf{u}^0(\mathbf{x});\boldsymbol{\alpha}^0\right]=\mathbf{Q}\left(\mathbf{x},\boldsymbol{\alpha}^0\right), \; \mathbf{x}\in\Omega_x, \tag{6.3}$$

$$\mathbf{B}\left[\mathbf{u}^0(\mathbf{x});\boldsymbol{\alpha}^0;\,\mathbf{x}\right]-\mathbf{B}^s\left(\mathbf{x},\boldsymbol{\alpha}^0\right)=\mathbf{0}, \; \mathbf{x}\in\partial\Omega_x\left(\boldsymbol{\alpha}^0\right). \tag{6.4}$$

The model response (result) is computed using the mathematical model represented by Eqs. (6.1) and (6.2) is a function of the model's dependent variables $\mathbf{u}(\mathbf{x})$ and parameters $\boldsymbol{\alpha}$ which can be represented formally as follows:

$$R(\mathbf{e})\triangleq R\left[\mathbf{u}(\mathbf{x});\boldsymbol{\alpha}\right]; \; \mathbf{e}\triangleq(\boldsymbol{\alpha},\mathbf{u})\in\mathscr{E}; \; \mathscr{E}=\mathscr{E}_\alpha\times\mathscr{E}_u. \tag{6.5}$$

Cacuci (1981a, 1981b) has shown that the most general definition of the sensitivity of an operator-valued model response $R(\mathbf{e})$, where $\mathbf{e}\triangleq(\boldsymbol{\alpha},\mathbf{u})\in\mathscr{E}$, to variations $\mathbf{h}\triangleq\left(\delta\boldsymbol{\alpha},\mathbf{v}^{(1)}\right)$ in the model parameters and state functions in a neighborhood around the nominal functions and parameter values $\mathbf{e}^0\triangleq\left(\boldsymbol{\alpha}^0,\mathbf{u}^0\right)\in\mathscr{E}$, is given by the 1st-order Gateaux- (G-) variation, which will be denoted as $\delta R\left(\mathbf{e}^0;\mathbf{h}\right)$ and is defined as follows:

$$\delta\mathbf{R}\left(\mathbf{e}^0;\mathbf{h}\right)\triangleq\left\{\frac{d}{d\varepsilon}\left[R\left(\mathbf{e}^0+\varepsilon\mathbf{h}\right)\right]\right\}_{\varepsilon=0}\triangleq\lim_{\varepsilon\to0}\left\{\frac{1}{\varepsilon}\left[R\left(\mathbf{e}^0+\varepsilon\mathbf{h}\right)-R\left(\mathbf{e}^0\right)\right]\right\}, \tag{6.6}$$

for a scalar $\varepsilon\in\mathscr{F}$ and for all (i.e., arbitrary) vectors $\mathbf{h}\in\mathscr{E}=\mathscr{E}_\alpha\times\mathscr{E}_u$ in a neighborhood $\left(\mathbf{e}^0+\varepsilon\mathbf{h}\right)$ around $\mathbf{e}^0=\left(\boldsymbol{\alpha}^0,\mathbf{u}^0\right)\in\mathscr{E}$. The G-variation $\delta R\left(\mathbf{e}^0;\mathbf{h}\right)$ is an operator defined on the same domain as $R(\mathbf{e})$ and has the same range as $R(\mathbf{e})$. The G-variation $\delta R\left(\mathbf{e}^0;\mathbf{h}\right)$ satisfies the relation: $R\left(\mathbf{e}^0+\varepsilon\mathbf{h}\right)-R\left(\mathbf{e}^0\right)=\delta R\left(\mathbf{e}^0;\mathbf{h}\right)+\Delta(\mathbf{h})$, with $\lim_{\varepsilon\to0}\left[\Delta(\varepsilon\mathbf{h})\right]/\varepsilon=0$. The existence of the G-variation $\delta R\left(\mathbf{e}^0;\mathbf{h}\right)$ does not guarantee its numerical computability. Numerical methods most often require that $\delta R\left(\mathbf{e}^0;\mathbf{h}\right)$ be linear in $\mathbf{h}\triangleq\left(\delta\boldsymbol{\alpha};\mathbf{v}^{(1)}\right)$ in a neighborhood $\left(\mathbf{e}^0+\varepsilon\mathbf{h}\right)$ around $\mathbf{e}^0=\left(\mathbf{u}^0,\boldsymbol{\alpha}^0\right)\in\mathscr{E}$. Formally, the necessary and sufficient conditions for the G-variation $\delta R\left(\mathbf{e}^0;\mathbf{h}\right)$ of a nonlinear operator $R(\mathbf{e})$ to be linear and continuous in $\mathbf{h}$ in a neighborhood $\left(\mathbf{e}^0+\varepsilon\mathbf{h}\right)$ around $\mathbf{e}^0=\left(\boldsymbol{\alpha}^0,\mathbf{u}^0\right)$ and thus admit a total 1st-order G-derivative, are as follows:

1. $R(\mathbf{e})$ satisfies a weak Lipschitz condition at $\mathbf{e}^0$, namely:

$$\left\|R\left(\mathbf{e}^0+\varepsilon\mathbf{h}\right)-R\left(\mathbf{e}^0\right)\right\|\le k\left\|\varepsilon\left(\mathbf{e}^0\right)\right\|, \; k<\infty. \tag{6.7}$$

2. $R(\mathbf{e})$ satisfies the following condition for a scalar $\varepsilon \in \mathcal{F}$ and vectors $\mathbf{h}_1, \mathbf{h}_2 \in \mathcal{E}$, where the symbol $\mathcal{F}$ denotes the underlying field of scalars:

$$R\left(\mathbf{e}^0 + \varepsilon \mathbf{h}_1 + \varepsilon \mathbf{h}_2\right) - R\left(\mathbf{e}^0 + \varepsilon \mathbf{h}_1\right) - R\left(\mathbf{e}^0 + \varepsilon \mathbf{h}_2\right) + R\left(\mathbf{e}^0\right) = o(\varepsilon). \tag{6.8}$$

In practice, the relations provided in Eqs. (6.7) and (6.8) are seldom used directly since the computation of the expression on the right side of Eq. (6.6) reveals if the respective expression is linear (or not) in $\mathbf{h} \triangleq \left(\delta\boldsymbol{\alpha}, \mathbf{v}^{(1)}\right)$ and, hence, in $\mathbf{v}^{(1)}(\mathbf{x})$.

Numerical methods (e.g., Newton's method and variants thereof) for solving Eqs. (6.1) and (6.2) also require the existence of the 1st-order G-derivatives of original model equations. Therefore, the conditions provided in Eqs. (6.7) and (6.8) are henceforth considered to be satisfied by the model responses and also by the operators underlying the physical system modeled by Eqs. (6.1) and (6.2).

When the 1st-order G-differential $\delta R\left(\mathbf{e}^0; \mathbf{h}\right) \equiv \left\{\delta R\left[\mathbf{u}(\mathbf{x}); \boldsymbol{\alpha}; \mathbf{v}^{(1)}(\mathbf{x}); \delta\boldsymbol{\alpha}\right]\right\}_{\boldsymbol{\alpha}^0}$ satisfies the conditions provided in Eqs. (6.7) and (6.8), it can be written as follows:

$$\begin{aligned}
&\left\{\delta R\left[\mathbf{u}(\mathbf{x}); \boldsymbol{\alpha}; \mathbf{v}^{(1)}(\mathbf{x}); \delta\boldsymbol{\alpha}\right]\right\}_{\boldsymbol{\alpha}^0} \\
&\triangleq \left\{\frac{d}{d\varepsilon} \int_{\lambda_1\left(\boldsymbol{\alpha}^0 + \varepsilon\delta\boldsymbol{\alpha}\right)}^{\omega_1\left(\boldsymbol{\alpha}^0 + \varepsilon\delta\boldsymbol{\alpha}\right)} \cdots \int_{\lambda_{TI}\left(\boldsymbol{\alpha}^0 + \varepsilon\delta\boldsymbol{\alpha}\right)}^{\omega_{TI}\left(\boldsymbol{\alpha}^0 + \varepsilon\delta\boldsymbol{\alpha}\right)} S\left(\mathbf{u}^0 + \varepsilon\mathbf{v}^{(1)}, \boldsymbol{\alpha}^0 + \varepsilon\delta\boldsymbol{\alpha}; \mathbf{x}\right) dx_1 \ldots dx_{TI}\right\} \\
&\triangleq \left\{\delta R\left[\mathbf{u}(\mathbf{x}); \boldsymbol{\alpha}; \delta\boldsymbol{\alpha}\right]\right\}_{dir} + \left\{\delta R\left[\mathbf{u}(\mathbf{x}); \boldsymbol{\alpha}; \mathbf{v}^{(1)}(\mathbf{x})\right]\right\}_{ind}.
\end{aligned} \tag{6.9}$$

In Eq. (6.9), the "direct effect" term $\left\{\delta R\left[\mathbf{u}(\mathbf{x}); \boldsymbol{\alpha}; \delta\boldsymbol{\alpha}\right]\right\}_{dir}$ comprises only dependencies on $\delta\boldsymbol{\alpha}$ and is defined as follows:

$$\left\{\delta R\left[\mathbf{u}(\mathbf{x}); \boldsymbol{\alpha}; \delta\boldsymbol{\alpha}\right]\right\}_{dir} \triangleq \left\{\frac{\partial R(\mathbf{u}; \boldsymbol{\alpha})}{\partial \boldsymbol{\alpha}}\right\}_{\boldsymbol{\alpha}^0} \delta\boldsymbol{\alpha} \triangleq \sum_{j_1=1}^{TP} \left\{R^{(1)}\left[j_1; \mathbf{u}(\mathbf{x}); \boldsymbol{\alpha}\right]\right\}_{dir} \delta\alpha_{j_1}, \tag{6.10}$$

where $\partial R(\mathbf{u}; \boldsymbol{\alpha}) / \partial \boldsymbol{\alpha}$ denotes the partial G-derivatives of $R(\mathbf{e})$ with respect to $\boldsymbol{\alpha}$, evaluated at the nominal parameter values, and where the following definitions were used:

$$\frac{\partial[\;]}{\partial \boldsymbol{\alpha}} \delta\boldsymbol{\alpha} \triangleq \sum_{i=1}^{TP} \frac{\partial[\;]}{\partial \alpha_i} \delta\alpha_i, \tag{6.11}$$

$$\begin{aligned}
\left\{R^{(1)}\left[j_1; \mathbf{u}(\mathbf{x}); \boldsymbol{\alpha}\right]\right\}_{dir} &\triangleq \left\{\int_{\lambda_1(\boldsymbol{\alpha})}^{\omega_1(\boldsymbol{\alpha})} \cdots \int_{\lambda_{TI}(\boldsymbol{\alpha})}^{\omega_{TI}(\boldsymbol{\alpha})} \frac{\partial S(\mathbf{u}; \boldsymbol{\alpha}; \boldsymbol{\alpha})}{\partial \alpha_{j_1}} dx_1 \ldots dx_{TI}\right\}_{\boldsymbol{\alpha}^0} \\
&+ \sum_{j=1}^{TI} \left\{\frac{\partial \omega_j(\boldsymbol{\alpha})}{\partial \alpha_{j_1}} \int_{\lambda_1}^{\omega_1} dx_1 .. \int_{\lambda_{j-1}}^{\omega_{j-1}} \int_{\lambda_{j+1}}^{\omega_{j+1}} .. \int_{\lambda_{TI}}^{\omega_{TI}} dx_{TI} S\left[\mathbf{u}\left(x_1, .., \omega_j(\boldsymbol{\alpha}), .., x_{N_x}\right); \boldsymbol{\alpha}\right]\right\}_{\boldsymbol{\alpha}^0}
\end{aligned} \tag{6.12}$$

$$-\sum_{j=1}^{TI}\left\{\frac{\partial\lambda_j(\boldsymbol{\alpha})}{\partial\alpha_{j_1}}\int_{\lambda_1}^{\omega_1}dx_1..\int_{\lambda_{j-1}}^{\omega_{j-1}}\int_{\lambda_{j+1}}^{\omega_{j+1}}..\int_{\lambda_{TI}}^{\omega_{TI}}dx_{TI}S\left[\mathbf{u}\left(x_1,..,\lambda_j(\boldsymbol{\alpha}),..,x_{N_x}\right);\boldsymbol{\alpha}\right]\right\}_{\boldsymbol{\alpha}^0}.$$

The notation on the left side of Eq. (6.11) comprises an implied transposed vector (since it represents an inner product), but the respective "dagger (†)" which indicates "transposition" has been omitted in order to keep the notation as simple as possible. "Daggers" indicating transposition will also be omitted in other inner products, whenever possible while avoiding ambiguities.

The direct-effect term can be computed once the nominal values $\mathbf{e}^0=\left(\mathbf{u}^0,\boldsymbol{\alpha}^0\right)$ are available. The notation $\{\ \}_{\boldsymbol{\alpha}^0}$ will be used to indicate that the quantity enclosed within the bracket is to be evaluated at the respective nominal parameter and state function values.

On the other hand, the quantity $\left\{\delta R\left[\mathbf{u}(\mathbf{x});\boldsymbol{\alpha};\mathbf{v}^{(1)}(\mathbf{x})\right]\right\}_{ind}$ in Eq. (6.9) comprises only variations in the state functions and is therefore called the "indirect-effect term," having the following expression:

$$\left\{\delta R\left[\mathbf{u}(\mathbf{x});\boldsymbol{\alpha};\mathbf{v}^{(1)}(\mathbf{x})\right]\right\}_{ind}\triangleq\int_{\lambda_1(\boldsymbol{\alpha}^0)}^{\omega_1(\boldsymbol{\alpha}^0)}dx_1..\int_{\lambda_{TI}(\boldsymbol{\alpha}^0)}^{\omega_{TI}(\boldsymbol{\alpha}^0)}dx_{TI}\left\{\frac{\partial S(\mathbf{u};\boldsymbol{\alpha};\mathbf{x})}{\partial\mathbf{u}}\mathbf{v}^{(1)}(\mathbf{x})\right\}_{\boldsymbol{\alpha}^0},\tag{6.13}$$

where

$$\frac{\partial[\ ]}{\partial\mathbf{u}}\mathbf{v}^{(1)}(\mathbf{x})\triangleq\sum_{i=1}^{TD}\frac{\partial[\ ]}{\partial u_i(\mathbf{x})}\delta u_i(\mathbf{x}).\tag{6.14}$$

The notation on the left side of Eq. (6.14) comprises an implied transposed vector (since it represents an inner product), but the respective "dagger (†)" which indicates "transposition" has been omitted to simplify the notation.

The "indirect effect" term induces variations in the response through the variations in the state functions, which are, in turn, caused by the parameter variations through the equations underlying the model. Evidently, the indirect-effect term can be quantified only after having determined the variations $\mathbf{v}^{(1)}(\mathbf{x})$ in terms of the variations $\delta\boldsymbol{\alpha}$. The "indirect effect" term induces variations in the response through the variations in the state functions, which are, in turn, caused by the parameter variations through the equations underlying the model. Evidently, the indirect-effect term can be quantified only after having determined the variations $\mathbf{v}^{(1)}(\mathbf{x})$ in terms of the variations $\delta\boldsymbol{\alpha}$.

The 1st-order relationship between the vectors $\mathbf{v}^{(1)}(\mathbf{x})$ and $\delta\boldsymbol{\alpha}$ is determined by solving the equations obtained by applying the definition of the G-differential to Eqs. (6.1) and (6.2), which yields the following equations:

$$\left\{\frac{d}{d\varepsilon}\mathbf{N}\left[\mathbf{u}^0+\varepsilon\mathbf{v}^{(1)}(\mathbf{x});\boldsymbol{\alpha}^0+\varepsilon\delta\boldsymbol{\alpha}\right]\right\}_{\varepsilon=0}=\left\{\frac{d}{d\varepsilon}\mathbf{Q}\left(\mathbf{x},\boldsymbol{\alpha}^0+\varepsilon\delta\boldsymbol{\alpha}\right)\right\}_{\varepsilon=0},\tag{6.15}$$

$$\left\{\frac{d}{d\varepsilon}\mathbf{B}\left[\mathbf{u}^0+\varepsilon\mathbf{v}^{(1)}(\mathbf{x});\boldsymbol{\alpha}^0+\varepsilon\delta\boldsymbol{\alpha}\right]\right\}_{\varepsilon=0}-\left\{\frac{d}{d\varepsilon}\mathbf{C}\left(\mathbf{x},\pm^0+\varepsilon\delta\boldsymbol{\alpha}\right)\right\}_{\varepsilon=0}=\mathbf{0}. \tag{6.16}$$

Carrying out the differentiations with respect to ε in Eqs. (6.15) and (6.16), and setting $\varepsilon=0$ in the resulting expressions yield the following equations:

$$\left\{\mathbf{N}^{(1)}(\mathbf{u};\boldsymbol{\alpha})\mathbf{v}^{(1)}(\mathbf{x})\right\}_{\boldsymbol{\alpha}^0}=\left\{\mathbf{q}_V^{(1)}(\mathbf{u};\boldsymbol{\alpha};\delta\boldsymbol{\alpha})\right\}_{\boldsymbol{\alpha}^0},\quad \mathbf{x}\in\Omega_x, \tag{6.17}$$

$$\left\{\mathbf{b}_V^{(1)}\left(\mathbf{u};\boldsymbol{\alpha};\mathbf{v}^{(1)};\delta\boldsymbol{\alpha}\right)\right\}_{\boldsymbol{\alpha}^0}=\mathbf{0},\quad \mathbf{x}\in\partial\Omega_x\left(\boldsymbol{\alpha}^0\right). \tag{6.18}$$

In Eqs. (6.17) and (6.18), the superscript "(1)" indicates "1st-Level" and the various quantities which appear in these equations are defined as follows:

$$\mathbf{N}^{(1)}(\mathbf{u};\boldsymbol{\alpha})\triangleq\left\{\frac{\partial\mathbf{N}(\mathbf{u};\boldsymbol{\alpha})}{\partial\mathbf{u}}\right\}\triangleq\begin{pmatrix}\dfrac{\partial N_1}{\partial u_1} & \cdots & \dfrac{\partial N_1}{\partial u_{TD}}\\ \vdots & \ddots & \vdots\\ \dfrac{\partial N_{TD}}{\partial u_1} & \cdots & \dfrac{\partial N_{TD}}{\partial u_{TD}}\end{pmatrix}; \tag{6.19}$$

$$\mathbf{q}_V^{(1)}(\mathbf{u};\boldsymbol{\alpha};\delta\boldsymbol{\alpha})\triangleq\frac{\partial\left[\mathbf{Q}(\boldsymbol{\alpha})-\mathbf{N}(\mathbf{u};\boldsymbol{\alpha})\right]}{\partial\boldsymbol{\alpha}}\delta\boldsymbol{\alpha}\triangleq\sum_{j_1=1}^{TP}\mathbf{s}_V^{(1)}(j_1;\mathbf{u};\boldsymbol{\alpha})\delta\alpha_{j_1}; \tag{6.20}$$

$$\mathbf{s}_V^{(1)}(j_1;\mathbf{u};\boldsymbol{\alpha})\triangleq\frac{\partial\left[\mathbf{Q}(\boldsymbol{\alpha})-\mathbf{N}(\mathbf{u};\boldsymbol{\alpha})\right]}{\partial\alpha_{j_1}}; \tag{6.21}$$

$$\left\{\mathbf{b}_V^{(1)}\left(\mathbf{u};\boldsymbol{\alpha};\mathbf{v}^{(1)};\delta\boldsymbol{\alpha}\right)\right\}_{\boldsymbol{\alpha}^0}\triangleq\left\{\frac{\partial\mathbf{B}(\mathbf{u};\boldsymbol{\alpha})}{\partial\mathbf{u}}\right\}_{\boldsymbol{\alpha}^0}\mathbf{v}^{(1)}+\left\{\frac{\partial\left[\mathbf{B}(\mathbf{u};\boldsymbol{\alpha})-\mathbf{C}(\boldsymbol{\alpha})\right]}{\partial\boldsymbol{\alpha}}\right\}_{\boldsymbol{\alpha}^0}\delta\boldsymbol{\alpha}. \tag{6.22}$$

The system that comprises Eqs. (6.17) and (6.18) is called the "1st-Level Variational Sensitivity System" (1st-LVSS). In order to determine the solutions of the 1st-LVSS that would correspond to every parameter variation $\delta\alpha_{j_1}$, $j_1=1,...,TP$, the 1st-LVSS would need to be solved TP times, with distinct right sides for each $\delta\alpha_{j_1}$, thus requiring TP large-scale computations. In other words, the actual form of the 1st-LVSS that would need to be solved in practice is as follows:

$$\left\{\mathbf{N}^{(1)}(\mathbf{u};\boldsymbol{\alpha})\mathbf{v}^{(1)}(j_1;\mathbf{x})\right\}_{\boldsymbol{\alpha}^0}=\left\{\mathbf{s}_V^{(1)}(j_1;\mathbf{u};\boldsymbol{\alpha})\right\}_{\boldsymbol{\alpha}^0},\quad j_1=1,...,TP;\ \mathbf{x}\in\Omega_x, \tag{6.23}$$

$$\left\{\mathbf{b}_V^{(1)}\left[\mathbf{u};\boldsymbol{\alpha};\mathbf{v}^{(1)}(j_1;\mathbf{x})\right]\right\}_{\boldsymbol{\alpha}^0}=\mathbf{0};\ j_1=1,...,TP;\ \mathbf{x}\in\partial\Omega_x\left(\boldsymbol{\alpha}^0\right). \tag{6.24}$$

In most practical situations, the number of model parameters significantly exceeds the number of functional responses of interest, in other words, $TR\ll TP$, so it would be advantageous to perform just TR (rather than TP) computations. The fundamental ideas underlying the 1st-CASAM-N are as conceived by Cacuci (1981a, 1981b), aiming at computing exactly and efficiently the "indirect effect term" without needing to compute explicitly the vectors $\mathbf{v}^{(1)}(j_1;\mathbf{x}),\ j_1=1,...,TP$. The qualifier "comprehensive" is meant to indicate that the 1st-CASAM-N considers that the internal and external boundaries $\partial\Omega_x(\boldsymbol{\alpha})$ of the phase-space domain depend on the uncertain model parameters $\boldsymbol{\alpha}$ and are thereby imprecisely known, subject to uncertainties. Thus, the 1st-CASAM-N represents a generalization of the pioneering works by Cacuci (1981a, 1981b) which had conceived the "adjoint sensitivity analysis methodology for nonlinear systems" but in which the domain boundary was considered to have been perfectly well-known, free of uncertainties.

The need for computing the vectors $\mathbf{v}^{(1)}(j_1;\mathbf{x}),\ j_1=1,...,TP$, is eliminated by expressing the indirect-effect term defined in Eq. (6.13) in terms of the solutions of the "1st-Level Adjoint Sensitivity System" (1st-LASS), the construction of which requires the introduction of adjoint operators. This is accomplished by introducing a real Hilbert space denoted as $\mathcal{H}_1(\Omega_x)$, endowed with an inner product of two vectors $\mathbf{u}^{(a)}(\mathbf{x})\in\mathcal{H}_1$ and $\mathbf{u}^{(b)}(\mathbf{x})\in\mathcal{H}_1$ denoted as $\left\langle \mathbf{u}^{(a)},\mathbf{u}^{(b)}\right\rangle_1$ and defined as follows:

$$\left\langle \mathbf{u}^{(a)},\mathbf{u}^{(b)}\right\rangle_1 \triangleq \left\{ \int_{\lambda_1(\boldsymbol{\alpha})}^{\omega_1(\boldsymbol{\alpha})} \cdots \int_{\lambda_{TI}(\boldsymbol{\alpha})}^{\omega_{TI}(\boldsymbol{\alpha})} \left[\mathbf{u}^{(a)}(\mathbf{x})\mathbf{u}^{(b)}(\mathbf{x})\right] dx_1 \ldots dx_{TI} \right\}_{\boldsymbol{\alpha}^0}, \tag{6.25}$$

where $\mathbf{u}^{(a)}(\mathbf{x})\mathbf{u}^{(b)}(\mathbf{x}) \triangleq \sum_{i=1}^{TD} u_i^{(a)}(\mathbf{x})u_i^{(b)}(\mathbf{x})$, and where the "dagger" (†), which indicates "transposition," has been omitted (to simplify the notation) in the representation of this scalar product.

Consider a vector $\mathbf{a}^{(1)}(\mathbf{x})\in\mathcal{H}_1$, which is an element in $\mathcal{H}_1(\Omega_x)$ but is otherwise arbitrary at this stage, and use Eq. (6.25) to form the inner product of $\mathbf{a}^{(1)}(\mathbf{x})\in\mathcal{H}_1$ with the relation provided in Eq. (6.17) to obtain:

$$\left\{\left\langle \mathbf{a}^{(1)},\mathbf{N}^{(1)}(\mathbf{u};\boldsymbol{\alpha})\mathbf{v}^{(1)}\right\rangle_1\right\}_{\boldsymbol{\alpha}^0} = \left\{\left\langle \mathbf{a}^{(1)},\mathbf{q}_V^{(1)}(\mathbf{u};\boldsymbol{\alpha};\delta\boldsymbol{\alpha})\right\rangle_1\right\}_{\boldsymbol{\alpha}^0}, \quad \mathbf{x}\in\Omega_x. \tag{6.26}$$

Using the definition of the adjoint operator in $\mathcal{H}_1(\Omega_x)$, the left side of Eq. (6.26) is transformed as follows:

$$\begin{aligned}\left\{\left\langle \mathbf{a}^{(1)},\mathbf{N}^{(1)}(\mathbf{u};\boldsymbol{\alpha})\mathbf{v}^{(1)}\right\rangle_1\right\}_{\boldsymbol{\alpha}^0} &= \left\{\left\langle \left[\mathbf{N}^{(1)}(\mathbf{u};\boldsymbol{\alpha})\right]^* \mathbf{a}^{(1)},\mathbf{v}^{(1)}\right\rangle_1\right\}_{\boldsymbol{\alpha}^0} \\ &+ \left\{\left[P^{(1)}\left(\mathbf{u};\boldsymbol{\alpha};\mathbf{a}^{(1)};\mathbf{v}^{(1)}\right)\right]_{\partial\Omega_x}\right\}_{\boldsymbol{\alpha}^0},\end{aligned} \tag{6.27}$$

where $\left[P^{(1)}\left(\mathbf{u};\boldsymbol{\alpha};\mathbf{a}^{(1)};\mathbf{v}^{(1)}\right)\right]_{\partial\Omega_x}$ denotes the associated bilinear concomitant evaluated on the space/time domain's boundary $\partial\Omega_x(\boldsymbol{\alpha}^0)$ and where $\left[\mathbf{N}^{(1)}(\mathbf{u};\boldsymbol{\alpha})\right]^*$ is the operator adjoint to $\mathbf{N}^{(1)}(\mathbf{u};\boldsymbol{\alpha})$. The symbol $[\ ]^*$ indicates the "adjoint" operator. In certain situations, it might be computationally advantageous to include certain boundary components of $\left[P^{(1)}\left(\mathbf{u};\boldsymbol{\alpha};\mathbf{a}^{(1)};\mathbf{v}^{(1)}\right)\right]_{\partial\Omega_x}$ into the components of $\left[\mathbf{N}^{(1)}(\mathbf{u};\boldsymbol{\alpha})\right]^*$.

The first term on the right side of Eq. (6.27) is required to represent the indirect-effect term defined in Eq. (6.13) by imposing the following relationship:

$$\left\{\left[\mathbf{N}^{(1)}(\mathbf{u};\boldsymbol{\alpha})\right]^*\mathbf{a}^{(1)}(\mathbf{x})\right\}_{\boldsymbol{\alpha}^0}=\left\{\partial S(\mathbf{u};\boldsymbol{\alpha})/\partial\mathbf{u}\right\}_{\boldsymbol{\alpha}^0}\triangleq\mathbf{q}_A^{(1)}\left[\mathbf{u}(\mathbf{x});\boldsymbol{\alpha}\right],\quad \mathbf{x}\in\Omega_x, \tag{6.28}$$

The domain of $\mathbf{A}^{(1)}(\mathbf{u};\boldsymbol{\alpha})$ is determined by selecting appropriate adjoint boundary and/or initial conditions, which will be denoted in operator form as:

$$\left\{\mathbf{b}_A^{(1)}\left(\mathbf{u};\mathbf{a}^{(1)};\boldsymbol{\alpha}\right)\right\}_{\boldsymbol{\alpha}^0}=\mathbf{0},\ \ \mathbf{x}\in\partial\Omega_x\left(\boldsymbol{\alpha}^0\right). \tag{6.29}$$

The previous boundary conditions for $\left[\mathbf{N}^{(1)}(\mathbf{u};\boldsymbol{\alpha})\right]^*$ are usually inhomogeneous, in other words, $\mathbf{b}_A^{(1)}(\mathbf{0};\mathbf{0};\boldsymbol{\alpha})\neq\mathbf{0}$, and are obtained by imposing the following requirements: (i) they must be independent of unknown values of $\mathbf{v}^{(1)}(\mathbf{x})$ and $\delta\boldsymbol{\alpha}$; (ii) the substitution of the boundary and/or initial conditions represented by Eqs. (6.24) and (6.29) into the expression of $\left\{\left[P^{(1)}\left(\mathbf{u};\boldsymbol{\alpha};\mathbf{a}^{(1)};\mathbf{v}^{(1)}\right)\right]_{\partial\Omega_x}\right\}_{\boldsymbol{\alpha}^0}$ must cause all terms containing unknown values of $\mathbf{v}^{(1)}(\mathbf{x})$ to vanish.

Constructing the adjoint initial and/or boundary conditions for $\left[\mathbf{N}^{(1)}(\mathbf{u};\boldsymbol{\alpha})\right]^*$ as previously described and implementing them together with the variational boundary and initial conditions represented by Eq. (6.24) into Eq. (6.27) reduces the bilinear concomitant $\left\{\left[P^{(1)}\left(\mathbf{u};\boldsymbol{\alpha};\mathbf{a}^{(1)};\mathbf{v}^{(1)}\right)\right]_{\partial\Omega_x}\right\}_{\boldsymbol{\alpha}^0}$ to a quantity denoted as $\left\{\left[\hat{P}^{(1)}\left(\mathbf{u};\boldsymbol{\alpha};\mathbf{a}^{(1)};\delta\boldsymbol{\alpha}\right)\right]_{\partial\Omega_x}\right\}_{\boldsymbol{\alpha}^0}$, which will contain boundary terms involving only known values of $\delta\boldsymbol{\alpha}$, $\boldsymbol{\alpha}^0$, $\mathbf{u}^0$, and $\boldsymbol{\psi}^{(1)}$ Since $\left\{\left[\hat{P}^{(1)}\left(\mathbf{u};\boldsymbol{\alpha};\mathbf{a}^{(1)};\delta\boldsymbol{\alpha}\right)\right]_{\partial\Omega_x}\right\}_{\boldsymbol{\alpha}^0}$ is linear in $\delta\boldsymbol{\alpha}$, it can be expressed in the following form: $\left[\hat{P}^{(1)}\left(\mathbf{u};\boldsymbol{\alpha};\mathbf{a}^{(1)};\delta\boldsymbol{\alpha}\right)\right]_{\partial\Omega_x}=\sum_{j_1=1}^{TP}\left[\partial\hat{P}^{(1)}\left(\mathbf{u};\boldsymbol{\alpha};\mathbf{a}^{(1)}\right)/\partial\alpha_{j_1}\right]\delta\alpha_{j_1}$.

The results obtained in Eqs. (6.27) and (6.28) are now replaced in Eq. (6.13) to obtain the following expression of the indirect-effect term as a function of $\mathbf{a}^{(1)}(\mathbf{x})$:

$$\begin{aligned}\left\{\delta R\left[\mathbf{u}(\mathbf{x});\boldsymbol{\alpha};\mathbf{v}^{(1)}(\mathbf{x})\right]\right\}_{ind}&=\left\{\left\langle\mathbf{a}^{(1)},\mathbf{q}_V^{(1)}(\mathbf{u};\boldsymbol{\alpha};\delta\boldsymbol{\alpha})\right\rangle_1\right\}_{\boldsymbol{\alpha}^0}\\&-\left\{\left[\hat{P}^{(1)}\left(\mathbf{u};\boldsymbol{\alpha};\mathbf{a}^{(1)};\delta\boldsymbol{\alpha}\right)\right]_{\partial\Omega_x}\right\}_{\boldsymbol{\alpha}^0}.\end{aligned} \tag{6.30}$$

Replacing in Eq. (6.10) the result obtained in Eq. (6.30) together with the expression for the direct-effect term provided in Eq. (6.12) yields the following expression for the first G-differential of the response $R\left[\mathbf{u}(\mathbf{x});\boldsymbol{\alpha}\right]$:

$$\begin{aligned}
&\left\{\delta R\left[\mathbf{u}(\mathbf{x});\boldsymbol{\alpha};\mathbf{v}^{(1)}(\mathbf{x});\delta\boldsymbol{\alpha}\right]\right\}_{\boldsymbol{\alpha}^0}=\left\{\delta R\left[\mathbf{u}(\mathbf{x});\boldsymbol{\alpha};\delta\boldsymbol{\alpha}\right]\right\}_{dir}\\
&+\left\{\left\langle \mathbf{a}^{(1)},\mathbf{q}_V^{(1)}(\mathbf{u};\boldsymbol{\alpha};\delta\boldsymbol{\alpha})\right\rangle_1\right\}_{\boldsymbol{\alpha}^0}\\
&-\left\{\left[\hat{P}^{(1)}\left(\mathbf{u};\boldsymbol{\alpha};\mathbf{a}^{(1)};\delta\boldsymbol{\alpha}\right)\right]_{\partial\Omega_x}\right\}_{\boldsymbol{\alpha}^0}\\
&\triangleq\sum_{j_1=1}^{TP}\left\{R^{(1)}\left[j_1;\mathbf{u}(\mathbf{x});\mathbf{a}^{(1)}(\mathbf{x});\boldsymbol{\alpha}\right]\right\}_{\boldsymbol{\alpha}^0}\delta\alpha_{j_1},
\end{aligned}\tag{6.31}$$

where, for each $j_1=1,\ldots,TP$, the quantity $R^{(1)}\left[j_1;\mathbf{u}(\mathbf{x});\mathbf{a}^{(1)}(\mathbf{x});\boldsymbol{\alpha}\right]$ denotes the 1st-order sensitivities of the response $R\left[\mathbf{u}(\mathbf{x});\boldsymbol{\alpha}\right]$ with respect to the model parameters α_{j_1} and has the following expression, for $j_1=1,...,TP$:

$$\begin{aligned}
&R^{(1)}\left[j_1;\mathbf{u}(\mathbf{x});\mathbf{a}^{(1)}(\mathbf{x});\boldsymbol{\alpha}\right]=-\frac{\partial\left[\hat{P}^{(1)}\left(\mathbf{u};\mathbf{a}^{(1)};\boldsymbol{\alpha};\right)\right]_{\partial\Omega_x}}{\partial\alpha_{j_1}}+\left\{R^{(1)}\left[j_1;\mathbf{u}(\mathbf{x});\boldsymbol{\alpha}\right]\right\}_{dir}\\
&+\int_{\lambda_1(\boldsymbol{\alpha})}^{\omega_1(\boldsymbol{\alpha})}dx_1\ldots\int_{\lambda_{TI}(\boldsymbol{\alpha})}^{\omega_{TI}(\boldsymbol{\alpha})}\mathbf{a}^{(1)}(\mathbf{x})\frac{\partial\left[\mathbf{Q}(\boldsymbol{\alpha})-\mathbf{N}(\mathbf{u};\boldsymbol{\alpha})\right]}{\partial\alpha_{j_1}}dx_{TI}\\
&\triangleq\int_{\lambda_1(\boldsymbol{\alpha})}^{\omega_1(\boldsymbol{\alpha})}\ldots\int_{\lambda_{TI}(\boldsymbol{\alpha})}^{\omega_{TI}(\boldsymbol{\alpha})}S^{(1)}\left[j_1;\mathbf{u}(\mathbf{x});\mathbf{a}^{(1)}(\mathbf{x});\boldsymbol{\alpha}\right]dx_1\ldots dx_{TI}\triangleq\frac{\partial R\left[\mathbf{u}(\mathbf{x});\boldsymbol{\alpha}\right]}{\partial\alpha_{j_1}}.
\end{aligned}\tag{6.32}$$

As indicated by Eq. (6.32), each of the 1st-order sensitivities $R^{(1)}\left[j_1;\mathbf{u}(\mathbf{x});\mathbf{a}^{(1)}(\mathbf{x});\boldsymbol{\alpha}\right]$ of the response $R\left[\mathbf{u}(\mathbf{x});\boldsymbol{\alpha}\right]$ with respect to the model parameters α_{j_1} (including boundary and initial conditions) can be computed inexpensively after having obtained the function $\mathbf{a}^{(1)}(\mathbf{x})\in\mathcal{H}_1$, using just quadrature formulas to evaluate the various inner products involving $\mathbf{a}^{(1)}(\mathbf{x})\in\mathcal{H}_1$. The function $\mathbf{a}^{(1)}(\mathbf{x})\in\mathcal{H}_1$ is obtained by solving numerically Eqs. (6.28) and (6.29), which is the only large-scale computation needed for obtaining all of the 1st-order sensitivities. The relations represented by Eqs. (6.28) and (6.29) are called the 1st-Level Adjoint Sensitivity System (1st-LASS); the solution, $\mathbf{a}^{(1)}(\mathbf{x})\in\mathcal{H}_1(\Omega_x)$, of the 1st-LASS is called the 1st-level adjoint function. It is important to note that the 1st-LASS is independent of parameter variation $\delta\alpha_{j_1}, j_1=1,...,TP$, and therefore needs to be solved only once, regardless of the number of model parameters under consideration. Furthermore, since the 1st-LASS is linear in $\mathbf{a}^{(1)}(\mathbf{x})$, solving it requires less computational effort than solving the original model, which is nonlinear in $\mathbf{u}(\mathbf{x})$. Finally, the function $S^{(1)}\left[j_1;\mathbf{u}(\mathbf{x});\mathbf{a}^{(1)}(\mathbf{x});\boldsymbol{\alpha}\right]$, which is constructed by using the left side of Eq. (6.32), will likely contain Dirac delta functions which are needed to represent the non-zero residual boundary terms contained in $\hat{P}^{(1)}\left(\mathbf{u};\mathbf{a}^{(1)};\boldsymbol{\alpha};\right)$ as a "volume integral over $dx_1\ldots dx_{TI}$." The function $S^{(1)}\left[j_1;\mathbf{u}(\mathbf{x});\mathbf{a}^{(1)}(\mathbf{x});\boldsymbol{\alpha}\right]$ will be used for determining the 2nd- and higher-order sensitivities of the response with respect to the model's parameters.

6.3 2ND-ORDER COMPREHENSIVE ADJOINT SENSITIVITY ANALYSIS METHODOLOGY

The 2nd-order sensitivities $\partial^2 R[\mathbf{u}(\mathbf{x});\boldsymbol{\alpha}]/\partial\alpha_{j_2}\partial\alpha_{j_1}$ are determined by treating them as the "1st-order sensitivities of the 1st-order sensitivities." Thus, the 1st-order G-differential of a 1st-order sensitivity $R^{(1)}\left[j_1;\mathbf{u}(\mathbf{x});\mathbf{a}^{(1)}(\mathbf{x});\boldsymbol{\alpha}\right]$, $j_1=1,...,TP$, is given by the following expression:

$$\begin{aligned}
&\left\{\delta R^{(1)}\left[j_1;\mathbf{u}(\mathbf{x});\mathbf{a}^{(1)}(\mathbf{x});\boldsymbol{\alpha};\mathbf{v}^{(1)}(\mathbf{x});\delta\mathbf{a}^{(1)}(\mathbf{x});\delta\boldsymbol{\alpha}\right]\right\}_{\boldsymbol{\alpha}^0}\\
&\triangleq\left\{\frac{d}{d\varepsilon}\delta R^{(1)}\left[j_1;\mathbf{u}(\mathbf{x})+\varepsilon\mathbf{v}^{(1)}(\mathbf{x});\mathbf{a}^{(1)}(\mathbf{x})+\varepsilon\delta\mathbf{a}^{(1)}(\mathbf{x});\boldsymbol{\alpha}+\varepsilon\delta\boldsymbol{\alpha}\right]\right\}_{\varepsilon=0}\\
&=\sum_{j_2=1}^{TP}\left\{\frac{\partial R^{(1)}\left[j_1;\mathbf{u};\mathbf{a}^{(1)};\boldsymbol{\alpha}\right]}{\partial\alpha_{j_2}}\right\}_{\boldsymbol{\alpha}^0}\delta\alpha_{j_2}+\left\{\delta R^{(1)}\left[j_1;\mathbf{u};\mathbf{a}^{(1)};\boldsymbol{\alpha};\mathbf{v}^{(1)}(\mathbf{x});\delta\mathbf{a}^{(1)}(\mathbf{x})\right]\right\}_{ind},
\end{aligned}\tag{6.33}$$

where the indirect-effect term $\left\{\delta R^{(1)}\left[j_1;\mathbf{u};\mathbf{a}^{(1)};\boldsymbol{\alpha};\mathbf{v}^{(1)}(\mathbf{x});\delta\mathbf{a}^{(1)}(\mathbf{x})\right]\right\}_{ind}$ comprises all dependencies on the vectors $\mathbf{v}^{(1)}(\mathbf{x})$ and $\delta\mathbf{a}^{(1)}(\mathbf{x})$ of variations in the state functions $\mathbf{u}(\mathbf{x})$ and $\mathbf{a}^{(1)}(\mathbf{x})$, respectively, and is defined as follows:

$$\begin{aligned}
&\left\{\delta R^{(1)}\left[j_1;\mathbf{u};\mathbf{a}^{(1)};\boldsymbol{\alpha};\mathbf{v}^{(1)};\delta\mathbf{a}^{(1)}\right]\right\}_{ind}\\
&\triangleq\int_{\lambda_1(\boldsymbol{\alpha})}^{\omega_1(\boldsymbol{\alpha})}dx_1...\int_{\lambda_{TI}(\boldsymbol{\alpha})}^{\omega_{TI}(\boldsymbol{\alpha})}dx_{TI}\left\{\frac{\partial S^{(1)}}{\partial\mathbf{u}}\mathbf{v}^{(1)}(\mathbf{x})+\frac{\partial S^{(1)}}{\partial\mathbf{a}^{(1)}}\delta\mathbf{a}^{(1)}(\mathbf{x})\right\}_{\boldsymbol{\alpha}^0}.
\end{aligned}\tag{6.34}$$

The functions $\mathbf{v}^{(1)}(\mathbf{x})$ and $\delta\mathbf{a}^{(1)}(\mathbf{x})$ are obtained by solving the following 2nd-Level Variational Sensitivity System (2nd-LVSS), which comprises the concatenation of the 1st-LVSS with the G-differentiated 1st-LASS:

$$\begin{aligned}
&\left\{\mathbf{VM}^{(2)}\left[2\times2;\mathbf{U}^{(2)}(2;\mathbf{x});\boldsymbol{\alpha}\right]\mathbf{V}^{(2)}(2;\mathbf{x})\right\}_{\boldsymbol{\alpha}^0}\\
&=\left\{\mathbf{Q}_V^{(2)}\left[2;\mathbf{U}^{(2)}(2;\mathbf{x});\boldsymbol{\alpha};\delta\boldsymbol{\alpha}\right]\right\}_{\boldsymbol{\alpha}^0},\quad \mathbf{x}\in\Omega_x,
\end{aligned}\tag{6.35}$$

$$\begin{aligned}
&\left\{\mathbf{B}_V^{(2)}\left[2;\mathbf{U}^{(2)}(2;\mathbf{x});\mathbf{V}^{(2)}(2;\mathbf{x});\boldsymbol{\alpha};\delta\boldsymbol{\alpha}\right]\right\}_{\boldsymbol{\alpha}^0}\\
&=\mathbf{0}[2],\ \mathbf{0}[2]\triangleq[\mathbf{0},\mathbf{0}]^{\dagger},\mathbf{x}\in\partial\Omega_x(\boldsymbol{\alpha}^0).
\end{aligned}\tag{6.36}$$

The argument "2" which appears in the list of arguments of the vector $\mathbf{U}^{(2)}(2;\mathbf{x})$ and the "variational vector" $\mathbf{V}^{(2)}(2;\mathbf{x})$ in Eq. (6.35) indicates that each of these vectors is a 2-block column vector, each block comprising a column vector of dimension TD, defined as follows:

$$\mathbf{U}^{(2)}(2;\mathbf{x}) \triangleq \begin{pmatrix} \mathbf{u}^{(1)}(\mathbf{x}) \\ \mathbf{a}^{(1)}(\mathbf{x}) \end{pmatrix}; \; \mathbf{V}^{(2)}(2;\mathbf{x}) \triangleq \delta\mathbf{U}^{(2)}(2;\mathbf{x}) \triangleq \begin{pmatrix} \mathbf{v}^{(2)}(1;\mathbf{x}) \\ \mathbf{v}^{(2)}(2;\mathbf{x}) \end{pmatrix} \triangleq \begin{pmatrix} \mathbf{v}^{(1)}(\mathbf{x}) \\ \delta\mathbf{a}^{(1)}(\mathbf{x}) \end{pmatrix}.$$

To distinguish block vectors from block matrices, two bold capital letters have been used (and will henceforth be used) to denote block matrices, as in the case of "the 2nd-level variational matrix" $\mathbf{VM}^{(2)}\left[2\times 2;\mathbf{u}^{(2)}(\mathbf{x});\boldsymbol{\alpha}\right]$. The "2nd-level" is indicated by the superscript "(2)". The argument "2×2", which appears in the list of arguments of $\mathbf{VM}^{(2)}\left[2\times 2;\mathbf{u}^{(2)}(\mathbf{x});\boldsymbol{\alpha}\right]$, indicates that this matrix is a 2×2-dimensional block matrix comprising four matrices, each of dimensions $TD\times TD$, having the following structure:

$$\mathbf{VM}^{(2)}\left[2\times 2;\mathbf{U}^{(2)}(2;\mathbf{x});\boldsymbol{\alpha}\right] \triangleq \begin{pmatrix} \mathbf{N}^{(1)} & \mathbf{0} \\ \mathbf{V}_{21}^{(2)} & \mathbf{V}_{22}^{(2)} \end{pmatrix}. \tag{6.37}$$

The other quantities which appear in Eqs. (6.35) and (6.36) are 2-block vectors having the same structure as $\mathbf{V}^{(2)}(2;\mathbf{x})$, and are defined as follows:

$$\mathbf{Q}_V^{(2)}\left[2;\mathbf{U}^{(2)}(2;\mathbf{x});\boldsymbol{\alpha};\delta\boldsymbol{\alpha}\right] \triangleq \begin{pmatrix} \mathbf{q}_V^{(2)}\left(1;\mathbf{U}^{(2)}(2;\mathbf{x});\boldsymbol{\alpha};\delta\boldsymbol{\alpha}\right) \\ \mathbf{q}_V^{(2)}\left(2;\mathbf{U}^{(2)}(2;\mathbf{x});\boldsymbol{\alpha};\delta\boldsymbol{\alpha}\right) \end{pmatrix} \triangleq \begin{pmatrix} \mathbf{q}_V^{(1)}(\mathbf{u};\boldsymbol{\alpha};\delta\boldsymbol{\alpha}) \\ \mathbf{q}_2^{(2)}\left(\mathbf{u};\mathbf{a}^{(1)};\boldsymbol{\alpha};\delta\boldsymbol{\alpha}\right) \end{pmatrix}; \tag{6.38}$$

$$\mathbf{q}_2^{(2)}\left(\mathbf{u};\boldsymbol{\alpha};\mathbf{a}^{(1)};\delta\boldsymbol{\alpha}\right) \triangleq \frac{\partial \mathbf{q}_A^{(1)}\left[\mathbf{u}(\mathbf{x});\boldsymbol{\alpha}\right]}{\partial\boldsymbol{\alpha}}\delta\boldsymbol{\alpha} - \frac{\partial\left\{\left[\mathbf{N}^{(1)}(\mathbf{u};\boldsymbol{\alpha})\right]^{*}\mathbf{a}^{(1)}(\mathbf{x})\right\}}{\partial\boldsymbol{\alpha}}\delta\boldsymbol{\alpha}; \tag{6.39}$$

$$\mathbf{q}_V^{(2)}\left[i;\mathbf{U}^{(2)}(2;\mathbf{x});\boldsymbol{\alpha};\delta\boldsymbol{\alpha}\right] \triangleq \sum_{j_2=1}^{TP}\mathbf{s}_V^{(2)}\left[i;j_2;\mathbf{U}^{(2)}(2;\mathbf{x});\boldsymbol{\alpha}\right]\delta\alpha_{j_2}, \quad i=1,2; \tag{6.40}$$

$$\mathbf{s}_V^{(2)}\left[1;j_2;\mathbf{U}^{(2)}(2;\mathbf{x});\boldsymbol{\alpha}\right] \triangleq \frac{\partial\left[\mathbf{Q}(\boldsymbol{\alpha}) - \mathbf{N}(\mathbf{u};\boldsymbol{\alpha})\right]}{\partial\alpha_{j_2}}; \tag{6.41}$$

$$\mathbf{s}_V^{(2)}\left[2;j_2;\mathbf{U}^{(2)}(2;\mathbf{x});\boldsymbol{\alpha}\right] \triangleq \frac{\partial \mathbf{q}_A^{(1)}\left[\mathbf{u}(\mathbf{x});\boldsymbol{\alpha}\right]}{\partial\alpha_{j_2}} - \frac{\partial\left\{\left[\mathbf{N}^{(1)}(\mathbf{u};\boldsymbol{\alpha})\right]^{*}\mathbf{a}^{(1)}(\mathbf{x})\right\}}{\partial\alpha_{j_2}}; \tag{6.42}$$

$$\begin{aligned} &\mathbf{B}_V^{(2)}\left[2;\mathbf{U}^{(2)}(2;\mathbf{x});\mathbf{V}^{(2)}(2;\mathbf{x});\boldsymbol{\alpha};\delta\boldsymbol{\alpha}\right] \\ &\triangleq \begin{pmatrix} \mathbf{b}_V^{(2)}\left[1;\mathbf{U}^{(2)}(2;\mathbf{x});\mathbf{V}^{(2)}(2;\mathbf{x});\boldsymbol{\alpha};\delta\boldsymbol{\alpha}\right] \\ \mathbf{b}_V^{(2)}\left[2;\mathbf{U}^{(2)}(2;\mathbf{x});\mathbf{V}^{(2)}(2;\mathbf{x});\boldsymbol{\alpha};\delta\boldsymbol{\alpha}\right] \end{pmatrix} \\ &\triangleq \begin{pmatrix} \mathbf{b}_V^{(1)}\left(\mathbf{u}^{(1)};\boldsymbol{\alpha};\delta\mathbf{u}^{(1)};\delta\boldsymbol{\alpha}\right) \\ \delta\mathbf{b}_A^{(1)}\left[\mathbf{U}^{(2)}(2;\mathbf{x});\mathbf{V}^{(2)}(2;\mathbf{x});\boldsymbol{\alpha};\delta\boldsymbol{\alpha}\right] \end{pmatrix}. \end{aligned} \tag{6.43}$$

$$\mathbf{V}_{21}^{(2)}\left(\mathbf{u};\mathbf{a}^{(1)};\boldsymbol{\alpha}\right)\triangleq\frac{\partial\left\{\left[\mathbf{N}^{(1)}\left(\mathbf{u};\boldsymbol{\alpha}\right)\right]^{*}\mathbf{a}^{(1)}\right\}}{\partial\mathbf{u}}-\frac{\mathbf{q}_{A}^{(1)}\left[\mathbf{u}\left(\mathbf{x}\right);\boldsymbol{\alpha}\right]}{\partial\mathbf{u}};\tag{6.44}$$

$$\mathbf{V}_{22}^{(2)}\left(\mathbf{u};\boldsymbol{\alpha}\right)\triangleq\left[\mathbf{N}^{(1)}\left(\mathbf{u};\boldsymbol{\alpha}\right)\right]^{*};\tag{6.45}$$

$$\delta\mathbf{b}_{A}^{(1)}\left(\mathbf{u};\mathbf{a}^{(1)};\boldsymbol{\alpha}\right)\triangleq\frac{\partial\mathbf{b}_{A}^{(1)}}{\partial\mathbf{u}}\mathbf{v}^{(1)}\left(\mathbf{x}\right)+\frac{\partial\mathbf{b}_{A}^{(1)}}{\partial\mathbf{a}^{(1)}}\delta\mathbf{a}^{(1)}\left(\mathbf{x}\right)+\frac{\partial\mathbf{b}_{A}^{(1)}}{\partial\boldsymbol{\alpha}}\delta\boldsymbol{\alpha}.\tag{6.46}$$

The argument "2" in the expression $\mathbf{0}[2]\triangleq[\mathbf{0},\mathbf{0}]^{\dagger}$ indicates that the quantity $\mathbf{0}[2]$ is a 2-block column vector comprising two vectors, each of which has TD components, all of which are zero-valued.

The need for solving the 2nd-LVSS is circumvented by deriving an alternative expression for the indirect-effect term $\left\{\delta R^{(1)}\left[j_1;\mathbf{u};\mathbf{a}^{(1)};\boldsymbol{\alpha};\mathbf{V}^{(2)}\left(2;\mathbf{x}\right)\right]\right\}_{ind}$ defined in Eq. (6.34) in which the function $\mathbf{V}^{(2)}\left(2;\mathbf{x}\right)$ is replaced by a 2nd-level adjoint function which is independent of variations in the model parameter and state functions. This 2nd-level adjoint function is the solution of a 2nd-Level Adjoint Sensitivity System (2nd-LASS), which will be constructed by using the same principles as employed for deriving the 1st-LASS. The 2nd-LASS is constructed in a Hilbert space, denoted as $\mathcal{H}_2\left(\Omega_x\right)$, which comprises as elements block vectors of the same form as $\mathbf{V}^{(2)}\left(2;\mathbf{x}\right)$, and is endowed with the following inner product of two vectors $\boldsymbol{\Psi}^{(2)}\left(2;\mathbf{x}\right)\triangleq\left[\boldsymbol{\psi}^{(2)}\left(1;\mathbf{x}\right),\boldsymbol{\psi}^{(2)}\left(2;\mathbf{x}\right)\right]^{\dagger}\in\mathcal{H}_2$ and $\boldsymbol{\Phi}^{(2)}\left(\mathbf{x}\right)\triangleq\left[\boldsymbol{\varphi}^{(2)}\left(1;\mathbf{x}\right),\boldsymbol{\varphi}^{(2)}\left(2;\mathbf{x}\right)\right]^{\dagger}\in\mathcal{H}_2$:

$$\left\langle\boldsymbol{\Psi}^{(2)}\left(2;\mathbf{x}\right),\boldsymbol{\Phi}^{(2)}\left(\mathbf{x}\right)\right\rangle_2\triangleq\sum_{i=1}^{2}\left\langle\boldsymbol{\psi}^{(2)}\left(i;\mathbf{x}\right),\boldsymbol{\varphi}^{(2)}\left(i;\mathbf{x}\right)\right\rangle_1.\tag{6.47}$$

The inner product defined in Eq. (6.47) is continuous in $\boldsymbol{\alpha}$ in a neighborhood of $\boldsymbol{\alpha}^0$ and will be used to construct the following 2nd-Level Adjoint Sensitivity System (2nd-LASS) for the 2nd-level adjoint function $\mathbf{A}^{(2)}\left(2;j_1;\mathbf{x}\right)\triangleq\left[\mathbf{a}^{(2)}\left(1;j_1;\mathbf{x}\right),\mathbf{a}^{(2)}\left(2;j_1;\mathbf{x}\right)\right]^{\dagger}\in\mathcal{H}_2\left(\Omega_x\right)$, for each $j_1=1,...,TP$:

$$\begin{aligned}&\left\{\mathbf{AM}^{(2)}\left[2\times2;\mathbf{U}^{(2)}\left(2;\mathbf{x}\right);\boldsymbol{\alpha}\right]\mathbf{A}^{(2)}\left(2;j_1;\mathbf{x}\right)\right\}_{\boldsymbol{\alpha}^0}\\&=\left\{\mathbf{Q}_{A}^{(2)}\left[2;j_1;\mathbf{U}^{(2)}\left(2;\mathbf{x}\right);\boldsymbol{\alpha}\right]\right\}_{\boldsymbol{\alpha}^0},\\&\quad for\ \ j_1=1,...,TP;\ \mathbf{x}\in\Omega_x,\end{aligned}\tag{6.48}$$

subject to boundary conditions represented as follows:

$$\begin{aligned}&\left\{\mathbf{B}_{A}^{(2)}\left[2;\mathbf{U}^{(2)}\left(2;\mathbf{x}\right);\mathbf{A}^{(2)}\left(2;j_1;\mathbf{x}\right);\boldsymbol{\alpha}\right]\right\}_{\boldsymbol{\alpha}^0}\\&=\mathbf{0}[2];\ \ j_1=1,...,TP;\ \mathbf{x}\in\partial\Omega_x\left(\boldsymbol{\alpha}^0\right),\end{aligned}\tag{6.49}$$

where:

$$\mathbf{Q}_A^{(2)}\left[2;j_1;\mathbf{U}^{(2)}(2;\mathbf{x});\boldsymbol{\alpha}\right]\triangleq\begin{pmatrix}\mathbf{q}_A^{(2)}\left(1;j_1;\mathbf{U}^{(2)};\boldsymbol{\alpha}\right)\\ \mathbf{q}_A^{(2)}\left(2;j_1;\mathbf{U}^{(2)};\boldsymbol{\alpha}\right)\end{pmatrix}$$
$$\triangleq\begin{pmatrix}\partial S^{(1)}\left[j_1;\mathbf{u}(\mathbf{x});\mathbf{a}^{(1)}(\mathbf{x});\boldsymbol{\alpha};\mathbf{v}^{(1)}(\mathbf{x})\right]/\partial\mathbf{u}\\ \partial S^{(1)}\left[j_1;\mathbf{u}(\mathbf{x});\mathbf{a}^{(1)}(\mathbf{x});\boldsymbol{\alpha};\mathbf{v}^{(1)}(\mathbf{x})\right]/\partial\mathbf{a}^{(1)}\end{pmatrix},\ j_1=1,\dots,TP. \tag{6.50}$$

$$\mathbf{AM}^{(2)}\left[2\times2;\mathbf{U}^{(2)}(2;\mathbf{x});\boldsymbol{\alpha}\right]\triangleq\left[\mathbf{VM}^{(2)}\left(2\times2;\mathbf{U}^{(2)}(2;\mathbf{x});\boldsymbol{\alpha}\right)\right]^*$$
$$=\begin{pmatrix}\left\{\left[\mathbf{N}^{(1)}(\mathbf{u};\boldsymbol{\alpha})\right]^*\right\}^\dagger & \left[\mathbf{V}_{21}^{(2)*}\right]^\dagger\\ \mathbf{0} & \left[\mathbf{V}_{22}^{(2)*}\right]^\dagger\end{pmatrix}. \tag{6.51}$$

The matrix $\mathbf{AM}^{(2)}\left[2\times2;\mathbf{u}^{(2)}(\mathbf{x});\boldsymbol{\alpha}\right]$ comprises (2×2) block matrices, each of dimensions TD^2, thus comprising a total of $(2\times2)TD^2$ components (or elements) and is obtained from the following relation:

$$\left\{\left\langle\mathbf{A}^{(2)}(2;\mathbf{x}),\mathbf{VM}^{(2)}\mathbf{V}^{(2)}(2;\mathbf{x})\right\rangle_2\right\}_{\boldsymbol{\alpha}^0}=\left\{\left[P^{(2)}\left(\mathbf{U}^{(2)};\mathbf{A}^{(2)};\mathbf{V}^{(2)};\boldsymbol{\alpha}\right)\right]_{\partial\Omega_x}\right\}_{\boldsymbol{\alpha}^0}$$
$$+\left\{\left\langle\mathbf{V}^{(2)}(2;\mathbf{x}),\mathbf{AM}^{(2)}\left[2\times2;\mathbf{U}^{(2)}(2;\mathbf{x});\boldsymbol{\alpha}\right]\mathbf{A}^{(2)}(2;\mathbf{x})\right\rangle_2\right\}_{\boldsymbol{\alpha}^0}, \tag{6.52}$$

where the quantity $\left\{\left[P^{(2)}\left(\mathbf{U}^{(2)};\mathbf{A}^{(2)};\mathbf{V}^{(2)};\boldsymbol{\alpha}\right)\right]_{\partial\Omega_x}\right\}_{\boldsymbol{\alpha}^0}$ denotes the corresponding bilinear concomitant defined on the domain's boundary, evaluated at the nominal values for the parameters and respective state functions.

The 2nd-level adjoint boundary/initial conditions represented by Eq. (6.49) are determined by requiring that: (i) they must be independent of unknown values of $\mathbf{V}^{(2)}(2;\mathbf{x})$; (ii) the substitution of the boundary and/or initial conditions represented by Eqs. (6.49) and (6.36) into the expression of $\left\{\left[P^{(2)}\left(\mathbf{U}^{(2)};\mathbf{A}^{(2)};\mathbf{V}^{(2)};\boldsymbol{\alpha}\right)\right]_{\partial\Omega_x}\right\}_{\boldsymbol{\alpha}^0}$ must cause all terms containing unknown values of $\mathbf{V}^{(2)}(2;\mathbf{x})$ to vanish.

Using in Eq. (6.34) the relations defining 2nd-LASS, the 2nd-LVSS and the relation provided in Eq. (6.52) yields the following alternative expression for the indirect-effect term in terms of the 2nd-level adjoint sensitivity function $\mathbf{A}^{(2)}(2;j_1;\mathbf{x})\triangleq\left[\mathbf{a}^{(2)}(1;j_1;\mathbf{x}),\mathbf{a}^{(2)}(2;j_1;\mathbf{x})\right]^\dagger$:

$$\left\{\delta R^{(1)}\left[j_1;\mathbf{u};\mathbf{a}^{(1)};\boldsymbol{\alpha};\mathbf{V}^{(2)}(2;\mathbf{x})\right]\right\}_{ind}=-\left\{\left[\hat{P}^{(2)}\left(\mathbf{U}^{(2)};\mathbf{A}^{(2)};\boldsymbol{\alpha};\delta\boldsymbol{\alpha}\right)\right]_{\partial\Omega_x}\right\}_{\boldsymbol{\alpha}^0}$$
$$+\left\{\left\langle\mathbf{A}^{(2)}(2;j_1;\mathbf{x}),\mathbf{Q}_V^{(2)}\left[2;\mathbf{U}^{(2)}(2;\mathbf{x});\boldsymbol{\alpha};\delta\boldsymbol{\alpha}\right]\right\rangle_2\right\}_{\boldsymbol{\alpha}^0}, \tag{6.53}$$

where $\left\{\left[\hat{P}^{(2)}\left(\mathbf{U}^{(2)};\mathbf{A}^{(2)};\boldsymbol{\alpha};\delta\boldsymbol{\alpha}\right)\right]_{\partial\Omega_x}\right\}_{\boldsymbol{\alpha}^0}$ denotes residual boundary terms which may not have vanished after having used the boundary and/or initial conditions represented by Eqs. (6.36) and (6.49).

Replacing the expression obtained in Eq. (6.53) into Eq. (6.33) yields the following expression:

$$\begin{aligned}&\left\{\delta R^{(1)}\left[j_1;\mathbf{U}^{(2)}(2;\mathbf{x});\mathbf{A}^{(2)}(2;j_1;\mathbf{x});\boldsymbol{\alpha};\delta\boldsymbol{\alpha}\right]\right\}_{\boldsymbol{\alpha}^0} = \\ &= \sum_{j_2=1}^{TP}\left\{R^{(2)}\left[j_2;j_1;\mathbf{U}^{(2)}(2;\mathbf{x});\mathbf{A}^{(2)}(2;j_1;\mathbf{x});\boldsymbol{\alpha}\right]\right\}_{\boldsymbol{\alpha}^0}\delta\alpha_{j_2},\ j_1=1,\dots,TP,\end{aligned} \tag{6.54}$$

where the quantity $R^{(2)}\left[j_2;j_1;\mathbf{U}^{(2)}(2;\mathbf{x});\mathbf{A}^{(2)}(2;j_1;\mathbf{x});\boldsymbol{\alpha}\right]$ denotes the 2nd-order sensitivity of the generic scalar-valued response $R\left[\mathbf{u}(\mathbf{x});\boldsymbol{\alpha}\right]$ with respect to the parameters α_{j_1} and α_{j_2} computed at the nominal values of the parameters and respective state functions, and has the following expression for $j_1, j_2 = 1,\dots,TP$:

$$\begin{aligned}&R^{(2)}\left[j_2;j_1;\mathbf{U}^{(2)}(2;\mathbf{x});\mathbf{A}^{(2)}(2;j_1;\mathbf{x});\boldsymbol{\alpha}\right] \triangleq \frac{\partial R^{(1)}\left[j_1;\mathbf{u}(\mathbf{x});\mathbf{a}^{(1)}(\mathbf{x});\boldsymbol{\alpha}\right]}{\partial\alpha_{j_2}} \\ &+\sum_{i=1}^{2}\left\langle \mathbf{a}^{(2)}(i;j_1;\mathbf{x}),\mathbf{s}_V^{(2)}\left[i;j_2;\mathbf{U}^{(2)}(2;\mathbf{x});\boldsymbol{\alpha}\right]\right\rangle_1 \\ &-\frac{\left\{\partial\hat{P}^{(2)}\left[\mathbf{U}^{(2)}(2;\mathbf{x});\mathbf{A}^{(2)}(2;j_1;\mathbf{x});\boldsymbol{\alpha}\right]\right\}_{\partial\Omega_x}}{\partial\alpha_{j_2}} \\ &\triangleq \int_{\lambda_1(\boldsymbol{\alpha})}^{\omega_1(\boldsymbol{\alpha})}\dots\int_{\lambda_{TI}(\boldsymbol{\alpha})}^{\omega_{TI}(\boldsymbol{\alpha})} S^{(2)}\left[j_2;j_1;\mathbf{U}^{(2)}(2;\mathbf{x});\mathbf{A}^{(2)}(2;j_1;\mathbf{x});\boldsymbol{\alpha}\right]dx_1\dots dx_{TI} \triangleq \frac{\partial^2 R\left[\mathbf{u}(\mathbf{x});\boldsymbol{\alpha}\right]}{\partial\alpha_{j_2}\partial\alpha_{j_1}}.\end{aligned} \tag{6.55}$$

Notably, the 2nd-LASS is independent of parameter variations $\delta\boldsymbol{\alpha}$ and variations $\mathbf{V}^{(2)}(2;\mathbf{x})$ in the respective state functions. It is also important to note that the $(2\times TD)^2$-dimensional matrix $\mathbf{AM}^{(2)}\left[2\times2;\mathbf{U}^{(2)}(2;\mathbf{x});\boldsymbol{\alpha}\right]$ is independent of the index j_1. Only the source-term $\mathbf{Q}_A^{(2)}\left[2;j_1;\mathbf{U}^{(2)}(2;\mathbf{x});\boldsymbol{\alpha}\right]$ depends on the index j_1. Therefore, the same solver can be used to invert the operator $\mathbf{AM}^{(2)}\left[2\times2;\mathbf{U}^{(2)}(2;\mathbf{x});\boldsymbol{\alpha}\right]$ in order to solve numerically the 2nd-LASS for each j_1-dependent source $\mathbf{Q}_A^{(2)}\left[2;j_1;\mathbf{U}^{(2)}(2;\mathbf{x});\boldsymbol{\alpha}\right]$ to obtain the corresponding j_1-dependent $2\times TD$-dimensional 2nd-level adjoint function (column vector) $\mathbf{A}^{(2)}(2;j_1;\mathbf{x}) \triangleq \left[\mathbf{a}^{(2)}(1;j_1;\mathbf{x}),\mathbf{a}^{(2)}(2;j_1;\mathbf{x})\right]^{\dagger}$. Computationally, it is efficient to store, if possible, the inverse matrix $\left\{\mathbf{AM}^{(2)}\left[2\times2;\mathbf{U}^{(2)}(2;\mathbf{x});\boldsymbol{\alpha}\right]\right\}^{-1}$, in order to multiply directly the inverse matrix $\left\{\mathbf{AM}^{(2)}\left[2\times2;\mathbf{U}^{(2)}(2;\mathbf{x});\boldsymbol{\alpha}\right]\right\}^{-1}$ with the corresponding source term $\mathbf{Q}_A^{(2)}\left[2;j_1;\mathbf{U}^{(2)}(2;\mathbf{x});\boldsymbol{\alpha}\right]$, for each index j_1, in order to obtain the corresponding j_1-dependent $2\times TD$-dimensional 2nd-level adjoint function $\mathbf{A}^{(2)}(2;j_1;\mathbf{x})\triangleq\left[\mathbf{a}^{(2)}(1;j_1;\mathbf{x}),\mathbf{a}^{(2)}(2;j_1;\mathbf{x})\right]^{\dagger}$. The two components $\mathbf{a}^{(2)}(1;j_1;\mathbf{x})$ and $\mathbf{a}^{(2)}(2;j_1;\mathbf{x})$ of the 2nd-level adjoint function $\mathbf{A}^{(2)}(2;j_1;\mathbf{x})\triangleq\left[\mathbf{a}^{(2)}(1;j_1;\mathbf{x}),\mathbf{a}^{(2)}(2;j_1;\mathbf{x})\right]^{\dagger}$ are distinguished from each other by the use of the numbers "1" and, respectively, "2" in the respective list of arguments. In this particularly simple case, the numbers "1" and "2" could have also been used as subscripts, in the customary notation for vector components but such a use would not lend itself to generalizations because the subscript notation would become unwieldy for the higher-level adjoint functions, as will be introduced in the Sections to follow.

Since the adjoint matrix $\mathbf{AM}^{(2)}\left[2\times 2;\mathbf{U}^{(2)}(2;\mathbf{x});\boldsymbol{\alpha}\right]$ is block diagonal, solving the 2nd-LASS is equivalent to solving two 1st-LASS, with two different source terms. Thus, the "solvers" and the computer program used for solving the 1st-LASS can also be used for solving the 2nd-LASS. The 2nd-LASS was designated as the "2nd-level" rather than the "2nd-order" adjoint sensitivity system since the 2nd-LASS does not involve any explicit 2nd-order G-derivatives of the operators underlying the original system but involves the inversion of the same operators that needed to be inverted for solving the 1st-LASS.

If the 2nd-LASS is solved TP-times, the 2nd-order mixed sensitivities $R^{(2)}\left[j_2;j_1;\mathbf{U}^{(2)}(2;\mathbf{x});\mathbf{A}^{(2)}(2;j_1;\mathbf{x});\boldsymbol{\alpha}\right]\equiv\partial^2 R/\partial\alpha_{j_2}\partial\alpha_{j_1}$ will be computed twice, in two different ways, in terms of two distinct 2nd-level adjoint functions. Consequently, the symmetry property $\partial^2 R\left[\mathbf{u}(\mathbf{x});\boldsymbol{\alpha}\right]/\partial\alpha_{j_2}\partial\alpha_{j_1}=\partial^2 R\left[\mathbf{u}(\mathbf{x});\boldsymbol{\alpha}\right]/\partial\alpha_{j_1}\partial\alpha_{j_2}$ provides an intrinsic (numerical) verification that the components of the 2nd-level adjoint function $\mathbf{A}^{(2)}(2;j_1;\mathbf{x})$, and the 1st-level adjoint function $\mathbf{a}^{(1)}(\mathbf{x})$ are all computed accurately.

The structure of the 2nd-LASS enables full flexibility for prioritizing the computation of the 2nd-order sensitivities. The computation of the 2nd-order sensitivities would logically be prioritized based on the relative magnitudes of the 1st-order sensitivities: the largest relative 1st-order response sensitivity should have the highest priority for computing the corresponding 2nd-order mixed sensitivities; then, the second largest relative 1st-order response sensitivity should be considered next, and so on. The unimportant 2nd-order sensitivities can be deliberately neglected while knowing the error incurred by neglecting them. Computing 2nd-order sensitivities that correspond to vanishing 1st-order sensitivities may also be of interest since vanishing 1st-order sensitivities may indicate critical points of the response in the phase-space of model parameters.

6.4 3RD-ORDER COMPREHENSIVE ADJOINT SENSITIVITY ANALYSIS METHODOLOGY

The 3rd-order sensitivities $\partial^3 R\left[\mathbf{u}(\mathbf{x});\boldsymbol{\alpha}\right]/\partial\alpha_{j_3}\partial\alpha_{j_2}\partial\alpha_{j_1}$ are computed by considering them to be the "1st-order sensitivities of a 2nd-order sensitivity." Thus, each of the 2nd-order sensitivity $R^{(2)}\left[j_2;j_1;\mathbf{U}^{(2)}(2;\mathbf{x});\mathbf{A}^{(2)}(2;j_1;\mathbf{x});\boldsymbol{\alpha}\right]\equiv\partial^2 R/\partial\alpha_{j_2}\partial\alpha_{j_1}$ is considered to be a "model response" which is assumed to satisfy the conditions stated in Eqs. (6.7) and (6.8) for each $j_1, j_2 = 1,...,TP$. Consequently, the 1st-order total G-differential of $R^{(2)}\left[j_2; j_1; \mathbf{U}^{(2)}(2;\mathbf{x}); \mathbf{A}^{(2)}(2; j_1; \mathbf{x}); \boldsymbol{\alpha}\right]$ will exist and will be linear in the variations $\mathbf{V}^{(2)}(2;\mathbf{x})$ and $\delta\mathbf{A}^{(2)}(2; j_1; \mathbf{x})$ in a neighborhood around the nominal values of the parameters and the respective state functions. By definition, the 1st-order total G-differential of $R^{(2)}\left[j_2; j_1; \mathbf{U}^{(2)}(2;\mathbf{x}); \mathbf{A}^{(2)}(2; j_1; \mathbf{x}); \boldsymbol{\alpha}\right]$, which will be denoted as $\left\{\delta R^{(2)}\left[j_2; j_1; \mathbf{U}^{(2)}(2;\mathbf{x}); \mathbf{A}^{(2)}(2; j_1; \mathbf{x}); \boldsymbol{\alpha}; \mathbf{V}^{(2)}(2;\mathbf{x}); \delta\mathbf{A}^{(2)}(2; j_1; \mathbf{x}); \delta\boldsymbol{\alpha}\right]\right\}_{\boldsymbol{\alpha}^0}$, is given by the following expression:

$$\begin{aligned}
&\left\{\delta R^{(2)}\left[j_2; j_1; \mathbf{U}^{(2)}(2;\mathbf{x}); \mathbf{A}^{(2)}(2; j_1; \mathbf{x}); \boldsymbol{\alpha}; \mathbf{V}^{(2)}(2;\mathbf{x}); \delta\mathbf{A}^{(2)}(2; j_1; \mathbf{x}); \delta\boldsymbol{\alpha}\right]\right\}_{\boldsymbol{\alpha}^0} \\
&\triangleq \left\{\frac{\partial R^{(2)}\left[j_2; j_1; \mathbf{U}^{(2)}(2;\mathbf{x}); \mathbf{A}^{(2)}(2; j_1; \mathbf{x}); \boldsymbol{\alpha}\right]}{\partial \boldsymbol{\alpha}} \delta\boldsymbol{\alpha}\right\}_{\boldsymbol{\alpha}^0} \\
&+\left\{\delta R^{(2)}\left[j_2; j_1; \mathbf{U}^{(2)}(2;\mathbf{x}); \mathbf{A}^{(2)}(2; j_1; \mathbf{x}); \boldsymbol{\alpha}; \mathbf{V}^{(2)}(2;\mathbf{x}); \delta\mathbf{A}^{(2)}(2; j_1; \mathbf{x})\right]\right\}_{ind},
\end{aligned} \tag{6.56}$$

where:

$$\begin{aligned}
&\left\{\delta R^{(2)}\left[j_2; j_1; \mathbf{U}^{(2)}(2;\mathbf{x}); \mathbf{A}^{(2)}(2; j_1; \mathbf{x}); \boldsymbol{\alpha}; \mathbf{V}^{(2)}(2;\mathbf{x}); \delta\mathbf{A}^{(2)}(2; j_1; \mathbf{x})\right]\right\}_{ind} \\
&\triangleq \int_{\lambda_1(\boldsymbol{\alpha})}^{\omega_1(\boldsymbol{\alpha})} \dots \int_{\lambda_{TI}(\boldsymbol{\alpha})}^{\omega_{TI}(\boldsymbol{\alpha})} \left\{\frac{\partial S^{(2)}\left[j_2; j_1; \mathbf{U}^{(2)}; \mathbf{A}^{(2)}; \boldsymbol{\alpha}\right]}{\partial \mathbf{U}^{(2)}(2;\mathbf{x})} \mathbf{V}^{(2)}(2;\mathbf{x})\right\}_{\boldsymbol{\alpha}^0} dx_1 \dots dx_{TI} \\
&+\int_{\lambda_1(\boldsymbol{\alpha})}^{\omega_1(\boldsymbol{\alpha})} \dots \int_{\lambda_{TI}(\boldsymbol{\alpha})}^{\omega_{TI}(\boldsymbol{\alpha})} \left\{\frac{\partial S^{(2)}\left[j_2; j_1; \mathbf{U}^{(2)}; \mathbf{A}^{(2)}; \boldsymbol{\alpha}\right]}{\partial \mathbf{A}^{(2)}(2; j_1; \mathbf{x})} \delta\mathbf{A}^{(2)}(2; j_1; \mathbf{x})\right\}_{\boldsymbol{\alpha}^0} dx_1 \dots dx_{TI}.
\end{aligned} \tag{6.57}$$

The indirect-effect term $\left\{\delta R^{(2)}\left[j_2; j_1; \mathbf{U}^{(2)}(2;\mathbf{x}); \mathbf{A}^{(2)}(2; j_1; \mathbf{x}); \boldsymbol{\alpha}; \delta\boldsymbol{\alpha}\right]\right\}_{dir}$ can be computed after having determined the vectors $\mathbf{V}^{(2)}(2;\mathbf{x})$ and $\delta\mathbf{A}^{(2)}(2; j_1; \mathbf{x})$, which are the solutions of the following 3rd-Level Variational Sensitivity System (3rd-LVSS):

$$\begin{aligned}
&\left\{\mathbf{VM}^{(3)}\left[4\times 4; \mathbf{U}^{(3)}(4; j_1; \mathbf{x}); \boldsymbol{\alpha}\right] \mathbf{V}^{(3)}(4; j_1; \mathbf{x})\right\}_{\boldsymbol{\alpha}^0} \\
&= \left\{\mathbf{Q}_V^{(3)}\left[4; j_1; \mathbf{U}^{(3)}(4; j_1; \mathbf{x}); \boldsymbol{\alpha}; \delta\boldsymbol{\alpha}\right]\right\}_{\boldsymbol{\alpha}^0}, \quad \mathbf{x} \in \Omega_x,
\end{aligned} \tag{6.58}$$

$$\left\{\mathbf{B}_V^{(3)}\left[4; \mathbf{U}^{(3)}(4; j_1; \mathbf{x}); \mathbf{V}^{(3)}(4; j_1; \mathbf{x}); \boldsymbol{\alpha}; \delta\boldsymbol{\alpha}\right]\right\}_{\boldsymbol{\alpha}^0} = \mathbf{0}[4]; \quad \mathbf{x} \in \partial\Omega_x\left(\boldsymbol{\alpha}^0\right); \tag{6.59}$$

where $\mathbf{0}[4] \triangleq [\mathbf{0}, \mathbf{0}, \mathbf{0}, \mathbf{0}]^\dagger$ and where:

$$\mathbf{VM}^{(3)}\left(4\times 4;\mathbf{U}^{(3)};\boldsymbol{\alpha}\right)\triangleq\begin{pmatrix}\mathbf{VM}^{(2)}\left(2\times 2\right) & \mathbf{0}\left[2\times 2\right]\\ \mathbf{VM}_{21}^{(3)}\left(2\times 2\right) & \mathbf{VM}_{22}^{(3)}\left(2\times 2\right)\end{pmatrix};\tag{6.60}$$

$$\mathbf{U}^{(3)}\left(4;j_1;\mathbf{x}\right)\triangleq\begin{pmatrix}\mathbf{U}^{(2)}\left(2;\mathbf{x}\right)\\ \mathbf{A}^{(2)}\left(2;j_1;\mathbf{x}\right)\end{pmatrix};$$

$$\mathbf{V}^{(3)}\left(4;j_1;\mathbf{x}\right)\triangleq\delta\mathbf{U}^{(3)}\left(4;j_1;\mathbf{x}\right)=\begin{pmatrix}\mathbf{V}^{(2)}\left(2;\mathbf{x}\right)\\ \delta\mathbf{A}^{(2)}\left(2;j_1;\mathbf{x}\right)\end{pmatrix}\tag{6.61}$$

$$=\left[\mathbf{v}^{(1)}\left(\mathbf{x}\right),\delta\mathbf{a}^{(1)}\left(\mathbf{x}\right),\delta\mathbf{a}^{(2)}\left(1;j_1;\mathbf{x}\right),\delta\mathbf{a}^{(2)}\left(2;j_1;\mathbf{x}\right)\right]^{\dagger};$$

$$\mathbf{VM}_{21}^{(3)}\left(2\times 2;\mathbf{x}\right)\triangleq\frac{\partial\left\{\mathbf{AM}^{(2)}\left[2\times 2;\mathbf{U}^{(2)}\left(2;\mathbf{x}\right);\boldsymbol{\alpha}\right]\mathbf{A}^{(2)}\left(2;j_1;\mathbf{x}\right)\right\}}{\partial\mathbf{U}^{(2)}\left(2;\mathbf{x}\right)}-\frac{\partial\mathbf{Q}_A^{(2)}\left[2;j_1;\mathbf{u}^{(2)}\left(\mathbf{x}\right);\boldsymbol{\alpha}\right]}{\partial\mathbf{U}^{(2)}\left(2;\mathbf{x}\right)};\tag{6.62}$$

$$\mathbf{VM}_{22}^{(3)}\left(2\times 2;\mathbf{x}\right)\triangleq\mathbf{AM}^{(2)}\left[2\times 2;\mathbf{U}^{(2)}\left(2;\mathbf{x}\right);\boldsymbol{\alpha}\right];\quad\mathbf{0}\left[2\times 2\right]\triangleq\begin{pmatrix}\mathbf{0} & \mathbf{0}\\ \mathbf{0} & \mathbf{0}\end{pmatrix};\tag{6.63}$$

$$\mathbf{Q}_V^{(3)}\left[4;j_1;\mathbf{U}^{(3)}\left(4;j_1;\mathbf{x}\right);\boldsymbol{\alpha};\delta\boldsymbol{\alpha}\right]\triangleq\begin{pmatrix}\mathbf{Q}_V^{(2)}\left[2;\mathbf{U}^{(2)}\left(2;\mathbf{x}\right);\boldsymbol{\alpha};\delta\boldsymbol{\alpha}\right]\\ \mathbf{Q}_2^{(3)}\left[2;j_1;\mathbf{U}^{(3)}\left(4;j_1;\mathbf{x}\right);\boldsymbol{\alpha};\delta\boldsymbol{\alpha}\right]\end{pmatrix}\tag{6.64}$$

$$\triangleq\left\{\mathbf{q}_V^{(3)}\left[1;j_1;\mathbf{U}^{(3)}\left(4;j_1;\mathbf{x}\right);\boldsymbol{\alpha};\delta\boldsymbol{\alpha}\right],\ldots,\mathbf{q}_V^{(3)}\left[4;j_1;\mathbf{U}^{(3)}\left(4;j_1;\mathbf{x}\right);\boldsymbol{\alpha};\delta\boldsymbol{\alpha}\right]\right\}^{\dagger};$$

$$\mathbf{q}_V^{(3)}\left[i;j_1;\mathbf{U}^{(3)}\left(4;j_1;\mathbf{x}\right);\boldsymbol{\alpha};\delta\boldsymbol{\alpha}\right]\triangleq\sum_{j_3=1}^{TP}\mathbf{s}_V^{(3)}\left[i;j_3;j_1;\mathbf{U}^{(3)}\left(4;j_1;\mathbf{x}\right);\boldsymbol{\alpha}\right]\delta\alpha_{j_3};\quad i=1,\ldots 4;\tag{6.65}$$

$$\mathbf{Q}_2^{(3)}\left[2;j_1;\mathbf{U}^{(3)}\left(4;j_1;\mathbf{x}\right);\boldsymbol{\alpha};\delta\boldsymbol{\alpha}\right]\triangleq\frac{\partial\mathbf{Q}_A^{(2)}\left[2;j_1;\mathbf{u}^{(2)}\left(\mathbf{x}\right);\boldsymbol{\alpha}\right]}{\partial\boldsymbol{\alpha}}\partial\boldsymbol{\alpha}-\frac{\partial\left\{\mathbf{AM}^{(2)}\left[2\times 2;\mathbf{U}^{(2)}\left(2;\mathbf{x}\right);\boldsymbol{\alpha}\right]\mathbf{A}^{(2)}\left(2;j_1;\mathbf{x}\right)\right\}}{\partial\boldsymbol{\alpha}}\partial\boldsymbol{\alpha};\tag{6.66}$$

$$\mathbf{B}_V^{(3)}\left[4;\mathbf{U}^{(3)}\left(4;j_1;\mathbf{x}\right);\mathbf{V}^{(3)}\left(4;j_1;\mathbf{x}\right);\boldsymbol{\alpha};\delta\boldsymbol{\alpha}\right]\triangleq\begin{pmatrix}\mathbf{B}_V^{(2)}\left[2;\mathbf{U}^{(2)}\left(2;\mathbf{x}\right);\mathbf{V}^{(2)}\left(2;\mathbf{x}\right);\boldsymbol{\alpha};\delta\boldsymbol{\alpha}\right]\\ \delta\mathbf{B}_A^{(2)}\left[2;\mathbf{U}^{(3)}\left(4;j_1;\mathbf{x}\right);\mathbf{V}^{(3)}\left(4;j_1;\mathbf{x}\right);\boldsymbol{\alpha};\delta\boldsymbol{\alpha}\right]\end{pmatrix}.\tag{6.67}$$

The variational matrix $\mathbf{VM}^{(3)}\left(4\times 4;\mathbf{U}^{(3)};\boldsymbol{\alpha}\right)$ comprises (4×4) block matrices, each comprising TD^2 components/elements; thus, the matrix $\mathbf{VM}^{(3)}\left(4\times 4;\mathbf{U}^{(3)};\boldsymbol{\alpha}\right)$ comprises a total of $(4\times 4)TD^2$ components/elements. Each of the vectors $\mathbf{V}^{(3)}(4;j_1;\mathbf{x})$, $\mathbf{Q}_V^{(3)}\left[4;j_1;\mathbf{U}^{(3)}(4;j_1;\mathbf{x});\boldsymbol{\alpha};\delta\boldsymbol{\alpha}\right]$ and $\mathbf{B}_V^{(3)}\left[4;\mathbf{U}^{(3)}(2;j_1;\mathbf{x});\mathbf{V}^{(3)}(4;j_1;\mathbf{x});\boldsymbol{\alpha};\delta\boldsymbol{\alpha}\right]$ comprises four TD-dimensional vectors, as shown in their respective definitions; thus, each of the vectors $\mathbf{V}^{(3)}(4;j_1;\mathbf{x})$, $\mathbf{Q}_V^{(3)}\left[4;j_1;\mathbf{U}^{(3)}(4;j_1;\mathbf{x});\boldsymbol{\alpha};\delta\boldsymbol{\alpha}\right]$ and $\mathbf{B}_V^{(3)}\left[4;\mathbf{U}^{(3)}(2;j_1;\mathbf{x});\mathbf{V}^{(3)}(4;j_1;\mathbf{x});\boldsymbol{\alpha};\delta\boldsymbol{\alpha}\right]$ comprises $(4\times TD)$ components/elements.

Solving the 3rd-LVSS would require TP^3 large-scale computations, which is unrealistic for large-scale systems comprising many parameters. The 3rd-CASAM-N circumvents the need for solving the 3rd-LVSS by deriving an alternative expression for the indirect-effect term defined in Eq. (6.57), in which the function $\mathbf{V}^{(3)}(4;j_1;\mathbf{x})$ is replaced by a 3rd-level adjoint function which is independent of parameter variations. This 3rd-level adjoint function is the solution of a 3rd-Level Adjoint Sensitivity System (3rd-LASS) which is constructed by applying the same principles as those used for constructing the 1st-LASS and the 2nd-LASS. The Hilbert space appropriate for constructing the 3rd-LASS, denoted as $\mathcal{H}_3(\Omega_x)$, comprises as elements block vectors of the same form as $\mathbf{V}^{(3)}(4;j_1;\mathbf{x})$. Thus, a generic block vector element of $\mathcal{H}_3(\Omega_x)$ will be denoted as $\boldsymbol{\Psi}^{(3)}(4;\mathbf{x})\triangleq\left[\boldsymbol{\psi}^{(3)}(1;\mathbf{x}),\boldsymbol{\psi}^{(3)}(2;\mathbf{x}),\boldsymbol{\psi}^{(3)}(3;\mathbf{x}),\boldsymbol{\psi}^{(3)}(4;\mathbf{x})\right]^\dagger\in\mathcal{H}_3(\Omega_x)$, comprising four TD-dimensional vector components of the form $\boldsymbol{\psi}^{(3)}(i;\mathbf{x})\triangleq\left[\psi_1^{(3)}(i;\mathbf{x}),...,\psi_{TD}^{(3)}(i;\mathbf{x})\right]^\dagger\in\mathcal{H}_1(\Omega_x)$, $i=1,2,3,4$, where each of these four components is a TD-dimensional column vector. The inner product of two vectors $\boldsymbol{\Psi}^{(3)}(4;\mathbf{x})\in\mathcal{H}_3(\Omega_x)$ and $\boldsymbol{\Phi}^{(3)}(4;\mathbf{x})\in\mathcal{H}_3(\Omega_x)$ in the Hilbert space $\mathcal{H}_3(\Omega_x)$ will be denoted as $\left\langle\boldsymbol{\Psi}^{(3)}(4;\mathbf{x}),\boldsymbol{\Phi}^{(3)}(4;\mathbf{x})\right\rangle_3$ and defined as follows:

$$\left\langle\boldsymbol{\Psi}^{(3)}(4;\mathbf{x}),\boldsymbol{\Phi}^{(3)}(4;\mathbf{x})\right\rangle_3\triangleq\sum_{i=1}^{4}\left\langle\boldsymbol{\psi}^{(3)}(i;\mathbf{x}),\boldsymbol{\varphi}^{(3)}(i;\mathbf{x})\right\rangle_1. \tag{6.68}$$

The inner product defined in Eq. (6.68) is continuous in $\boldsymbol{\alpha}$, in a neighborhood around $\boldsymbol{\alpha}^0$. The steps for constructing the 3rd-LASS are conceptually similar to those underlying the construction of the 1st-LASS and 2nd-LASS. Using the definition of the inner product defined in Eq. (6.68), construct the inner product of Eq. (6.58) with a vector $\mathbf{A}^{(3)}(4;\mathbf{x})\triangleq\left[\mathbf{a}^{(3)}(1;\mathbf{x}),\mathbf{a}^{(3)}(2;\mathbf{x}),\mathbf{a}^{(3)}(3;\mathbf{x}),\mathbf{a}^{(3)}(4;\mathbf{x})\right]^\dagger\in\mathcal{H}_3(\Omega_x)$ to obtain the following relation:

$$\begin{aligned}&\left\{\left\langle\mathbf{A}^{(3)}(4;\mathbf{x}),\mathbf{VM}^{(3)}\left[4\times 4;\mathbf{U}^{(3)}(4;j_1;\mathbf{x});\boldsymbol{\alpha}\right]\mathbf{V}^{(3)}(4;j_1;\mathbf{x})\right\rangle_3\right\}_{\boldsymbol{\alpha}^0}\\&=\left\{\left\langle\mathbf{A}^{(3)}(4;\mathbf{x}),\mathbf{Q}_V^{(3)}\left[4;j_1;\mathbf{U}^{(3)}(4;j_1;\mathbf{x});\boldsymbol{\alpha};\delta\boldsymbol{\alpha}\right]\right\rangle_3\right\}_{\boldsymbol{\alpha}^0},\quad\mathbf{x}\in\Omega_x.\end{aligned} \tag{6.69}$$

The inner product on the left side of Eq. (6.69) is further transformed by using the definition of the adjoint operator to obtain the following relation:

$$
\begin{aligned}
&\left\{\left\langle \mathbf{A}^{(3)}(4;\mathbf{x}), \mathbf{VM}^{(3)}\left[4\times 4;\mathbf{U}^{(3)}(4;j_1;\mathbf{x});\boldsymbol{\alpha}\right]\mathbf{V}^{(3)}(4;j_1;\mathbf{x})\right\rangle_3\right\}_{\boldsymbol{\alpha}^0} \\
&=\left\{\left\langle \mathbf{V}^{(3)}(4;j_1;\mathbf{x}), \mathbf{AM}^{(3)}\left[4\times 4;\mathbf{U}^{(3)}(4;j_1;\mathbf{x});\boldsymbol{\alpha}\right]\mathbf{A}^{(3)}(4;\mathbf{x})\right\rangle_3\right\}_{\boldsymbol{\alpha}^0} \\
&+\left\{\left[P^{(3)}\left(\mathbf{U}^{(3)};\mathbf{A}^{(3)};\mathbf{V}^{(3)};\boldsymbol{\alpha}\right)\right]_{\partial\Omega_x}\right\}_{\boldsymbol{\alpha}^0},
\end{aligned}
\tag{6.70}
$$

where:

$$
\begin{aligned}
&\mathbf{AM}^{(3)}\left[4\times 4;\mathbf{U}^{(3)}(4;j_1;\mathbf{x});\boldsymbol{\alpha}\right] \triangleq \left[\mathbf{VM}^{(3)}\left(4\times 4;\mathbf{U}^{(3)};\boldsymbol{\alpha}\right)\right]^* \\
&=\begin{pmatrix}
\left\{\left[\mathbf{VM}^{(2)}(2\times 2)\right]^*\right\}^\dagger & \left\{\left[\mathbf{VM}_{21}^{(3)}(2\times 2)\right]^*\right\}^\dagger \\
\mathbf{0}[2\times 2] & \left\{\left[\mathbf{VM}_{22}^{(3)}(2\times 2)\right]^*\right\}^\dagger
\end{pmatrix},
\end{aligned}
\tag{6.71}
$$

and where $\left\{\left[P^{(3)}\left(\mathbf{U}^{(3)};\mathbf{A}^{(3)};\mathbf{V}^{(3)};\boldsymbol{\alpha}\right)\right]_{\partial\Omega_x}\right\}_{\boldsymbol{\alpha}^0}$ denotes the corresponding bilinear concomitant on the domain's boundary, evaluated at the nominal values for the parameters and respective state functions. The adjoint matrix $\mathbf{AM}^{(3)}\left[4\times 4;\mathbf{U}^{(3)}(4;j_1;\mathbf{x});\boldsymbol{\alpha}\right]$ comprises $(4\times TD)^2$ components/elements, while the adjoint function $\mathbf{A}^{(3)}(4;\mathbf{x}) \in \mathcal{H}_3(\Omega_x)$ comprises $(4\times TD)$ components/elements.

The domain of the adjoint matrix operator $\mathbf{AM}^{(3)}\left[4\times 4;\mathbf{U}^{(3)}(4;j_1;\mathbf{x});\boldsymbol{\alpha}\right]$ is specified by requiring that the function $\mathbf{A}^{(3)}(4;\mathbf{x}) \triangleq \left[\mathbf{a}^{(3)}(1;\mathbf{x}),\mathbf{a}^{(3)}(2;\mathbf{x}),\mathbf{a}^{(3)}(3;\mathbf{x}),\mathbf{a}^{(3)}(4;\mathbf{x})\right]^\dagger$ $\in \mathcal{H}_3(\Omega_x)$ satisfies adjoint boundary/initial conditions denoted as follows:

$$
\left\{\mathbf{B}_A^{(3)}\left[4;\mathbf{U}^{(3)}(4;j_1;\mathbf{x});\mathbf{A}^{(3)}(4;\mathbf{x});\boldsymbol{\alpha}\right]\right\}_{\boldsymbol{\alpha}^0} = \mathbf{0}[4], \quad \mathbf{x}\in\partial\Omega_x\left(\boldsymbol{\alpha}^0\right). \tag{6.72}
$$

The 3rd-level adjoint boundary/initial conditions represented by Eq. (6.72) are determined by requiring that: (i) they must be independent of unknown values of $\mathbf{V}^{(3)}(4;j_1;\mathbf{x})$; (ii) the substitution of the boundary and/or initial conditions represented by Eqs. (6.72) and (6.59) into the expression of $\left\{\left[P^{(3)}\left(\mathbf{U}^{(3)};\mathbf{A}^{(3)};\mathbf{V}^{(3)};\boldsymbol{\alpha}\right)\right]_{\partial\Omega_x}\right\}_{\boldsymbol{\alpha}^0}$ must cause all terms containing unknown values of $\mathbf{V}^{(3)}(4;j_1;\mathbf{x})$ to vanish.

Implementing the boundary/initial conditions represented by Eqs. (6.72) and (6.59) into Eq. (6.70) will transform the latter into the following form:

$$
\begin{aligned}
&\left\{\left\langle \mathbf{V}^{(3)}(4;j_1;\mathbf{x}), \mathbf{AM}^{(3)}\left[4\times 4;\mathbf{U}^{(3)}(4;j_1;\mathbf{x});\boldsymbol{\alpha}\right]\mathbf{A}^{(3)}(4;\mathbf{x})\right\rangle_3\right\}_{\boldsymbol{\alpha}^0} \\
&=\left\{\left\langle \mathbf{A}^{(3)}(4;\mathbf{x}), \mathbf{VM}^{(3)}\left[4\times 4;\mathbf{U}^{(3)}(4;j_1;\mathbf{x});\boldsymbol{\alpha}\right]\mathbf{V}^{(3)}(4;j_1;\mathbf{x})\right\rangle_3\right\}_{\boldsymbol{\alpha}^0} \\
&-\left\{\left[\hat{P}^{(3)}\left(\mathbf{U}^{(3)};\mathbf{A}^{(3)};\delta\boldsymbol{\alpha}\right)\right]_{\partial\Omega_x}\right\}_{\boldsymbol{\alpha}^0},
\end{aligned}
\tag{6.73}
$$

where $\left\{\left[\hat{P}^{(3)}\left(\mathbf{U}^{(3)};\mathbf{A}^{(3)};\delta\boldsymbol{\alpha}\right)\right]_{\partial\Omega_x}\right\}_{\boldsymbol{\alpha}^0}$ denotes residual boundary terms that may not vanish automatically. The right side of Eq. (6.69) is now used to replace the first term on the right side of Eq. (6.73) to obtain the following relation:

$$
\begin{aligned}
&\left\{\left\langle \mathbf{V}^{(3)}(4;j_1;\mathbf{x}),\mathbf{AM}^{(3)}\left[4\times 4;\mathbf{U}^{(3)}(4;j_1;\mathbf{x});\boldsymbol{\alpha}\right]\mathbf{A}^{(3)}(4;\mathbf{x})\right\rangle_3\right\}_{\boldsymbol{\alpha}^0} \\
&=\left\{\left\langle \mathbf{A}^{(3)}(4;\mathbf{x}),\mathbf{Q}_V^{(3)}\left[4;j_1;\mathbf{U}^{(3)}(4;j_1;\mathbf{x});\boldsymbol{\alpha};\delta\boldsymbol{\alpha}\right]\right\rangle_3\right\}_{\boldsymbol{\alpha}^0} \\
&-\left\{\left[\hat{P}^{(3)}\left(\mathbf{U}^{(3)};\mathbf{A}^{(3)};\delta\boldsymbol{\alpha}\right)\right]_{\partial\Omega_x}\right\}_{\boldsymbol{\alpha}^0}.
\end{aligned}
\tag{6.74}
$$

The definition of the 3rd-level adjoint function $\mathbf{A}^{(3)}(4;\mathbf{x})\in\mathcal{H}_3(\Omega_x)$ is now completed by requiring that the left side of Eq. (6.74) and the right side of Eq. (6.57) represent the "indirect effect term" $\left\{\delta R^{(2)}\left[j_2;j_1;\mathbf{U}^{(2)}(2;\mathbf{x});\mathbf{A}^{(2)}(2;j_1;\mathbf{x});\boldsymbol{\alpha};\mathbf{V}^{(2)}(2;\mathbf{x});\delta\mathbf{A}^{(2)}(2;j_1;\mathbf{x})\right]\right\}$, for each of the indices $j_1=1,...,TP;\ j_2=1,....,j_1$. Hence, there will be $TP(TP+1)/2$ distinct 3rd-level adjoint functions $\mathbf{A}^{(3)}(4;j_2;j_1;\mathbf{x})\triangleq\left[\mathbf{a}^{(3)}(1;j_2;j_1;\mathbf{x}),\mathbf{a}^{(3)}(2;j_2;j_1;\mathbf{x}),\mathbf{a}^{(3)}(3;j_2;j_1;\mathbf{x}),\mathbf{a}^{(3)}(4;j_2;j_1;\mathbf{x})\right]^{\dagger}$, corresponding to the indices $j_1=1,...,TP;\ j_2=1,....,j_1$. Each of these distinct 3rd-level adjoint functions will correspond to a specific (j_1,j_2)-dependent indirect-effect term.

The left side of Eq. (6.74) will be identical to the right side of Eq. (6.57) by requiring that the following relation be satisfied by the 3rd-level adjoint functions $\mathbf{A}^{(3)}(4;j_2;j_1;\mathbf{x})\triangleq\left[\mathbf{a}^{(3)}(1;j_2;j_1;\mathbf{x}),\mathbf{a}^{(3)}(2;j_2;j_1;\mathbf{x}),\mathbf{a}^{(3)}(3;j_2;j_1;\mathbf{x}),\mathbf{a}^{(3)}(4;j_2;j_1;\mathbf{x})\right]^{\dagger}$:

$$
\begin{aligned}
&\left\{\mathbf{AM}^{(3)}\left[4\times 4;\mathbf{U}^{(3)}(4;j_1;\mathbf{x});\boldsymbol{\alpha}\right]\mathbf{A}^{(3)}(4;j_2;j_1;\mathbf{x})\right\}_{\boldsymbol{\alpha}^0} \\
&=\left\{\mathbf{Q}_A^{(3)}\left[4;j_2;j_1;\mathbf{U}^{(3)}(4;j_1;\mathbf{x});\boldsymbol{\alpha}\right]\right\}_{\boldsymbol{\alpha}^0},\ j_1=1,...,TP;\ j_2=1,....,j_1,
\end{aligned}
\tag{6.75}
$$

where:

$$
\begin{aligned}
&\mathbf{Q}_A^{(3)}\left[4;j_2;j_1;\mathbf{U}^{(3)}(4;j_1;\mathbf{x});\boldsymbol{\alpha}\right] \\
&\triangleq\left[\mathbf{q}_A^{(3)}\left(1;j_2;j_1;\mathbf{U}^{(3)};\boldsymbol{\alpha}\right),....,\mathbf{q}_A^{(3)}\left(1;j_2;j_1;\mathbf{U}^{(3)};\boldsymbol{\alpha}\right)\right]^{\dagger};
\end{aligned}
\tag{6.76}
$$

$$
\mathbf{q}_A^{(3)}\left(1;j_2;j_1;\mathbf{U}^{(3)};\boldsymbol{\alpha}\right)\triangleq\partial S^{(2)}\left[j_2;j_1;\mathbf{u}^{(2)};\mathbf{a}^{(2)};\boldsymbol{\alpha}\right]/\partial\mathbf{u}^{(1)};
\tag{6.77}
$$

$$
\mathbf{q}_A^{(3)}\left(2;j_2;j_1;\mathbf{U}^{(3)};\boldsymbol{\alpha}\right)\triangleq\partial S^{(2)}\left[j_2;j_1;\mathbf{u}^{(2)};\mathbf{a}^{(2)};\boldsymbol{\alpha}\right]/\partial\mathbf{a}^{(1)};
\tag{6.78}
$$

$$
\mathbf{q}_A^{(3)}\left(3;j_2;j_1;\mathbf{U}^{(3)};\boldsymbol{\alpha}\right)\triangleq\partial S^{(2)}\left[j_2;j_1;\mathbf{u}^{(2)};\mathbf{a}^{(2)};\boldsymbol{\alpha}\right]/\partial\mathbf{a}^{(2)}(1;j_1;\mathbf{x});
\tag{6.79}
$$

$$
\mathbf{q}_A^{(3)}\left(4;j_2;j_1;\mathbf{U}^{(3)};\boldsymbol{\alpha}\right)\triangleq\partial S^{(2)}\left[j_2;j_1;\mathbf{u}^{(2)};\mathbf{a}^{(2)};\boldsymbol{\alpha}\right]/\partial\mathbf{a}^{(2)}(2;j_1;\mathbf{x}).
\tag{6.80}
$$

The boundary conditions to be satisfied by each of the 3rd-level adjoint functions $\mathbf{A}^{(3)}(4;j_2;j_1;\mathbf{x}) \triangleq \left[\mathbf{a}^{(3)}(1;j_2;j_1;\mathbf{x}),\mathbf{a}^{(3)}(2;j_2;j_1;\mathbf{x}),\mathbf{a}^{(3)}(3;j_2;j_1;\mathbf{x}),\mathbf{a}^{(3)}(4;j_2;j_1;\mathbf{x})\right]^{\dagger}$ are those represented by Eq. (6.72), namely:

$$\begin{aligned}&\left\{\mathbf{B}_{A}^{(3)}\left[4;\mathbf{U}^{(3)}(4;j_1;\mathbf{x});\mathbf{A}^{(3)}(4;j_2;j_1;\mathbf{x});\boldsymbol{\alpha}\right]\right\}_{\boldsymbol{\alpha}^0}=\mathbf{0}[4];\\&j_1=1,\dots,TP;\, j_2=1,\dots,j_1;\ \mathbf{x}\in\partial\Omega_x\left(\boldsymbol{\alpha}^0\right).\end{aligned}\tag{6.81}$$

Since the source $\mathbf{Q}_{A}^{(3)}\left[4;j_2;j_1;\mathbf{U}^{(3)}(4;j_1;\mathbf{x});\boldsymbol{\alpha}\right]$ may contain distributions, the equality in Eq. (6.75) is considered to hold in the weak sense.

The system of equations represented by Eqs. (6.75) and (6.81) will be called the 3rd-Level Adjoint Sensitivity System (3rd-LASS); its solution, $\mathbf{A}^{(3)}(4;j_2;j_1;\mathbf{x})$, will be called the 3rd-level adjoint function.

Using the equations underlying the 3rd-LASS together with Eqs. (6.74) in (6.57) yields the following expression for the indirect-effect term, for $j_1=1,\dots,TP;\ j_2=1,\dots,j_1$:

$$\begin{aligned}&\left\{\delta R^{(2)}\left[j_2;j_1;\mathbf{U}^{(2)}(2;\mathbf{x});\mathbf{A}^{(2)}(2;j_1;\mathbf{x});\boldsymbol{\alpha};\mathbf{V}^{(2)}(2;\mathbf{x});\delta\mathbf{A}^{(2)}(2;j_1;\mathbf{x})\right]\right\}_{ind}\\&=\left\{\left\langle\mathbf{A}^{(3)}(4;j_2;j_1;\mathbf{x}),\mathbf{Q}_{V}^{(3)}\left[4;j_1;\mathbf{U}^{(3)}(4;j_1;\mathbf{x});\boldsymbol{\alpha};\delta\boldsymbol{\alpha}\right]\right\rangle_3\right\}_{\boldsymbol{\alpha}^0}\\&-\left\{\left[\hat{P}^{(3)}\left(\mathbf{U}^{(3)};\mathbf{A}^{(3)};\delta\boldsymbol{\alpha}\right)\right]_{\partial\Omega_x}\right\}_{\boldsymbol{\alpha}^0}\\&\equiv\left\{\delta R^{(2)}\left[j_2;j_1;\mathbf{U}^{(3)}(4;j_1;\mathbf{x});\mathbf{A}^{(3)}(4;j_2;j_1;\mathbf{x});\boldsymbol{\alpha}\right]\right\}_{ind}.\end{aligned}\tag{6.82}$$

As the identity in the last line of Eq. (6.82) indicates, the dependence of the indirect-effect term on the variational function $\mathbf{V}^{(3)}(4;j_1;\mathbf{x})$ has been replaced by the dependence on the adjoint function $\mathbf{A}^{(3)}(4;j_2;j_1;\mathbf{x})$, for each $j_1=1,\dots,TP;\ j_2=1,\dots,j_1$.

Replacing the expression obtained in Eq. (6.82) for the indirect-effect term into Eq. (6.56) yields the following expression:

$$\begin{aligned}&\left\{\delta R^{(2)}\left[j_2;j_1;\mathbf{U}^{(3)}(4;j_1;\mathbf{x});\mathbf{A}^{(3)}(4;j_2;j_1;\mathbf{x});\boldsymbol{\alpha};\delta\boldsymbol{\alpha}\right]\right\}_{\boldsymbol{\alpha}^0}\\&=\left\{\frac{\partial R^{(2)}\left[j_2;j_1;\mathbf{U}^{(3)};\boldsymbol{\alpha}\right]}{\partial\boldsymbol{\alpha}}\delta\boldsymbol{\alpha}\right\}_{\boldsymbol{\alpha}^0}-\left\{\left[\hat{P}^{(3)}\left(\mathbf{U}^{(3)};\mathbf{A}^{(3)};\delta\boldsymbol{\alpha}\right)\right]_{\partial\Omega_x}\right\}_{\boldsymbol{\alpha}^0}\\&+\left\{\left\langle\mathbf{A}^{(3)}(4;j_2;j_1;\mathbf{x}),\mathbf{Q}_{V}^{(3)}\left[4;j_1;\mathbf{U}^{(3)};\boldsymbol{\alpha};\delta\boldsymbol{\alpha}\right]\right\rangle_3\right\}_{\boldsymbol{\alpha}^0}\end{aligned}\tag{6.83}$$

In component form, the total differential expressed by Eq. (6.83) has the following expression, for $j_1;j_2=1,\dots,TP$:

$$\left\{\delta R^{(2)}\left[j_2;j_1;\mathbf{U}^{(3)}(4;j_1;\mathbf{x});\mathbf{A}^{(3)}(4;j_2;j_1;\mathbf{x});\boldsymbol{\alpha};\delta\boldsymbol{\alpha}\right]\right\}_{\boldsymbol{\alpha}^0}$$
$$=\sum_{j_3=1}^{TP}\left\{R^{(3)}\left[j_3;j_2;j_1;\mathbf{U}^{(3)}(4;j_1;\mathbf{x});\mathbf{A}^{(3)}(4;j_2;j_1;\mathbf{x});\boldsymbol{\alpha}\right]\right\}_{\boldsymbol{\alpha}^0}\delta\alpha_{j_3} \tag{6.84}$$

where the quantity $R^{(3)}\left[j_3;j_2;j_1;\mathbf{U}^{(3)}(4;j_1;\mathbf{x});\mathbf{A}^{(3)}(4;j_2;j_1;\mathbf{x});\boldsymbol{\alpha}\right]$ denotes the 3rd-order sensitivity of the generic scalar-valued response $R[\mathbf{u}(\mathbf{x});\boldsymbol{\alpha}]$ with respect to any three model parameters α_{j_1}, α_{j_2}, α_{j_3}, and has the following expression for $j_1,j_2,j_3=1,\dots,TP$:

$$R^{(3)}\left[j_3;j_2;j_1;\mathbf{U}^{(3)}(4;j_1;\mathbf{x});\mathbf{A}^{(3)}(4;j_2;j_1;\mathbf{x});\boldsymbol{\alpha}\right]$$
$$\triangleq\frac{\partial R^{(2)}\left[j_2;j_1;\mathbf{U}^{(3)}(4;j_1;\mathbf{x});\boldsymbol{\alpha}\right]}{\partial\alpha_{j_3}}-\frac{\left[\partial\hat{P}^{(3)}\left(\mathbf{U}^{(3)};\mathbf{A}^{(3)};\delta\pm\right)\right]_{\partial\Omega_x}}{\partial\alpha_{j_3}}$$
$$+\sum_{i=1}^{4}\left\langle\mathbf{a}^{(3)}(i;j_2;j_1;\mathbf{x}),\mathbf{s}_V^{(3)}\left[i;j_3;j_1;\mathbf{U}^{(3)}(4;j_1;\mathbf{x});\boldsymbol{\alpha}\right]\right\rangle_1 \tag{6.85}$$
$$\triangleq\int_{\lambda_1(\boldsymbol{\alpha})}^{\omega_1(\boldsymbol{\alpha})}dx_1..\int_{\lambda_{TI}(\boldsymbol{\alpha})}^{\omega_{TI}(\boldsymbol{\alpha})}dx_{TI}S^{(3)}\left[j_3;j_2;j_1;\mathbf{U}^{(3)}(4;j_1;\mathbf{x});\mathbf{A}^{(3)}(4;j_2;j_1;\mathbf{x});\boldsymbol{\alpha}\right]$$
$$\triangleq\frac{\partial^3 R[\mathbf{u}(\mathbf{x});\boldsymbol{\alpha}]}{\partial\alpha_{j_3}\partial\alpha_{j_2}\partial\alpha_{j_1}}.$$

The matrix $\mathbf{AM}^{(3)}\left[4\times 4;\mathbf{U}^{(3)}(4;j_1;\mathbf{x});\boldsymbol{\alpha}\right]$ is block diagonal; therefore, solving the 3rd-LASS is equivalent to solving three 1st-LASS, with different source terms. The 3rd-LASS was designated as "the 3rd-level" rather than "3rd-order" adjoint sensitivity system since the 3rd-LASS does not involve any explicit 2nd-order and/or 3rd-order G-derivatives of the operators underlying the original system but involves the inversion of the same operators that needed to be inverted for solving the 1st-LASS.

By solving the 3rd-LASS $TP(TP+1)/2$ times, the 3rd-order mixed sensitivities $\left\{\partial^3 R(\boldsymbol{\alpha},\mathbf{u})/\partial\alpha_{j_3}\partial\alpha_{j_2}\partial\alpha_{j_1}\right\}_{(\boldsymbol{\alpha}^0)}$ will be computed three times, in three different ways. Consequently, the multiple symmetries intrinsic to the 3rd-order sensitivities provide an intrinsic numerical verification that the components of the 1st-, 2nd-, and 3rd-level adjoint functions are computed accurately.

The structure of the 3rd-LASS enables full flexibility for prioritizing the computation of the 3rd-order sensitivities. The computation of the 3rd-order sensitivities would logically be prioritized based on the relative magnitudes of the 2nd-order sensitivities so that the unimportant 3rd-order sensitivities can be deliberately neglected while knowing the error incurred by neglecting them.

6.5 4TH-ORDER COMPREHENSIVE ADJOINT SENSITIVITY ANALYSIS METHODOLOGY

The 3rd-order sensitivities $\partial^3 R[\mathbf{u}(\mathbf{x});\boldsymbol{\alpha}]/\partial\alpha_{j_3}\partial\alpha_{j_2}\partial\alpha_{j_1}$ will be assumed to satisfy the conditions stated in Eqs. (6.7) and (6.8) for each $j_1, j_2, j_3 = 1,...,TP$, so that the 1st-order total G-differential of $R^{(3)}\left[j_3;j_2;j_1;\mathbf{U}^{(3)};\mathbf{A}^{(3)};\boldsymbol{\alpha}\right]$ will exist and will be linear in the variations $\mathbf{V}^{(3)}(4;j_1;\mathbf{x}) \triangleq \delta\mathbf{U}^{(3)}(4;j_1;\mathbf{x})$ and $\delta\mathbf{A}^{(3)}(4;j_2;j_1;\mathbf{x})$ in a neighborhood around the nominal values of the parameters and the respective state functions. By definition, the 1st-order total G-differential of $R^{(3)}\left[j_3;j_2;j_1;\mathbf{U}^{(3)}(4;j_1;\mathbf{x});\mathbf{A}^{(3)}(4;j_2;j_1;\mathbf{x});\boldsymbol{\alpha}\right]$, which will be denoted as $\left\{\delta R^{(3)}\left[j_3;j_2;j_1;\mathbf{U}^{(3)};\mathbf{A}^{(3)};\boldsymbol{\alpha};\mathbf{V}^{(3)}(4;j_1;\mathbf{x});\delta\mathbf{A}^{(3)}(4;j_2;j_1;\mathbf{x});\delta\boldsymbol{\alpha}\right]\right\}_{\boldsymbol{\alpha}^0}$, is given by the following expression:

$$
\begin{aligned}
&\left\{\delta R^{(3)}\left[j_3;j_2;j_1;\mathbf{U}^{(3)};\mathbf{A}^{(3)};\boldsymbol{\alpha};\mathbf{V}^{(3)}(4;j_1;\mathbf{x});\delta\mathbf{A}^{(3)}(4;j_2;j_1;\mathbf{x});\delta\boldsymbol{\alpha}\right]\right\}_{\boldsymbol{\alpha}^0} \\
&\triangleq \left\{\frac{d}{d\varepsilon} R^{(3)}\left[j_3;j_2;j_1;\mathbf{U}^{(3)}(4;j_1;\mathbf{x})+\varepsilon\mathbf{V}^{(3)}(4;j_1;\mathbf{x});\right.\right. \\
&\qquad \left.\left.\mathbf{A}^{(3)}(4;j_2;j_1;\mathbf{x})+\varepsilon\delta\mathbf{A}^{(3)}(4;j_2;j_1;\mathbf{x});\boldsymbol{\alpha}+\varepsilon\delta\boldsymbol{\alpha}\right]\right\}_{\varepsilon=0} \\
&\triangleq \left\{\delta R^{(3)}\left[j_3;j_2;j_1;\mathbf{U}^{(3)};\mathbf{A}^{(3)};\boldsymbol{\alpha};\delta\boldsymbol{\alpha}\right]\right\}_{dir} \\
&\quad + \left\{\delta R^{(3)}\left[j_3;j_2;j_1;\mathbf{U}^{(3)};\mathbf{A}^{(3)};\boldsymbol{\alpha};\mathbf{V}^{(3)}(4;j_1;\mathbf{x});\delta\mathbf{A}^{(3)}(4;j_2;j_1;\mathbf{x})\right]\right\}_{ind}.
\end{aligned}
\tag{6.86}
$$

In Eq. (6.86), the quantity $\left\{\delta R^{(3)}\left[j_3;j_2;j_1;\mathbf{U}^{(3)};\mathbf{A}^{(3)};\boldsymbol{\alpha};\delta\boldsymbol{\alpha}\right]\right\}_{dir}$ denotes the "direct-effect term," which comprises all of the dependencies on the vector $\delta\boldsymbol{\alpha}$ of parameter variations and has the following expression:

$$
\begin{aligned}
&\left\{\delta R^{(3)}\left[j_3;j_2;j_1;\mathbf{U}^{(3)}(4;j_1;\mathbf{x});\mathbf{A}^{(3)}(4;j_2;j_1;\mathbf{x});\boldsymbol{\alpha};\delta\boldsymbol{\alpha}\right]\right\}_{dir} \triangleq \\
&\left\{\frac{\partial R^{(3)}\left[j_3;j_2;j_1;\mathbf{U}^{(3)}(4;j_1;\mathbf{x});\mathbf{A}^{(3)}(4;j_2;j_1;\mathbf{x});\boldsymbol{\alpha}\right]}{\partial\boldsymbol{\alpha}}\delta\boldsymbol{\alpha}\right\}_{\boldsymbol{\alpha}^0} \\
&= \sum_{j_4=1}^{TP}\left\{\frac{\partial R^{(3)}\left[j_3;j_2;j_1;\mathbf{U}^{(3)}(4;j_1;\mathbf{x});\mathbf{A}^{(3)}(4;j_2;j_1;\mathbf{x});\boldsymbol{\alpha}\right]}{\partial\alpha_{j_4}}\right\}_{\boldsymbol{\alpha}^0}\delta\alpha_{j_4}.
\end{aligned}
\tag{6.87}
$$

In Eq. (6.86), the quantity $\left\{\delta R^{(3)}\left[j_3;j_2;j_1;\mathbf{U}^{(3)};\mathbf{A}^{(3)};\boldsymbol{\alpha};\mathbf{V}^{(3)};\delta\mathbf{A}^{(3)}\right]\right\}_{ind}$ denotes the "indirect-effect term," which comprises all of the dependencies on the vectors $\mathbf{V}^{(3)}(4;j_1;\mathbf{x})$ and $\delta\mathbf{A}^{(3)}(4;j_2;j_1;\mathbf{x})$ of variations in the state functions $\mathbf{U}^{(3)}(4;j_1;\mathbf{x})$ and $\mathbf{A}^{(3)}(4;j_2;j_1;\mathbf{x})$; this indirect-effect term is defined as follows:

$$\left\{\delta R^{(3)}\left[j_3;j_2;j_1;\mathbf{U}^{(3)};\mathbf{A}^{(3)};\boldsymbol{\alpha};\mathbf{V}^{(3)}(4;j_1;\mathbf{x});\delta\mathbf{A}^{(3)}(4;j_2;j_1;\mathbf{x})\right]\right\}_{ind}$$

$$\triangleq\left\{\frac{\partial R^{(3)}\left[j_3;j_2;j_1;\mathbf{U}^{(3)};\mathbf{A}^{(3)};\boldsymbol{\alpha}\right]}{\partial\mathbf{U}^{(3)}(4;j_1;\mathbf{x})}\right\}_{\boldsymbol{\alpha}^0}\mathbf{V}^{(3)}(4;j_1;\mathbf{x})$$
$$+\left\{\frac{\partial R^{(3)}\left[j_3;j_2;j_1;\mathbf{U}^{(3)};\mathbf{A}^{(3)};\boldsymbol{\alpha}\right]}{\partial\mathbf{A}^{(3)}(4;j_2;j_1;\mathbf{x})}\right\}_{\boldsymbol{\alpha}^0}\delta\mathbf{A}^{(3)}(4;j_2;j_1;\mathbf{x}), \tag{6.88}$$

where:

$$\frac{\partial R^{(3)}\left[j_3;j_2;j_1;\mathbf{U}^{(3)};\mathbf{A}^{(3)};\boldsymbol{\alpha}\right]}{\partial\mathbf{U}^{(3)}(4;j_1;\mathbf{x})}\mathbf{V}^{(3)}(4;j_1;\mathbf{x})$$
$$\triangleq\frac{\partial R^{(3)}\left[j_3;j_2;j_1;\mathbf{U}^{(3)};\mathbf{A}^{(3)};\boldsymbol{\alpha}\right]}{\partial\mathbf{u}(\mathbf{x})}\mathbf{v}^{(1)}(\mathbf{x})$$
$$+\frac{\partial R^{(3)}\left[j_3;j_2;j_1;\mathbf{U}^{(3)};\mathbf{A}^{(3)};\boldsymbol{\alpha}\right]}{\partial\mathbf{a}^{(1)}(\mathbf{x})}\delta\mathbf{a}^{(1)}(\mathbf{x})$$
$$+\sum_{i=1}^{2}\frac{\partial R^{(3)}\left[j_3;j_2;j_1;\mathbf{U}^{(3)};\mathbf{A}^{(3)};\boldsymbol{\alpha}\right]}{\partial\mathbf{a}^{(2)}(i;j_1;\mathbf{x})}\delta\mathbf{a}^{(2)}(i;j_1;\mathbf{x}), \tag{6.89}$$

and

$$\frac{\partial R^{(3)}\left[j_3;j_2;j_1;\mathbf{U}^{(3)};\mathbf{A}^{(3)};\boldsymbol{\alpha}\right]}{\partial\mathbf{A}^{(3)}(4;j_2;j_1;\mathbf{x})}\delta\mathbf{A}^{(3)}(4;j_2;j_1;\mathbf{x})$$
$$\triangleq\sum_{i=1}^{4}\frac{\partial R^{(3)}\left[j_3;j_2;j_1;\mathbf{U}^{(3)};\mathbf{A}^{(3)};\boldsymbol{\alpha}\right]}{\partial\mathbf{a}^{(3)}(i;j_2;j_1;\mathbf{x})}\delta\mathbf{a}^{(3)}(i;j_2;j_1;\mathbf{x}). \tag{6.90}$$

The direct-effect term $\left\{\delta R^{(3)}\left[j_3;j_2;j_1;\mathbf{U}^{(3)};\mathbf{A}^{(3)};\boldsymbol{\alpha};\delta\boldsymbol{\alpha}\right]\right\}_{dir}$ can be computed immediately; however, the indirect-effect term $\left\{\delta R^{(3)}\left[j_3;j_2;j_1;\mathbf{U}^{(3)};\mathbf{A}^{(3)};\boldsymbol{\alpha};\mathbf{V}^{(3)};\delta\mathbf{A}^{(3)}\right]\right\}_{ind}$ can be computed only after having determined the vectors $\mathbf{V}^{(3)}(4;j_1;\mathbf{x})$ and $\delta\mathbf{A}^{(3)}(4;j_2;j_1;\mathbf{x})$. The vector $\mathbf{V}^{(3)}(4;j_1;\mathbf{x})$ is the solution of the 3rd-LVSS defined by Eqs. (6.58) and (6.59). The vector $\delta\mathbf{A}^{(3)}(4;j_2;j_1;\mathbf{x})$ is the solution of the G-differentiated 3rd-LASS. Taking the G-differential of Eqs. (6.75) and (6.81) yields the following system of equations for $\delta\mathbf{A}^{(3)}(4;j_2;j_1;\mathbf{x})$, for $j_1=1,...,TP;\ j_2=1,...,j_1$:

$$
\begin{aligned}
&\frac{\partial\left\{\mathbf{AM}^{(3)}\left[4\times 4;\mathbf{U}^{(3)}(4;j_1;\mathbf{x});\boldsymbol{\alpha}\right]\mathbf{A}^{(3)}(4;j_2;j_1;\mathbf{x})\right\}}{\partial \mathbf{U}^{(3)}(4;j_1;\mathbf{x})}\mathbf{V}^{(3)}(4;j_1;\mathbf{x})\\
&-\frac{\partial \mathbf{Q}_A^{(3)}\left[4;j_2;j_1;\mathbf{U}^{(3)}(4;j_1;\mathbf{x});\boldsymbol{\alpha}\right]}{\partial \mathbf{U}^{(3)}(4;j_1;\mathbf{x})}\mathbf{V}^{(3)}(4;j_1;\mathbf{x})\\
&+\mathbf{AM}^{(3)}\left[4\times 4;\mathbf{U}^{(3)}(4;j_1;\mathbf{x});\boldsymbol{\alpha}\right]\delta\mathbf{A}^{(3)}(4;j_2;j_1;\mathbf{x})\\
&=\frac{\partial \mathbf{Q}_A^{(3)}\left[4;j_2;j_1;\mathbf{U}^{(3)}(4;j_1;\mathbf{x});\boldsymbol{\alpha}\right]}{\partial\boldsymbol{\alpha}}\partial\boldsymbol{\alpha}\\
&-\frac{\partial\left\{\mathbf{AM}^{(3)}\left[4\times 4;\mathbf{U}^{(3)}(4;j_1;\mathbf{x});\boldsymbol{\alpha}\right]\mathbf{A}^{(3)}(4;j_2;j_1;\mathbf{x})\right\}}{\partial\boldsymbol{\alpha}}\partial\boldsymbol{\alpha},\quad \mathbf{x}\in\Omega_x,
\end{aligned}
\tag{6.91}
$$

$$
\left\{\delta\mathbf{B}_A^{(3)}\left[4;\mathbf{U}^{(3)}(4;j_1;\mathbf{x});\mathbf{A}^{(3)}(4;j_2;j_1;\mathbf{x});\boldsymbol{\alpha}\right]\right\}_{\boldsymbol{\alpha}^0}=\mathbf{0}[4];\ \mathbf{x}\in\partial\Omega_x\left(\boldsymbol{\alpha}^0\right). \tag{6.92}
$$

The quantities that appear in Eq. (6.91) are evaluated at the nominal values of the parameters and respective state functions, but the notation $\{\ \}_{\pm^0}$, which indicates this evaluation, has been omitted in order to simplify the notation.

Concatenating Eqs. (6.91) and (6.92) with the 3rd-LVSS yields the following system of equations, which will be called the "4th-order variational sensitivity system" (4th-LVSS), for determining the vectors $\mathbf{V}^{(3)}(4;j_1;\mathbf{x})$ and $\delta\mathbf{A}^{(3)}(4;j_2;j_1;\mathbf{x})$, for $j_1=1,\ldots,TP;\ j_2=1,\ldots,j_1$:

$$
\begin{aligned}
&\left\{\mathbf{VM}^{(4)}\left[8\times 8;\mathbf{U}^{(4)}(8;j_2;j_1;\mathbf{x});\boldsymbol{\alpha}\right]\mathbf{V}^{(4)}(8;j_2;j_1;\mathbf{x})\right\}_{\boldsymbol{\alpha}^0}\\
&=\left\{\mathbf{Q}_V^{(4)}\left[8;j_1;\mathbf{U}^{(4)}(8;j_2;j_1;\mathbf{x});\boldsymbol{\alpha};\delta\boldsymbol{\alpha}\right]\right\}_{\boldsymbol{\alpha}^0},\ \mathbf{x}\in\Omega_x,
\end{aligned}
\tag{6.93}
$$

$$
\left\{\mathbf{B}_V^{(4)}\left[8;\mathbf{U}^{(4)}(8;j_1;\mathbf{x});\mathbf{V}^{(4)}(8;j_1;\mathbf{x});\boldsymbol{\alpha};\delta\boldsymbol{\alpha}\right]\right\}_{\boldsymbol{\alpha}^0}=\mathbf{0}[8];\ \mathbf{x}\in\partial\Omega_x\left(\boldsymbol{\alpha}^0\right); \tag{6.94}
$$

where $\mathbf{0}[8]\triangleq[\mathbf{0},\mathbf{0},\mathbf{0},\mathbf{0},\mathbf{0},\mathbf{0},\mathbf{0},\mathbf{0}]^\dagger$, and where:

$$
\begin{aligned}
&\mathbf{VM}^{(4)}\left[8\times 8;\mathbf{U}^{(4)}(8;j_2;j_1;\mathbf{x});\boldsymbol{\alpha}\right]\\
&\triangleq\begin{pmatrix}\mathbf{VM}^{(3)}\left[4\times 4;\mathbf{U}^{(3)}(4;j_1;\mathbf{x});\boldsymbol{\alpha}\right] & \mathbf{0}[4\times 4]\\ \mathbf{VM}_{21}^{(4)}(4\times 4;\mathbf{x}) & \mathbf{VM}_{22}^{(4)}(4\times 4;\mathbf{x})\end{pmatrix};
\end{aligned}
\tag{6.95}
$$

$$
\mathbf{U}^{(4)}(8;j_2;j_1;\mathbf{x})\triangleq\begin{pmatrix}\mathbf{U}^{(3)}(4;j_1;\mathbf{x})\\ \mathbf{A}^{(3)}(4;j_2;j_1;\mathbf{x})\end{pmatrix}; \tag{6.96}
$$

$$\mathbf{V}^{(4)}(8;j_2;j_1;\mathbf{x}) \triangleq \delta\mathbf{U}^{(4)}(8;j_2;j_1;\mathbf{x}) = \begin{pmatrix} \mathbf{V}^{(3)}(4;j_1;\mathbf{x}) \\ \delta\mathbf{A}^{(3)}(4;j_2;j_1;\mathbf{x}) \end{pmatrix}$$
$$= \left[\mathbf{v}^{(1)}(\mathbf{x}), \delta\mathbf{a}^{(1)}(\mathbf{x}), \delta\mathbf{a}^{(2)}(1;j_1;\mathbf{x}), \delta\mathbf{a}^{(2)}(2;j_1;\mathbf{x}), \right. \tag{6.97}$$
$$\left. \delta\mathbf{a}^{(3)}(1;j_2;j_1;\mathbf{x}),...,\delta\mathbf{a}^{(3)}(4;j_2;j_1;\mathbf{x})\right]^{\dagger};$$

$$\mathbf{VM}_{21}^{(4)}(4\times4;\mathbf{x}) \triangleq -\frac{\partial \mathbf{Q}_A^{(3)}\left[4;j_2;j_1;\mathbf{U}^{(3)}(4;j_1;\mathbf{x});\boldsymbol{\alpha}\right]}{\partial \mathbf{U}^{(3)}(4;j_1;\mathbf{x})}$$
$$+\frac{\partial\left\{\mathbf{AM}^{(3)}\left[4\times4;\mathbf{U}^{(3)}(4;j_1;\mathbf{x});\boldsymbol{\alpha}\right]\mathbf{A}^{(3)}(4;j_2;j_1;\mathbf{x})\right\}}{\partial \mathbf{U}^{(3)}(4;j_1;\mathbf{x})}; \tag{6.98}$$

$$\mathbf{VM}_{22}^{(4)}(4\times4;\mathbf{x}) \triangleq \mathbf{AM}^{(3)}\left[4\times4;\mathbf{U}^{(3)}(4;j_1;\mathbf{x});\boldsymbol{\alpha}\right]; \tag{6.99}$$

$$\mathbf{Q}_V^{(4)}\left[8;j_1;\mathbf{U}^{(4)}(4;j_2;j_1;\mathbf{x});\boldsymbol{\alpha};\delta\boldsymbol{\alpha}\right] \triangleq \begin{pmatrix} \mathbf{Q}_V^{(3)}\left[4;j_1;\mathbf{U}^{(3)};\boldsymbol{\alpha};\delta\boldsymbol{\alpha}\right] \\ \mathbf{Q}_2^{(4)}\left[4;j_1;\mathbf{U}^{(4)};\boldsymbol{\alpha};\delta\boldsymbol{\alpha}\right] \end{pmatrix} \tag{6.100}$$
$$\equiv \left\{\mathbf{q}_V^{(4)}\left[1;j_1;\mathbf{U}^{(4)};\boldsymbol{\alpha};\delta\boldsymbol{\alpha}\right],...,\mathbf{q}_V^{(4)}\left[8;j_1;\mathbf{U}^{(4)};\boldsymbol{\alpha};\delta\boldsymbol{\alpha}\right]\right\}^{\dagger};$$

$$\mathbf{Q}_2^{(4)}\left[4;j_1;\mathbf{U}^{(4)}(4;j_2;j_1;\mathbf{x});\boldsymbol{\alpha};\delta\boldsymbol{\alpha}\right] \triangleq \frac{\partial \mathbf{Q}_A^{(3)}\left[4;j_2;j_1;\mathbf{U}^{(3)};\boldsymbol{\alpha}\right]}{\partial\boldsymbol{\alpha}}\partial\boldsymbol{\alpha}$$
$$-\frac{\partial\left\{\mathbf{AM}^{(3)}\left[4\times4;\mathbf{U}^{(3)}(4;j_1;\mathbf{x});\boldsymbol{\alpha}\right]\mathbf{A}^{(3)}(4;j_2;j_1;\mathbf{x})\right\}}{\partial\boldsymbol{\alpha}}\partial\boldsymbol{\alpha}; \tag{6.101}$$

$$\mathbf{B}_V^{(4)}\left[8;\mathbf{U}^{(4)}(8;j_2;j_1;\mathbf{x});\mathbf{V}^{(4)}(8;j_2;j_1;\mathbf{x});\boldsymbol{\alpha};\delta\boldsymbol{\alpha}\right]$$
$$\triangleq \begin{pmatrix} \mathbf{B}_V^{(3)}\left[4;\mathbf{U}^{(3)}(4;j_1;\mathbf{x});\mathbf{V}^{(3)}(4;j_1;\mathbf{x});\boldsymbol{\alpha};\delta\boldsymbol{\alpha}\right] \\ \delta\mathbf{B}_A^{(2)}\left[2;\mathbf{U}^{(2)}(2;\mathbf{x});\mathbf{A}^{(2)}(2;j_1;\mathbf{x});\boldsymbol{\alpha}\right] \end{pmatrix}. \tag{6.102}$$

For subsequent reference, it is noted that the quantities $\mathbf{q}_V^{(4)}\left[i;j_2;j_1;\mathbf{U}^{(4)}(4;j_2;j_1;\mathbf{x});\boldsymbol{\alpha};\delta\boldsymbol{\alpha}\right]$ are linear in the parameter variations $\delta\alpha_i$, $i=1,...,TP$, and can therefore be written in the following form:

$$\mathbf{q}_V^{(4)}\left[i;j_2;j_1;\mathbf{U}^{(4)}(4;j_2;j_1;\mathbf{x});\boldsymbol{\alpha};\delta\boldsymbol{\alpha}\right]$$
$$\triangleq \sum_{j_4=1}^{TP}\mathbf{s}_V^{(3)}\left[i;j_4;j_2;j_1;\mathbf{U}^{(4)}(4;\mathbf{x});\boldsymbol{\alpha}\right]\delta\alpha_{j_4};\ i=1,...,8. \tag{6.103}$$

The variational matrix $\mathbf{VM}^{(4)}\left[8\times 8;\mathbf{U}^{(4)}(8;j_2;j_1;\mathbf{x});\boldsymbol{\alpha}\right]$ comprises (8×8) block matrices, each comprising TD^2 components/elements; thus, the matrix $\mathbf{VM}^{(4)}\left[8\times 8;\mathbf{U}^{(4)}(8;j_2;j_1;\mathbf{x});\boldsymbol{\alpha}\right]$ comprises a total of $(8\times 8)TD^2$ components/elements. Each of the vectors $\mathbf{V}^{(4)}(8;j_2;j_1;\mathbf{x})$, $\mathbf{Q}_V^{(4)}\left[8;j_1;\mathbf{U}^{(3)}(4;j_1;\mathbf{x});\boldsymbol{\alpha};\delta\boldsymbol{\alpha}\right]$, and $\mathbf{B}_V^{(4)}\left[8;\mathbf{U}^{(4)}(8;j_2;j_1;\mathbf{x});\mathbf{V}^{(4)}(8;j_2;j_1;\mathbf{x});\boldsymbol{\alpha};\delta\boldsymbol{\alpha}\right]$ comprises eight TD-dimensional vectors, as shown in their respective definitions. Thus, each of the vectors $\mathbf{B}_V^{(4)}\left[8;\mathbf{U}^{(4)}(8;j_2;j_1;\mathbf{x});\mathbf{V}^{(4)}(8;j_2;j_1;\mathbf{x});\boldsymbol{\alpha};\delta\boldsymbol{\alpha}\right],\mathbf{V}^{(4)}(8;j_2;j_1;\mathbf{x}),\mathbf{Q}_V^{(4)}\left[8;j_1;\mathbf{U}^{(3)}(4;j_1;\mathbf{x});\boldsymbol{\alpha};\delta\boldsymbol{\alpha}\right]$ comprises $(8\times TD)$ components/elements.

Solving the 4th-LVSS would require TP^4 large-scale computations, which is unrealistic for large-scale systems comprising many parameters. The 4th-CASAM-N circumvents the need for solving the 3rd-LVSS by deriving an alternative expression for the indirect-effect term defined in Eq. (6.88), in which the function $\mathbf{V}^{(4)}(8;j_2;j_1;\mathbf{x})$ is replaced by a 4th-level adjoint function which is independent of parameter variations. This 4th-level adjoint function will be the solution of a 4th-Level Adjoint Sensitivity System (4th-LASS) which will be constructed by applying the same principles as those used for constructing the 1st-LASS, the 2nd-LASS, and the 3rd-LASS. The Hilbert space appropriate for constructing the 4th-LASS will be denoted as $\mathcal{H}_4(\Omega_x)$ and comprises as elements block vectors of the same form as $\mathbf{V}^{(4)}(8;j_2;j_1;\mathbf{x})$. Thus, a generic block vector in $\mathcal{H}_4(\Omega_x)$ will have the structure $\boldsymbol{\Psi}^{(4)}(8;\mathbf{x})\triangleq\left[\boldsymbol{\psi}^{(4)}(1;\mathbf{x}),...,\boldsymbol{\psi}^{(4)}(8;\mathbf{x})\right]^\dagger\in\mathcal{H}_4(\Omega_x)$, comprising eight TD-dimensional vectors of the form $\boldsymbol{\psi}^{(4)}(i;\mathbf{x})\triangleq\left[\psi_1^{(4)}(i;\mathbf{x}),...,\psi_{TD}^{(4)}(i;\mathbf{x})\right]^\dagger\in\mathcal{H}_1(\Omega_x)$, $i=1,...,8$. The inner product of two vectors $\boldsymbol{\Psi}^{(4)}(8;\mathbf{x})\in\mathcal{H}_4(\Omega_x)$ and $\boldsymbol{\Phi}^{(4)}(8;\mathbf{x})\in\mathcal{H}_4(\Omega_x)$ in the Hilbert space $\mathcal{H}_4(\Omega_x)$ will be denoted as $\left\langle\boldsymbol{\Psi}^{(4)}(8;\mathbf{x}),\boldsymbol{\Phi}^{(4)}(8;\mathbf{x})\right\rangle_4$ and defined as follows:

$$\left\langle\boldsymbol{\Psi}^{(4)}(8;\mathbf{x}),\boldsymbol{\Phi}^{(4)}(8;\mathbf{x})\right\rangle_4\triangleq\sum_{i=1}^{8}\left\langle\boldsymbol{\psi}^{(3)}(i;\mathbf{x}),\boldsymbol{\varphi}^{(3)}(i;\mathbf{x})\right\rangle_1. \tag{6.104}$$

The inner product defined in Eq. (6.104) is continuous in the model parameters $\boldsymbol{\alpha}$ in a neighborhood around $\boldsymbol{\alpha}^0$.

Forming the inner product of Eq. (6.93) with a vector $\mathbf{A}^{(4)}(8;\mathbf{x})\triangleq\left[\mathbf{a}^{(4)}(1;\mathbf{x}),...,\mathbf{a}^{(4)}(8;\mathbf{x})\right]^\dagger\in\mathcal{H}_4(\Omega_x)$ yields the following relation:

$$\begin{aligned}&\left\{\left\langle\mathbf{A}^{(4)}(8;\mathbf{x}),\mathbf{VM}^{(4)}\left[8\times 8;\mathbf{U}^{(4)}(8;j_2;j_1;\mathbf{x});\boldsymbol{\alpha}\right]\mathbf{V}^{(4)}(8;j_2;j_1;\mathbf{x})\right\rangle_3\right\}_{\boldsymbol{\alpha}^0}\\&=\left\{\left\langle\mathbf{A}^{(4)}(8;\mathbf{x}),\mathbf{Q}_V^{(4)}\left[8;j_1;\mathbf{U}^{(4)}(8;j_2;j_1;\mathbf{x});\boldsymbol{\alpha};\delta\boldsymbol{\alpha}\right]\right\rangle_3\right\}_{\boldsymbol{\alpha}^0},\quad\mathbf{x}\in\Omega_x.\end{aligned} \tag{6.105}$$

The inner product on the left side of Eq. (6.105) is further transformed by using the definition of the adjoint operator to obtain the following relation:

$$\left\{ \left\langle \mathbf{A}^{(4)}(8;\mathbf{x}), \mathbf{VM}^{(4)}\left[8\times 8;\mathbf{U}^{(4)}(8;j_2;j_1;\mathbf{x});\boldsymbol{\alpha}\right]\mathbf{V}^{(4)}(8;j_2;j_1;\mathbf{x}) \right\rangle_3 \right\}_{\boldsymbol{\alpha}^0}$$
$$= \left\{ \left\langle \mathbf{V}^{(4)}(8;j_2;j_1;\mathbf{x}), \mathbf{AM}^{(4)}\left[8\times 8;\mathbf{U}^{(4)}(8;j_2;j_1;\mathbf{x});\boldsymbol{\alpha}\right]\mathbf{A}^{(4)}(8;\mathbf{x}) \right\rangle_3 \right\}_{\boldsymbol{\alpha}^0} \tag{6.106}$$
$$+ \left\{ \left[P^{(4)}\left(\mathbf{U}^{(4)};\mathbf{A}^{(4)};\mathbf{V}^{(4)};\boldsymbol{\alpha}\right) \right]_{\partial\Omega_x} \right\}_{\boldsymbol{\alpha}^0},$$

where:

$$\mathbf{AM}^{(4)}\left[8\times 8;\mathbf{U}^{(4)}(8;j_2;j_1;\mathbf{x});\boldsymbol{\alpha}\right] \triangleq \left[\mathbf{VM}^{(4)}\left(8\times 8;\mathbf{U}^{(4)};\boldsymbol{\alpha}\right)\right]^*$$
$$= \begin{pmatrix} \left\{\left[\mathbf{VM}^{(3)}(4\times 4)\right]^*\right\}^\dagger & \left\{\left[\mathbf{VM}_{21}^{(4)}(4\times 4)\right]^*\right\}^\dagger \\ \mathbf{0}[4\times 4] & \left\{\left[\mathbf{VM}_{22}^{(3)}(4\times 4)\right]^*\right\}^\dagger \end{pmatrix}, \tag{6.107}$$

and where $\left\{ \left[P^{(4)}\left(\mathbf{U}^{(4)};\mathbf{A}^{(4)};\mathbf{V}^{(4)};\boldsymbol{\alpha}\right) \right]_{\partial\Omega_x} \right\}_{\boldsymbol{\alpha}^0}$ denotes the corresponding bilinear concomitant on the domain's boundary, evaluated at the nominal values for the parameters and respective state functions. The adjoint matrix $\mathbf{AM}^{(4)}\left[8\times 8;\mathbf{U}^{(4)}(8;j_2;j_1;\mathbf{x});\boldsymbol{\alpha}\right]$ comprises $(8\times 8)TD^2$ components/elements, while the adjoint function $\mathbf{A}^{(4)}(8;\mathbf{x}) \in \mathcal{H}_4(\Omega_x)$ comprises $(8\times TD)$ components/elements.

The domain of the adjoint matrix-operator $\mathbf{AM}^{(4)}\left[8\times 8;\mathbf{U}^{(4)}(8;j_2;j_1;\mathbf{x});\boldsymbol{\alpha}\right]$ is specified by requiring that the function $\mathbf{A}^{(4)}(8;\mathbf{x}) \triangleq \left[\mathbf{a}^{(4)}(1;\mathbf{x}),...,\mathbf{a}^{(4)}(8;\mathbf{x})\right]^\dagger \in \mathcal{H}_4(\Omega_x)$ satisfies adjoint boundary/initial conditions denoted as follows:

$$\left\{ \mathbf{B}_A^{(4)}\left[8;\mathbf{U}^{(4)}(8;j_2;j_1;\mathbf{x});\mathbf{A}^{(4)}(8;\mathbf{x});\boldsymbol{\alpha}\right] \right\}_{\boldsymbol{\alpha}^0} = \mathbf{0}[8], \quad \mathbf{x} \in \partial\Omega_x(\boldsymbol{\alpha}^0). \tag{6.108}$$

The 4th-level adjoint boundary/initial conditions represented by Eq. (6.108) are determined by requiring that: (i) they must be independent of unknown values of $\mathbf{V}^{(4)}(8;j_2;j_1;\mathbf{x})$; (ii) the substitution of the boundary and/or initial conditions represented by Eqs. (6.108) and (6.94) into the expression of $\left\{ \left[P^{(4)}\left(\mathbf{U}^{(4)};\mathbf{A}^{(4)};\mathbf{V}^{(4)};\boldsymbol{\alpha}\right) \right]_{\partial\Omega_x} \right\}_{\boldsymbol{\alpha}^0}$ must cause all terms containing unknown values of $\mathbf{V}^{(4)}(8;j_2;j_1;\mathbf{x})$ to vanish.

Implementing the boundary/initial conditions represented by Eqs. (6.108) and (6.94) into Eq. (6.106) will transform the later relation into the following form:

$$\left\{ \left\langle \mathbf{V}^{(4)}(8;j_2;j_1;\mathbf{x}), \mathbf{AM}^{(4)}\left[8\times 8;\mathbf{U}^{(4)}(8;j_2;j_1;\mathbf{x});\boldsymbol{\alpha}\right]\mathbf{A}^{(4)}(8;\mathbf{x}) \right\rangle_3 \right\}_{\boldsymbol{\alpha}^0}$$
$$= \left\{ \left\langle \mathbf{A}^{(4)}(8;\mathbf{x}), \mathbf{VM}^{(4)}\left[8\times 8;\mathbf{U}^{(4)}(8;j_2;j_1;\mathbf{x});\boldsymbol{\alpha}\right]\mathbf{V}^{(4)}(8;j_2;j_1;\mathbf{x}) \right\rangle_3 \right\}_{\boldsymbol{\alpha}^0} \tag{6.109}$$
$$- \left\{ \left[\hat{P}^{(4)}\left(\mathbf{U}^{(4)};\mathbf{A}^{(4)};\delta\boldsymbol{\alpha}\right) \right]_{\partial\Omega_x} \right\}_{\boldsymbol{\alpha}^0},$$

where $\left\{\left[\hat{P}^{(4)}\left(\mathbf{U}^{(4)};\mathbf{A}^{(4)};\delta\boldsymbol{\alpha}\right)\right]_{\partial\Omega_x}\right\}_{\boldsymbol{\alpha}^0}$ denotes residual boundary terms that may not vanish automatically.

The right side of Eq. (6.105) is now used in the first term on the right side of Eq. (6.109) to obtain the following relation:

$$\begin{aligned}
&\left\{\left\langle \mathbf{V}^{(4)}\left(8;j_2;j_1;\mathbf{x}\right),\mathbf{AM}^{(4)}\left[8\times 8;\mathbf{U}^{(4)}\left(8;j_2;j_1;\mathbf{x}\right);\boldsymbol{\alpha}\right]\mathbf{A}^{(4)}\left(8;\mathbf{x}\right)\right\rangle_3\right\}_{\boldsymbol{\alpha}^0} \\
&=\left\{\left\langle \mathbf{A}^{(4)}\left(8;\mathbf{x}\right),\mathbf{Q}_V^{(4)}\left[8;j_1;\mathbf{U}^{(4)}\left(8;j_2;j_1;\mathbf{x}\right);\boldsymbol{\alpha};\delta\boldsymbol{\alpha}\right]\right\rangle_3\right\}_{\boldsymbol{\alpha}^0} \\
&-\left\{\left[\hat{P}^{(4)}\left(\mathbf{U}^{(4)};\mathbf{A}^{(4)};\delta\boldsymbol{\alpha}\right)\right]_{\partial\Omega_x}\right\}_{\boldsymbol{\alpha}^0}.
\end{aligned} \tag{6.110}$$

The definition of the 4th-level adjoint function $\mathbf{A}^{(4)}\left(8;\mathbf{x}\right)\triangleq\left[\mathbf{a}^{(1)}\left(1;\mathbf{x}\right),\ldots,\mathbf{a}^{(4)}\left(8;\mathbf{x}\right)\right]^{\dagger}$ $\in\mathcal{H}_4\left(\Omega_x\right)$ is now completed by requiring that the left side of Eq. (6.110) and the right side of Eq. (6.88) represent the "indirect effect term" $\left\{\delta R^{(3)}\left[j_3;j_2;j_1;\mathbf{U}^{(3)};\mathbf{A}^{(3)};\boldsymbol{\alpha};\mathbf{V}^{(3)};\delta\mathbf{A}^{(3)}\right]\right\}_{ind}$, for each of the indices $j_1=1,\ldots,TP$; $j_2=1,\ldots.,j_1;j_3=1,\ldots.,j_2$. Hence, there will be $TP\left(TP+1\right)\left(TP+2\right)/6$ distinct 4th-level adjoint functions $\mathbf{A}^{(4)}\left(8;\mathbf{x}\right)\in\mathcal{H}_4\left(\Omega_x\right)$, each corresponding to one combination of the indices $j_1=1,\ldots,TP;j_2=1,\ldots.,j_1;j_3=1,\ldots.,j_2$. Each of these distinct 4th-level adjoint functions will correspond to a specific $\left(j_1,j_2,j_3\right)$-dependent indirect-effect term.

The left side of Eq. (6.110) will be identical to the right side of Eq. (6.88) by requiring that the following relation be satisfied by the 4th-level adjoint functions $\mathbf{A}^{(4)}\left(8;j_3;j_2;j_1;\mathbf{x}\right)\triangleq\left[\mathbf{a}^{(4)}\left(1;j_3;j_2;j_1;\mathbf{x}\right),\ldots,\mathbf{a}^{(4)}\left(8;j_3;j_2;j_1;\mathbf{x}\right)\right]^{\dagger}\in\mathcal{H}_4\left(\Omega_x\right)$, for each value of the indices $j_1=1,\ldots,TP;j_2=1,\ldots.,j_1;j_3=1,\ldots.,j_2$:

$$\begin{aligned}
&\mathbf{AM}^{(4)}\left[8\times 8;\mathbf{U}^{(4)}\left(8;j_2;j_1;\mathbf{x}\right);\boldsymbol{\alpha}\right]\mathbf{A}^{(4)}\left(8;j_3;j_2;j_1;\mathbf{x}\right) \\
&=\mathbf{Q}_A^{(4)}\left[8;j_3;j_2;j_1;\mathbf{U}^{(4)}\left(4;j_1;\mathbf{x}\right);\boldsymbol{\alpha}\right],
\end{aligned} \tag{6.111}$$

where:

$$\begin{aligned}
&\mathbf{Q}_A^{(4)}\left[8;j_3;j_2;j_1;\mathbf{U}^{(4)}\left(4;j_1;\mathbf{x}\right);\boldsymbol{\alpha}\right]\triangleq\left\{\mathbf{q}_A^{(4)}\left[1;j_3;j_2;j_1;\mathbf{U}^{(4)}\left(4;j_1;\mathbf{x}\right);\boldsymbol{\alpha}\right],\right. \\
&\left.\ldots,\mathbf{q}_A^{(4)}\left[8;j_3;j_2;j_1;\mathbf{U}^{(4)}\left(4;j_1;\mathbf{x}\right);\boldsymbol{\alpha}\right]\right\}^{\dagger},
\end{aligned} \tag{6.112}$$

$$\mathbf{q}_A^{(4)}\left[1;j_3;j_2;j_1;\mathbf{U}^{(4)}\left(4;j_1;\mathbf{x}\right);\boldsymbol{\alpha}\right]\triangleq\frac{\partial R^{(3)}\left[j_3;j_2;j_1;\mathbf{U}^{(3)};\mathbf{A}^{(3)};\boldsymbol{\alpha}\right]}{\partial\mathbf{u}\left(\mathbf{x}\right)}; \tag{6.113}$$

$$\mathbf{q}_A^{(4)}\left[2;j_3;j_2;j_1;\mathbf{U}^{(4)}(4;j_1;\mathbf{x});\boldsymbol{\alpha}\right]\triangleq\frac{\partial R^{(3)}\left[j_3;j_2;j_1;\mathbf{U}^{(3)};\mathbf{A}^{(3)};\boldsymbol{\alpha}\right]}{\partial \mathbf{a}^{(1)}(\mathbf{x})}; \quad (6.114)$$

$$\begin{aligned}&\mathbf{q}_A^{(4)}\left[2+i;j_3;j_2;j_1;\mathbf{U}^{(4)}(4;j_1;\mathbf{x});\boldsymbol{\alpha}\right]\\&\triangleq\frac{\partial R^{(3)}\left[j_3;j_2;j_1;\mathbf{U}^{(3)};\mathbf{A}^{(3)};\boldsymbol{\alpha}\right]}{\partial \mathbf{a}^{(2)}(i;j_1;\mathbf{x})};\; i=1,2;\end{aligned} \quad (6.115)$$

$$\begin{aligned}&\mathbf{q}_A^{(4)}\left[4+i;j_3;j_2;j_1;\mathbf{U}^{(4)}(4;j_1;\mathbf{x});\boldsymbol{\alpha}\right]\\&\triangleq\frac{\partial R^{(3)}\left[j_3;j_2;j_1;\mathbf{U}^{(3)};\mathbf{A}^{(3)};\boldsymbol{\alpha}\right]}{\partial \mathbf{a}^{(3)}(i;j_2;j_1;\mathbf{x})};\; i=1,2,3,4;\end{aligned} \quad (116)$$

The boundary conditions to be satisfied by each of the 4th-level adjoint functions $\mathbf{A}^{(4)}(8;j_3;j_2;j_1;\mathbf{x})\triangleq\left[\mathbf{a}^{(4)}(1;j_3;j_2;j_1;\mathbf{x}),...,\mathbf{a}^{(4)}(8;j_3;j_2;j_1;\mathbf{x})\right]^{\dagger}\in\mathcal{H}_4(\Omega_x)$ are those represented by Eq. (6.108), namely:

$$\begin{aligned}&\left\{\mathbf{B}_A^{(4)}\left[8;\mathbf{U}^{(4)}(8;j_2;j_1;\mathbf{x});\mathbf{A}^{(4)}(8;j_3;j_2;j_1;\mathbf{x});\boldsymbol{\alpha}\right]\right\}_{\boldsymbol{\alpha}^0}=\mathbf{0}[8],\;\; \mathbf{x}\in\partial\Omega_x(\boldsymbol{\alpha}^0);\\&j_1=1,...,TP;\; j_2=1,...,j_1;\; j_3=1,...,j_2.\end{aligned} \quad (6.117)$$

Since the source $\mathbf{Q}_A^{(4)}\left[8;j_3;j_2;j_1;\mathbf{U}^{(4)}(4;j_1;\mathbf{x});\boldsymbol{\alpha}\right]$ may contain distributions, the equality in Eq. (6.111) is considered to hold in the weak sense, and the Riesz representation theorem ensures that this weak equality holds uniquely.

The system of equations represented by Eqs. (6.111) and (6.117) will be called the 4th-Level Adjoint Sensitivity System (4th-LASS); its solution, $\mathbf{A}^{(4)}(8;j_3;j_2;j_1;\mathbf{x})$, will be called the 4th-level adjoint function. Using the equations underlying the 4th-LASS and Eq. (6.110) in Eq. (6.88) yields the following expression for the indirect-effect term:

$$\begin{aligned}&\left\{\delta R^{(3)}\left[j_3;j_2;j_1;\mathbf{U}^{(3)};\mathbf{A}^{(3)};\boldsymbol{\alpha};\mathbf{V}^{(3)};\delta\mathbf{A}^{(3)}\right]\right\}_{ind}\\&=\left\{\left\langle\mathbf{A}^{(4)},\mathbf{Q}_V^{(4)}\left[8;j_1;\mathbf{U}^{(4)};\boldsymbol{\alpha};\delta\boldsymbol{\alpha}\right]\right\rangle_3\right\}_{\boldsymbol{\alpha}^0}-\left\{\left[\hat{P}^{(4)}\left(\mathbf{U}^{(4)};\mathbf{A}^{(4)};\delta\boldsymbol{\alpha}\right)\right]_{\partial\Omega_x}\right\}_{\boldsymbol{\alpha}^0}\\&\equiv\left\{\delta R^{(3)}\left[j_3;j_2;j_1;\mathbf{U}^{(4)}(8;j_2;j_1;\mathbf{x});\mathbf{A}^{(4)}(8;j_3;j_2;j_1;\mathbf{x});\boldsymbol{\alpha}\right]\right\}_{ind}.\end{aligned} \quad (6.118)$$

As the identity in the last line of Eq. (6.118) indicates, the dependence of the indirect-effect term on the function $\mathbf{V}^{(4)}(8;j_2;j_1;\mathbf{x})$ has been replaced by the dependence on the adjoint function $\mathbf{A}^{(4)}(8;j_3;j_2;j_1;\mathbf{x})$, for each $j_1=1,...,TP;\; j_2=1,...,j_1; j_3=1,...,j_2$.

Replacing the expression obtained in Eq. (6.118) for the indirect-effect term together with the expression for the direct-effect term provided in Eq. (6.87) yields the following expression for the total differential defined in Eq. (6.86):

$$\begin{aligned}
&\left\{\delta R^{(3)}\left[j_3;j_2;j_1;\mathbf{U}^{(4)}(8;j_2;j_1;\mathbf{x});\mathbf{A}^{(4)}(8;j_3;j_2;j_1;\mathbf{x});\delta\boldsymbol{\alpha}\right]\right\}_{\boldsymbol{\alpha}^0} \\
&=\left\{\frac{\partial R^{(3)}\left[j_3;j_2;j_1;\mathbf{U}^{(3)}(4;j_1;\mathbf{x});\mathbf{A}^{(3)}(4;j_2;j_1;\mathbf{x});\boldsymbol{\alpha}\right]}{\partial\boldsymbol{\alpha}}\delta\boldsymbol{\alpha}\right\}_{\boldsymbol{\alpha}^0} \\
&-\left\{\left[\hat{P}^{(4)}\left(\mathbf{U}^{(4)};\mathbf{A}^{(4)};\delta\boldsymbol{\alpha}\right)\right]_{\partial\Omega_x}\right\}_{\boldsymbol{\alpha}^0} \\
&+\left\{\left\langle\mathbf{A}^{(4)}(8;j_3;j_2;j_1;\mathbf{x}),\mathbf{Q}_V^{(4)}\left[8;j_1;\mathbf{U}^{(4)}(8;j_2;j_1;\mathbf{x});\boldsymbol{\alpha};\delta\boldsymbol{\alpha}\right]\right\rangle_3\right\}_{\boldsymbol{\alpha}^0}
\end{aligned} \tag{6.119}$$

In component form, the total differential expressed by Eq. (6.119) can be written in the following form, for each $j_1=1,...,TP;\ j_2=1,...,j_1;j_3=1,...,j_2$:

$$\begin{aligned}
&\left\{\delta R^{(3)}\left[j_3;j_2;j_1;\mathbf{U}^{(4)}(8;j_2;j_1;\mathbf{x});\mathbf{A}^{(4)}(8;j_3;j_2;j_1;\mathbf{x});\delta\boldsymbol{\alpha}\right]\right\}_{\boldsymbol{\alpha}^0} \\
&=\sum_{j_4=1}^{TP}\left\{R^{(4)}\left[j_4;j_3;j_2;j_1;\mathbf{U}^{(4)};\mathbf{A}^{(4)};\boldsymbol{\alpha}\right]\right\}_{\boldsymbol{\alpha}^0}\delta\alpha_{j_4},
\end{aligned} \tag{6.120}$$

where the quantity $R^{(4)}\left[j_4;j_3;j_2;j_1;\mathbf{U}^{(4)}(8;j_2;j_1;\mathbf{x});\mathbf{A}^{(4)}(8;j_3;j_2;j_1;\mathbf{x});\boldsymbol{\alpha}\right]$ denotes the 4th-order sensitivity of the generic scalar-valued response $R\left[\mathbf{u}(\mathbf{x});\boldsymbol{\alpha}\right]$ with respect to any four model parameters α_{j_1}, α_{j_2}, α_{j_3}, α_{j_3}, and has the following expression:

$$\begin{aligned}
&R^{(4)}\left[j_4;j_3;j_2;j_1;\mathbf{U}^{(4)}(8;j_2;j_1;\mathbf{x});\mathbf{A}^{(4)}(8;j_3;j_2;j_1;\mathbf{x});\boldsymbol{\alpha}\right] \\
&\triangleq\frac{\partial R^{(3)}\left[j_3;j_2;j_1;\mathbf{U}^{(3)};\mathbf{A}^{(3)};\boldsymbol{\alpha}\right]}{\partial\alpha_{j_4}}-\frac{\left[\partial\hat{P}^{(4)}\left(\mathbf{U}^{(4)};\mathbf{A}^{(4)};\delta\boldsymbol{\alpha}\right)\right]_{\partial\Omega_x}}{\partial\alpha_{j_4}} \\
&+\sum_{i=1}^{8}\left\langle\mathbf{a}^{(4)}(i;j_3;j_2;j_1;\mathbf{x}),\mathbf{s}_V^{(3)}\left[i;j_4;j_2;j_1;\mathbf{U}^{(4)}(4;\mathbf{x});\boldsymbol{\alpha}\right]\right\rangle_1 \\
&\equiv\frac{\partial^4 R\left[\mathbf{u}(\mathbf{x});\boldsymbol{\alpha}\right]}{\partial\alpha_{j_1}\partial\alpha_{j_3}\partial\alpha_{j_3}\partial\alpha_{j_4}};\ j_1,j_2,j_3,j_4=1,...,j_3=1,...,TP.
\end{aligned} \tag{6.121}$$

As Eqs. (6.111) and (6.117) indicate, solving the 4th-LASS once provides the 4th-level adjoint function $\mathbf{A}^{(4)}(8;j_3;j_2;j_1;\mathbf{x})$, for each of the indices $j_1=1,...,TP$; $j_2=1,...,j_1$; $j_3=1,...,j_2$. In turn, the availability of $\mathbf{A}^{(4)}(8;j_3;j_2;j_1;\mathbf{x})$ enables the exact and efficient computation of the G-differential $\delta R^{(3)}\left[j_3;j_2;j_1;\mathbf{U}^{(4)}(8;j_2;j_1;\mathbf{x});\right.$

$\mathbf{A}^{(4)}(8; j_3; j_2; j_1; \mathbf{x}); \delta\boldsymbol{\alpha}\big]$. Thus, the exact computation of all of the partial 4th-order sensitivities $\partial^4 R[\mathbf{u}(\mathbf{x}); \boldsymbol{\alpha}] / \partial\alpha_{j_1}\partial\alpha_{j_3}\partial\alpha_{j_3}\partial\alpha_{j_4}$ requires, at most, $TP(TP+1)(TP+2)/6$ large-scale (adjoint) computations using the 4th-LASS, rather than, at least, $O(TP^4)$ large-scale computations as would be required by forward methods.

The adjoint matrix $\mathbf{AM}^{(4)}\left[8\times 8; \mathbf{U}^{(4)}(8; j_2; j_1; \mathbf{x}); \boldsymbol{\alpha}\right]$ is block diagonal; therefore, solving the 4th-LASS is equivalent to solving four 1st-LASS, with different source terms. The 4th-LASS was designated as the "4th-level" rather than "4th-order" adjoint sensitivity system since the 3rd-LASS does not involve any explicit 2nd-order, 3rd-order, and/or 4th-order G-derivatives of the operators underlying the original system but involves the inversion of the operators similar to those that needed to be inverted for solving the 1st-LASS.

By solving the 4th-LASS $TP(TP+1)(TP+2)/6$ times, the 4th-order mixed sensitivities $\partial^4 R[\mathbf{u}(\mathbf{x}); \boldsymbol{\alpha}] / \partial\alpha_{j_1}\partial\alpha_{j_3}\partial\alpha_{j_3}\partial\alpha_{j_4}$ will be computed four times, in four different ways using distinct adjoint functions. Consequently, the multiple symmetries intrinsic to the 4th-order sensitivities provide an intrinsic numerical verification that the components of the 1st-, 2nd-, 3rd-, and 4th-level adjoint functions are computed accurately.

The structure of the 4th-LASS enables full flexibility for prioritizing the computation of the 3rd-order sensitivities. The computation of the 4th-order sensitivities would be prioritized based on the relative magnitudes of the 3rd-order sensitivities so that the unimportant 4th-order sensitivities can be deliberately neglected while knowing the error incurred by neglecting them.

6.6 CONCLUDING REMARKS

The 1st- through 4th-order adjoint sensitivity analysis methodology presented in this chapter is applicable to nonlinear models and to linear models involving model responses that only depend on the forward dependent (state) functions. The PERP benchmark is modeled by the linear neutron transport equation and its neutron leakage response only depends on the neutron flux, which is the forward-dependent (state) function for this benchmark. Consequently, the general 4th-order comprehensive adjoint sensitivity analysis methodology presented in this chapter provides the generic framework for performing the sensitivity analysis of the PERP benchmark. The specific forms taken on by the respective 1st-LASS through 4th-LASS, and the specific forms taken on by the sensitivities of the PERP benchmark's leakage response with respect to the PERP benchmark's microscopic total cross-sections, from 1st-order through 4th-order, will be presented in Chapter 7. The numerical values of these sensitivities were used in the illustrative applications of the 2nd-BERRU-PM and 4th-BERRU-PM methodologies presented in the foregoing Chapters.

7 Polyethylene Reflected Plutonium (PERP) Reactor Physics Benchmark

Sensitivities of the Leakage Response to Total Cross-Sections

7.1 INTRODUCTION

The material presented in this chapter is based on the work of Cacuci (2023a) and summarizes the mathematical framework underlying the derivation of the exact expressions for the 1st-, 2nd-, 3rd, and 4th-order sensitivities of the PERP benchmark's leakage response with respect to the benchmark's 180 group-averaged microscopic total cross-sections. These expressions were used for the most efficient computation of the numerical values of these sensitivities, which were used in the illustrative applications of the 2nd-BERRU-PM and 4th-BERRU-PM methodologies presented in the preceding chapters.

Section 7.2 presents the derivation of the computational expressions for the 180 1st-order sensitivities of the PERP leakage response with respect to the group-averaged total microscopic cross-sections. Section 7.3 presents the derivation of the computational expressions for the 32,400 2nd-order sensitivities (of which 16,290 are distinct) of the PERP leakage response with respect to the group-averaged total microscopic cross-sections. The derivation of the computational expressions for the 5,832,000 3rd-order sensitivities (of which 988,260 are distinct) is presented in Section 7.4, while Section 7.5 presents the derivation of the computational expressions for the 1,049,760,000 4th-order sensitivities (of which 45,212,895 are distinct). Computational considerations are summarized in Section 7.6 and concluding remarks are offered in Section 7.7.

7.2 COMPUTATION OF 1ST-ORDER SENSITIVITIES $\partial L(\alpha)/\partial t_j$

This section presents the computational expressions of the 1st-order sensitivities of the PERP benchmark's leakage response with respect to the PERP model's multigroup total microscopic cross-sections. The multigroup form of the neutron transport equation can be written in matrix form as follows:

DOI: 10.1201/9781003478119-7

$$\mathbf{B}(\boldsymbol{\alpha})\boldsymbol{\varphi}(r,\boldsymbol{\Omega}) = \mathbf{Q}(\boldsymbol{\alpha}), \tag{7.1}$$

$$\boldsymbol{\varphi}(r,\boldsymbol{\Omega}) = \mathbf{0}, \; r = r_s \in S_b, \; \boldsymbol{\Omega} \bullet \mathbf{n} < 0, \tag{7.2}$$

where

$$\begin{gathered}
\boldsymbol{\varphi}(r,\boldsymbol{\Omega}) \triangleq \begin{pmatrix} \varphi^1 \\ \bullet \\ \varphi^g \\ \bullet \\ \varphi^G \end{pmatrix}; \; \mathbf{Q}(\boldsymbol{\alpha}) \triangleq \begin{pmatrix} Q^1 \\ \bullet \\ Q^g \\ \bullet \\ Q^G \end{pmatrix}; \\
\mathbf{B}(\boldsymbol{\alpha}) \triangleq \begin{pmatrix} B_{11}(\boldsymbol{\alpha}) & \bullet & B_{1g}(\boldsymbol{\alpha}) & \bullet & B_{1G}(\boldsymbol{\alpha}) \\ \bullet & \bullet & \bullet & \bullet & \bullet \\ B_{g1}(\boldsymbol{\alpha}) & \bullet & B_{gg}(\boldsymbol{\alpha}) & \bullet & B_{gG}(\boldsymbol{\alpha}) \\ \bullet & \bullet & \bullet & \bullet & \bullet \\ B_{G1}(\boldsymbol{\alpha}) & \bullet & B_{Gg}(\boldsymbol{\alpha}) & \bullet & B_{GG}(\boldsymbol{\alpha}) \end{pmatrix};
\end{gathered} \tag{7.3}$$

with components defined as follows:

$$\begin{gathered}
B_{gh}(\boldsymbol{\alpha}) \triangleq \delta_{gh} B_{gg}^0(\boldsymbol{\alpha}) - B_{gh}^1(\boldsymbol{\alpha}); \; g,h = 1,\ldots,G; \\
B_{gg}^0(\boldsymbol{\alpha})\varphi^g(r,\boldsymbol{\Omega}) \triangleq \left[\boldsymbol{\Omega} \bullet \nabla + \Sigma_t^g(\boldsymbol{\alpha};r)\right]\varphi^g(r,\boldsymbol{\Omega}); \\
B_{gh}^1(\boldsymbol{\alpha})\varphi^h(r,\boldsymbol{\Omega}') \triangleq \int_{4\pi} d\boldsymbol{\Omega}'\left[\Sigma_s^{h\to g}(\boldsymbol{\alpha};r,\boldsymbol{\Omega}' \bullet \boldsymbol{\Omega}) + \chi^g(\boldsymbol{\alpha};r)\left(\nu\Sigma_f\right)^h(\boldsymbol{\alpha};r)\right]\varphi^h(r,\boldsymbol{\Omega}').
\end{gathered} \tag{7.4}$$

The notation δ_{gh} denotes the Kronecker delta function, which is defined as usual, in other words, $\delta_{gh} = 1, \, if \;\, g = h$ and $\delta_{gh} = 0, \, if \;\, g \neq h$.

The multigroup form of the PERP leakage response is defined as follows:

$$L(\boldsymbol{\alpha};\boldsymbol{\varphi}) \triangleq \int_{S_b} dS \sum_{g=1}^{G} \int_{\boldsymbol{\Omega}\cdot\mathbf{n}>0} d\boldsymbol{\Omega}\, \boldsymbol{\Omega}\cdot\mathbf{n}\varphi^g(r,\boldsymbol{\Omega}). \tag{7.5}$$

When the response sensitivities with respect to the boundaries of the phase-space domain are *not* considered, the G-differential, denoted as $\delta L(\boldsymbol{\alpha};\delta\varphi^g)$, of the leakage response defined in Eq. (7.5) for a variation $\delta\mathbf{u} \triangleq (\delta\boldsymbol{\varphi};\delta\boldsymbol{\alpha})$ around the nominal values $\mathbf{u}^0 \triangleq (\boldsymbol{\varphi}^0;\boldsymbol{\alpha}^0)$ is obtained, by definition, as follows:

$$\begin{aligned}
\delta L(\boldsymbol{\alpha};\delta\boldsymbol{\varphi}) &\triangleq \frac{d}{d\varepsilon}\left\{\int_{S_b} dS \sum_{g=1}^{G} \int_{\boldsymbol{\Omega}\cdot\mathbf{n}>0} d\boldsymbol{\Omega}\, \boldsymbol{\Omega}\cdot\mathbf{n}\left[\varphi^{g,0}(r,\boldsymbol{\Omega}) + \varepsilon\delta\varphi^g(r,\boldsymbol{\Omega})\right]\right\}_{\varepsilon=0} \\
&= \int_{S_b} dS \sum_{g=1}^{G} \int_{\boldsymbol{\Omega}\cdot\mathbf{n}>0} d\boldsymbol{\Omega}\, \boldsymbol{\Omega}\cdot\mathbf{n}\left[\delta\varphi^g(r,\boldsymbol{\Omega})\right].
\end{aligned} \tag{7.6}$$

In general, the total G-variation of the response comprises two contributions, one arising directly from the parameter variations $\delta\boldsymbol{\alpha}$ (this contribution is called the "direct-effect term") and another contribution, stemming from the variations $\delta\boldsymbol{\varphi}$ in the state functions (this contribution is called the "indirect-effect term"). In the particular case of the leakage response considered in this work, the variation $\delta L\left(\boldsymbol{\alpha};\delta\varphi^g\right)$ does not depend directly on the parameter variations $\delta\boldsymbol{\alpha}$ but depends only on variations $\delta\boldsymbol{\varphi}$ in the state functions. Therefore, $\delta L\left(\boldsymbol{\alpha};\delta\varphi^g\right)$ coincides with the "indirect-effect term."

The variations $\delta\varphi^g\left(r,\boldsymbol{\Omega}\right)$ are the solutions of the 1st-order G-differentials of the model described by Eqs. (7.1) and (7.2). Applying the definition of the G-differential to these equations yields the following relations:

$$
\begin{aligned}
&\frac{d}{d\varepsilon}\Big\{\boldsymbol{\Omega}\bullet\nabla\left[\varphi^{g,0}\left(r,\boldsymbol{\Omega}\right)+\varepsilon\delta\varphi^g\left(r,\boldsymbol{\Omega}\right)\right]\\
&+\Sigma_t^g\left(\boldsymbol{\alpha}^0+\varepsilon\delta\boldsymbol{\alpha};r\right)\left[\varphi^{g,0}\left(r,\boldsymbol{\Omega}\right)+\varepsilon\delta\varphi^g\left(r,\boldsymbol{\Omega}\right)\right]\\
&-\sum_{h=1}^{G}\int_{4\pi}d\boldsymbol{\Omega}'\left[\varphi^{h,0}\left(r,\boldsymbol{\Omega}\right)+\varepsilon\delta\varphi^h\left(r,\boldsymbol{\Omega}\right)\right]\Big[\Sigma_s^{h\to g}\left(\boldsymbol{\alpha}^0+\varepsilon\delta\boldsymbol{\alpha};r,\boldsymbol{\Omega}'\bullet\boldsymbol{\Omega}\right)\\
&+\chi^g\left(\boldsymbol{\alpha}^0+\varepsilon\delta\boldsymbol{\alpha};r\right)\left(\nu\Sigma_f\right)^h\left(\boldsymbol{\alpha}^0+\varepsilon\delta\boldsymbol{\alpha};r\right)\Big]-Q^g\left(\boldsymbol{\alpha}^0+\varepsilon\delta\boldsymbol{\alpha}\right)\Big\}_{\varepsilon=0}=0,
\end{aligned}
\tag{7.7}
$$

$$
\frac{d}{d\varepsilon}\left\{\left[\varphi^{g,0}\left(r,\boldsymbol{\Omega}\right)+\varepsilon\delta\varphi^g\left(r,\boldsymbol{\Omega}\right)\right]\right\}_{\varepsilon=0}=0,\ r=r_s\in S_b,\ \boldsymbol{\Omega}\bullet\mathbf{n}<0,\ g=1,...,G. \tag{7.8}
$$

Carrying out the differentiations with respect to ε in Eqs. (7.7) and (7.8), and setting $\varepsilon=0$ in the resulting equations yields the following system of equations for the variations $\delta\varphi^g\left(\mathbf{r},\boldsymbol{\Omega}\right)$ in terms of the parameter variations $\delta\boldsymbol{\alpha}$:

$$
B^g\left(\boldsymbol{\alpha}^0\right)\left[\delta\varphi^g\left(\mathbf{r},\boldsymbol{\Omega}\right)\right]=Q^{(1)}\left(g;\boldsymbol{\alpha}^0,\boldsymbol{\varphi}^0;\delta\boldsymbol{\alpha}\right),\ g=1,...,G, \tag{7.9}
$$

$$
\delta\varphi^g\left(r,\boldsymbol{\Omega}\right)=0,\ r=r_s\in S_b,\ \boldsymbol{\Omega}\bullet\mathbf{n}<0,\ g=1,...,G, \tag{7.10}
$$

where:

$$
Q^{(1)}\left(g;\boldsymbol{\alpha}^0,\boldsymbol{\varphi}^0;\delta\boldsymbol{\alpha}\right)\triangleq\left\{\frac{\partial Q^g\left(\boldsymbol{\alpha}\right)}{\partial\boldsymbol{\alpha}}\delta\boldsymbol{\alpha}-\frac{\partial\left\{B^g\left(\boldsymbol{\alpha}\right)\left[\varphi^g\left(r,\boldsymbol{\Omega}\right)\right]\right\}}{\partial\boldsymbol{\alpha}}\delta\boldsymbol{\alpha}\right\}_{\boldsymbol{\alpha}^0}, \tag{7.11}
$$

$$
\begin{aligned}
&\frac{\partial\left\{B^g\left(\boldsymbol{\alpha}\right)\left[\varphi^g\left(r,\boldsymbol{\Omega}\right)\right]\right\}}{\partial\boldsymbol{\alpha}}\delta\boldsymbol{\alpha}\triangleq\varphi^g\left(r,\boldsymbol{\Omega}\right)\frac{\partial\Sigma_t^g\left(\boldsymbol{\alpha}\right)}{\partial\boldsymbol{\alpha}}\delta\boldsymbol{\alpha}\\
&-\sum_{h=1}^{G}\int_{4\pi}d\boldsymbol{\Omega}'\varphi^h\left(r,\boldsymbol{\Omega}'\right)\left\{\frac{\partial\Sigma_s^{h\to g}\left(\boldsymbol{\alpha};\boldsymbol{\Omega}'\bullet\boldsymbol{\Omega}\right)}{\partial\boldsymbol{\alpha}}\delta\boldsymbol{\alpha}+\frac{\partial\left[\chi^g\left(\boldsymbol{\alpha}\right)\left(\nu\Sigma_f\right)^h\left(\boldsymbol{\alpha}\right)\right]}{\partial\boldsymbol{\alpha}}\delta\boldsymbol{\alpha}\right\}.
\end{aligned}
\tag{7.12}
$$

In matrix vector form, Eqs. (7.9) and (7.10) can be written as follows:

$$\mathbf{B}(\boldsymbol{\alpha})\left[\delta\boldsymbol{\varphi}(r,\boldsymbol{\Omega})\right]=\mathbf{Q}^{(1)}(\boldsymbol{\alpha};\delta\boldsymbol{\alpha}), \tag{7.13}$$

$$\delta\boldsymbol{\varphi}(r,\boldsymbol{\Omega})=\mathbf{0},\ \ r=r_s\in S_b,\ \boldsymbol{\Omega}\bullet\mathbf{n}<0, \tag{7.14}$$

where:

$$\delta\boldsymbol{\varphi}(r,\boldsymbol{\Omega})\triangleq\begin{pmatrix}\delta\varphi^1(\mathbf{r},\boldsymbol{\Omega})\\ \bullet\\ \delta\varphi^g(\mathbf{r},\boldsymbol{\Omega})\\ \bullet\\ \delta\varphi^G(\mathbf{r},\boldsymbol{\Omega})\end{pmatrix};\ \mathbf{Q}^{(1)}(\boldsymbol{\alpha};\delta\boldsymbol{\alpha})\triangleq\begin{pmatrix}Q^{(1)}(1;\boldsymbol{\alpha},\boldsymbol{\varphi};\delta\boldsymbol{\alpha})\\ \bullet\\ Q^{(1)}(g;\boldsymbol{\alpha},\boldsymbol{\varphi};\delta\boldsymbol{\alpha})\\ \bullet\\ Q^{(1)}(G;\boldsymbol{\alpha},\boldsymbol{\varphi};\delta\boldsymbol{\alpha})\end{pmatrix}. \tag{7.15}$$

The relations represented by Eqs. (7.13) and (7.14) constitute the 1st-Level Variational Sensitivity System (1st-LVSS), which could be solved, in principle, to obtain the components of the function $\delta\boldsymbol{\varphi}$ for every parameter variation of interest but this procedure is extremely expensive computationally. The objective of the 1st-CASAM-L (Cacuci, 2022) is to construct an alternative expression for the "indirect-effect term" $\delta L(\boldsymbol{\alpha};\delta\boldsymbol{\varphi})$ which would not depend on $\delta\boldsymbol{\varphi}$. This objective can be achieved by implementing the following sequence of steps, which are all to be carried out at the nominal parameter values $\boldsymbol{\alpha}^0$ although this will not be explicitly indicated in the derivations to follow in order to keep the notation as simple as possible:

1. Consider that the square-integrable functions $\boldsymbol{\varphi}\in\mathcal{H}_{(1)}$ and $\delta\boldsymbol{\varphi}\in\mathcal{H}_{(1)}$ are elements of a Hilbert space denoted as $\mathcal{H}_{(1)}$, in which the inner product between two elements $\mathbf{u}(r,\boldsymbol{\Omega})\in\mathcal{H}_{(1)}$ and $\mathbf{v}(r,\boldsymbol{\Omega})\in\mathcal{H}_{(1)}$ is denoted as $\langle\mathbf{u}(r,\boldsymbol{\Omega}),\mathbf{v}(r,\boldsymbol{\Omega})\rangle_{(1)}$ and is defined as follows:

$$\langle\mathbf{u}(r,\boldsymbol{\Omega}),\mathbf{v}(r,\boldsymbol{\Omega})\rangle_{(1)}\triangleq\sum_{g=1}^{G}\int_0^{r_s}4\pi r^2dr\int_{4\pi}d\boldsymbol{\Omega}\,u^g(r,\boldsymbol{\Omega})v^g(r,\boldsymbol{\Omega}). \tag{7.16}$$

2. Form the inner product of Eq. (7.13) with a yet undefined function $\boldsymbol{\psi}^{(1)}(r,\boldsymbol{\Omega})\triangleq\left[\psi^{(1),g}(r,\boldsymbol{\Omega}),\dots,\psi^{(1),G}(r,\boldsymbol{\Omega})\right]^{\dagger}\in\mathcal{H}_{(1)}$ to obtain the following relation:

$$\begin{aligned}&\left\langle\boldsymbol{\psi}^{(1)}(r,\boldsymbol{\Omega}),\mathbf{B}(\boldsymbol{\alpha})\left[\delta\boldsymbol{\varphi}(r,\boldsymbol{\Omega})\right]\right\rangle_{(1)}=\left\langle\boldsymbol{\psi}^{(1)}(r,\boldsymbol{\Omega}),\mathbf{Q}^{(1)}(\boldsymbol{\alpha};\delta\boldsymbol{\alpha})\right\rangle_{(1)}\\&=\sum_{g=1}^{G}\int_0^{r_s}4\pi r^2dr\int_{4\pi}d\boldsymbol{\Omega}\,\psi^{(1),g}(r,\boldsymbol{\Omega})Q^{(1)}(g;\boldsymbol{\alpha}^0,\boldsymbol{\varphi}^0;\delta\boldsymbol{\alpha}).\end{aligned} \tag{7.17}$$

3. Using the inner product defined in (7.16) and the definition of the adjoint operator in $\mathcal{H}_{(1)}$, recast the left side of Eq. (7.17) into the following equivalent form:

$$\left\langle \psi^{(1)}(r,\boldsymbol{\Omega}), \mathbf{B}(\boldsymbol{\alpha})\left[\delta\boldsymbol{\varphi}(r,\boldsymbol{\Omega})\right]\right\rangle_{(1)}$$
$$= \left\langle \delta\boldsymbol{\varphi}(r,\boldsymbol{\Omega}), \mathbf{A}(\boldsymbol{\alpha})\psi^{(1)}(r,\boldsymbol{\Omega})\right\rangle_{(1)} + P\left[\delta\boldsymbol{\varphi},\psi^{(1)}\right], \tag{7.18}$$

where

$$P\left[\delta\boldsymbol{\varphi},\psi^{(1)}\right] \triangleq \int_{S_b} dS \sum_{g=1}^{G} \int_{\boldsymbol{\Omega}\cdot\mathbf{n}<0} \left|\boldsymbol{\Omega}\bullet\mathbf{n}\right| \left[\delta\varphi^{g}(r,\boldsymbol{\Omega})\psi^{(1),g}(r,\boldsymbol{\Omega})\right] d\boldsymbol{\Omega}$$
$$-\int_{S_b} dS \sum_{g=1}^{G} \int_{\boldsymbol{\Omega}\cdot\mathbf{n}>0} \boldsymbol{\Omega}\bullet\mathbf{n}\left[\delta\varphi^{g}(r,\boldsymbol{\Omega})\psi^{(1),g}(r,\boldsymbol{\Omega})\right] d\boldsymbol{\Omega}, \tag{7.19}$$

denotes the bilinear concomitant evaluated on the PERP sphere's surface S_b and where $\mathbf{A}(\boldsymbol{\alpha})$ denotes the operator adjoint to $\mathbf{B}(\boldsymbol{\alpha})$, having components $A_{gh}(\boldsymbol{\alpha}) \triangleq \left[B_{hg}(\boldsymbol{\alpha})\right]^{*}$, where the symbol $[\ ]^{*}$ indicates "formal adjoint operator."

4. Require the first term on the right side of Eq. (7.18) to represent the functional defined in Eq. (7.6), which is achieved by imposing the following relation:

$$\mathbf{A}(\boldsymbol{\alpha})\psi^{(1)}(r,\boldsymbol{\Omega}) = \boldsymbol{\Omega}\bullet\mathbf{n}\delta(r-r_s). \tag{7.20}$$

5. Use the boundary condition given in Eq. (7.14) to eliminate the corresponding boundary terms in the bilinear concomitant $P\left[\delta\boldsymbol{\varphi},\psi^{(1)}\right]$ in Eq. (7.19) and eliminate the appearance of the remaining boundary terms in Eq. (7.19) by imposing the following (adjoint) boundary condition:

$$\psi^{(1)}(r,\boldsymbol{\Omega}) = \mathbf{0},\ r = r_s \in S_b,\ \boldsymbol{\Omega}\bullet\mathbf{n} > 0. \tag{7.21}$$

The relations given in Eqs. (7.20) and (7.21) constitute the 1st-Level Adjoint Sensitivity System (1st-LASS), which has the following component form for $g = 1,...,G$:

$$-\boldsymbol{\Omega}\bullet\nabla\psi^{(1),g}(r,\boldsymbol{\Omega}) + \Sigma_t^{g}(\boldsymbol{\alpha})\psi^{(1),g}(r,\boldsymbol{\Omega})$$
$$-\sum_{h=1}^{G}\int_{4\pi} d\boldsymbol{\Omega}'\,\psi^{(1),h}(r,\boldsymbol{\Omega}')\left[\Sigma_s^{g\to h}(\boldsymbol{\alpha};\boldsymbol{\Omega}\bullet\boldsymbol{\Omega}') + (\nu\Sigma_f)^{g}(\boldsymbol{\alpha})\chi^{h}(\boldsymbol{\alpha})\right] \tag{7.22}$$
$$= \boldsymbol{\Omega}\bullet\mathbf{n}\delta(r-r_s);$$

$$\psi^{(1),g}(r,\boldsymbol{\Omega}) = 0,\ r = r_s \in S_b,\ \boldsymbol{\Omega}\bullet\mathbf{n} > 0. \tag{7.23}$$

It is important to note that the relations presented in Eqs. (7.20) and (7.21) hold at and in a neighborhood of, the nominal values $\mathbf{u}^0 \triangleq \left(\boldsymbol{\varphi}^0;\boldsymbol{\alpha}^0\right)$. In order to simplify the notation, the superscript "zero" denoting nominal values has been omitted in the foregoing derivations and will occasionally be omitted henceforth, as well. This simplification should not cause any loss of clarity since it will become clear from the context which quantities are to be evaluated/computed using the nominal values for the model parameters.

6. Use the relations underlying the 1st-LVSS and 1st-LASS in Eq. (7.6) to obtain the following expression for the "indirect-effect term":

$$\delta L(\boldsymbol{\alpha};\delta\boldsymbol{\varphi}) = \left\langle \boldsymbol{\psi}^{(1)}(r,\boldsymbol{\Omega}), \mathbf{Q}^{(1)}(\boldsymbol{\alpha};\delta\boldsymbol{\alpha})\right\rangle_{(1)} = \sum_{g=1}^{G}\int_{0}^{r_d} 4\pi r^2 dr \int_{4\pi} d\boldsymbol{\Omega}\, \psi^{(1),g} (r,\boldsymbol{\Omega}) Q^{(1)}\left(g;\boldsymbol{\alpha}^0,\boldsymbol{\varphi}^0;\delta\boldsymbol{\alpha}\right) \triangleq \sum_{j=1}^{J_\alpha} \frac{\partial L(\boldsymbol{\alpha})}{\partial \alpha_j}\delta\alpha_j. \tag{7.24}$$

7. Replacing in (7.24) the expression of $Q^{(1)}\left(g;\boldsymbol{\alpha}^0,\boldsymbol{\varphi}^0;\delta\boldsymbol{\alpha}\right)$ from Eq. (7.11) and identifying the expressions that multiply the arbitrary parameter variations α_j yields the expression of each partial sensitivity $\partial L\left(\boldsymbol{\alpha},\boldsymbol{\varphi};\boldsymbol{\psi}^{(1)}\right)/\partial\alpha_j$ for all α_j, $j=1,...,J_\alpha$. Thus, the computation of all of the J_α partial sensitivities $\partial L\left(\boldsymbol{\alpha},\boldsymbol{\varphi};\boldsymbol{\psi}^{(1)}\right)/\partial\alpha_j$, $j=1,...,J_\alpha$, requires just a single large-scale computation in order to determine $\boldsymbol{\psi}^{(1)}(r,\boldsymbol{\Omega})$, followed by J_α inexpensive computations to perform each integration (quadrature) involving the 1st-level adjoint function $\boldsymbol{\psi}^{(1)}(r,\boldsymbol{\Omega})$. In particular, the expression for computing the 1st-order partial sensitivity with respect to the total cross-sections is as follows:

$$\frac{\partial L\left(\boldsymbol{\alpha},\boldsymbol{\varphi};\boldsymbol{\psi}^{(1)}\right)}{\partial t_j} = -\sum_{g=1}^{G}\int_{0}^{r_s} 4\pi r^2 dr \int_{4\pi} d\boldsymbol{\Omega}\, \psi^{(1),g}(\mathbf{r},\boldsymbol{\Omega})\varphi^g(\mathbf{r},\boldsymbol{\Omega}) \frac{\partial \Sigma_t^g(\boldsymbol{\alpha})}{\partial t_j}, \quad for\ j=1,...,J_{\sigma t}=180. \tag{7.25}$$

The following relations hold when the parameters t_j designate the group-averaged total microscopic cross-sections, in other words, $t_j \equiv \sigma_{t,i_j}^{g_j}$ and $t_k \equiv \sigma_{t,i_k}^{g_k}$, respectively:

$$\frac{\partial \Sigma_t^g(\mathbf{t})}{\partial t_j} \triangleq \frac{\partial \Sigma_t^g(\mathbf{t})}{\partial \sigma_{t,i_j}^{g_j}} = \frac{\partial\left(\sum_{m=1}^{M}\sum_{i=1}^{I} N_{i,m}\sigma_{t,i}^{g}\right)}{\partial \sigma_{t,i_j}^{g_j}} = \delta_{g_j g} N_{i_j,m_j}, \tag{7.26}$$

$$\frac{\partial^2 \Sigma_t^g(\mathbf{t})}{\partial t_j \partial t_k} \triangleq \frac{\partial^2 \Sigma_t^g(\mathbf{t})}{\partial \sigma_{t,i_j}^{g_j}\partial \sigma_{t,i_k}^{g_k}} = \frac{\partial\left(\delta_{g_j g} N_{i_j,m_j}\right)}{\partial \sigma_{t,i_k}^{g_k}} = 0, \tag{7.27}$$

where the subscripts i_j, g_j and m_j denote the isotope, the energy group, and material associated with the parameter $\sigma_{t,i_j}^{g_j}$, respectively; the subscripts i_k, g_k, and m_k denote the isotope, the energy group, and material associated with the parameter

$\sigma_{t,i_k}^{g_k}$, respectively; and $\delta_{g_j g}$ denotes the Kronecker delta function (e.g., $\delta_{g_j g}=1$ if $g_j=g$; $\delta_{g_j g}=0$ if $g_j\neq g$).

7.3 COMPUTATION OF 2ND-ORDER SENSITIVITIES $\partial^2 L(\boldsymbol{\alpha})/\partial t_{j_2}\partial t_{j_1}$

The 2nd-order sensitivities which involve *only* the total microscopic cross-sections, are obtained by G-differentiating the expression provided in Eq. (7.25) for the 1st-order sensitivities of the leakage response with respect to the group-averaged total microscopic cross-sections. Taking the G-differential of Eq. (7.25) yields the following result:

$$\delta\left\{\frac{\partial L\left(\boldsymbol{\alpha},\boldsymbol{\varphi};\boldsymbol{\psi}^{(1)}\right)}{\partial t_{j_1}}\right\}_{\boldsymbol{\alpha}^0}=\frac{d}{d\varepsilon}\left\{-\sum_{g=1}^{G}\int_0^{r_s}4\pi r^2 dr\int_{4\pi}d\boldsymbol{\Omega}\left[\psi^{(1),g}(\mathbf{r},\boldsymbol{\Omega})\right.\right.$$
$$\left.\left.+\varepsilon\delta\psi^{(1),g}(\mathbf{r},\boldsymbol{\Omega})\right]\left[\varphi^{g}(\mathbf{r},\boldsymbol{\Omega})+\varepsilon\delta\varphi^{g}(\mathbf{r},\boldsymbol{\Omega})\right]\frac{\partial\left[\Sigma_t^g(\boldsymbol{\alpha})+\varepsilon\delta\Sigma_t^g(\boldsymbol{\alpha})\right]}{\partial t_{j_1}}\right\}_{\varepsilon=0}, \tag{7.28}$$
$$for\ j_1=1,\dots,J_{\sigma t}=180.$$

Carrying out the differentiations with respect to ε in Eq. (7.28) and subsequently setting $\varepsilon=0$ in the resulting equations yields the following relation:

$$\delta\left\{\frac{\partial L\left(\boldsymbol{\alpha},\boldsymbol{\varphi};\boldsymbol{\psi}^{(1)}\right)}{\partial t_{j_1}}\right\}_{\boldsymbol{\alpha}^0}=-\sum_{g=1}^{G}\int_0^{r_s}4\pi r^2 dr\int_{4\pi}d\boldsymbol{\Omega}\,\delta\psi^{(1),g}(\mathbf{r},\boldsymbol{\Omega})$$
$$\times\left\{\varphi^{g}(\mathbf{r},\boldsymbol{\Omega})\frac{\partial\Sigma_t^g(\boldsymbol{\alpha})}{\partial t_{j_1}}\right\}_{\boldsymbol{\alpha}^0}-\sum_{g=1}^{G}\int_0^{r_s}4\pi r^2 dr\int_{4\pi}d\boldsymbol{\Omega}\,\delta\varphi^{g}(\mathbf{r},\boldsymbol{\Omega})$$
$$\times\left\{\psi^{(1),g}(\mathbf{r},\boldsymbol{\Omega})\frac{\partial\Sigma_t^g(\boldsymbol{\alpha})}{\partial t_{j_1}}\right\}_{\boldsymbol{\alpha}^0}-\sum_{g=1}^{G}\int_0^{r_s}4\pi r^2 dr\int_{4\pi}d\boldsymbol{\Omega} \tag{7.29}$$
$$\times\left\{\psi^{(1),g}(\mathbf{r},\boldsymbol{\Omega})\varphi^{g}(\mathbf{r},\boldsymbol{\Omega})\right\}_{\boldsymbol{\alpha}^0}\frac{\partial}{\partial t_{j_1}}\delta\Sigma_t^g(\boldsymbol{\alpha});\ j_1=1,\dots,J_{\sigma t}=180.$$

Taking into account the facts that: (i) the parameter t_{j_1} represents the group-averaged total microscopic cross-sections $\sigma_{t,i}^{g}$ with $j_1=(i-1)G+g$, $i=1,\dots,I=6$, $g=1,\dots,G=30$; and (ii) $\Sigma_t^g(\mathbf{t})=\sum_{m=1}^{M=2}\sum_{i=1}^{I=6}N_{i,m}\sigma_{t,i}^{g}$, it follows that the last term on the left side of Eq. (7.29) is identically zero.

In Eq. (7.29), the variations $\delta\varphi^{g}(r,\boldsymbol{\Omega})$ are the solutions of Eqs. (7.13) and (7.14) while the variations $\delta\psi^{(1),g}(\mathbf{r},\boldsymbol{\Omega})$ are the solutions of the G-differentiated 1st-LASS, namely:

$$\left\{A^g(\alpha)\delta\psi^{(1),g}(r,\boldsymbol{\Omega})\right\}_{\alpha^0}=\left\{Q^{(2)}\left(g;\alpha,\boldsymbol{\psi}^{(1)};\delta\alpha\right)\right\}_{\alpha^0},\quad g=1,\ldots,G, \tag{7.30}$$

$$\left\{\delta\psi^{(1),g}(r,\boldsymbol{\Omega})\right\}_{\alpha^0}=0,\ r=r_s,\ \boldsymbol{\Omega}\bullet\mathbf{n}>0,\ g=1,\ldots,G, \tag{7.31}$$

where:

$$\begin{aligned}Q^{(2)}\left(g;\alpha,\psi^{(1)};\delta\alpha\right)&\triangleq-\frac{\partial\left\{A^g(\alpha)\left[\psi^{(1),g}(r,\boldsymbol{\Omega})\right]\right\}}{\partial\alpha}\delta\alpha\\&=-\psi^{(1),g}(r,\boldsymbol{\Omega})\frac{\partial\Sigma_t^g(\alpha)}{\partial\alpha}\delta\alpha+\sum_{h=1}^{G}\int_{4\pi}d\boldsymbol{\Omega}'\,\psi^{(1),h}(r,\boldsymbol{\Omega}')\\&\times\left\{\frac{\partial\Sigma_s^{g\to h}(\alpha;\boldsymbol{\Omega}\bullet\boldsymbol{\Omega}')}{\partial\alpha}\delta\alpha+\frac{\partial\left[(\nu\Sigma_f)^g(\alpha)\chi^h(\alpha)\right]}{\partial\alpha}\delta\alpha\right\},\ g=1,\ldots,G.\end{aligned} \tag{7.32}$$

In matrix vector form, which will be used in the sequel, Eqs. (7.30) and (7.31) are expressed as follows:

$$\left\{\mathbf{A}(\alpha)\left[\delta\boldsymbol{\psi}^{(1)}(r,\boldsymbol{\Omega})\right]\right\}_{\alpha^0}=\left\{\mathbf{Q}^{(2)}\left(\alpha,\boldsymbol{\psi}^{(1)};\delta\alpha\right)\right\}_{\alpha^0}, \tag{7.33}$$

$$\delta\boldsymbol{\psi}^{(1)}(r,\boldsymbol{\Omega})=\mathbf{0},\ r=r_s,\ \boldsymbol{\Omega}\bullet\mathbf{n}>0. \tag{7.34}$$

where:

$$\delta\boldsymbol{\psi}^{(1)}(r,\boldsymbol{\Omega})\triangleq\begin{pmatrix}\delta\psi^{(1),1}(r,\boldsymbol{\Omega})\\ \bullet\\ \delta\psi^{(1),G}(r,\boldsymbol{\Omega})\end{pmatrix},\quad \mathbf{Q}^{(2)}\left(\alpha,\boldsymbol{\psi}^{(1)};\delta\alpha\right)\triangleq\begin{pmatrix}Q^{(2)}\left(1;\alpha,\boldsymbol{\psi}^{(1)};\delta\alpha\right)\\ \bullet\\ Q^{(2)}\left(G;\alpha,\boldsymbol{\psi}^{(1)};\delta\alpha\right)\end{pmatrix}. \tag{7.35}$$

The 2nd-order sensitivities expressed by Eq. (7.29) could be computed, in principle, if the functions $\delta\varphi^g(r,\boldsymbol{\Omega})$ and $\delta\psi^{(1),g}(\mathbf{r},\boldsymbol{\Omega})$ were available. However, determining the functions $\delta\varphi^g(r,\boldsymbol{\Omega})$ would require solving the 1st-LVSS, while determining the functions $\delta\psi^{(1),g}(r,\boldsymbol{\Omega})$ would require solving Eqs. (7.33) and (7.34), which would altogether amount to twice as many large-scale computations as there are parameter variations. To avoid the need for performing so many large-scale computations, the right side of Eq. (7.29) will be recast in terms of the solutions of a 2nd-Level Adjoint Sensitivity System (2nd-LASS), which will be constructed so as to eliminate the appearance of the functions $\delta\varphi^g(r,\boldsymbol{\Omega})$ and $\delta\psi^{(1),g}(r,\boldsymbol{\Omega})$. The requisite 2nd-LASS is constructed by applying the 4th-CASAM-L (Cacuci, 2022), comprising the following sequence of steps which are to be carried out at the nominal parameter values α^0:

1. Consider a Hilbert space, denoted as $\mathcal{H}_{(2)}$, comprising block vector elements $\mathbf{u}^{(2)}(r,\mathbf{\Omega}) \in \mathcal{H}_{(2)}$ having the structure: $\mathbf{u}^{(2)}(r,\mathbf{\Omega}) \triangleq \left[\mathbf{u}_1^{(2)}(r,\mathbf{\Omega}), \mathbf{u}_2^{(2)}(r,\mathbf{\Omega})\right]^{\dagger}$, with components of the form $\mathbf{u}_i^{(2)}(r,\mathbf{\Omega}) \triangleq \left[u_i^{(2),1}(r,\mathbf{\Omega}),\dots,u_i^{(2),g}(r,\mathbf{\Omega}),\dots,u_i^{(2),G}(r,\mathbf{\Omega})\right]^{\dagger}$, $i=1,2$.
2. Define the inner product in $\mathcal{H}_{(2)}$, denoted as $\left\langle \mathbf{u}^{(2)}(r,\mathbf{\Omega}), \mathbf{v}^{(2)}(r,\mathbf{\Omega}) \right\rangle_{(2)}$, of two arbitrary elements $\mathbf{u}^{(2)}(r,\mathbf{\Omega}) \in \mathcal{H}_{(2)}$ and $\mathbf{v}^{(2)}(r,\mathbf{\Omega}) \in \mathcal{H}_{(2)}$ in $\mathcal{H}_{(2)}$, where $\mathbf{v}^{(2)}(r,\mathbf{\Omega}) \triangleq \left[\mathbf{v}_1^{(2)}(r,\mathbf{\Omega}), \mathbf{v}_2^{(2)}(r,\mathbf{\Omega})\right]^{\dagger}$, $\mathbf{v}_i^{(2)}(r,\mathbf{\Omega}) \triangleq \left[v_i^{(2),1}(r,\mathbf{\Omega}),\dots,v_i^{(2),G}(r,\mathbf{\Omega})\right]^{\dagger}$, $i=1,2$, as follows:

$$
\begin{aligned}
\left\langle \mathbf{u}^{(2)}(r,\mathbf{\Omega}), \mathbf{v}^{(2)}(r,\mathbf{\Omega}) \right\rangle_{(2)} &\triangleq \sum_{i=1}^{2} \left\langle \mathbf{u}_i^{(2)}(r,\mathbf{\Omega}), \mathbf{v}_i^{(2)}(r,\mathbf{\Omega}) \right\rangle_{(1)} \\
&= \sum_{i=1}^{2} \sum_{g=1}^{G} \int_0^{r_s} 4\pi r^2 dr \int_{4\pi} d\mathbf{\Omega}\, u_i^{(2),g}(r,\mathbf{\Omega})\, v_i^{(2),g}(r,\mathbf{\Omega}).
\end{aligned}
\tag{7.36}
$$

3. For a 2 × 2 block matrix linear operator $\mathbf{L}^{(2)} \triangleq \begin{pmatrix} L_{11}^{(2)} & L_{12}^{(2)} \\ L_{21}^{(2)} & L_{22}^{(2)} \end{pmatrix}$ in the Hilbert space $\mathcal{H}_{(2)}$, use the inner product defined in Eq. (7.36) to define the formal adjoint operator, denoted as $\mathbf{A}^{(2)} \triangleq \begin{pmatrix} A_{11}^{(2)} & A_{12}^{(2)} \\ A_{21}^{(2)} & A_{22}^{(2)} \end{pmatrix}$, of $\mathbf{L}^{(2)}$, through the following relationship:

$$
\left\langle \mathbf{v}^{(2)}, \mathbf{L}^{(2)}\mathbf{u}^{(2)} \right\rangle_{(2)} = \left\langle \mathbf{u}^{(2)}, \mathbf{A}^{(2)}\mathbf{v}^{(2)} \right\rangle_{(2)} + P^{(2)}\left[\mathbf{u}^{(2)}, \mathbf{v}^{(2)}\right], \tag{7.37}
$$

where $P^{(2)}\left[\mathbf{u}^{(2)}, \mathbf{v}^{(2)}\right]$ denotes the corresponding bilinear concomitant evaluated on the domain's boundary.
4. Apply the definition provided in Eq. (7.37) to form the inner product of Eqs. (7.13) and (7.33), respectively, with a yet undefined function $\boldsymbol{\psi}^{(2)}(j_1;r,\mathbf{\Omega}) \triangleq \left[\psi_1^{(2)}(j_1;r,\mathbf{\Omega}), \psi_2^{(2)}(j_1;r,\mathbf{\Omega})\right]^{\dagger} \in \mathcal{H}_{(2)}$ to obtain the following relation, evaluated at $\boldsymbol{\alpha}^0$:

$$
\begin{aligned}
&\left\{ \left\langle \boldsymbol{\psi}^{(2)}(j_1;r,\mathbf{\Omega}), \begin{pmatrix} \mathbf{B}(\boldsymbol{\alpha})\left[\delta\varphi(r,\mathbf{\Omega})\right] \\ \mathbf{A}(\boldsymbol{\alpha})\left[\delta\psi^{(1)}(r,\mathbf{\Omega})\right] \end{pmatrix} \right\rangle_{(2)} \right\}_{\boldsymbol{\alpha}^0} \\
&= \left\{ \left\langle \boldsymbol{\psi}^{(2)}(j_1;r,\mathbf{\Omega}), \begin{pmatrix} \mathbf{Q}^{(1)}(\boldsymbol{\alpha};\delta\boldsymbol{\alpha}) \\ \mathbf{Q}^{(2)}(\boldsymbol{\alpha},\boldsymbol{\psi}^{(1)};\delta\boldsymbol{\alpha}) \end{pmatrix} \right\rangle_{(2)} \right\}_{\boldsymbol{\alpha}^0}.
\end{aligned}
\tag{7.38}
$$

5. Use the relation shown in Eq. (7.37) to recast the left side of Eq. (7.38) into the following equivalent form:

$$\left\{\left\langle \psi^{(2)}(j_1;r,\mathbf{\Omega}), \begin{pmatrix} \mathbf{B}(\alpha)[\delta\varphi(r,\mathbf{\Omega})] \\ \mathbf{A}(\alpha)\left[\delta\psi^{(1)}(r,\mathbf{\Omega})\right] \end{pmatrix} \right\rangle_{(2)}\right\}_{\alpha^0}$$
$$= \left\{\left\langle \left[\delta\varphi, \delta\psi^{(1)}\right], \begin{pmatrix} \mathbf{A}(\alpha)\psi_1^{(2)}(j_1;r,\mathbf{\Omega}) \\ \mathbf{B}(\alpha)\psi_2^{(2)}(j_1;r,\mathbf{\Omega}) \end{pmatrix} \right\rangle_{(2)}\right\}_{\alpha^0} \tag{7.39}$$
$$+P^{(2)}\left[\delta\varphi(r,\mathbf{\Omega}), \delta\psi^{(1)}(r,\mathbf{\Omega}); \psi_1^{(2)}(j_1;r,\mathbf{\Omega}), \psi_2^{(2)}(j_1;r,\mathbf{\Omega})\right],$$

where $P^{(2)}\left[\delta\varphi(r,\mathbf{\Omega}), \delta\psi^{(1)}(r,\mathbf{\Omega}); \psi_1^{(2)}(j_1;r,\mathbf{\Omega}), \psi_2^{(2)}(j_1;r,\mathbf{\Omega})\right]$ denotes the corresponding bilinear concomitant evaluated on the PERP sphere's outer boundary S_b at $r = r_s$.

6. Require the first term on the right side of Eq. (7.39) to represent the same function as the right side of Eq. (7.29). This requirement will be satisfied by requiring that the following relations be satisfied by the components of the function $\psi^{(2)}(j_1;r,\mathbf{\Omega}) \triangleq \left[\psi_1^{(2)}(j_1;r,\mathbf{\Omega}), \psi_2^{(2)}(j_1;r,\mathbf{\Omega})\right]^\dagger \in \mathcal{H}_{(2)}$ for $j_1 = 1,...,J_{\sigma t}$:

$$\left\{\mathbf{A}(\alpha)\psi_1^{(2)}(j_1;r,\mathbf{\Omega})\right\}_{\alpha^0} = -\left\{\mathbf{S}(j_1;\alpha)\psi^{(1)}(\mathbf{r},\mathbf{\Omega})\right\}_{\alpha^0}, \; j_1 = 1,...,J_{\sigma t}; \tag{7.40}$$

$$\left\{\mathbf{B}(\alpha)\psi_2^{(2)}(j_1;r,\mathbf{\Omega})\right\}_{\alpha^0} = -\left\{\mathbf{S}(j_1;\alpha)\varphi(\mathbf{r},\mathbf{\Omega})\right\}_{\alpha^0}; \; j_1 = 1,...,J_{\sigma t}, \tag{7.41}$$

where, for each $j_1 = 1,...,J_{\sigma t}$, $\mathbf{S}(j_1;\alpha)$ is a $G \times G$ diagonal matrix having non-zero elements on its diagonal of the form $\partial\Sigma_t^g(\alpha)/\partial t_{j_1}$, $g = 1,...,G$, in other words,

$$\mathbf{S}(j_1;\alpha) \triangleq \begin{pmatrix} \dfrac{\partial\Sigma_t^1(\alpha)}{\partial t_{j_1}} & \bullet & 0 \\ \bullet & \bullet & \bullet \\ 0 & \bullet & \dfrac{\partial\Sigma_t^1(\alpha)}{\partial t_{j_1}} \end{pmatrix} \triangleq Diag\left[\frac{\partial\Sigma_t^g(\alpha)}{\partial t_{j_1}}\right], \; g = 1,...,G. \tag{7.42}$$

7. Use in Eq. (7.39) the boundary conditions shown in Eqs. (7.14) and (7.34). Furthermore, set to zero the remaining terms in the bilinear concomitant $P^{(2)}\left[\delta\varphi(r,\mathbf{\Omega}), \delta\psi^{(1)}(r,\mathbf{\Omega}); \psi_1^{(2)}(j_1;r,\mathbf{\Omega}), \psi_2^{(2)}(j_1;r,\mathbf{\Omega})\right]$ in Eq. (7.39) by requiring that the components of the function $\psi^{(2)}(j_1;r,\mathbf{\Omega}) \in \mathcal{H}_{(2)}$ to satisfy the following boundary conditions:

$$\psi_1^{(2)}(j_1;r,\mathbf{\Omega}) = \mathbf{0}, \; r = r_s, \; \mathbf{\Omega}\bullet\mathbf{n} > 0; \; j_1 = 1,...,J_{\sigma t}; \tag{7.43}$$

$$\psi_2^{(2)}(j_1;r,\mathbf{\Omega}) = \mathbf{0},\ r = r_s,\ \mathbf{\Omega}\bullet\mathbf{n} < 0;\ j_1 = 1,...,J_{\sigma t}. \tag{7.44}$$

The previous boundary conditions complete the well-posed definition of the 2nd-level adjoint function $\psi^{(2)}(j_1;r,\mathbf{\Omega}) \triangleq \left[\psi_1^{(2)}(j_1;r,\mathbf{\Omega}),\psi_2^{(2)}(j_1;r,\mathbf{\Omega})\right]^\dagger \in \mathsf{H}_{(2)}$ as the solution of the system of equations shown in (7.40)–(7.44), which are called "the 2nd-Level Adjoint Sensitivity System (2nd-LASS)." Recall that the reason for calling this system "2nd-Level" (as opposed to "2nd-Order") stems from the fact that this system does not involve any "2nd-order" differential or derivatives of the dependent variables (i.e., state functions) even though the solution of the 2nd-LASS is used for computing, efficiently and exactly, the 2nd-order sensitivities of the response with respect to the model parameters.

8. Use Eqs. (7.39)–(7.44) in Eq. (7.29) to obtain the following alternative expression for the differential $\delta\left\{\partial L\left(\alpha,\varphi;\psi^{(1)}\right)/\partial t_{j_1}\right\}_{\alpha^0}$ in terms of the 2nd-level adjoint function $\psi^{(2)}(j_1;r,\mathbf{\Omega}) \triangleq \left[\psi_1^{(2)}(j_1;r,\mathbf{\Omega}),\psi_2^{(2)}(j_1;r,\mathbf{\Omega})\right]^\dagger \in \mathcal{H}_{(2)}$:

$$\begin{aligned}
\delta\left\{\frac{\partial L\left(\alpha,\varphi;\psi^{(1)};\psi^{(2)}\right)}{\partial t_{j_1}}\right\}_{\pm^0} &= \left\{\left\langle \psi^{(2)}(j_1;r,\mathbf{\Omega}),\begin{pmatrix}\mathbf{Q}^{(1)}(\alpha;\delta\alpha)\\ \mathbf{Q}^{(2)}\left(\alpha,\psi^{(1)};\delta\alpha\right)\end{pmatrix}\right\rangle_{(2)}\right\}_{\alpha^0}\\
&= \left\{\sum_{g=1}^{G}\int_0^{r_s}4\pi r^2dr\int_{4\pi}d\mathbf{\Omega}\,\psi_1^{(2),g}(j_1;r,\mathbf{\Omega})Q^{(1)}(g;\alpha,\varphi;\delta\alpha)\right\}_{\alpha^0}\\
&+\left\{\sum_{g=1}^{G}\int_0^{r_s}4\pi r^2dr\int_{4\pi}d\mathbf{\Omega}\,\psi_2^{(2),g}(j_1;r,\mathbf{\Omega})Q^{(2)}\left(g;\alpha,\psi^{(1)};\delta\alpha\right)\right\}_{\alpha^0}\\
&= \sum_{j_2=1}^{J_\alpha}\left\{\frac{\partial^2 L(\alpha)}{\partial\alpha_{j_2}\partial t_{j_1}}\delta\alpha_{j_2}\right\}_{\alpha^0},\quad j_1 = 1,...J_{\sigma t}.
\end{aligned}\tag{7.45}$$

It is important to note that the differential expression provided in Eq. (7.45) comprises *all* of the 2nd-order partial sensitivities that involve the group-averaged microscopic total cross-sections. The 2nd-order partial sensitivities $\left\{\partial^2 L\left(\alpha,\varphi;\psi^{(1)},\psi^{(2)}\right)/\partial t_{j_1}\partial\alpha_{j_2}\right\}_{\alpha^0}$, $j_1 = 1,...,J_{\sigma t}$, $j_2 = 1,...,J_\alpha$, which involve the 180 total cross-sections (t_{j_1}) and the other parameters (α_{j_2}), are obtained by replacing in Eq. (7.45) the expressions of $Q^{(1)}(g;\alpha,\varphi;\delta\alpha)$ and $Q^{(2)}\left(g;\alpha,\psi^{(1)};\delta\alpha\right)$ by their definitions provided in Eqs. (7.11) and (7.32), respectively, and subsequently identifying the expressions that multiply the corresponding parameter variations.

In particular, identifying in Eq. (7.45) the quantities that multiply *only* the variation in the total microscopic cross-sections provides the following expression for the 2nd-order response sensitivities that involve *only* the group-averaged total microscopic cross-sections:

$$\frac{\partial^2 L\left(\boldsymbol{\alpha},\boldsymbol{\varphi};\boldsymbol{\psi}^{(1)};\boldsymbol{\psi}^{(2)}\right)}{\partial t_{j_2}\partial t_{j_1}} = -\sum_{g=1}^{G}\int_0^{r_s} 4\pi r^2 dr \int_{4\pi} d\boldsymbol{\Omega}\left[\psi_1^{(2),g}\left(j_1;r,\boldsymbol{\Omega}\right)\varphi^g\left(r,\boldsymbol{\Omega}\right)\right.$$
$$\left.+\psi_2^{(2),g}\left(j_1;r,\boldsymbol{\Omega}\right)\psi^{(1),g}\left(r,\boldsymbol{\Omega}\right)\right]\frac{\partial \Sigma_t^g\left(\boldsymbol{\alpha}\right)}{\partial t_{j_2}},\ j_1 = 1,...,J_{\sigma t};\ j_2 = 1,...j_1. \tag{7.46}$$

The aforementioned expression is to be evaluated at the nominal parameter values, but the corresponding indication $\{\ \}_{\boldsymbol{\alpha}^0}$ has been omitted, for simplicity.

7.4 COMPUTATION OF 3RD-ORDER SENSITIVITIES $\partial^3 L(\boldsymbol{\alpha})/\partial t_{j_3}\partial t_{j_2}\partial t_{j_1}$

The 3rd-order sensitivities $\partial^3 L\left(\boldsymbol{\alpha},\boldsymbol{\varphi};\boldsymbol{\psi}^{(1)};\boldsymbol{\psi}^{(2)};\boldsymbol{\psi}^{(3)}\right)/\partial t_{j_3}\partial t_{j_2}\partial t_{j_1}$ of the PERP benchmark's leakage response with respect to the group-averaged microscopic total cross-sections are obtained by applying the 4th-CASAM-L (Cacuci, 2022) to the expression of the 2nd-order sensitivities obtained in (7.46). The first step in the application of the 4th-CASAM-L is to determine the total G-differential of the expression given in Eq. (7.46), which yields the following result for $j_1 = 1,...,J_{\sigma t}$; and $j_2 = 1,...,j_1$:

$$\delta\left\{\frac{\partial^2 L\left(\boldsymbol{\alpha},\boldsymbol{\varphi};\boldsymbol{\psi}^{(1)};\boldsymbol{\psi}^{(2)}\right)}{\partial t_{j_2}\partial t_{j_1}}\right\}_{\boldsymbol{\alpha}^0} \triangleq \frac{d}{d\varepsilon}\left\{-\sum_{g=1}^{G}\int_0^{r_s} 4\pi r^2 dr \int_{4\pi} d\boldsymbol{\Omega}\right.$$
$$\times\left\{\left[\psi_1^{(2),g}\left(j_1;r,\boldsymbol{\Omega}\right)+\varepsilon\delta\psi_1^{(2),g}\left(j_1;r,\boldsymbol{\Omega}\right)\right]\left[\varphi^g\left(\mathbf{r},\boldsymbol{\Omega}\right)+\varepsilon\delta\varphi^g\left(\mathbf{r},\boldsymbol{\Omega}\right)\right]\right.$$
$$\left.+\left[\psi_2^{(2),g}\left(j_1;r,\boldsymbol{\Omega}\right)+\varepsilon\delta\psi_2^{(2),g}\left(j_1;r,\boldsymbol{\Omega}\right)\right]\left[\psi^{(1),g}\left(\mathbf{r},\boldsymbol{\Omega}\right)+\varepsilon\delta\psi^{(1),g}\left(\mathbf{r},\boldsymbol{\Omega}\right)\right]\right\}$$
$$\left.\times\frac{\partial\left[\Sigma_t^g\left(\boldsymbol{\alpha}\right)+\varepsilon\delta\Sigma_t^g\left(\boldsymbol{\alpha}\right)\right]}{\partial t_{j_2}}\right\}_{\varepsilon=0} = -\left\{\sum_{g=1}^{G}\int_0^{r_s} 4\pi r^2 dr \int_{4\pi} d\boldsymbol{\Omega}\left[\delta\psi_1^{(2),g}\left(j_1;r,\boldsymbol{\Omega}\right)\varphi^g\left(r,\boldsymbol{\Omega}\right)\right.\right.$$
$$+\psi_1^{(2),g}\left(j_1;r,\boldsymbol{\Omega}\right)\delta\varphi^g\left(r,\boldsymbol{\Omega}\right)+\delta\psi_2^{(2),g}\left(j_1;r,\boldsymbol{\Omega}\right)\psi^{(1),g}\left(r,\boldsymbol{\Omega}\right)$$
$$\left.\left.+\psi_2^{(2),g}\left(j_1;r,\boldsymbol{\Omega}\right)\delta\psi^{(1),g}\left(r,\boldsymbol{\Omega}\right)\right]\frac{\partial\Sigma_t^g\left(\boldsymbol{\alpha}\right)}{\partial t_{j_2}}\right\}_{\boldsymbol{\alpha}^0}, \tag{7.47}$$

where: (i) the functions $\delta\varphi^g\left(r,\boldsymbol{\Omega}\right)$ are the solutions of Eqs. (7.13) and (7.14); (ii) the functions $\delta\psi^{(1),g}\left(r,\boldsymbol{\Omega}\right)$ are the solutions of Eqs. (7.30) and (7.31); (iii) the functions $\delta\psi_1^{(2),g}\left(j_1;r,\boldsymbol{\Omega}\right)$ and $\delta\psi_2^{(2),g}\left(j_1;r,\boldsymbol{\Omega}\right)$ are the solutions of the G-differentiated 2nd-LASS defined by Eqs. (7.40)–(7.44), namely:

$$\left\{\mathbf{A}\left(\boldsymbol{\alpha}\right)\delta\boldsymbol{\psi}_1^{(2)}\left(j_1;r,\boldsymbol{\Omega}\right)+\mathbf{S}\left(j_1;\boldsymbol{\alpha}\right)\delta\boldsymbol{\psi}^{(1)}\left(\mathbf{r},\boldsymbol{\Omega}\right)\right\}_{\boldsymbol{\alpha}^0}$$
$$=\left\{\mathbf{Q}_1^{(3)}\left[\boldsymbol{\alpha};\boldsymbol{\psi}^{(1)}\left(\mathbf{r},\boldsymbol{\Omega}\right);\boldsymbol{\psi}_1^{(2)}\left(j_1;r,\boldsymbol{\Omega}\right);\delta\boldsymbol{\alpha}\right]\right\}_{\boldsymbol{\alpha}^0}, \tag{7.48}$$

$$\left\{\mathbf{B}(\alpha)\delta\psi_2^{(2)}(j_1;r,\boldsymbol{\Omega})+\mathbf{S}(j_1;\pm)\delta\varphi(\mathbf{r},\boldsymbol{\Omega})\right\}_{\alpha^0} \\ =\left\{\mathbf{Q}_2^{(3)}\left[\alpha;\varphi(\mathbf{r},\boldsymbol{\Omega});\psi_2^{(2)}(j_1;r,\boldsymbol{\Omega});\delta\alpha\right]\right\}_{\alpha^0}, \tag{7.49}$$

$$\delta\psi_1^{(2)}(j_1;r,\boldsymbol{\Omega})=\mathbf{0},\ r=r_s,\ \boldsymbol{\Omega}\bullet\mathbf{n}>0;\ j_1=1,\dots,J_{\sigma t}, \tag{7.50}$$

$$\delta\psi_2^{(2)}(j_1;r,\boldsymbol{\Omega})=\mathbf{0},\ r=r_s,\ \boldsymbol{\Omega}\bullet\mathbf{n}<0;\ j_1=1,\dots,J_{\sigma t}, \tag{7.51}$$

where:

$$\mathbf{Q}_1^{(3)}\left[\alpha;\psi^{(1)}(r,\boldsymbol{\Omega});\psi_1^{(2)}(j_1;r,\boldsymbol{\Omega});\delta\alpha\right] \\ \triangleq-\left\{\left[\frac{\partial\mathbf{S}(j_1;\alpha)}{\partial\alpha}\delta\alpha\right]\psi^{(1)}(r,\boldsymbol{\Omega})\right\}_{\alpha^0}-\left\{\left[\frac{\partial\mathbf{A}(\alpha)}{\partial\alpha}\delta\alpha\right]\psi_1^{(2)}(j_1;r,\boldsymbol{\Omega})\right\}_{\alpha^0} \\ \triangleq\sum_{j_3=1}^{J_\alpha}\frac{\partial\mathbf{Q}_1^{(3)}\left[\alpha;\psi^{(1)}(r,\boldsymbol{\Omega});\psi_1^{(2)}(j_1;r,\boldsymbol{\Omega})\right]}{\partial\alpha_{j_3}}\delta\alpha_{j_3}, \tag{7.52}$$

$$\mathbf{Q}_2^{(3)}\left[\alpha;\varphi(r,\boldsymbol{\Omega});\psi_2^{(2)}(j_1;r,\boldsymbol{\Omega});\delta\alpha\right]\triangleq-\left\{\left[\frac{\partial\mathbf{S}(j_1;\alpha)}{\partial\alpha}\delta\alpha\right]\varphi(r,\boldsymbol{\Omega})\right\}_{\alpha^0} \\ -\left\{\left[\frac{\partial\mathbf{B}(\alpha)}{\partial\alpha}\delta\alpha\right]\psi_2^{(2)}(j_1;r,\boldsymbol{\Omega})\right\}_{\alpha^0}=\sum_{j_3=1}^{J_\alpha}\frac{\partial\mathbf{Q}_2^{(3)}}{\partial\alpha_{j_3}}\delta\alpha_{j_3}. \tag{7.53}$$

Thus, in order to compute the value of the differential shown in Eq. (7.47), it would be first necessary to solve the following system of matrix equations subject to the boundary conditions shown in Eqs. (7.14), (7.34), (7.50) and (7.51):

$$\left\{\begin{pmatrix}\mathbf{B}(\alpha)&\mathbf{0}&\mathbf{0}&\mathbf{0}\\\mathbf{0}&\mathbf{A}(\alpha)&\mathbf{0}&\mathbf{0}\\\mathbf{0}&\mathbf{S}(j_1;\alpha)&\mathbf{A}(\alpha)&\mathbf{0}\\\mathbf{S}(j_1;\alpha)&\mathbf{0}&\mathbf{0}&\mathbf{B}(\alpha)\end{pmatrix}\right\}_{\alpha^0}\begin{pmatrix}\delta\varphi\\\delta\psi^{(1)}\\\delta\psi_1^{(2)}(j_1)\\\delta\psi_2^{(2)}(j_1)\end{pmatrix} \\ =\left\{\begin{pmatrix}\mathbf{Q}^{(1)}(\alpha;\delta\alpha)\\\mathbf{Q}^{(2)}\left(\alpha,\psi^{(1)};\delta\alpha\right)\\\mathbf{Q}_1^{(3)}\left[\alpha;\psi^{(1)};\psi_1^{(2)}(j_1);\delta\alpha\right]\\\mathbf{Q}_2^{(3)}\left[\alpha;\varphi;\psi_2^{(2)}(j_1);\delta\alpha\right]\end{pmatrix}\right\}_{\alpha^0}. \tag{7.54}$$

It is evident that solving Eq. (7.54) amounts to impractically many large-scale computations, which can be avoided by recasting the right side of Eq. (7.47) in terms the solutions of a 3rd-Level Adjoint Sensitivity System (3rd-LASS), which is constructed by applying the 4th-CASAM-L (Cacuci, 2022) comprising the following sequence of operations, all of which are to be carried out at the nominal parameter values $\boldsymbol{\alpha}^0$:

1. Consider a Hilbert space, denoted as $\mathcal{H}_{(3)}$, comprising elements denoted generically as $\mathbf{u}^{(3)}(r,\boldsymbol{\Omega}) \in \mathcal{H}_{(3)}$, which have a "*four-component block-vector*" structure of the form $\mathbf{u}^{(3)}(r,\boldsymbol{\Omega}) \triangleq \left[\mathbf{u}_1^{(3)}(r,\boldsymbol{\Omega}), \mathbf{u}_2^{(3)}(r,\boldsymbol{\Omega}), \mathbf{u}_3^{(3)}(r,\boldsymbol{\Omega}), \mathbf{u}_4^{(3)}(r,\boldsymbol{\Omega})\right]^\dagger$, with each of the four vector components having the following structure: $\mathbf{u}_i^{(3)}(r,\boldsymbol{\Omega}) \triangleq \left[u_i^{(3),1}(r,\boldsymbol{\Omega}),...,u_i^{(3),g}(r,\boldsymbol{\Omega}),...,u_i^{(3),G}(r,\boldsymbol{\Omega})\right]^\dagger$, $i = 1,2,3,4$.
2. Define the inner product in $\mathcal{H}_{(3)}$, denoted as $\left\langle \mathbf{u}^{(3)}(r,\boldsymbol{\Omega}), \mathbf{v}^{(3)}(r,\boldsymbol{\Omega}) \right\rangle_{(3)}$, of two arbitrary elements in $\mathcal{H}_{(3)}$, denoted as $\mathbf{u}^{(3)}(r,\boldsymbol{\Omega}) \in \mathcal{H}_{(3)}$ and $\mathbf{v}^{(3)}(r,\boldsymbol{\Omega}) \in \mathcal{H}_{(3)}$, with $\mathbf{v}^{(3)}(r,\boldsymbol{\Omega}) \triangleq \left[\mathbf{v}_1^{(3)}(r,\boldsymbol{\Omega}), \mathbf{v}_2^{(3)}(r,\boldsymbol{\Omega}), \mathbf{v}_3^{(3)}(r,\boldsymbol{\Omega}), \mathbf{v}_4^{(3)}(r,\boldsymbol{\Omega})\right]^\dagger$, and $\mathbf{v}_i^{(3)}(r,\boldsymbol{\Omega})$ $\left[\mathbf{v}_i^{(3),1}(r,\boldsymbol{\Omega}),...,\mathbf{v}_i^{(3),g}(r,\boldsymbol{\Omega}),...,\mathbf{v}_i^{(3),G}(r,\boldsymbol{\Omega})\right]^\dagger$, $i = 1,2,3,4$, as follows:

$$\begin{aligned}\left\langle \mathbf{u}^{(3)}(r,\boldsymbol{\Omega}), \mathbf{v}^{(3)}(r,\boldsymbol{\Omega}) \right\rangle_{(3)} &\triangleq \sum_{i=1}^{4} \left\langle \mathbf{u}_i^{(3)}(r,\boldsymbol{\Omega}), \mathbf{v}_i^{(3)}(r,\boldsymbol{\Omega}) \right\rangle_{(1)} \\ &= \sum_{i=1}^{4}\sum_{g=1}^{G} \int_0^{r_s} 4\pi r^2 dr \int_{4\pi} d\boldsymbol{\Omega}\, u_i^{(3),g}(r,\boldsymbol{\Omega})\, v_i^{(3),g}(r,\boldsymbol{\Omega}).\end{aligned} \tag{7.55}$$

3. Apply the definition provided in Eq. (7.55) to form the inner product of Eq. (7.54) with a function $\boldsymbol{\psi}^{(3)}(j_2;j_1;r,\boldsymbol{\Omega}) \triangleq \left[\boldsymbol{\psi}_1^{(3)}(j_2;j_1;r,\boldsymbol{\Omega}),...,\boldsymbol{\psi}_4^{(3)}(j_2;j_1;r,\boldsymbol{\Omega})\right]^\dagger$ $\in \mathcal{H}_{(3)}$, which is just required to be square integrable at this stage but which remains to be fully defined, to obtain the following relation, evaluated at $\boldsymbol{\alpha}^0$:

$$\begin{aligned}&\left\langle \boldsymbol{\psi}^{(3)}(j_2;j_1;r,\boldsymbol{\Omega}), \begin{pmatrix} \mathbf{B}(\boldsymbol{\alpha}) & \mathbf{0} & \mathbf{0} & \mathbf{0} \\ \mathbf{0} & \mathbf{A}(\boldsymbol{\alpha}) & \mathbf{0} & \mathbf{0} \\ \mathbf{0} & \mathbf{S}(j_1;\boldsymbol{\alpha}) & \mathbf{A}(\boldsymbol{\alpha}) & \mathbf{0} \\ \mathbf{S}(j_1;\boldsymbol{\alpha}) & \mathbf{0} & \mathbf{0} & \mathbf{B}(\boldsymbol{\alpha}) \end{pmatrix} \begin{pmatrix} \delta\boldsymbol{\varphi} \\ \delta\boldsymbol{\psi}^{(1)} \\ \delta\boldsymbol{\psi}_1^{(2)}(j_1) \\ \delta\boldsymbol{\psi}_2^{(2)}(j_1) \end{pmatrix} \right\rangle_{(3)} \\ &= \left\langle \boldsymbol{\psi}^{(3)}(j_2;j_1;r,\boldsymbol{\Omega}), \begin{pmatrix} \mathbf{Q}^{(1)}(\boldsymbol{\alpha};\delta\boldsymbol{\alpha}) \\ \mathbf{Q}^{(2)}\left(\boldsymbol{\alpha},\boldsymbol{\psi}^{(1)};\delta\boldsymbol{\alpha}\right) \\ \mathbf{Q}_1^{(3)}\left[\boldsymbol{\alpha};\boldsymbol{\psi}^{(1)};\boldsymbol{\psi}_1^{(2)}(j_1);\delta\boldsymbol{\alpha}\right] \\ \mathbf{Q}_2^{(3)}\left[\boldsymbol{\alpha};\boldsymbol{\varphi};\boldsymbol{\psi}_2^{(2)}(j_1);\delta\boldsymbol{\alpha}\right] \end{pmatrix} \right\rangle_{(3)}.\end{aligned} \tag{7.56}$$

4. Use the customary definition of the adjoint operator in $\mathcal{H}_{(3)}$, endowed with the inner product defined in Eq. (7.55) to recast the left side of Eq. (7.56) into the following equivalent form:

$$\left\langle \psi^{(3)}(j_2;j_1;r,\mathbf{\Omega}), \begin{pmatrix} \mathbf{B}(\alpha) & \mathbf{0} & \mathbf{0} & \mathbf{0} \\ \mathbf{0} & \mathbf{A}(\alpha) & \mathbf{0} & \mathbf{0} \\ \mathbf{0} & \mathbf{S}(j_1;\alpha) & \mathbf{A}(\alpha) & \mathbf{0} \\ \mathbf{S}(j_1;\alpha) & \mathbf{0} & \mathbf{0} & \mathbf{B}(\alpha) \end{pmatrix} \begin{pmatrix} \delta\varphi \\ \delta\psi^{(1)} \\ \delta\psi_1^{(2)}(j_1) \\ \delta\psi_2^{(2)}(j_1) \end{pmatrix} \right\rangle_{(3)}$$
$$= \left\langle \begin{pmatrix} \delta\varphi \\ \delta\psi^{(1)} \\ \delta\psi_1^{(2)}(j_1) \\ \delta\psi_2^{(2)}(j_1) \end{pmatrix}^{\dagger}, \begin{pmatrix} \mathbf{A}(\alpha) & \mathbf{0} & \mathbf{0} & \mathbf{S}(j_1;\alpha) \\ \mathbf{0} & \mathbf{B}(\alpha) & \mathbf{S}(j_1;\alpha) & \mathbf{0} \\ \mathbf{0} & \mathbf{0} & \mathbf{B}(\alpha) & \mathbf{0} \\ \mathbf{0} & \mathbf{0} & \mathbf{0} & \mathbf{A}(\alpha) \end{pmatrix} \begin{pmatrix} \psi_1^{(3)}(j_2;j_1) \\ \psi_2^{(3)}(j_2;j_1) \\ \psi_3^{(3)}(j_2;j_1) \\ \psi_4^{(3)}(j_2;j_1) \end{pmatrix} \right\rangle_{(3)} \tag{7.57}$$
$$+P^{(3)}\left[\delta\varphi,\delta\psi^{(1)};\delta\psi^{(2)}(j_1);\psi^{(3)}(j_2;j_1)\right],$$

where $P^{(3)}\left[\delta\varphi,\delta\psi^{(1)};\delta\psi^{(2)}(j_1);\psi^{(3)}(j_2;j_1)\right]$ denotes the corresponding bilinear concomitant evaluated on the PERP sphere's outer boundary S_b at $r=r_s$.

5. Require the first term on the right side of Eq. (7.57) to represent the same function as the right side of Eq. (7.47). This requirement will be satisfied by requiring that the following relations be satisfied by the components of the function $\psi^{(3)}(j_2;j_1;r,\mathbf{\Omega}) \triangleq \left[\psi_1^{(3)}(j_2;j_1;r,\mathbf{\Omega}),...,\psi_4^{(3)}(j_2;j_1;r,\mathbf{\Omega})\right]^{\dagger} \in \mathcal{H}_{(3)}$:

$$\left\{ \begin{pmatrix} \mathbf{A}(\alpha) & \mathbf{0} & \mathbf{0} & \mathbf{S}(j_1;\alpha) \\ \mathbf{0} & \mathbf{B}(\alpha) & \mathbf{S}(j_1;\alpha) & \mathbf{0} \\ \mathbf{0} & \mathbf{0} & \mathbf{B}(\alpha) & \mathbf{0} \\ \mathbf{0} & \mathbf{0} & \mathbf{0} & \mathbf{A}(\alpha) \end{pmatrix} \begin{pmatrix} \psi_1^{(3)}(j_2;j_1;r,\mathbf{\Omega}) \\ \psi_2^{(3)}(j_2;j_1;r,\mathbf{\Omega}) \\ \psi_3^{(3)}(j_2;j_1;r,\mathbf{\Omega}) \\ \psi_4^{(3)}(j_2;j_1;r,\mathbf{\Omega}) \end{pmatrix} \right\}_{\alpha^0}$$
$$= -\left\{ \begin{pmatrix} \mathbf{S}(j_2;\alpha)\psi_1^{(2)}(j_1;r,\mathbf{\Omega}) \\ \mathbf{S}(j_2;\alpha)\psi_2^{(2)}(j_1;r,\mathbf{\Omega}) \\ \mathbf{S}(j_2;\alpha)\varphi(r,\mathbf{\Omega}) \\ \mathbf{S}(j_2;\alpha)\psi^{(1)}(r,\mathbf{\Omega}) \end{pmatrix} \right\}_{\alpha^0}, \; for \; j_1 = 1,...,J_{\sigma t}; \; j_2 = 1,...,j_1. \tag{7.58}$$

6. Use in Eq. (7.57) the boundary conditions provided in Eqs. (7.14), (7.34), (7.50), and (7.51). Furthermore, set to zero the remaining terms in the bilinear concomitant $P^{(3)}\left[\delta\varphi,\delta\psi^{(1)};\delta\psi^{(2)}(j_1);\psi^{(3)}(j_2;j_1)\right]$ in

Eq. (7.57) by requiring the components of the function $\psi^{(3)}(j_2;j_1;r,\mathbf{\Omega}) \triangleq \left[\psi_1^{(3)}(j_2;j_1;r,\mathbf{\Omega}),...,\psi_4^{(3)}(j_2;j_1;r,\mathbf{\Omega})\right]^{\dagger} \in \mathcal{H}_{(3)}$ to satisfy the following boundary conditions:

$$\begin{aligned}&\psi_i^{(3)}(j_2;j_1;r,\mathbf{\Omega}) = \mathbf{0}, r = r_s; \mathbf{\Omega}\bullet\mathbf{n} < 0; i = 2,3; j_1 = 1,...,J_{\sigma t};\\ &j_2 = 1,...,j_1; \psi_i^{(3)}(j_2;j_1;r,\mathbf{\Omega}) = \mathbf{0}, r = r_s; \mathbf{\Omega}\bullet\mathbf{n} > 0; i = 1,4;\\ &j_1 = 1,...,J_{\sigma t}; j_2 = 1,...,j_1.\end{aligned} \tag{7.59}$$

7. The system of equations comprising Eq. (7.58) together with the boundary conditions provided in Eq. (7.59) are called "the *3rd-Level Adjoint Sensitivity System* (3rd-LASS)" for the 3rd-level adjoint function $\psi^{(3)}(j_2;j_1;r,\mathbf{\Omega}) \in \mathcal{H}_{(3)}$. The reason for calling this system "3rd-*Level*" (as opposed to "3rd-*Order*") stems from the fact that this system does not involve any 2nd- and/or 3rd-order differentials or derivatives of the dependent variables (i.e., state functions) even though the solution of the 3rd-LASS is used for computing, efficiently and exactly, the 3rd-order sensitivities of the response with respect to the model parameters. In component form, the 3rd-LASS is solved in the following order for $j_1 = 1,...,J_{\sigma t}; j_2 = 1,...,j_1$:

$$\mathbf{A}(\boldsymbol{\alpha})\psi_4^{(3)}(j_2;j_1;r,\mathbf{\Omega}) = -\mathbf{S}(j_2;\boldsymbol{\alpha})\psi^{(1)}(r,\mathbf{\Omega}); \tag{7.60}$$

$$\psi_4^{(3)}(j_2;j_1;r,\mathbf{\Omega}) = \mathbf{0},\ r = r_s,\ \mathbf{\Omega}\bullet\mathbf{n} > 0; \tag{7.61}$$

$$\mathbf{B}(\boldsymbol{\alpha})\psi_3^{(3)}(j_2;j_1;r,\mathbf{\Omega}) = -\mathbf{S}(j_2;\boldsymbol{\alpha})\varphi(r,\mathbf{\Omega}); \tag{7.62}$$

$$\psi_3^{(3)}(j_2;j_1;r,\mathbf{\Omega}) = \mathbf{0},\ r = r_s,\ \mathbf{\Omega}\bullet\mathbf{n} < 0; \tag{7.63}$$

$$\begin{aligned}\mathbf{B}(\boldsymbol{\alpha})\psi_2^{(3)}(j_2;j_1;r,\mathbf{\Omega}) &= -\mathbf{S}(j_2;\boldsymbol{\alpha})\psi_2^{(2)}(j_1;r,\mathbf{\Omega})\\ &-\mathbf{S}(j_1;\boldsymbol{\alpha})\psi_3^{(3)}(j_2;j_1;r,\mathbf{\Omega});\end{aligned} \tag{7.64}$$

$$\psi_2^{(3)}(j_2;j_1;r,\mathbf{\Omega}) = \mathbf{0},\ r = r_s,\ \mathbf{\Omega}\bullet\mathbf{n} < 0; \tag{7.65}$$

$$\begin{aligned}\mathbf{A}(\boldsymbol{\alpha})\psi_1^{(3)}(j_2;j_1;r,\mathbf{\Omega}) &= -\mathbf{S}(j_2;\boldsymbol{\alpha})\psi_1^{(2)}(j_1;r,\mathbf{\Omega})\\ &-\mathbf{S}(j_1;\boldsymbol{\alpha})\psi_4^{(3)}(j_2;j_1;r,\mathbf{\Omega});\end{aligned} \tag{7.66}$$

$$\psi_1^{(3)}(j_2;j_1;r,\mathbf{\Omega}) = \mathbf{0},\ r = r_s,\ \mathbf{\Omega}\bullet\mathbf{n} > 0. \tag{7.67}$$

8. Using the equations underlying the 3rd-LASS in Eq. (7.47) leads to the following alternative expression for the differential $\delta\left\{\partial^2 L(\boldsymbol{\alpha},\varphi;\psi^{(1)})/\partial t_{j_1}\partial t_{j_2}\right\}_{\boldsymbol{\alpha}^0}$ in terms of the 3rd-level adjoint function $\psi^{(3)}(j_2;j_1;r,\mathbf{\Omega}) \in \mathcal{H}_{(3)}$:

$$\delta\left\{\frac{\partial^2 L\left(\boldsymbol{\alpha},\boldsymbol{\varphi};\boldsymbol{\psi}^{(1)};\boldsymbol{\psi}^{(2)}\right)}{\partial t_{j_2}\partial t_{j_1}}\right\}_{\boldsymbol{\alpha}^0}=\left\{\left\langle \boldsymbol{\psi}_1^{(3)}\left(j_2;j_1;r,\boldsymbol{\Omega}\right),\mathbf{Q}^{(1)}\left(\boldsymbol{\alpha};\delta\boldsymbol{\alpha}\right)\right\rangle_{(1)}\right\}_{\boldsymbol{\alpha}^0}$$
$$+\left\{\left\langle \boldsymbol{\psi}_2^{(3)}\left(j_2;j_1;r,\boldsymbol{\Omega}\right),\mathbf{Q}^{(2)}\left(\boldsymbol{\alpha},\boldsymbol{\psi}^{(1)};\delta\boldsymbol{\alpha}\right)\right\rangle_{(1)}\right\}_{\boldsymbol{\alpha}^0}$$
$$+\left\{\left\langle \boldsymbol{\psi}_3^{(3)}\left(j_2;j_1;r,\boldsymbol{\Omega}\right),\mathbf{Q}_1^{(3)}\left[\boldsymbol{\alpha};\boldsymbol{\psi}^{(1)};\boldsymbol{\psi}_1^{(2)}\left(j_1\right);\delta\boldsymbol{\alpha}\right]\right\rangle_{(1)}\right\}_{\boldsymbol{\alpha}^0} \tag{7.68}$$
$$+\left\{\left\langle \boldsymbol{\psi}_4^{(3)}\left(j_2;j_1;r,\boldsymbol{\Omega}\right),\mathbf{Q}_2^{(3)}\left[\boldsymbol{\alpha};\boldsymbol{\psi}^{(1)};\boldsymbol{\psi}_2^{(2)}\left(j_1\right);\delta\boldsymbol{\alpha}\right]\right\rangle_{(1)}\right\}_{\boldsymbol{\alpha}^0}$$
$$=\sum_{j_3=1}^{J_\alpha}\left\{\frac{\partial^3 L\left(\boldsymbol{\alpha},\boldsymbol{\varphi};\boldsymbol{\psi}^{(1)};\boldsymbol{\psi}^{(2)};\boldsymbol{\psi}^{(3)}\right)}{\partial\alpha_{j_3}\partial t_{j_2}\partial t_{j_1}}\delta\alpha_{j3}\right\}_{\boldsymbol{\alpha}^0},\ j_1=1,\ldots J_{\sigma t};\ j_2=1,\ldots,j_1.$$

It is important to note that the expression provided in Eq. (7.68) comprises *all* of the 3rd-order partial sensitivities that involve the group-averaged microscopic total cross-sections, in other words, the 3rd-order partial sensitivities $\partial^3 L\left(\boldsymbol{\alpha},\boldsymbol{\varphi};\boldsymbol{\psi}^{(1)};\boldsymbol{\psi}^{(2)};\boldsymbol{\psi}^{(3)}\right)/\partial\alpha_{j_3}\partial t_{j_2}\partial t_{j_1}$, $j_1=1,\ldots,J_{\sigma t};\ j_2=1,\ldots,j_1,\ j_3=1,\ldots,J_\alpha$, describes a total of $\left[180(180+1)/2\right]\times 7{,}477$ 3rd-order response sensitivities with respect to the microscopic total cross-sections (t_{j_1} and t_{j_2}) and also with respect to all of the other parameters (α_{j_3}).

In particular, identifying in Eq. (7.68) the quantities that multiply *only* the variations in the total microscopic cross-sections will reduce the expressions provided in Eqs. (7.52) and (7.53) as follows:

$$\mathbf{Q}_1^{(3)}\left[\boldsymbol{\alpha};\boldsymbol{\psi}^{(1)}\left(\mathbf{r},\boldsymbol{\Omega}\right);\boldsymbol{\psi}_1^{(2)}\left(j_1;r,\boldsymbol{\Omega}\right);\delta\boldsymbol{\alpha}\right]=-\sum_{j_3=1}^{J_{\sigma t}}\mathbf{S}\left(j_3;\boldsymbol{\alpha}\right)\boldsymbol{\psi}_1^{(2)}\left(j_1;r,\boldsymbol{\Omega}\right)\delta t_{j_3}, \tag{7.69}$$

$$\mathbf{Q}_2^{(3)}\left[\boldsymbol{\alpha};\boldsymbol{\varphi}\left(\mathbf{r},\boldsymbol{\Omega}\right);\boldsymbol{\psi}_2^{(2)}\left(j_1;r,\boldsymbol{\Omega}\right);\delta\boldsymbol{\alpha}\right]\triangleq-\sum_{j_3=1}^{J_{\sigma t}}\mathbf{S}\left(j_3;\boldsymbol{\alpha}\right)\boldsymbol{\psi}_2^{(2)}\left(j_1;r,\boldsymbol{\Omega}\right)\delta t_{j_3}, \tag{7.70}$$

where $\mathbf{S}\left(j_3;\boldsymbol{\alpha}\right)$ has the same definition as in Eq. (7.42), except that the index j_1, which appears in Eq. (7.42), is replaced by the index j_3.

Inserting the results from Eqs. (7.69) and (7.70) into Eq. (7.68) and identifying the specific quantity that multiplies the specific variation δt_{j_3} provides the following expression for the 3rd-order response sensitivities that involve *only* the group-averaged total microscopic cross-sections:

$$\left\{\frac{\partial^3 L\left(\boldsymbol{\alpha},\boldsymbol{\varphi};\boldsymbol{\psi}^{(1)};\boldsymbol{\psi}^{(2)};\boldsymbol{\psi}^{(3)}\right)}{\partial t_{j_3}\partial t_{j_2}\partial t_{j_1}}\right\}_{\boldsymbol{\alpha}^0}=-\left\{\left\langle \boldsymbol{\psi}_1^{(3)}\left(j_2;j_1;r,\boldsymbol{\Omega}\right),\mathbf{S}\left(j_3;\boldsymbol{\alpha}\right)\boldsymbol{\varphi}\left(r,\boldsymbol{\Omega}\right)\right\rangle_{(1)}\right\}_{\boldsymbol{\alpha}^0}$$

$$
\begin{aligned}
&-\left\{\left\langle \psi_2^{(3)}(j_2;j_1;r,\boldsymbol{\Omega}),\mathbf{S}(j_3;\alpha)\psi^{(1)}(r,\boldsymbol{\Omega})\right\rangle_{(1)}\right\}_{\alpha^0}\\
&-\left\{\left\langle \psi_3^{(3)}(j_2;j_1;r,\boldsymbol{\Omega}),\mathbf{S}(j_3;\alpha)\psi_1^{(2)}(j_1;r,\boldsymbol{\Omega})\right\rangle_{(1)}\right\}_{\alpha^0}\\
&-\left\{\left\langle \psi_4^{(3)}(j_2;j_1;r,\boldsymbol{\Omega}),\mathbf{S}(j_3;\alpha)\psi_2^{(2)}(j_1;r,\boldsymbol{\Omega})\right\rangle_{(1)}\right\}_{\alpha^0},\\
&for\ j_1=1,\dots J_{\sigma t};\ j_2=1,\dots,j_1;\ j_3=1,\dots,j_2.
\end{aligned}
\tag{7.71}
$$

The expression provided in Eq. (7.71) can be written in component form as follows:

$$
\begin{aligned}
&\frac{\partial^3 L\left(\alpha,\varphi;\psi^{(1)};\psi^{(2)};\psi^{(3)}\right)}{\partial t_{j_3}\partial t_{j_2}\partial t_{j_1}}=-\sum_{g=1}^{G}\int_0^{r_s}4\pi r^2 dr\int_{4\pi}d\boldsymbol{\Omega}\left[\psi_1^{(3),g}(j_2;j_1;r,\boldsymbol{\Omega})\varphi^g(r,\boldsymbol{\Omega})\right.\\
&+\psi_2^{(3),g}(j_2;j_1;r,\boldsymbol{\Omega})\psi^{(1),g}(r,\boldsymbol{\Omega})+\psi_3^{(3),g}(j_2;j_1;r,\boldsymbol{\Omega})\psi_1^{(2),g}(j_1;r,\boldsymbol{\Omega})\\
&\left.+\psi_4^{(3),g}(j_2;j_1;r,\boldsymbol{\Omega})\psi_2^{(2),g}(j_1;r,\boldsymbol{\Omega})\right]\frac{\partial\Sigma_t^g(\alpha)}{\partial t_{j_3}},\\
&for\ j_1=1,\dots JTX;\ j_2=1,\dots j_1;\ j_3=1,\dots,j_2.
\end{aligned}
\tag{7.72}
$$

All of the quantities appearing in Eq. (7.72) are to be evaluated at the nominal parameter values but the indication $\{\ \}_{\alpha^0}$ has been omitted in order to simplify the notation.

7.5 COMPUTATION OF 4TH-ORDER SENSITIVITIES $\partial^4 L(\alpha)/\partial t_{j_4}\partial t_{j_3}\partial t_{j_2}\partial t_{j_1}$

The 4th-order sensitivities of the PERP leakage response with respect to the group-averaged microscopic total cross-sections are obtained by determining the total 1st-order G-differential of the expression of the 3rd-order sensitivities provided in Eq. (7.71). Applying the definition of the G-differential to Eq. (7.71) yields the following expression:

$$
\begin{aligned}
&\delta\left\{\frac{\partial^3 L\left(\alpha,\varphi;\psi^{(1)};\psi^{(2)};\psi^{(3)}\right)}{\partial t_{j_3}\partial t_{j_2}\partial t_{j_1}}\right\}_{\alpha^0}\\
&=\frac{d}{d\varepsilon}\left\{\frac{\partial^3 L\left(\alpha+\varepsilon\delta\alpha,\varphi+\varepsilon\delta\varphi;\psi^{(1)}+\varepsilon\delta\psi^{(1)};\psi^{(2)}+\varepsilon\delta\psi^{(2)};\psi^{(3)}+\varepsilon\delta\psi^{(3)}\right)}{\partial t_{j_3}\partial t_{j_2}\partial t_{j_1}}\right\}_{\alpha^0}\\
&=\delta\left\{\frac{\partial^3 L\left(\alpha,\varphi;\psi^{(1)};\psi^{(2)};\psi^{(3)}\right)}{\partial t_{j_3}\partial t_{j_2}\partial t_{j_1}}\right\}_{dir}+\delta\left\{\frac{\partial^3 L\left(\alpha,\varphi;\psi^{(1)};\psi^{(2)};\psi^{(3)}\right)}{\partial t_{j_3}\partial t_{j_2}\partial t_{j_1}}\right\}_{ind},\\
&for\ j_1=1,\dots J_{\sigma t};j_2=1,\dots j_1;\ j_3=1,\dots,j_2.
\end{aligned}
\tag{7.73}
$$

In Eq. (7.73), the "direct-effect" term is defined as follows:

$$\delta\left\{\frac{\partial^3 L\left(\boldsymbol{\alpha},\boldsymbol{\varphi};\boldsymbol{\psi}^{(1)};\boldsymbol{\psi}^{(2)};\boldsymbol{\psi}^{(3)}\right)}{\partial t_{j_3}\partial t_{j_2}\partial t_{j_1}}\right\}_{dir} \triangleq -\left\{\left\langle \boldsymbol{\psi}_1^{(3)}\left(j_2;j_1;r,\boldsymbol{\Omega}\right),\left[\frac{\partial \mathbf{S}\left(j_3;\boldsymbol{\alpha}\right)}{\partial \boldsymbol{\alpha}}\delta\boldsymbol{\alpha}\right]\boldsymbol{\varphi}\left(r,\boldsymbol{\Omega}\right)\right\rangle_{(1)}\right\}_{\boldsymbol{\alpha}^0}$$
$$-\left\{\left\langle \boldsymbol{\psi}_2^{(3)}\left(j_2;j_1;r,\boldsymbol{\Omega}\right),\left[\frac{\partial \mathbf{S}\left(j_3;\boldsymbol{\alpha}\right)}{\partial \boldsymbol{\alpha}}\delta\boldsymbol{\alpha}\right]\boldsymbol{\psi}^{(1)}\left(r,\boldsymbol{\Omega}\right)\right\rangle_{(1)}\right\}_{\boldsymbol{\alpha}^0}$$
$$-\left\{\left\langle \boldsymbol{\psi}_3^{(3)}\left(j_2;j_1;r,\boldsymbol{\Omega}\right),\left[\frac{\partial \mathbf{S}\left(j_3;\boldsymbol{\alpha}\right)}{\partial \boldsymbol{\alpha}}\delta\boldsymbol{\alpha}\right]\boldsymbol{\psi}_1^{(2)}\left(j_1;r,\boldsymbol{\Omega}\right)\right\rangle_{(1)}\right\}_{\boldsymbol{\alpha}^0} \tag{7.74}$$
$$-\left\{\left\langle \boldsymbol{\psi}_4^{(3)}\left(j_2;j_1;r,\boldsymbol{\Omega}\right),\left[\frac{\partial \mathbf{S}\left(j_3;\boldsymbol{\alpha}\right)}{\partial \boldsymbol{\alpha}}\delta\boldsymbol{\alpha}\right]\boldsymbol{\psi}_2^{(2)}\left(j_1;r,\boldsymbol{\Omega}\right)\right\rangle_{(1)}\right\}_{\boldsymbol{\alpha}^0},$$

while the "indirect-effect" term is defined as follows:

$$\delta\left\{\frac{\partial^3 L\left(\boldsymbol{\alpha},\boldsymbol{\varphi};\boldsymbol{\psi}^{(1)};\boldsymbol{\psi}^{(2)};\boldsymbol{\psi}^{(3)}\right)}{\partial t_{j_3}\partial t_{j_2}\partial t_{j_1}}\right\}_{ind} \triangleq \left\{\sum_{i=1}^{8}\left\langle \mathbf{v}_i^{(4)}\left(j_2;j_1;r,\boldsymbol{\Omega}\right),\mathbf{r}_i^{(4)}\left(j_3;j_2;j_1;r,\boldsymbol{\Omega}\right)\right\rangle_{(1)}\right\}_{\boldsymbol{\alpha}^0}. \tag{7.75}$$

The quantities appearing on the right side of Eq. (7.75) are defined as follows:

$$\mathbf{v}^{(4)}\left(j_2;j_1;r,\boldsymbol{\Omega}\right) \triangleq \left[\mathbf{v}_1^{(4)}\left(j_2;j_1;r,\boldsymbol{\Omega}\right),...,\mathbf{v}_8^{(4)}\left(j_2;j_1;r,\boldsymbol{\Omega}\right)\right]^\dagger$$
$$\triangleq \left[\delta\boldsymbol{\varphi},\delta\boldsymbol{\psi}^{(1)},\delta\boldsymbol{\psi}_1^{(2)}\left(j_1\right),\delta\boldsymbol{\psi}_2^{(2)}\left(j_1\right),\delta\boldsymbol{\psi}_1^{(3)}\left(j_2;j_1\right),\delta\boldsymbol{\psi}_2^{(3)}\left(j_2;j_1\right),\right. \tag{7.76}$$
$$\left.\delta\boldsymbol{\psi}_3^{(3)}\left(j_2;j_1\right),\delta\boldsymbol{\psi}_4^{(3)}\left(j_2;j_1\right)\right]^\dagger,$$

$$\mathbf{r}^{(4)}\left(j_3;j_2;j_1;r,\boldsymbol{\Omega}\right) \triangleq \left[\mathbf{r}_1^{(4)}\left(j_3;j_2;j_1;r,\boldsymbol{\Omega}\right),...,\mathbf{r}_8^{(4)}\left(j_3;j_2;j_1;r,\boldsymbol{\Omega}\right)\right]^\dagger, \tag{7.77}$$

with components defined as follows:

$$\mathbf{r}_1^{(4)}\left(j_3;j_2;j_1;r,\boldsymbol{\Omega}\right) \triangleq -\mathbf{S}\left(j_3;\boldsymbol{\alpha}\right)\boldsymbol{\psi}_1^{(3)}\left(j_2;j_1;r,\boldsymbol{\Omega}\right), \tag{7.78}$$

$$\mathbf{r}_2^{(4)}\left(j_3;j_2;j_1;r,\boldsymbol{\Omega}\right) \triangleq -\mathbf{S}\left(j_3;\boldsymbol{\alpha}\right)\boldsymbol{\psi}_2^{(3)}\left(j_2;j_1;r,\boldsymbol{\Omega}\right), \tag{7.79}$$

$$\mathbf{r}_3^{(4)}\left(j_3;j_2;j_1;r,\boldsymbol{\Omega}\right) \triangleq -\mathbf{S}\left(j_3;\boldsymbol{\alpha}\right)\boldsymbol{\psi}_3^{(3)}\left(j_2;j_1;r,\boldsymbol{\Omega}\right), \tag{7.80}$$

$$\mathbf{r}_4^{(4)}\left(j_3;j_2;j_1;r,\boldsymbol{\Omega}\right) \triangleq -\mathbf{S}\left(j_3;\boldsymbol{\alpha}\right)\boldsymbol{\psi}_4^{(3)}\left(j_2;j_1;r,\boldsymbol{\Omega}\right), \tag{7.81}$$

$$\mathbf{r}_5^{(4)}(j_3;j_2;j_1;r,\mathbf{\Omega}) \triangleq -\mathbf{S}(j_3;\boldsymbol{\alpha})\varphi(r,\mathbf{\Omega}), \tag{7.82}$$

$$\mathbf{r}_6^{(4)}(j_3;j_2;j_1;r,\mathbf{\Omega}) \triangleq -\mathbf{S}(j_3;\boldsymbol{\alpha})\psi^{(1)}(r,\mathbf{\Omega}), \tag{7.83}$$

$$\mathbf{r}_7^{(4)}(j_3;j_2;j_1;r,\mathbf{\Omega}) \triangleq -\mathbf{S}(j_3;\boldsymbol{\alpha})\psi_1^{(2)}(j_1;r,\mathbf{\Omega}), \tag{7.84}$$

$$\mathbf{r}_8^{(4)}(j_3;j_2;j_1;r,\mathbf{\Omega}) \triangleq -\mathbf{S}(j_3;\boldsymbol{\alpha})\psi_2^{(2)}(j_1;r,\mathbf{\Omega}). \tag{7.85}$$

The components of the vector $\mathbf{v}^{(4)}(j_2;j_1;r,\mathbf{\Omega})$ defined in Eq. (7.76) are the solutions of the following 4th-Level Variational Sensitivity System (4th-LVSS):

$$\mathbf{V}^{(4)}(j_2;j_1;r,\mathbf{\Omega})\mathbf{v}^{(4)}(j_2;j_1;r,\mathbf{\Omega}) = \mathbf{q}^{(4)}(j_2;j_1;r,\mathbf{\Omega}), \tag{7.86}$$

satisfying the following boundary conditions:

$$\delta\varphi(r,\mathbf{\Omega}) = \mathbf{0}, \ \ r = r_s, \ \ \mathbf{\Omega}\bullet\mathbf{n} < 0, \tag{7.87}$$

$$\delta\psi^{(1)}(r,\mathbf{\Omega}) = \mathbf{0}, \ \ r = r_s, \ \ \mathbf{\Omega}\bullet\mathbf{n} > 0, \tag{7.88}$$

$$\delta\psi_1^{(2)}(j_1;r,\mathbf{\Omega}) = \mathbf{0}, \ r = r_s, \ \mathbf{\Omega}\bullet\mathbf{n} > 0; \ j_1 = 1,\dots,J_{\sigma t}, \tag{7.89}$$

$$\delta\psi_2^{(2)}(j_1;r,\mathbf{\Omega}) = \mathbf{0}, \ \ r = r_s, \ \ \mathbf{\Omega}\bullet\mathbf{n} < 0; \ j_1 = 1,\dots,J_{\sigma t}, \tag{7.90}$$

$$\begin{aligned} &\delta\psi_i^{(3)}(j_2;j_1;r,\mathbf{\Omega}) = \mathbf{0}, \ r = r_s; \ \mathbf{\Omega}\bullet\mathbf{n} > 0; \\ &\textit{for } i = 1,4; \ j_1 = 1,\dots,J_{\sigma t}; \ j_2 = 1,\dots,j_1; \end{aligned} \tag{7.91}$$

$$\begin{aligned} &\delta\psi_i^{(3)}(j_2;j_1;r,\mathbf{\Omega}) = \mathbf{0}, \ r = r_s; \ \mathbf{\Omega}\bullet\mathbf{n} < 0; \\ &\textit{for } \ i = 2,3; \ j_1 = 1,\dots,J_{\sigma t}; \ j_2 = 1,\dots,j_1. \end{aligned} \tag{7.92}$$

In Eq. (7.86), the block matrix $\mathbf{V}^{(4)}(j_2;j_1;r,\mathbf{\Omega})$ and the block vector $\mathbf{q}^{(4)}(j_2;j_1;r,\mathbf{\Omega})$ are defined as follows:

$$\mathbf{V}^{(4)}(j_2;j_1;r,\mathbf{\Omega}) \triangleq \begin{pmatrix} \mathbf{B} & \bullet & \bullet & \bullet & \bullet & \bullet & \bullet & \bullet \\ \bullet & \mathbf{A} & \bullet & \bullet & \bullet & \bullet & \bullet & \bullet \\ \bullet & \mathbf{S}(j_1) & \mathbf{A} & \bullet & \bullet & \bullet & \bullet & \bullet \\ \mathbf{S}(j_1) & \bullet & \bullet & \mathbf{B} & \bullet & \bullet & \bullet & \bullet \\ \bullet & \bullet & \mathbf{S}(j_2) & \bullet & \mathbf{A} & \bullet & \bullet & \mathbf{S}(j_1) \\ \bullet & \bullet & \bullet & \mathbf{S}(j_2) & \bullet & \mathbf{B} & \mathbf{S}(j_1) & \bullet \\ \mathbf{S}(j_2) & \bullet & \bullet & \bullet & \bullet & \bullet & \mathbf{B} & \bullet \\ \bullet & \mathbf{S}(j_2) & \bullet & \bullet & \bullet & \bullet & \bullet & \mathbf{A} \end{pmatrix}, \tag{7.93}$$

$$\mathbf{q}^{(4)}(j_2;j_1;r,\boldsymbol{\Omega}) \triangleq \Big\{\mathbf{Q}^{(1)}(\delta\boldsymbol{\alpha}),\mathbf{Q}^{(2)}\big(\boldsymbol{\psi}^{(1)};\delta\boldsymbol{\alpha}\big),\mathbf{Q}_1^{(3)}\big[\boldsymbol{\psi}^{(1)};\boldsymbol{\psi}_1^{(2)}(j_1);\delta\boldsymbol{\alpha}\big],$$
$$\mathbf{Q}_2^{(3)}\big[\boldsymbol{\varphi};\boldsymbol{\psi}_2^{(2)}(j_1);\delta\boldsymbol{\alpha}\big],\mathbf{Q}_1^{(4)}\big[\boldsymbol{\psi}^{(1)};\boldsymbol{\psi}^{(2)}(j_1);\boldsymbol{\psi}^{(3)}(j_2;j_1);\delta\boldsymbol{\alpha}\big],$$
$$\mathbf{Q}_2^{(4)}\big[\boldsymbol{\psi}^{(1)};\boldsymbol{\psi}^{(2)}(j_1);\boldsymbol{\psi}^{(3)}(j_2;j_1);\delta\boldsymbol{\alpha}\big],\mathbf{Q}_3^{(4)}\big[\boldsymbol{\psi}^{(1)};\boldsymbol{\psi}^{(2)}(j_1);\boldsymbol{\psi}^{(3)}(j_2;j_1);\delta\boldsymbol{\alpha}\big], \tag{7.94}$$
$$\mathbf{Q}_4^{(4)}\big[\boldsymbol{\psi}^{(1)};\boldsymbol{\psi}^{(2)}(j_1);\boldsymbol{\psi}^{(3)}(j_2;j_1);\delta\boldsymbol{\alpha}\big]\Big\}^{\dagger}.$$

In order to enhance the visibility of the structure of the 8×8 matrix $\mathbf{V}^{(4)}(j_2;j_1;r,\boldsymbol{\Omega})$ defined in Eq. (7.93), the zero-valued block matrix components of this matrix have been represented by dots, while the dependence of the non-zero matrices on the vector of parameters $\boldsymbol{\alpha}$ and on the independent variables have been suppressed. In Eq. (7.94), the source terms $\mathbf{Q}^{(1)}(\delta\boldsymbol{\alpha})$, $\mathbf{Q}^{(2)}\big(\boldsymbol{\psi}^{(1)};\delta\boldsymbol{\alpha}\big)$, $\mathbf{Q}_i^{(3)}\big[\boldsymbol{\alpha};\boldsymbol{\psi}^{(1)}(r,\boldsymbol{\Omega});\boldsymbol{\psi}_i^{(2)}(j_1;r,\boldsymbol{\Omega});\delta\boldsymbol{\alpha}\big]$, $i=1,2,$ are as defined in Eqs. (7.15), (7.35), (7.52), and (7.53), respectively, while the source terms $\mathbf{Q}_i^{(4)}\big[\boldsymbol{\psi}^{(1)};\boldsymbol{\psi}^{(2)}(j_1);\boldsymbol{\psi}^{(3)}(j_2;j_1);\delta\boldsymbol{\alpha}\big]$, $i=1,2,3,4,$ are defined as follows:

$$\mathbf{Q}_1^{(4)}\big[\boldsymbol{\psi}^{(1)};\boldsymbol{\psi}^{(2)}(j_1);\boldsymbol{\psi}^{(3)}(j_2;j_1);\delta\boldsymbol{\alpha}\big] \triangleq -\left[\frac{\partial \mathbf{A}(\boldsymbol{\alpha})}{\partial\boldsymbol{\alpha}}\delta\boldsymbol{\alpha}\right]\boldsymbol{\psi}_1^{(3)}(j_2;j_1)$$
$$-\left[\frac{\partial \mathbf{S}(j_1;\boldsymbol{\alpha})}{\partial\boldsymbol{\alpha}}\delta\boldsymbol{\alpha}\right]\boldsymbol{\psi}_4^{(3)}(j_2;j_1)-\left[\frac{\partial \mathbf{S}(j_2;\boldsymbol{\alpha})}{\partial\boldsymbol{\alpha}}\delta\boldsymbol{\alpha}\right]\boldsymbol{\psi}_1^{(2)}(j_1;r,\boldsymbol{\Omega}), \tag{7.95}$$

$$\mathbf{Q}_2^{(4)}\big[\boldsymbol{\psi}^{(1)};\boldsymbol{\psi}^{(2)}(j_1);\boldsymbol{\psi}^{(3)}(j_2;j_1);\delta\boldsymbol{\alpha}\big] \triangleq -\left[\frac{\partial \mathbf{B}(\boldsymbol{\alpha})}{\partial\boldsymbol{\alpha}}\delta\boldsymbol{\alpha}\right]\boldsymbol{\psi}_2^{(3)}(j_2;j_1)$$
$$-\left[\frac{\partial \mathbf{S}(j_1;\boldsymbol{\alpha})}{\partial\boldsymbol{\alpha}}\delta\boldsymbol{\alpha}\right]\boldsymbol{\psi}_3^{(3)}(j_2;j_1)-\left[\frac{\partial \mathbf{S}(j_2;\boldsymbol{\alpha})}{\partial\boldsymbol{\alpha}}\delta\boldsymbol{\alpha}\right]\boldsymbol{\psi}_2^{(2)}(j_1;r,\boldsymbol{\Omega}), \tag{7.96}$$

$$\mathbf{Q}_3^{(4)}\big[\boldsymbol{\psi}^{(1)};\boldsymbol{\psi}^{(2)}(j_1);\boldsymbol{\psi}^{(3)}(j_2;j_1);\delta\boldsymbol{\alpha}\big]$$
$$\triangleq -\left[\frac{\partial \mathbf{B}(\boldsymbol{\alpha})}{\partial\boldsymbol{\alpha}}\delta\boldsymbol{\alpha}\right]\boldsymbol{\psi}_3^{(3)}(j_2;j_1)-\left[\frac{\partial \mathbf{S}(j_2;\boldsymbol{\alpha})}{\partial\boldsymbol{\alpha}}\delta\boldsymbol{\alpha}\right]\boldsymbol{\varphi}(r,\boldsymbol{\Omega}), \tag{7.97}$$

$$\mathbf{Q}_4^{(4)}\big[\boldsymbol{\psi}^{(1)};\boldsymbol{\psi}^{(2)}(j_1);\boldsymbol{\psi}^{(3)}(j_2;j_1);\delta\boldsymbol{\alpha}\big]$$
$$\triangleq -\left[\frac{\partial \mathbf{A}(\boldsymbol{\alpha})}{\partial\boldsymbol{\alpha}}\delta\boldsymbol{\alpha}\right]\boldsymbol{\psi}_4^{(3)}(j_2;j_1)-\left[\frac{\partial \mathbf{S}(j_2;\boldsymbol{\alpha})}{\partial\boldsymbol{\alpha}}\delta\boldsymbol{\alpha}\right]\boldsymbol{\psi}^{(1)}(r,\boldsymbol{\Omega}). \tag{7.98}$$

When only the total microscopic cross-sections are considered, the source terms $\mathbf{Q}^{(1)}(\delta\boldsymbol{\alpha})$, $\mathbf{Q}^{(2)}\big(\boldsymbol{\psi}^{(1)};\delta\boldsymbol{\alpha}\big)$, $\mathbf{Q}_i^{(3)}\big[\boldsymbol{\alpha};\boldsymbol{\psi}^{(1)}(r,\boldsymbol{\Omega});\boldsymbol{\psi}_i^{(2)}(j_1;r,\boldsymbol{\Omega});\delta\boldsymbol{\alpha}\big]$, $i=1,2,$ and $\mathbf{Q}_i^{(4)}\big[\boldsymbol{\psi}^{(1)};\boldsymbol{\psi}^{(2)}(j_1);\boldsymbol{\psi}^{(3)}(j_2;j_1);\delta\boldsymbol{\alpha}\big]$, $i=1,2,3,4,$ reduce to the following expressions:

$$\mathbf{Q}^{(1)}(\delta\boldsymbol{\alpha}) \triangleq -\sum_{j_4=1}^{J_{\sigma t}} \mathbf{S}(j_4;\boldsymbol{\alpha})\varphi(r,\boldsymbol{\Omega})\delta t_{j_4}, \tag{7.99}$$

$$\mathbf{Q}^{(2)}\left(\boldsymbol{\psi}^{(1)};\delta\boldsymbol{\alpha}\right) \triangleq -\sum_{j_4=1}^{J_{\sigma t}} \mathbf{S}(j_4;\boldsymbol{\alpha})\boldsymbol{\psi}^{(1)}(r,\boldsymbol{\Omega})\delta t_{j_4}, \tag{7.100}$$

$$\mathbf{Q}_1^{(3)}\left[\boldsymbol{\alpha};\boldsymbol{\psi}^{(1)}(r,\boldsymbol{\Omega});\boldsymbol{\psi}_1^{(2)}(j_1;r,\boldsymbol{\Omega});\delta\boldsymbol{\alpha}\right] \triangleq -\sum_{j_4=1}^{J_{\sigma t}} \mathbf{S}(j_4;\boldsymbol{\alpha})\boldsymbol{\psi}_1^{(2)}(j_1;r,\boldsymbol{\Omega})\delta t_{j_4}, \tag{7.101}$$

$$\mathbf{Q}_2^{(3)}\left[\boldsymbol{\alpha};\varphi(r,\boldsymbol{\Omega});\boldsymbol{\psi}_2^{(2)}(j_1;r,\boldsymbol{\Omega});\delta\boldsymbol{\alpha}\right] \triangleq -\sum_{j_4=1}^{J_{\sigma t}} \mathbf{S}(j_4;\boldsymbol{\alpha})\boldsymbol{\psi}_2^{(2)}(j_1;r,\boldsymbol{\Omega})\delta t_{j_4}, \tag{7.102}$$

$$\mathbf{Q}_1^{(4)}\left[\boldsymbol{\psi}^{(1)};\boldsymbol{\psi}^{(2)}(j_1);\boldsymbol{\psi}^{(3)}(j_2;j_1);\delta\boldsymbol{\alpha}\right] \triangleq -\sum_{j_4=1}^{J_{\sigma t}} \mathbf{S}(j_4;\boldsymbol{\alpha})\boldsymbol{\psi}_1^{(3)}(j_2;j_1)\delta t_{j_4}, \tag{7.103}$$

$$\mathbf{Q}_2^{(4)}\left[\boldsymbol{\psi}^{(1)};\boldsymbol{\psi}^{(2)}(j_1);\boldsymbol{\psi}^{(3)}(j_2;j_1);\delta\boldsymbol{\alpha}\right] \triangleq -\sum_{j_4=1}^{J_{\sigma t}} \mathbf{S}(j_4;\boldsymbol{\alpha})\boldsymbol{\psi}_2^{(3)}(j_2;j_1)\delta t_{j_4}, \tag{7.104}$$

$$\mathbf{Q}_3^{(4)}\left[\boldsymbol{\psi}^{(1)};\boldsymbol{\psi}^{(2)}(j_1);\boldsymbol{\psi}^{(3)}(j_2;j_1);\delta\boldsymbol{\alpha}\right] \triangleq -\sum_{j_4=1}^{J_{\sigma t}} \mathbf{S}(j_4;\boldsymbol{\alpha})\boldsymbol{\psi}_3^{(3)}(j_2;j_1)\delta t_{j_4}, \tag{7.105}$$

$$\mathbf{Q}_4^{(4)}\left[\boldsymbol{\psi}^{(1)};\boldsymbol{\psi}^{(2)}(j_1);\boldsymbol{\psi}^{(3)}(j_2;j_1);\delta\boldsymbol{\alpha}\right] \triangleq -\sum_{j_4=1}^{J_{\sigma t}} \mathbf{S}(j_4;\boldsymbol{\alpha})\boldsymbol{\psi}_4^{(3)}(j_2;j_1)\delta t_{j_4}. \tag{7.106}$$

It is impractical to solve the 4th-LVSS, as will be demonstrated by discussing (at the conclusion of this subsection) the CPU time that would be needed to solve repeatedly the 4th-LVSS.

To avoid the need for solving the 4th-LVSS, the appearance of the vector $\mathbf{v}^{(4)}(j_2;j_1;r,\boldsymbol{\Omega})$ will be eliminated from appearing in the expression of the indirect-effect term expressed by the right side of (7.75). This elimination will be accomplished by recasting the inner products that involve $\mathbf{v}^{(4)}(j_2;j_1;r,\boldsymbol{\Omega})$ in terms of equivalent inner products, which will involve the solutions of a 4th-Level Adjoint Sensitivity System (4th-LASS) instead of the components of $\mathbf{v}^{(4)}(j_2;j_1;r,\boldsymbol{\Omega})$. Furthermore, the solutions of the 4th-LASS will not depend on the parameter variations. The 4th-LASS will be constructed by applying the 4th-CASAM-L (Cacuci, 2022) comprising the following sequence of steps.

1. Consider a Hilbert space, denoted as $\mathcal{H}_{(4)}$, comprising elements denoted generically as $\mathbf{u}^{(4)}(r,\boldsymbol{\Omega}) \in \mathcal{H}_{(4)}$, which have an "*eight-component block-vector*" structure of the form $\mathbf{u}^{(4)}(r,\boldsymbol{\Omega}) \triangleq \left[\mathbf{u}_1^{(4)}(r,\boldsymbol{\Omega}),...,\mathbf{u}_8^{(4)}(r,\boldsymbol{\Omega})\right]^\dagger$, with each of the eight vector components having the following structure: $\mathbf{u}_i^{(4)}(r,\boldsymbol{\Omega}) \triangleq \left[u_i^{(4),1}(r,\boldsymbol{\Omega}),...,u_i^{(4),g}(r,\boldsymbol{\Omega}),...,u_i^{(4),G}(r,\boldsymbol{\Omega})\right]^\dagger$, $i = 1,...,8$.

2. Define the inner product in $\mathcal{H}_{(4)}$, which will be denoted as $\left\langle \mathbf{u}^{(4)}(r,\boldsymbol{\Omega}), \mathbf{w}^{(4)}(r,\boldsymbol{\Omega}) \right\rangle_{(4)}$ of two elements $\mathbf{u}^{(4)}(r,\boldsymbol{\Omega}) \in \mathcal{H}_{(4)}$ and $\mathbf{w}^{(4)}(r,\boldsymbol{\Omega}) \in \mathcal{H}_{(4)}$ in $\mathcal{H}_{(4)}$ having the structure $\mathbf{w}^{(4)}(r,\boldsymbol{\Omega}) \triangleq \left[\mathbf{w}_1^{(4)}(r,\boldsymbol{\Omega}),...,\mathbf{w}_8^{(4)}(r,\boldsymbol{\Omega})\right]^\dagger$, $\mathbf{w}_i^{(4)}(r,\boldsymbol{\Omega}) \triangleq \left[w_i^{(4),1}(r,\boldsymbol{\Omega}),...,w_i^{(4),G}(r,\boldsymbol{\Omega})\right]^\dagger$, $i=1,...,8$, as follows:

$$\begin{aligned}\left\langle \mathbf{u}^{(4)}(r,\boldsymbol{\Omega}), \mathbf{w}^{(4)}(r,\boldsymbol{\Omega}) \right\rangle_{(4)} &\triangleq \sum_{i=1}^{8} \left\langle \mathbf{u}_i^{(4)}(r,\boldsymbol{\Omega}), \mathbf{w}_i^{(4)}(r,\boldsymbol{\Omega}) \right\rangle_{(1)} \\ &= \sum_{i=1}^{8} \sum_{g=1}^{G} \int_0^{r_s} 4\pi r^2 dr \int_{4\pi} d\boldsymbol{\Omega}\, u_i^{(4),g}(r,\boldsymbol{\Omega})\, w_i^{(4),g}(r,\boldsymbol{\Omega}).\end{aligned} \tag{7.107}$$

3. Apply the definition provided in Eq. (7.107) to form the inner product of a vector $\boldsymbol{\psi}^{(4)}(j_3;j_2;j_1;r,\boldsymbol{\Omega}) \triangleq \left[\psi_1^{(4)}(j_3;j_2;j_1;r,\boldsymbol{\Omega}),...,\psi_8^{(4)}(j_3;j_2;j_1;r,\boldsymbol{\Omega})\right]^\dagger \in \mathcal{H}_{(4)}$ with Eq. (7.86) to obtain the following relation, evaluated at $\boldsymbol{\alpha}^0$:

$$\begin{aligned}&\left\{ \left\langle \boldsymbol{\psi}^{(4)}(j_3;j_2;j_1;r,\boldsymbol{\Omega}), \mathbf{V}^{(4)}(j_2;j_1;r,\boldsymbol{\Omega})\, \mathbf{v}^{(4)}(j_2;j_1;r,\boldsymbol{\Omega}) \right\rangle_{(4)} \right\}_{\boldsymbol{\alpha}^0} \\ &= \left\{ \left\langle \boldsymbol{\psi}^{(4)}(j_3;j_2;j_1;r,\boldsymbol{\Omega}),\ \mathbf{q}^{(4)}(j_2;j_1;r,\boldsymbol{\Omega}) \right\rangle_{(4)} \right\}_{\boldsymbol{\alpha}^0}.\end{aligned} \tag{7.108}$$

4. Use the customary definition of the adjoint operator in $\mathcal{H}_{(4)}$, endowed with the inner product defined in Eq. (7.107) to recast the left side of Eq. (7.108) into the following equivalent form:

$$\begin{aligned}&\left\{ \left\langle \boldsymbol{\psi}^{(4)}(j_3;j_2;j_1;r,\boldsymbol{\Omega}), \mathbf{V}^{(4)}(j_2;j_1;r,\boldsymbol{\Omega})\, \mathbf{v}^{(4)}(j_2;j_1;r,\boldsymbol{\Omega}) \right\rangle_{(4)} \right\}_{\boldsymbol{\alpha}^0} \\ &= \left\{ \left\langle \mathbf{v}^{(4)}(j_2;j_1;r,\boldsymbol{\Omega}), \left[\mathbf{V}^{(4)}(j_2;j_1;r,\boldsymbol{\Omega})\right]^* \boldsymbol{\psi}^{(4)}(j_3;j_2;j_1;r,\boldsymbol{\Omega}) \right\rangle_{(4)} \right\}_{\boldsymbol{\alpha}^0} \\ &+ P^{(4)}\left[\delta\boldsymbol{\varphi}, \delta\boldsymbol{\psi}^{(1)}; \delta\boldsymbol{\psi}^{(2)}(j_1); \delta\boldsymbol{\psi}^{(3)}(j_2;j_1); \boldsymbol{\psi}^{(4)}(j_3;j_2;j_1)\right],\end{aligned} \tag{7.109}$$

where $P^{(4)}\left[\delta\boldsymbol{\varphi}, \delta\boldsymbol{\psi}^{(1)}; \delta\boldsymbol{\psi}^{(2)}(j_1); \delta\boldsymbol{\psi}^{(3)}(j_2;j_1); \boldsymbol{\psi}^{(4)}(j_3;j_2;j_1)\right]$ denotes the corresponding bilinear concomitant evaluated on the PERP sphere's outer boundary S_b at $r=r_s$ and where the block matrix operator $\left[\mathbf{V}^{(4)}(j_2;j_1;r,\boldsymbol{\Omega})\right]^*$ is the formal adjoint of the block matrix operator $\mathbf{V}^{(4)}(j_2;j_1;r,\boldsymbol{\Omega})$, having the following structure:

$$\left[\mathbf{V}^{(4)}\left(j_2;j_1;r,\boldsymbol{\Omega}\right)\right]^* \triangleq \begin{pmatrix} \mathbf{A} & \bullet & \bullet & \mathbf{S}(j_1) & \bullet & \bullet & \mathbf{S}(j_2) & \bullet \\ \bullet & \mathbf{B} & \mathbf{S}(j_1) & \bullet & \bullet & \bullet & \bullet & \mathbf{S}(j_2) \\ \bullet & \bullet & \mathbf{B} & \bullet & \mathbf{S}(j_2) & \bullet & \bullet & \bullet \\ \bullet & \bullet & \bullet & \mathbf{A} & \bullet & \mathbf{S}(j_2) & \bullet & \bullet \\ \bullet & \bullet & \bullet & \bullet & \mathbf{B} & \bullet & \bullet & \bullet \\ \bullet & \bullet & \bullet & \bullet & \bullet & \mathbf{A} & \bullet & \bullet \\ \bullet & \bullet & \bullet & \bullet & \bullet & \mathbf{S}(j_1) & \mathbf{A} & \bullet \\ \bullet & \bullet & \bullet & \bullet & \mathbf{S}(j_1) & \bullet & \bullet & \mathbf{B} \end{pmatrix}. \quad (7.110)$$

5. Require the first term on the right side of Eq. (7.109) to represent the same function as the indirect-effect term defined in Eq. (7.75). This requirement will be accomplished by requiring that the following relations be satisfied by the components of the function $\boldsymbol{\psi}^{(4)}\left(j_3;j_2;j_1;r,\boldsymbol{\Omega}\right)$:

$$\left\{\left[\mathbf{V}^{(4)}\left(j_2;j_1;r,\boldsymbol{\Omega}\right)\right]^* \boldsymbol{\psi}^{(4)}\left(j_3;j_2;j_1;r,\boldsymbol{\Omega}\right)\right\}_{\alpha^0} = \mathbf{r}^{(4)}\left(j_3;j_2;j_1;r,\boldsymbol{\Omega}\right), \quad (7.111)$$
$$for\ j_1 = 1,...,J_{\sigma t};\ j_2 = 1,...,j_1;\ j_3 = 1,...,j_2.$$

6. Use in Eq. (7.109) the boundary conditions shown in Eqs. (7.87)–(7.92). Furthermore, set to zero the remaining terms in the bilinear concomitant $P^{(4)}\left[\delta\boldsymbol{\varphi},\delta\boldsymbol{\psi}^{(1)};\delta\boldsymbol{\psi}^{(2)}(j_1);\delta\boldsymbol{\psi}^{(3)}(j_2;j_1);\boldsymbol{\psi}^{(4)}(j_3;j_2;j_1)\right]$ in Eq. (7.109) by requiring the components of the function $\boldsymbol{\psi}^{(4)}\left(j_3;j_2;j_1;r,\boldsymbol{\Omega}\right)$ to satisfy the following boundary conditions:

$$\begin{aligned} &\psi_i^{(4)}\left(j_3;j_2;j_1;r,\boldsymbol{\Omega}\right) = \mathbf{0},\ r = r_s;\ \boldsymbol{\Omega}\bullet\mathbf{n} < 0; \\ &i = 2,3,5,8;\ j_1 = 1,...,J_{\sigma t};\ j_2 = 1,...,j_1;\ j_3 = 1,...,j_2; \\ &\psi_i^{(4)}\left(j_3;j_2;j_1;r,\boldsymbol{\Omega}\right) = \mathbf{0},\ r = r_s;\ \boldsymbol{\Omega}\bullet\mathbf{n} > 0; \\ &i = 1,4,6,7;\ j_1 = 1,...,J_{\sigma t};\ j_2 = 1,...,j_1;\ j_3 = 1,...,j_2. \end{aligned} \quad (7.112)$$

7. The boundary conditions are shown in Eq. (7.112) to complete the well-posed definition of the 4th-level adjoint function $\boldsymbol{\psi}^{(4)}\left(j_3;j_2;j_1;r,\boldsymbol{\Omega}\right) \in \mathcal{H}_{(4)}$ as the solution of Eqs. (7.111) and (7.112), which are called "the *4th-Level Adjoint Sensitivity System* (4th-LASS)." The reason for calling this system "4th-*Level*" (as opposed to "4th-*Order*") stems from the fact that this system does not involve any 2nd-, 3rd-, and/or 4th-order differentials or derivatives of the dependent variables (i.e., state functions) even though the solution of the 4th-LASS is used for computing, efficiently and exactly, the 4th-order sensitivities of the response with respect to the model parameters.

8. Use the equations underlying the 4th-LVSS and 4th-LASS in Eq. (7.75) to obtain the following alternative expression for the indirect-effect term $\delta\left\{\partial^3 L\left(\alpha,\varphi;\psi^{(1)};\psi^{(2)};\psi^{(3)}\right)/\partial t_{j_3}\partial t_{j_2}\partial t_{j_1}\right\}_{ind}$ in terms of the 4th-level adjoint function $\psi^{(4)}(j_3;j_2;j_1;r,\mathbf{\Omega})\in\mathcal{H}_{(4)}$:

$$\delta\left\{\frac{\partial^3 L\left(\alpha,\varphi;\psi^{(1)};\psi^{(2)};\psi^{(3)}\right)}{\partial t_{j_3}\partial t_{j_2}\partial t_{j_1}}\right\}_{ind}$$
$$=\sum_{i=1}^{8}\left\{\left\langle\psi_i^{(4)}(j_3;j_2;j_1;r,\mathbf{\Omega}),\mathbf{q}_i^{(4)}(j_2;j_1;r,\mathbf{\Omega})\right\rangle_{(1)}\right\}_{\alpha^0}, \tag{7.113}$$
$$for\ j_1=1,\ldots J_{\sigma t};\ j_2=1,\ldots j_1;\ j_3=1,\ldots j_2.$$

The differential expression for the indirect-effect term $\delta\left\{\partial^3 L\left(\alpha,\varphi;\psi^{(1)};\psi^{(2)};\psi^{(3)}\right)/\partial t_{j_3}\partial t_{j_2}\partial t_{j_1}\right\}_{ind}$ provided in Eq. (7.113) comprises *all* of the 4th-order partial sensitivities that involve the group-averaged microscopic total cross-sections. In particular, if *only* the variations in the total microscopic cross-sections are considered, then the direct-effect term defined in Eq. (7.74) vanishes and the indirect-effect term in Eq. (7.113) reduces to the following particular form:

$$\delta\left\{\frac{\partial^3 L\left(\alpha,\varphi;\psi^{(1)};\psi^{(2)};\psi^{(3)}\right)}{\partial t_{j_3}\partial t_{j_2}\partial t_{j_1}}\right\}_{\alpha^0}=\delta\left\{\frac{\partial^3 L\left(\alpha,\varphi;\psi^{(1)};\psi^{(2)};\psi^{(3)}\right)}{\partial t_{j_3}\partial t_{j_2}\partial t_{j_1}}\right\}_{ind}$$
$$=\sum_{i=1}^{8}\left\{\left\langle\psi^{(4)}(j_3;j_2;j_1;r,\mathbf{\Omega}),\mathbf{q}_i^{(4)}(j_2;j_1;r,\mathbf{\Omega})\right\rangle_{(1)}\right\}_{\alpha^0} \tag{7.114}$$
$$=\sum_{j_4=1}^{J_{\sigma t}}\left\{\frac{\partial^4 L\left(\alpha,\varphi;\psi^{(1)};\psi^{(2)};\psi^{(3)};\psi^{(4)}\right)}{\partial t_{j_4}\partial t_{j_3}\partial t_{j_2}\partial t_{j_1}}\delta t_{j_4}\right\}_{\alpha^0},$$
$$j_1=1,\ldots J_{\sigma t};\ j_2=1,\ldots,j_1;\ j_3=1,\ldots,j_2.$$

Identifying the specific quantity that multiplies the specific variation δt_{j_4} provides the following expression for the 4th-order response sensitivities which involve *only* the group-averaged total microscopic cross-sections:

$$\frac{\partial^4 L\left(\alpha,\varphi;\psi^{(1)};\psi^{(2)};\psi^{(3)};\psi^{(4)}\right)}{\partial t_{j_4}\partial t_{j_3}\partial t_{j_2}\partial t_{j_1}}=-\left\langle\psi_1^{(4)}(j_3;j_2;j_1;r,\mathbf{\Omega}),\mathbf{S}(j_4;\alpha)\varphi(r,\mathbf{\Omega})\right\rangle_{(1)}$$
$$-\left\langle\psi_2^{(4)}(j_3;j_2;j_1;r,\mathbf{\Omega}),\mathbf{S}(j_4;\alpha)\psi^{(1)}(r,\mathbf{\Omega})\right\rangle_{(1)} \tag{7.115}$$
$$-\left\langle\psi_3^{(4)}(j_3;j_2;j_1;r,\mathbf{\Omega}),\mathbf{S}(j_4;\alpha)\psi_1^{(2)}(j_1;r,\mathbf{\Omega})\right\rangle_{(1)}$$

$$-\left\langle \psi_4^{(4)}(j_3;j_2;j_1;r,\mathbf{\Omega}),\mathbf{S}(j_4;\alpha)\psi_2^{(2)}(j_1;r,\mathbf{\Omega})\right\rangle_{(1)}$$
$$-\left\langle \psi_5^{(4)}(j_3;j_2;j_1;r,\mathbf{\Omega}),\mathbf{S}(j_4;\alpha)\psi_1^{(3)}(j_2;j_1;r,\mathbf{\Omega})\right\rangle_{(1)}$$
$$-\left\langle \psi_6^{(4)}(j_3;j_2;j_1;r,\mathbf{\Omega}),\mathbf{S}(j_4;\alpha)\psi_2^{(3)}(j_2;j_1;r,\mathbf{\Omega})\right\rangle_{(1)}$$
$$-\left\langle \psi_7^{(4)}(j_3;j_2;j_1;r,\mathbf{\Omega}),\mathbf{S}(j_4;\alpha)\psi_3^{(3)}(j_2;j_1;r,\mathbf{\Omega})\right\rangle_{(1)}$$
$$-\left\langle \psi_8^{(4)}(j_3;j_2;j_1;r,\mathbf{\Omega}),\mathbf{S}(j_4;\alpha)\psi_4^{(3)}(j_2;j_1;r,\mathbf{\Omega})\right\rangle_{(1)},$$
$$for\ j_1=1,...,J_{\sigma t};\ j_2=1,...,j_1;\ j_3=1,...,j_2;\ j_4=1,...,j_3.$$

In component form, the 4th-LASS is solved to determine the components of the 4th-level adjoint function $\psi^{(4)}(j_3;j_2;j_1;r,\mathbf{\Omega})$ in the following sequence:

$$\mathbf{B}(\alpha)\psi_5^{(4)}(j_3;j_2;j_1;r,\mathbf{\Omega})=-\mathbf{S}(j_3;\alpha)\varphi(r,\mathbf{\Omega}), \tag{7.116}$$

$$\psi_5^{(4)}(j_3;j_2;j_1;r,\mathbf{\Omega})=\mathbf{0},\ r=r_s;\ \mathbf{\Omega}\bullet\mathbf{n}<0;$$
$$j_1=1,...,J_{\sigma t};\ j_2=1,...,j_1;\ j_3=1,...,j_2, \tag{7.117}$$

$$\mathbf{A}(\alpha)\psi_6^{(4)}(j_3;j_2;j_1;r,\mathbf{\Omega})=-\mathbf{S}(j_3;\alpha)\psi^{(1)}(r,\mathbf{\Omega}), \tag{7.118}$$

$$\psi_6^{(4)}(j_3;j_2;j_1;r,\mathbf{\Omega})=\mathbf{0},\ r=r_s;\ \mathbf{\Omega}\bullet\mathbf{n}>0;$$
$$j_1=1,...,J_{\sigma t};j_2=1,...,j_1;\ \ j_3=1,...,j_2, \tag{7.119}$$

$$\mathbf{A}(\alpha)\psi_7^{(4)}(j_3;j_2;j_1;r,\mathbf{\Omega})=-\mathbf{S}(j_1;\alpha)\psi_6^{(4)}(j_3;j_2;j_1;r,\mathbf{\Omega})$$
$$-\mathbf{S}(j_3;\alpha)\psi_1^{(2)}(j_1;r,\mathbf{\Omega}), \tag{7.120}$$

$$\psi_7^{(4)}(j_3;j_2;j_1;r,\mathbf{\Omega})=\mathbf{0},\ r=r_s;\ \mathbf{\Omega}\bullet\mathbf{n}>0;$$
$$j_1=1,...,J_{\sigma t};j_2=1,...,j_1;\ \ j_3=1,...,j_2, \tag{7.121}$$

$$\mathbf{B}(\alpha)\psi_8^{(4)}(j_3;j_2;j_1;r,\mathbf{\Omega})=-\mathbf{S}(j_1;\alpha)\psi_5^{(4)}(j_3;j_2;j_1;r,\mathbf{\Omega})$$
$$-\mathbf{S}(j_3;\alpha)\psi_2^{(2)}(j_1;r,\mathbf{\Omega}), \tag{7.122}$$

$$\psi_8^{(4)}(j_3;j_2;j_1;r,\mathbf{\Omega})=\mathbf{0},\ r=r_s;\ \mathbf{\Omega}\bullet\mathbf{n}<0;$$
$$j_1=1,...,J_{\sigma t};\ j_2=1,...,j_1;\ j_3=1,...,j_2, \tag{7.123}$$

$$\mathbf{B}(\alpha)\psi_3^{(4)}(j_3;j_2;j_1;r,\mathbf{\Omega})=-\mathbf{S}(j_2;\alpha)\psi_5^{(4)}(j_3;j_2;j_1;r,\mathbf{\Omega})$$
$$-\mathbf{S}(j_3;\alpha)\psi_3^{(3)}(j_2;j_1;r,\mathbf{\Omega}), \tag{7.124}$$

$$\psi_3^{(4)}(j_3;j_2;j_1;r,\boldsymbol{\Omega}) = \mathbf{0},\ r = r_s;\ \boldsymbol{\Omega}\bullet\mathbf{n} < 0;$$
$$j_1 = 1,\dots,J_{\sigma t};\ j_2 = 1,\dots,j_1;\ j_3 = 1,\dots,j_2, \tag{7.125}$$

$$\mathbf{A}(\boldsymbol{\alpha})\psi_4^{(4)}(j_3;j_2;j_1;r,\boldsymbol{\Omega}) = -\mathbf{S}(j_2;\boldsymbol{\alpha})\psi_6^{(4)}(j_3;j_2;j_1;r,\boldsymbol{\Omega})$$
$$-\mathbf{S}(j_3;\boldsymbol{\alpha})\psi_4^{(3)}(j_2;j_1;r,\boldsymbol{\Omega}), \tag{7.126}$$

$$\psi_4^{(4)}(j_3;j_2;j_1;r,\boldsymbol{\Omega}) = \mathbf{0},\ r = r_s;\ \boldsymbol{\Omega}\bullet\mathbf{n} > 0;$$
$$j_1 = 1,\dots,J_{\sigma t};\ j_2 = 1,\dots,j_1;\ j_3 = 1,\dots,j_2, \tag{7.127}$$

$$\mathbf{A}(\boldsymbol{\alpha})\psi_1^{(4)}(j_3;j_2;j_1;r,\boldsymbol{\Omega}) = -\mathbf{S}(j_1;\boldsymbol{\alpha})\psi_4^{(4)}(j_3;j_2;j_1;r,\boldsymbol{\Omega})$$
$$-\mathbf{S}(j_2;\boldsymbol{\alpha})\psi_7^{(4)}(j_3;j_2;j_1;r,\boldsymbol{\Omega}) - \mathbf{S}(j_3;\boldsymbol{\alpha})\psi_1^{(3)}(j_2;j_1;r,\boldsymbol{\Omega}), \tag{7.128}$$

$$\psi_1^{(4)}(j_3;j_2;j_1;r,\boldsymbol{\Omega}) = \mathbf{0},\ r = r_s;\ \boldsymbol{\Omega}\bullet\mathbf{n} > 0;$$
$$j_1 = 1,\dots,J_{\sigma t};\ j_2 = 1,\dots,j_1;\ j_3 = 1,\dots,j_2, \tag{7.129}$$

$$\mathbf{B}(\boldsymbol{\alpha})\psi_2^{(4)}(j_3;j_2;j_1;r,\boldsymbol{\Omega}) = -\mathbf{S}(j_1;\boldsymbol{\alpha})\psi_3^{(4)}(j_3;j_2;j_1;r,\boldsymbol{\Omega})$$
$$-\mathbf{S}(j_2;\boldsymbol{\alpha})\psi_8^{(4)}(j_3;j_2;j_1;r,\boldsymbol{\Omega}) - \mathbf{S}(j_3;\boldsymbol{\alpha})\psi_2^{(3)}(j_2;j_1;r,\boldsymbol{\Omega}), \tag{7.130}$$

$$\psi_2^{(4)}(j_3;j_2;j_1;r,\boldsymbol{\Omega}) = \mathbf{0},\ r = r_s;\ \boldsymbol{\Omega}\bullet\mathbf{n} < 0;$$
$$j_1 = 1,\dots,J_{\sigma t};\ j_2 = 1,\dots,j_1;\ j_3 = 1,\dots,j_2. \tag{7.131}$$

The quantities in Eqs. (7.115)–(7.131) are to be evaluated at the nominal parameter values but the corresponding indicator $\{\ \}_{\boldsymbol{\alpha}^0}$ has been omitted in order to simplify the respective notations.

7.6 COMPUTATIONAL CONSIDERATIONS

The number of sensitivities of the PERP leakage response which involve solely the group-averaged total microscopic cross-sections are as follows: (i) 180 1st-order sensitivities; (ii) 32,400 2nd-order sensitivities, of which 16,290 are distinct; and (iii) 5,832,000 3rd-order sensitivities, of which 988,260 are distinct; (iv) 1,049,760,000 4th-order sensitivities, of which 45,212,895 are distinct. Numerical results for these sensitivities have been obtained by using the code PARTISN (Alcouffe et al., 2008) with a S_{32} angular quadrature (ISN=32). The computer used for these computations was a DELL computer with an 8-core processor (AMD FX-8350). The CPU time for a typical adjoint or forward computation, in other words, to solve Eqs. (7.60) or (7.62), using a S_{32} angular quadrature (ISN=32) is ca. 20 seconds while the CPU time for computing the integrals over the various adjoint functions which appear in the definition of the respective sensitivity, see Eq. (7.72), is ca. 0.002 seconds.

A total of 33,301 adjoint computations are needed to obtain all of the *distinct* 3rd-order sensitivities $\partial^3 L(\boldsymbol{\alpha})/\partial t_{j_3}\partial t_{j_2}\partial t_{j_1}$, comprising the following large-scale computations:

1. One adjoint PARTISN computation to obtain $\psi^{(1)}(r,\mathbf{\Omega})$;
2. 180 adjoint-like PARTISN computations to obtain $\psi_1^{(2)}(j_1;r,\mathbf{\Omega})$;
3. 180 forward-like PARTISN computations to obtain $\psi_2^{(2)}(j_1;r,\mathbf{\Omega})$;
4. $180(180+1)/2=16{,}290$ adjoint-like PARTISN computations to obtain $\psi_1^{(3)}(j_2;j_1;r,\mathbf{\Omega})$ since the source terms for the equations satisfied by these functions depend on both indices j_1 and j_2;
5. $180(180+1)/2=16{,}290$ forward-like PARTISN computations to obtain $\psi_2^{(3)}(j_2;j_1;r,\mathbf{\Omega})$ since the source terms for the equations satisfied by these functions depend on both indices j_1 and j_2;
6. 180 forward-like PARTISN computations to obtain $\psi_3^{(3)}(j_2;j_1;r,\mathbf{\Omega})$ since the source terms for the equations satisfied by these functions depend only on the index j_2;
7. 180 adjoint-like PARTISN computations to obtain $\psi_4^{(3)}(j_2;j_1;r,\mathbf{\Omega})$ since the source terms for the equations satisfied by these functions depend only on the index j_2.

A total number of 2,075,341 adjoint-like or forward-like large-scale computations are needed to obtain all of the *distinct* 4th-order sensitivities $\partial^4 L(\boldsymbol{\alpha})/\partial t_{j4}\partial t_{j3}\partial t_{j2}\partial t_{j1}$ using the 4th-CASAM-L method, as follows:

1. 988,260 adjoint-like PARTISN computations for obtaining $\psi_1^{(4)}(j_3;j_2;j_1;r,\mathbf{\Omega})$;
2. 988,260 forward-like PARTISN computations for obtaining $\psi_2^{(4)}(j_3;j_2;j_1;r,\mathbf{\Omega})$;
3. 16,290 forward-like PARTISN computations for obtaining $\psi_3^{(4)}(j_3;j_2;j_1;r,\mathbf{\Omega})$;
4. 16,290 adjoint-like PARTISN computations for obtaining $\psi_4^{(4)}(j_3;j_2;j_1;r,\mathbf{\Omega})$;
5. 180 forward-like PARTISN computations for obtaining $\psi_5^{(4)}(j_3;j_2;j_1;r,\mathbf{\Omega})$;
6. 180 adjoint-like PARTISN computations for obtaining $\psi_6^{(4)}(j_3;j_2;j_1;r,\mathbf{\Omega})$;
7. 16,290 adjoint-like PARTISN computations for obtaining $\psi_7^{(4)}(j_3;j_2;j_1;r,\mathbf{\Omega})$;
8. 16,290 forward-like PARTISN computations for obtaining $\psi_8^{(4)}(j_3;j_2;j_1;r,\mathbf{\Omega})$;
9. 33,301 adjoint computations for obtaining all of the *distinct* 3rd-order sensitivities $\partial^3 L(\boldsymbol{\alpha})/\partial t_{j_3}\partial t_{j_2}\partial t_{j_1}$.

On the other hand, if the *distinct* 4th-order sensitivities $\partial^4 L(\boldsymbol{\alpha})/\partial t_{j_4}\partial t_{j_3}\partial t_{j_2}\partial t_{j_1}$ were computed by solving the 4th-LVSS, which comprises Eq. (7.86) and the corresponding boundary conditions, a total of $46{,}217{,}625 = 180+180\times 181/2+180\times 181\times 182/3!+180\times 181\times 182\times 183/4!$ forward computations would be needed.

To compute the 180 1st-order sensitivities, one adjoint large-scale computation is needed in order to obtain the 1st-level adjoint function $\psi^{(1)}(r,\boldsymbol{\Omega})$. Thus, the CPU time needed is ca. 24 seconds for computing $\psi^{(1)}$ plus ca. 1 second for computing the integrals over this adjoint function. By comparison, ca. 270 minutes would be needed to compute these 1st-order sensitivities using the FD formula, assuming that the optimal step size would be guessed *ab initio.*

To compute the $180(180+1)/2=16{,}290$ distinct 2nd-order sensitivities, $180\times 2=360$, adjoint large-scale computations are needed to obtain the 2nd-level adjoint functions $\psi_1^{(2)}(j_1;r,\boldsymbol{\Omega})$ and $\psi_2^{(2)}(j_1;r,\boldsymbol{\Omega})$, $j_1=1,\ldots,180$. Thus, the CPU time needed is ca. 2.4 hours for computing $\psi_1^{(2)}(j_1;r,\boldsymbol{\Omega})$ and $\psi_2^{(2)}(j_1;r,\boldsymbol{\Omega})$, $j_1=1,\ldots,180$. An additional CPU time of ca. 3 minutes is needed for computing the integrals over these adjoint functions which provide the values of both the 1st-order and 2nd-order sensitivities. The 2nd-order mixed sensitivities would be computed twice, using two distinct adjoint functions. By comparison, ca. 810 CPU hours would be needed to directly compute the 2nd-order sensitivities (only) using the FD formula, assuming that the optimal step sizes would be guessed *ab initio.*

To compute the $180(180+1)(180+2)/3!=988{,}260$ distinct 3rd-order sensitivities, 32,940 adjoint large-scale computations are needed to obtain the 3rd-level adjoint functions $\psi_i^{(3)}(j_1;j_2;r,\boldsymbol{\Omega})$, $i=1,\ldots,4;\ j_1,j_2=1,\ldots,180$, which requires ca. 220 hours CPU time. An additional CPU time of ca. 0.6 hours would be needed for computing the integrals over these adjoint functions to obtain all of the 1st-order, 2nd-order, and 3rd-order sensitivities. The mixed 2nd-order sensitivities would be computed twice, while the mixed-sensitivities would be computed three times, using distinct adjoint functions in all cases. By comparison, 98,817 hours would be needed to compute these 3rd-order sensitivities using the FD formula, assuming that the optimal step sizes would be guessed *ab initio.*

To compute the $180(180+1)(180+2)(180+3)/4!=45{,}212{,}985$ distinct 4th-order sensitivities would require 2,042,040 adjoint large-scale computations to obtain the 4th-level adjoint functions $\psi_i^{(4)}(j_1;j_2;j_3;r,\boldsymbol{\Omega})$, $i=1,\ldots,8$; $j_1,j_2,j_3=1,\ldots,180$, which would amount to ca. 25,525 CPU hours; additionally, ca. 50 CPU hours would be needed for computing the integrals over the various adjoint functions to obtain all of the 1st-, 2nd-, 3rd-, and 4th-order. By comparison, ca. 1015 years would be needed to compute just the 4th-order sensitivities using the FD formula, assuming that the optimal step sizes would be guessed ab initio. Evidently, finite-difference schemes are rendered impractical by the "curse of dimensionality" (Bellman, 1957).

7.7 CONCLUDING REMARKS

The expressions presented in this chapter were used to compute the values presented in Table 2.4, which were used, in turn, in all of the illustrative applications of the 2nd-BERRU-PM and 2nd-BERRU-PM methodologies presented in the previous Chapters of this book. The expressions for computing the sensitivities (of 1st- through 4th-orders) of the PERP benchmark's leakage response with respect to all 7,477 PERP parameters that have non-zero nominal values, as described in

Chapter 2, are available in the book by Cacuci and Fang (2023), along with the numerical values of these sensitivities.

The traditional methods of sensitivity analysis aim at computing sensitivities of responses directly to the *primary* parameters (i.e., microscopic cross-sections, isotopic number densities, etc.) involved in the computational model of the physical system under consideration. Although the sensitivities to the primary model parameters are ultimately of interest for subsequent use in predictive modeling activities (which includes the quantification of the uncertainties induced in responses by uncertainties in the primary model parameters, assimilation of experimental data for calibrating the model's parameters, and improving the model's predictions), the primary parameters seldom appear explicitly in the equations underlying the model. For example, the primary model parameters (e.g., microscopic cross-sections, atomic number densities) do not appear explicitly in the forward and adjoint transport equations modeling the OECD/NEA reactor physics benchmark, as evidenced by the derivations presented in this chapter. What appears explicitly in these equations are the group-averaged *macroscopic* cross-sections, which are *functions* of the primary model parameters (i.e., isotopic number densities and microscopic cross-sections) and which can be considered to be *features* of the neutron transport equation. This fact has motivated the conception and development by Cacuci (2024a, 2024b, 2024c) of the "n^{th}-Order Features Adjoint Sensitivity Analysis Methodology for Nonlinear Systems (n^{th}-FASAM-N)," which significantly reduces the computational effort for computing efficiently and exactly *sensitivities of model responses to model features* (i.e., functions of the primary model responses), and subsequently compute the sensitivities of responses to the primary model parameters by using the sensitivities to the model features. The ongoing development of this novel methodology is expected to provide the ultimate tool of unparalleled computational efficiency for arbitrarily high-order exact sensitivity analysis of large-scale computational models.

References

Abramov, R. (2009). The multidimensional moment-constrained maximum entropy problem: A BFGS algorithm with constraint scaling. *J. Comput. Phys.*, **228**, 96–108.

Abramov, R., Majda, A. and Kleeman, R. (2005). Information theory and predictability for low frequency variability. *J. Atmos. Sci.*, **62**, 65–87.

Alcouffe, R. E., Baker, R. S., Dahl, J. A., Turner, S. A. and Ward, R. (2008). *PARTISN: A Time-Dependent, Parallel Neutral Particle Transport Code System* (LA-UR-08-07258). Los Alamos, NM: Los Alamos National Laboratory.

Anthes, R. A. (1974). Data assimilation and initialization of hurricane prediction models. *J. Atmos. Sci.*, **31**, 702.

Bandyopadhyay, K., Bhattacharya, A., Biswas, P. and Drabold, D. (2005). Maximum entropy and the problem of moments: A stable algorithm. *Phys. Rev. E*, **71**.

Barhen, J., Cacuci, D. G., Wagschal, J. J., Bjerke, M. A. and Mullins, C. B. (1982). Uncertainty analysis of time-dependent nonlinear systems: Theory and application to transient thermal hydraulics. *Nucl. Sci. Eng.*, **81**, 23–44.

Bellman, R. E. (1957). *Dynamic Programming*. Rand Corporation; Princeton, NJ, USA: Princeton University Press.

Bernardo, J. M. and Smith, A. F. M. (2006). *Bayesian Theory*. Chichester; West Sussex: John Wiley & Sons.

Cacuci, D. G. (1981a). Sensitivity theory for nonlinear systems: I. Nonlinear functional analysis approach. *J. Math. Phys.*, **22**, 2794–2812.

Cacuci, D. G. (1981b). Sensitivity theory for nonlinear systems: II. Extensions to additional classes of responses. *J. Math. Phys.*, **22**, 2803–2812.

Cacuci, D. G. (2014). Predictive modeling of coupled multi-physics systems: I. Theory. *Ann. Nucl. Energy*, **70**, 266–278.

Cacuci, D. G. (2018). *The Second-Order Adjoint Sensitivity Analysis Methodology*. Boca Raton, FL: CRC Press.

Cacuci, D. G. (2019). *BERRU Predictive Modeling: Best-Estimate Results with Reduced Uncertainties*. Berlin: Springer.

Cacuci, D. G. (2022). *The nth-Order Comprehensive Adjoint Sensitivity Analysis Methodology: Overcoming the Curse of Dimensionality. Volume I: Linear Systems*. Cham: Springer Nature.

Cacuci, D. G. and Fang, R. (2023). *The nth-Order Comprehensive Adjoint Sensitivity Analysis Methodology (nth-CASAM): Overcoming the Curse of Dimensionality in Sensitivity and Uncertainty Analysis, Volume II: Application to a Large-Scale System, 463 pages*. Springer Nature Switzerland, Cham, 2023; https://doi.org/10.1007/978-3-031-19635-5

Cacuci, D. G. (2023a). *The nth-Order Comprehensive Adjoint Sensitivity Analysis Methodology (nth-CASAM): Overcoming the Curse of Dimensionality in Sensitivity and Uncertainty Analysis, Volume III: Nonlinear Systems*. Cham: Springer Nature, 369 pages. https://doi.org/10.1007/978-3-031-22757-8

Cacuci, D. G. (2023b). Second-order MaxEnt predictive modelling methodology. I: Deterministically incorporated computational model (2nd-BERRU-PMD). *Am. J. Comput. Math.*, **13**, 236–266. https://doi.org/10.4236/ajcm.2023.132013

Cacuci, D. G. (2023c). Second-order MaxEnt predictive modelling methodology. II: Probabilistically incorporated computational model (2nd-BERRU-PMP). *Am. J. Comput. Math.*, **13**, 267–294. https://doi.org/10.4236/ajcm.2023.132014

Cacuci, D. G. (2023d). Fourth-order predictive modelling: I. General-purpose closed-form fourth-order moments-constrained MaxEnt distribution. *Am. J. Comput. Math.*, **13**, 413–438. https://doi.org/10.4236/ajcm.2023.134024

Cacuci, D. G. (2023e). Fourth-order predictive modelling: II. 4th-BERRU-PM methodology for combining measurements with computations to obtain best-estimate results with reduced uncertainties. *Am. J. Comput. Math.*, **13**, 439–475. https://doi.org/10.4236/ajcm.2023.134025

Cacuci, D. G. (2024a). Second-order FASAM. *Am. J. Comput. Math.*, **13**, 267–294. https://doi.org/10.4236/ajcm.2023.132014

Cacuci, D. G. (2024b). Introducing the nth-order features adjoint sensitivity analysis methodology for nonlinear systems (nth-FASAM-N): I. Mathematical framework. *Am. J. Comput. Math.*, **14**, 11–42. https://doi.org/10.4236/ajcm.2024.141002

Cacuci, D. G. (2024c). Introducing the nth-order features adjoint sensitivity analysis methodology for nonlinear systems (nth-FASAM-N): II. Illustrative example. *Am. J. Comput. Math.*, **14**, 43–95. https://doi.org/10.4236/ajcm.2024.141003

Cacuci, D. G. and Ionescu-Bujor, M. (2010). Model calibration and best-estimate prediction through experimental data assimilation: I. Mathematical framework. *Nucl. Sci. Eng.*, **165**, 18–44.

Cacuci, D. G., Navon, M. I. and Ionescu-Bujor, M. (2014). *Computational Methods for Data Evaluation and Assimilation*. Boca Raton, FL: Chapman & Hall/CRC.

Cecchini, I., Farinelli, U., Gandini, A. and Salvatores, M. (1964). A/Conf.28/M. In *Proceedings of the 3rd International Conference on the Peaceful Uses of Atomic Energy*, Geneva, Vol. 2, p. 627.

Chadwick, M. B., Herman, M., Obložinský, P., Dunn, M. E., Danon, Y., Kahler, A. C., Smith, D. L., Pritychenko, B., Arbanas, G., Brewer, R., Brown, D. A., Capote, R., Carlson, A. D., Cho, Y. S., Derrien, H., Guber, K., Hale, G. M., Hoblit, S., Holloway, S. and Young, P. G. (2011). ENDF/B-VII.1: Nuclear data for science and technology: Cross sections, covariances, fission product yields and decay data. *Nucl. Data Sheets*, **112**, 2887–2996. https://doi.org/10.1016/j.nds.2011.11.002

Conlin, J. L., Parsons, D. K., Gardiner, S. J., Gray, M., Lee, M. B. and White, M. C. (2013). *MENDF71X: Multigroup Neutron Cross-Section Data Tables Based upon ENDF/B-VII.1X*. Los Alamos National Laboratory Report LA-UR-15-29571. Los Alamos, NM: Los Alamos National Laboratory.

Derber, J. C. (1985). The variational four-dimensional assimilation of analysis using filtered models as constraints. PhD Thesis, University of Wisconsin-Madison.

Derber, J. C. (1987). Variational four-dimensional analysis using the quasigeostrophic constraint. *Mon. Weather Rev.*, **115**, 998.

Derber, J. C., Parrish, D. F. and Lord, S. J. (1991). The new global operational analysis system at the National Meteorological Center. *Weather Forecast.*, **6**, 538.

Dragt, J. B., Dekker, J. W. M., Gruppelaar, H. and Janssen, A. J. (1977). Methods of adjustment and error evaluation of neutron capture cross sections. *Nucl. Sci. Eng.*, **62**, 11.

Evangelou, E., Zhu, Z. and Smith, R. L. (2011). Estimation and prediction for spatial generalized linear mixed models using high order Laplace approximation. *J. Stat. Plan. Infer.*, **141**(11), 3564–3577. https://doi.org/10.1016/j.jspi.2011.05.008

Faragó, I., Havasi, Á. and Zlatev, Z. (eds) (2013). *Advanced Numerical Methods for Complex Environmental Models: Needs and Availability*. Bussum: Bentham Science Publishers.

Gandin, L. S. (1963). *Objective Analysis of Meteorological Fields*. Leningrad: Gridromet (in Russian). English translation by the Israel Program for Scientific Translations, 1965.

Gandini, A. and Petilli, M. (1973). *AMARA: A Code Using the Lagrange Multipliers Method for Nuclear Data Adjustment* (CNEN-RI/FI(73)39). Casaccia/Rome: Comitato Nazionale Energia Nucleare.

Hall, M. C. G. and Cacuci, D. G. (1983). Physical interpretation of the adjoint functions for sensitivity analysis of atmospheric models. *J. Atmos. Sci.*, **40**, 2537–2546.

Hall, M. C. G., Cacuci, D. G. and Schlesinger, M. E. (1982). Sensitivity analysis of a radiative-convective model by the adjoint method. *J. Atmos. Sci.*, **39**, 2038–2050.

Haven, K., Majda, A. and Abramov, R. (2005). Quantifying predictability through information theory: Small sample estimation in a non-Gaussian framework. *J. Comput. Phys.*, **206**, 334–362.

Hoefer, A., Buss, O. and Neuber, J. C. (2011a). How confident can we be in confidence intervals for the computational bias obtained with the generalized linear least squares methodology? – A toy model analysis. In Proceedings of the International Conference on Nuclear Criticality, ICNC 2011, Edinburgh, September 19–22.

Hoefer, A., Buss, O. and Neuber, J. C. (2011b). Limitations of the generalized linear least squares methodology for bias estimation in criticality safety analysis. In IAEA International Workshop on Burnup Credit Criticality Calculation Methods and Applications, Beijing, October. https://www.researchgate.net/publication/276271639

Hoffmann, R. N. (1986). A four-dimensional analysis exactly satisfying equations of motion. *Mon. Weather Rev.*, **114**(2), 388.

Humi, I., Wagschal, J. J. and Yeivin, Y. (1964). Multi-group constants from integral data. In Proceedings of the 3rd International Conference on the Peaceful Uses of Atomic Energy, Geneva, Vol. 2, p. 398.

Jaynes, E. T. (1957). Information theory and statistical mechanics. *Phys. Rev.*, **106**, 620.

Kuroi, H. and Mitani, H. (1975). Adjustment to cross-section data to fit integral experiments by least squares method. *J. Nucl. Sci. Technol.*, **12**, 663.

Lahoz, W., Khattatov, B. and Ménard, R. (eds) (2010). *Data Assimilation: Making Sense of Observations*. Heidelberg: Springer Verlag.

Lewins, J. (1965). *IMPORTANCE: The Adjoint Function*. Oxford: Pergamon Press.

Lewis, J. M., Lakshmivarahan, S. and Dhall, S. K. (2006). *Dynamic Data Assimilation: A Least Square Approach*. Cambridge: Cambridge University Press.

Lorenc, A. C. (1986). Analysis-methods for numerical weather prediction. *Q. J. R. Meteorol. Soc.*, **112**(474), 1177.

Luo, Z., Wang, X. and Liu, D. (2020). Prediction on the static response of structures with large-scale uncertain-but-bounded parameters based on the adjoint sensitivity analysis. *Struct. Multidiscip. Optim.*, **61**, 123–139. https://doi.org/10.1007/s00158-019-02349-w

MacFarlane, R., Muir, D. W., Boicourt, R. M., Kahler, A. C. and Conlin, J. L. (2017). *The NJOY Nuclear Data Processing System, Version 2016* (LA-UR-17-20093). Los Alamos, NM: Los Alamos National Laboratory.

Mahfouf, J. F., Buizza, R. and Errico, R. M. (1996). Strategy for including physical processes in the ECMWF variational data assimilation system. In ECMWF Workshop on Non-Linear Aspects of Data Assimilation, Reading, September 9–11.

Muller, I. and Ruggeri, T. (1993). *Extended Thermodynamics, Springer Tracts in Natural Philosophy*. Heidelberg: Springer.

Navon, I. M. (1986). A review of variational and optimization methods in meteorology. In Sasaki, Y. K. (ed.), *Festive Volume of the International Symposium on Variational Methods in Geosciences* (Developments in Geo-Mathematics, vol. 5). New York, USA: Elsevier Science Pub. Co., 29.

Navon, I. M. (1998). Practical and theoretical aspects of adjoint parameter estimation and identifiability in meteorology and oceanography. *Dyn. Atmos. Oceans* (Special Issue in honor of Richard Pfeffer), **27**(1–4), 55.

Práger, T. and Kelemen, F. D. (2013). Adjoint methods and their application in earth sciences. In Faragó, I., Havasi, Á. and Zlatev, Z. (eds), *Advanced Numerical Methods for Complex Environmental Models: Needs and Availability*. Bussum: Bentham Science Publishers, 203–275.

Rabier, F. (2005). Overview of global data assimilation developments in numerical weather-prediction centers. *Q. J. R. Meteorol. Soc.*, **131**(613), 3215.

Roulston, M. and Smith, L. (2002). Evaluating probabilistic forecasts using information theory. *Mon. Weather Rev.*, **130**, 1653–1660.

Rowlands, J., et al. (1973). The production and performance of the adjusted cross-section set FGL5. In Proceedings of the International Symposium on Physics of Fast Reactors, ROOT: CERN, Tokyo. http://www.cern.ch/root.

Sasaki, Y. K. (1955). A fundamental study of the numerical prediction based on the variational principle. *J. Meteorol. Soc. Japan*, **33**, 262.

Sasaki, Y. K. (1958). An objective analysis based on the variational method. *J. Meteorol. Soc. Japan*, **36**(3), 77.

SCALE. (2009). *A Modular Code System for Performing Standardized Computer Analyses for Licensing Evaluation* (ORNL/TM 2005/39, Version 6). Oak Ridge, TN: Oak Ridge National Laboratory.

Shannon, C. E. (1948). A mathematical theory of communication. *Bell Syst. Tech. J.*, **27**, 379–423.

Shun, Z. and McCullagh, P. (1995). Laplace approximation of high dimensional integrals. *J. R. Stat. Soc. Series B Stat. Methodol.*, **57**(4), 749–760.

Stacey, W. M. (1974). *Variational Methods in Nuclear Reactor Physics*. New York: Academic Press.

Tang, Y. and Reid, N. (2023). Laplace and saddle point approximations in high dimensions. arXiv:2107.10885v3. https://doi.org/10.48550/arXiv.2107.10885

Tukey, J. W. (1957). *The Propagation of Errors, Fluctuations and Tolerances* (Technical Reports No. 10-12). Princeton, NJ: Princeton University Press.

Usachhev, L. N. (1964). Perturbation theory for the breeding ratio and for other number ratios pertaining to various reactor processes. *J. Nucl. Energy Part A/B*, **18**, 571.

Valentine, T. E. (2006). *Polyethylene-Reflected Plutonium Metal Sphere Subcritical Noise Measurements. SUB-PU-METMIXED-001, International Handbook of Evaluated Criticality Safety Benchmark Experiments* (NEA/NSC/DOC(95)03/I-IX). Paris: Organization for Economic Co-Operation and Development; Nuclear Energy Agency.

Venard, C., Santamarina, A., Leclainche, A. and Mournier, C. (2009). The R.I.B. tool for the determination of computational bias and associated uncertainty in the CRISTAL criticality safety package. In Proceedings of ANS Nuclear Criticality Safety Division Topical Meeting (NCSD 2009), Richland, WA.

Weisbin, C. R., et al. (1978). Application of sensitivity and uncertainty methodology to fast reactor integral experiment analysis. *Nucl. Sci. Eng.*, **66**, 307.

Wilson, W. B., Perry, R. T., Shores, E. F., Charlton, W. S., Parish, T. A., Estes, G. P., Brown, T. H., Arthur, E. D., Bozoian, M., England, T. R., Madland, D. G. and Stewart, J. E. (2002). SOURCES4C: A code for calculating (α,n), spontaneous fission, and delayed neutron sources and spectra. In Proceedings of the American Nuclear Society/Radiation Protection and Shielding Division 12th Biennial Topical Meeting, Santa Fe, NM, April 14–18.

Wu, X. (2003). Calculation of maximum entropy densities with application to income distribution. *J. Econ.*, **115**, 347–354.

Index